Unsolved!

Unsolved!

The History and Mystery
of the World's Greatest Ciphers
from Ancient Egypt
to Online Secret Societies

CRAIG P. BAUER

Princeton University Press | Princeton and Oxford

Library of Congress Cataloging-in-Publication Data
Names: Bauer, Craig P., author.
Title: Unsolved! : the history and mystery of the world's greatest ciphers
from ancient Egypt to online secret societies / Craig Bauer.
Description: Princeton : Princeton University Press, [2017] | Includes
bibliographical references and index.
Identifiers: LCCN 2016027673 | ISBN 9780691167671 (hardback)
Subjects: LCSH: Ciphers—History. | Ciphers—Problems, exercises, etc. |
BISAC: COMPUTERS / Security / Cryptography. | COMPUTERS / Computer
Science. | MATHEMATICS / Recreations & Games.
Classification: LCC Z104 .B33 2017 | DDC 005.8/2409—dc23 LC record
available at https://lccn.loc.gov/2016027673

British Library Cataloging-in-Publication Data is available
This book has been composed in Bodoni Std, Brill and Din Pro
Printed on acid-free paper. ∞
Printed in the United States of America

1 3 5 7 9 10 8 6 4 2

To René Stein and Klaus Schmeh, both of whom I value greatly for the

help they've provided, as well as for their friendship.

Contents

Preface

A message that has been hidden through the use of a code or cipher is automatically of greater interest than one that can be instantly read. Perhaps the disguise obscures something of no interest whatsoever, but on the other hand, almost anything might be revealed, including the location of buried treasure worth millions (chapter 9), the identity of a murderer or serial killer (chapters 4, 5, and 6), or information that rewrites history (chapters 1 and 2). Other ciphers seem unlikely to reveal anything significant, but the decades during which they've escaped attempts at solution guarantee fame for any would-be solver (chapters 3, 7, and 8). Then there are the ciphers whose solutions will garner the solver much attention because they likely require mathematical breakthroughs or approaches so creative that no one else could manage to find them (chapters 10 and 11).

In the past, unsolved ciphers have been cracked by scholars with decades of experience in the field (James Gillogly, Kent Boklan), as well as by amateurs (Donald Harden and his wife Bettye). I think it's likely that several of the ciphers in the pages to follow will be cracked by readers before much time has passed.

A chronological approach to the ciphers that remain unsolved may seem the most logical, but a little must be learned about how ciphers can be broken in order to appreciate how the present examples have resisted all attempts thus far. So, in order to provide the necessary background first, the chronological approach has been avoided.

To create a bit of suspense and prevent the reader from concluding that every cipher remains unsolved, I mixed in some that actually have solutions, which are shown after they're discussed. These are not chosen at random, but rather selected for the insight they provide into the unsolved ciphers.

The unsolved ciphers themselves were also selected carefully. There are far too many to cover all of them thoroughly in a single volume, so instead I picked ones that offer as much variety as possible, while avoiding those that would require long technical discussions. If your favorite wasn't included, feel free to suggest it for a potential future volume.

Of necessity, some references are online only. If a link becomes broken, you may sometimes still find the information by using https://archive.org/web/ to find an archived copy of the page.

All of my books come with good user support. If you wish to ask me anything, please send an email to cryptoauthor@gmail.com.

I hope you enjoy the show!

Acknowledgments

Early work on this book was supported by the National Security Agency. I was honored to be allowed to serve as the 2011–2012 Scholar-in-Residence at their Center for Cryptologic History. A one-semester sabbatical granted by York College of Pennsylvania helped me to bring the book to completion. René Stein, National Cryptologic Museum librarian, was of tremendous help. Klaus Schmeh led me to many examples I hadn't previously encountered. Some of Klaus's papers are available in the pages of *Cryptologia*, in English, but most of his work is available currently only in German. These works include his excellent books and his blog (http://scienceblogs.de/klausis-krypto-kolumne/author/kschmeh/). Readers who know German are encouraged to check out his writings.

Others I am grateful to for their help include Carlos Alvarado, Gordon L. Anderson, Jeanne Anderson, Chris Bennett, Paula Bérard, Betsy Blumenthal, Paolo Bonavoglia, Colleen Boyle, Lauren Bucca, Karen Carter, Chris Christensen, Paul M. Clemens, John W. Dawson, Ralph Erskine, Sue Fairchild, Gerry Feltus, Mary Ann Folter, Stuart Freed, Tony Gaffney, John Gallehawk, Joscelyn Godwin, Patrik Granholm, Carolyn Graves-Brown, Tina Hampson, Logan Harris, David Hatch, Brian Heinhold, Lucy Hughes, Liam Hurd, Meghan Kanabay, Dimitri Karetnikov, Vickie Kearn, Kevin Knight, Oliver Knörzer, Benedek Láng, Donald P. Levasseur, Robert Lewand, Greg Link, Simon Martin, Adrienne Mayor, Beáta Megyesi, Anne Marie Menta, Dante Molle, Brent Morris, David Oranchak, James Ramm, Sravana Reddy, Jim Reeds, Dirk Rijmenants, Eleanor Robson, Moshe Rubin, Gordon Rugg, CaseyAnn Salanova, David Saunders, Bruce Schneier, Claryn Spies, Joseph Splawski, Claiborne Thompson, Kyle R. Triplett, Jim Tucker, Michael Tymn, David Visco, Hunter Willingham, Rene Zandbergen, and Lucy Zhou.

If I forgot anyone, I apologize. All of the help I received was truly appreciated!

Unsolved!

A King's Quest

The Thirst for Knowledge

King Rudolf II, Holy Roman Emperor from 1576 to 1612, was widely regarded as insane. The Catholic Church thought the only solution was an exorcism. He devoted minimal energy to ruling, instead pouring himself into occult subjects like alchemy. One of his goals was to find the philosopher's stone, which was said to allow its possessor to change lesser metals into gold and even transform the possessor himself so that he might live forever. Toward this end, he turned his castle in Prague into the world center for such alchemical and occult studies. He brought in as many of the top experts as he could and spared no expense in purchasing (or even stealing) books and manuscripts that might help him in his quest.

One book that attracted Rudolf's attention is known as the *Devil's Bible* or *Codex Gigas* (giant book), for it weighed in at about 160 pounds (72.6 kg) and is believed to be the largest manuscript in the world. Dating to the late thirteenth century, the contents are an odd assembly, including an entire Latin Bible, Hippocrates's medical writings, *The Chronicle of Bohemia* by Cosmas of Prague, magic formulas, and a work on exorcism. According to legend, it was created by a condemned monk in a single night with help from the devil. The devil appears in the manuscript on a page that has become darker than the rest, as if infernal forces were at work.

The mainstream view today is that the page that the devil appears on, as well as the facing page, have simply become darker from receiving far more use (and hence sunlight) than the other pages. Although I consider the mystery solved in this instance, Rudolf purchased another strange manuscript around which a much greater mystery arose.

While it measured only approximately 6 by 9 inches, and he could carry it around easily, Rudolf could not even begin to read the small manuscript.

FIG. 1.1 Emperor Rudolf II

Missing a cover, it consisted of more than two hundred pages of odd symbols, not matching any known alphabet, and it could be hiding anything. A large number of images were assumed to be connected with the text, but they might have been intended to misdirect the uninitiated from the manuscript's true subject matter.

For old manuscripts like this, collectors refer to "leaves" rather than pages. Each leaf has a front (recto) and a back (verso) and can be referred to by a leaf number followed by "r" for recto or a "v" for verso. In the case of Rudolf's manuscript, some leaves were larger but had multiple folds, so they would fit in with the rest. It may well have been the illustrations that led Rudolf to purchase the volume, so they'll be examined first.

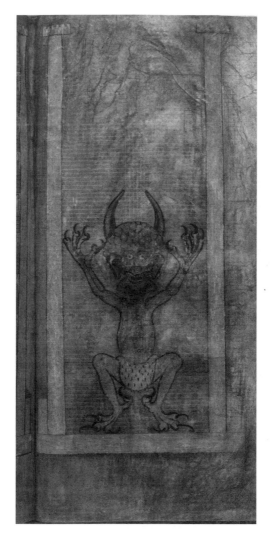

FIG. 1.2 A page from Rudolf II's largest book—*Codex Gigas*

A Picture Is Worth a Thousand Words

The manuscript begins with sketches of various plants. In many cases, the text wraps around the drawings, indicating that the drawings were made first. An example of this appears in fig. 1.3. As you look at this image, take a moment to try to figure out what sort of plant was being depicted. Many people think the top portion of the plant resembles a sunflower. But sunflowers

FIG. 1.3 Leaf 33v of the manuscript (herbal/botanical?)

certainly don't have the sort of root system depicted above. The roots look more like potatoes!

For Brian Heinhold, a professor of mathematics and computer science at Mount St. Mary's University, another plant came to mind. He wrote, "The 'weird' plant looks to me like a Jerusalem artichoke. It grows in this part of the world and you can sometimes find its tubers sold in grocery store produce sections."[1]

In any case, the various plant images have caused this portion of the manuscript to be labeled by researchers as herbal or botanical.

The manuscript continues with illustrations that are considered to be astronomical or astrological. In the one reproduced in fig. 1.4 (a page that folds

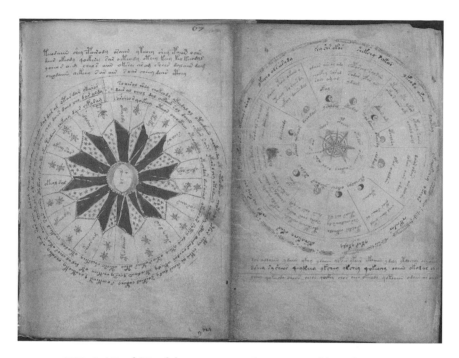

FIG. 1.4 Leaf 67r of the manuscript (astronomical/astrological?)

out), the left-hand side shows an object that could represent the sun with twelve rays emanating off of it. Between each of the rays, we have a solid triangle and another triangle with stars in it. The number of stars in each varies, and the significance of this variation is unclear. However, the number twelve, in connection with the sun, could well be a reference to the twelve signs of the zodiac.

We then come to a section that has been tentatively labeled biological. It contains many images of naked women standing in barrels of some sort. Water, or another fluid, appears to be flowing. Could this be a depiction of the human circulatory system, or is it supposed to be some sort of plumbing system?

What is perhaps the strangest portion of the manuscript comes next. In the image below, the left illustration seems to be centered on a sun, while the right image may have a sun or a crescent moon at the center. The label cosmological is reasonable for this section, but then what are we to make of the large foldout that follows? This illustration has met with a wide range of interpretations!

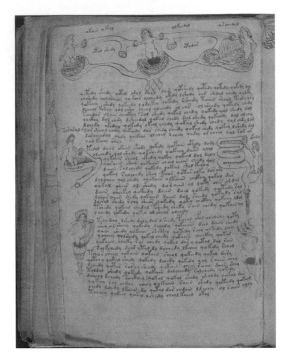

FIG. 1.5 Leaf 77v of the manuscript (biological?)

FIG. 1.6 Leaves 85r (part) and 86v (part)
(part of 85–86 foldout) of the manuscript (cosmological?)

FIG. 1.7 Leaves 85v and 86r (foldout) of the manuscript (cosmological?)

The next section is more straightforward. We once again have plants, but typically many per page, as opposed to the one-at-a-time approach of the herbal/botanical section. An example of this is shown in fig. 1.8. These pages are often referred to as the pharmaceutical section.

Space considerations forced me to make hard choices in selecting which pages to reproduce here. Fortunately, readers interested in seeing the entire manuscript can find it online at http://brbl-dl.library.yale.edu/vufind/Record/3519597.

There's yet another section that closes out the manuscript, but it doesn't contain any illustrations. It's referred to as the "text only" portion and has been conjectured to consist of recipes, an appendix, or a postscript.

These divisions are subject to debate. Indeed, William Romaine Newbold, a professor at the University of Pennsylvania, and Robert S. Brumbaugh, a professor at Yale, divided it into five sections,[2] whereas others, such as I, posit six.[3] Assuming that six is accurate, we recap the divisions here:

FIG. 1.8 Leaf 99v of the manuscript (pharmaceutical?)

1. Herbal/Botanical
2. Astronomical/Astrological
3. Biological
4. Cosmological
5. Pharmaceutical
6. Text Only (Recipes? Appendix? Postscript?)

But remember, it's possible that the pictures (our only initial clue as to the subject matter) are not actually related to the text.

Expert Opinions

Rudolf II pursued the best of whatever caught his interest. His scientific collections ranged from astronomy to zoology. His interest in astronomy is responsible for bringing Tycho Brahe and Johannes Kepler together—a pairing that eventually led to the discovery that the orbital paths of the planets are ellipses. Rudolf II also built a first-class art collection containing works by Titian, Holbein, Dürer, and others. His garden was filled with rare and exotic flowers, and he had a wide selection of wild animals, including lions, tigers, leopards, panthers, bears, and a lion cub named Otakar, who was sometimes allowed to roam the halls freely, terrifying guests.

When Rudolf wanted to look upon objects whose beauty was unsurpassed anywhere in the world, he could do so. When he wanted expert opinions, they were available. He was relentless in his quest for knowledge. On one occasion, he singed his beard as he attempted to turn lead into gold. With the hope of uncovering occult wisdom hidden within his strange manuscript, he turned to an expert in cryptanalysis. But before revealing the result, let's take a look for ourselves.

A MASCed Text?

The text itself appears (before serious analysis) to be an incredibly simple cipher, the sort in which each letter is replaced consistently with some other letter. Julius Caesar made the replacements easy to remember by always using the letter that came three positions later in the alphabet. Using our modern alphabet, this means that A is replaced with D, B is replaced with E, C is replaced with F, etc. We only begin to have problems when we get to X. There is no letter three positions later. To fix this problem, we loop back to the start of the alphabet and use A. Similarly, Y is replaced with B, and Z is replaced with C. The original alphabet (plaintext) and the cipher alphabet (ciphertext) can both be written out for handy reference like so:

```
ABCDEFGHIJKLMNOPQRSTUVWXYZ (Plaintext)
DEFGHIJKLMNOPQRSTUVWXYZABC (Ciphertext)
```

The message "THE EVIL THAT MEN DO LIVES AFTER THEM. THE GOOD IS OFT INTERRED WITH THEIR BONES." would be enciphered as

"WKH HYLO WKDW PHQ GR OLYHV DIWHU WKHP. WKH JRRG LV RIW LQWHUUHG ZLWK WKHLU ERQHV." with this method.[4]

Other encipherers chose substitutions that weren't so neatly ordered. Indeed, the ciphertext alphabet was sometimes random, like this:

```
ABCDEFGHIJKLMNOPQRSTUVWXYZ (Plaintext)
JCTBUQAFMOSXWYGDLIVEKRPHZN (Ciphertext)
```

Between these two extremes, a keyword (or words), such as IRON MAIDEN can be used to partially scramble the ciphertext alphabet:

```
ABCDEFGHIJKLMNOPQRSTUVWXYZ (Plaintext)
IRONMADEBCFGHJKLPQSTUVWXYZ (Ciphertext)
```

Because we do not want both A and G to be enciphered as I, we deleted the second I in IRON MAIDEN, when aligning it with our original alphabet. The second N is also deleted. The letters not appearing in our key then follow alphabetically. A longer keyword would offer a more thorough scrambling.

Ciphers constructed in this way, regardless of how the substitutions are determined, are known as monoalphabetic substitution ciphers. Because we are substituting other letters for the ones we start with, it makes sense to call it a substitution cipher. The monoalphabetic part just means that only one ciphertext alphabet is ever used. The cipher disguises the original message, just as a mask hides the identity of a superhero, so we abbreviate monoalphabetic substitution cipher as MASC and refer to MASCed text.

The enciphered manuscript Rudolf II purchased had a very fancy MASC. Typically, an encipherer MASCs his or her messages using his or her own alphabet. That is, Cyrillic ciphertext usually hides Russian plaintext, and Hebrew ciphertext usually hides Hebrew plaintext.

But as the images above demonstrate, Rudolf's acquisition used a strange set of characters to represent the alphabet of the original plaintext. This twist makes it harder to guess the original language. The basic characters are shown in fig. 1.9.

If a MASC was used to disguise the text that Rudolf bought, it should be easy to remove. But before we get into the techniques that allow this sort of cipher to be broken, we first give some evidence that a MASC really was used. For even as far back as the early 1600s, many different types of ciphers had seen use.

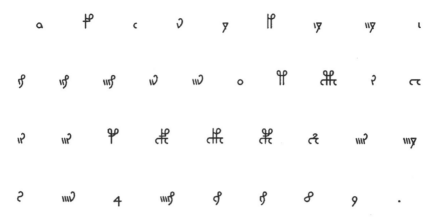

FIG. 1.9 Basic character set

Evidence of a MASC

There is one extremely important character in the cipher Rudolf had his cryptanalyst contemplating that was not listed in the table above, namely the blank space! The blank was actually the most common symbol on the two hundred–plus pages of manuscript. It shouldn't be taken for granted because early manuscripts often ran all of the words together. It wasn't until the fourteenth century that all European works clearly separated words.

Knowing where words begin and end is a huge aid to recovering the original message. It can even help us determine what language the message was in before we begin deciphering. For example, in English the average word length has been estimated at values such as 4.5 letters and 5.1 letters, whereas for German text we have an average of about 5.9 letters. The greater length for German is due, in part, to that language's tendency to combine existing words when a neologism is required, rather than inventing a brand new word. The longest example that is "in use" is, according to the *Guinness Book of World Records*, Rechtsschutzversicherungsgesellschaften. In English, it is rendered as five words: insurance companies providing legal protection.

The average word length for the enciphered manuscript is between four and five letters. The reason a single value cannot be given definitively is detailed later. But even such an imprecise range rules out German. This is some help, but we cannot immediately conclude that the underlying language is English, for example, as there are other languages with an average word length in this range.

The strongest evidence that a MASC was used is repetitions of groups of letters and even entire ciphertext words. A more complicated cipher would lead to much more variation in the ciphertext. At the extreme end, an ideal cipher is indistinguishable from completely random text. The massive amount of repetition argues against any cipher more complex than a MASC.

How to Reveal MASCed Text

If we suspect a MASC, we may make use of various patterns in the given language to recover the message. A quick demonstration in English serves to illustrate this approach. Consider the following ciphertext, which disguises a quote from a 1606 proposition of the archdukes in Vienna.[5]

```
GNX LBURXHW NX NYHRIRXHRD KYJW NY ANPBIDX, BJVGWLNXHX,
MBFFBJNXHX BYD HGR JNMR, XSBINYO YK RCSRYXR HK QNYD
BJJ MNYDX KQ HIRBXTIRX, JRBIY XRVIRHX BYD TXR
XVBYDBJKTX ABWX KQ GBILNYO GNX RYRLNRX … GR BJXK GBX
B AGKJR JNFIBIW KQ LBONV FKKMX. GR XHINERX BJJ HGR
HNLR HK RJNLNYBHR OKD VKLSJRHRJW XK HGBH GR LBW NY
QTHTIR XRIER B DNQQRIRYH LBXHRI.
```

We can start by counting how often each letter appears in the ciphertext.

A = 3	G = 12	M = 4	S = 3	Y = 17
B = 26	H = 20	N = 22	T = 5	Z = 0
C = 1	I = 15	O = 4	U = 1	
D = 9	J = 16	P = 1	V = 5	
E = 2	K = 15	Q = 7	W = 7	
F = 4	L = 10	R = 35	X = 30	

We can covert these to probabilities by dividing each number by the total number of letters, namely 264. We get

A = .0113636	G = .0454545	M = .0151515	S = .0113636	Y = .0643939
B = .0984848	H = .0757575	N = .0833333	T = .0189393	Z = .00000
C = .0037878	I = .0568181	O = .0151515	U = .0037878	

D = .0340909   J = .060606   P = .0037878   V = .0189393

E = .0075757   K = .0568181   Q = .0265151   W = .0265151

F = .0151515   L = .0378787   R = .1325757   X = .1136363

It's often preferable to work with probabilities, but if you prefer "percent chance," you may multiply all of the above values by 100. For example, R makes up about 13.26 percent of the ciphertext.

We now compare these to the probabilities of the letters in normal (plaintext) English.

Table 1.1 Probabilities of Letters in English

Letter	Probability	Letter	Probability
A	.08167	N	.06749
B	.01492	O	.07507
C	.02782	P	.01929
D	.04253	Q	.00095
E	.12702	R	.05987
F	.02228	S	.06327
G	.02015	T	.09056
H	.06094	U	.02758
I	.06966	V	.00978
J	.00153	W	.02360
K	.00772	X	.00150
L	.04025	Y	.01974
M	.02406	Z	.00074

Source: Beutelspacher, A., *Cryptology* (Washington, DC: Mathematical Association of America, 1994), 10.

As you can see, in normal English, the letter E stands out as being the most common (probable). The most frequent letter in our ciphertext was R, so we make the tentative assumption that R represents E.

We don't have to limit ourselves to looking at the frequencies of letters. We can also look at the frequencies of two-letter combinations, three-letter combinations, and even words. The most common two letter combination is TH. This is due, in part, to the great popularity of the word THE in English. We can hunt for the word THE by looking for the most common three-letter word in the ciphertext. Unfortunately, we have a four-way tie! The three-letter ciphertext words GNX, BYD, HGR, and BJJ each occur twice. However, we can quickly eliminate BJJ as a possibility because it doesn't have the right form. An enciphered version of THE must contain three distinct letters. Now, we suspected that E was replaced with R (because of the high frequency of R), and the possibility HGR ends with R, making it look like the best match for THE. By comparison, if GNX represents THE, we have X, a symbol with probability = .1136363 representing E. If BYD represents THE, we have D, a symbol with probability = .0340909 representing E. This latter probability is way too small. We cannot eliminate X as standing for E because its probability is close to that of E, but R still seems like the best fit. So we tentatively identify HGR as THE. Filling the cipher letters HGR in with their proposed equivalents, we get the following:

```
H      E T      TE E TE                ,    H    T ,
GNX LBURXHW NX NYHRIRXHRD KYJW NY ANPBIDX, BJVGWLNXHX,


        T      THE      E,             E   E    E T
MBFFBJNXHX BYD HGR JNMR,  XSBINYO YK RCSRYXR HK QNYD


            T E    E , E    E ET          E
BJJ MNYDX KQ HIRBXTIRX, JRBIY XRVIRHX BYD TXR


                  H       H    E E E ... HE      H
XVBYDBJKTX ABWX KQ GBILNYO GNX RYRLNRX ... GR BJXK GBX


    H E                      . HE  T  E      THE
B AGKJR JNFIBIW KQ LBONV FKKMX. GR XHINERX BJJ HGR


T E T E      TE        ETE      TH T HE
HNLR HK RJNLNYBHR OKD VKLSJRHRJW XK HGBH GR LBW NY


    T  E  E  E      E E T    TE .
QTHTIR XRIER B DNQQRIRYH LBXHRI.
```

While we cannot yet read the message, the work so far seems okay. If making the substitutions indicated had led to words like EHE or HHE that don't exist in

English, we would backtrack and try other substitutions. The words that remain partial (i.e., still have letters missing) offer possibilities. For example, HGBH, which we identified as TH-T with only the third letter unknown, must be THAT. If we had a situation where it was impossible to fill in missing letters to make a valid English word, we would have to backtrack and try different substitutions.

Now, identifying B as A, we can fill in all appearances of B in the ciphertext and look for other easily identifiable words.

```
H    A E T       TE E TE           A  , A H    T ,
GNX LBURXHW NX NYHRIRXHRD KYJW NY ANPBIDX, BJVGWLNXHX,

 A  A   T A   THE    E,  A        E  E  E T
MBFFBJNXHX BYD HGR JNMR, XSBINYO YK RCSRYXR HK QNYD

A            T EA   E ,  EA   E  ET A      E
BJJ MNYDX KQ HIRBXTIRX, JRBIY XRVIRHX BYD TXR

  A A    A      HA    H   E E  E ... HE A    H
XVBYDBJKTX ABWX KQ GBILNYO GNX RYRLNRX ... GR BJXK GBX

A H E    A      A       . HE  T   E A    THE
B AGKJR JNFIBIW KQ LBONV FKKMX. GR XHINERX BJJ HGR

T  E T  E    ATE         ETE      THAT HE A
HNLR HK RJNLNYBHR OKD VKLSJRHRJW XK HGBH GR LBW NY

   T  E  E EA    E ET A TE .
QTHTIR XRIER B DNQQRIRYH LBXHRI.
```

Once again, things look good. Our substitution has not introduced any impossible words, and we have the one-letter word A appearing twice. There are many possible routes to solving a MASC. Another person attacking this cipher might have begun by noticing that the cipher letter B appeared by itself twice, and therefore must represent A or I, the only one-letter words in English.

Similarly, we can continue our decipherment in a number of ways. Notice that we have a two-letter word that begins with T. The only possibility is TO, so we now replace the previously unknown letter K with O everywhere.

```
H     A E T        TE E TE  O          A   , A H    T ,
GNX LBURXHW NX NYHRIRXHRD KYJW NY ANPBIDX, BJVGWLNXHX,


 A   A  T A   THE     E,   A      O E  E  E TO
MBFFBJNXHX BYD HGR JNMR, XSBINYO YK RCSRYXR HK QNYD


A         O  T EA   E , EA    E  ET A     E
BJJ MNYDX KQ HIRBXTIRX, JRBIY XRVIRHX BYD TXR


   A   A O    A   O  HA      H   E E  E … HE A  O H
XVBYDBJKTX ABWX KQ GBILNYO GNX RYRLNRX … GR BJXK GBX


A  HO E      A   O   A      OO . HE  T   E  A    THE
B AGKJR JNFIBIW KQ LBONV FKKMX. GR XHINERX BJJ HGR


T   E TO E      ATE  O  O    ETE    O THAT HE A
HNLR HK RJNLNYBHR OKD VKLSJRHRJW XK HGBH GR LBW NY


   T   E   E  EA      E E T  A TE .
QTHTIR XRIER B DNQQRIRYH LBXHRI.
```

When attacking such ciphers, progress tends to become more and more rapid as letters are recovered. Feel free at this point to stop reading and complete the decipherment yourself.

Your path to the solution may differ from mine. The next thing that jumps out at me is the word A?? where the second and third letters must be the same. Running through the alphabet, the only possibilities are ADD, ALL, and ASS. (Remember, the message is from 1606, so APP won't work!) There's no need to try all three possibilities substituted throughout the message. We simply look at the word in context. Which do you like best:

ADD THE T--E   or   ALL THE T--E   or   ASS THE T--E

Though it is not guaranteed, the second choice looks the best to me. My mind automatically fills in the missing letters to get the phrase ALL THE TIME. One can imagine ADD THE TIME as a solution, perhaps as part of a mathematical equation written out in words, but replacing J with D everywhere makes it impossible to complete some other words when restricting ourselves to valid English. So, once again we fill in a newly uncovered letter, L, and look for more easily identified words. We'll leave T - - E as is for the moment, in case the phrase was ALL THE TIDE or some other unlikely possibility.

```
H   A E T       TE E TE  O L      A  , AL H   T ,
GNX LBURXHW NX NYHRIRXHRD KYJW NY ANPBIDX, BJVGWLNXHX,

 A  AL T A  THE L  E,  A      O E  E  E TO
MBFFBJNXHX BYD HGR JNMR, XSBINYO YK RCSRYXR HK QNYD

ALL      O  T EA   E , LEA   E ET A     E
BJJ MNYDX KQ HIRBXTIRX, JRBIY XRVIRHX BYD TXR

   A  ALO   A  O  HA     H  E E  E ... HE AL O H
XVBYDBJKTX ABWX KQ GBILNYO GNX RYRLNRX ... GR BJXK GBX

A  HOLE L  A  O  A     OO . HE  T  E  ALL THE
B AGKJR JNFIBIW KQ LBONV FKKMX. GR XHINERX BJJ HGR

T  E TO EL    ATE O  O  LETEL   O THAT HE  A
HNLR HK RJNLNYBHR OKD VKLSJRHRJW XK HGBH GR LBW NY

   T  E  E EA     E ET  A TE .
QTHTIR XRIER B DNQQRIRYH LBXHRI.
```

At this stage, we can recognize AL-O as ALSO and -HOLE as WHOLE. You might recognize other words as well. If not, filling in the newly identified S and W will help.

```
H S A EST  S  TE ESTE  O L     W A  S, AL H   STS,
GNX LBURXHW NX NYHRIRXHRD KYJW NY ANPBIDX, BJVGWLNXHX,

 A  AL TS A  THE L  E,  A      O E  E SE TO
MBFFBJNXHX BYD HGR JNMR, XSBINYO YK RCSRYXR HK QNYD

ALL    S O  T EAS  ES, LEA   SE  ETS A    SE
BJJ MNYDX KQ HIRBXTIRX, JRBIY XRVIRHX BYD TXR
```

```
S A  ALO S  WA S  O   HA        H S E  E   ES  …  HE ALSO HAS
XVBYDBJKTX ABWX KQ GBILNYO GNX RYRLNRX … GR BJXK GBX
```

```
A WHOLE L   A    O   A      OO S. HE ST   ES ALL THE
B AGKJR JNFIBIW KQ LBONV FKKMX. GR XHINERX BJJ HGR
```

```
T  E TO EL    ATE  O   O  LETEL  SO THAT HE  A
HNLR HK RJNLNYBHR OKD VKLSJRHRJW XK HGBH GR LBW NY
```

```
   T  E SE  E A     E E T  ASTE .
QTHTIR XRIER B DNQQRIRYH LBXHRI.
```

We could continue recognizing partial words, but there is a powerful tool that we haven't made use of yet. Various people have, over the years, gone through the trouble to categorize all of the words, in various size English dictionaries, by their patterns. At first this was done by hand, but now computer programs allow such lists to be formed rapidly. With such a list in hand, or running on a computer, it's easy to look at all words with a given pattern. For example, we have -O--LETEL-, where the missing letters are all distinct. One pattern word program, using a dictionary of 751,322 words[6] yields only a single possibility, COMPLETELY.

Using such programs (which return results in the blink of an eye and are freely available online) saves much thought. We may even begin our attack with this approach. If a ciphertext contains a word with a rare pattern, we might have a unique solution without knowing any letters, or at least with a short list of possibilities to consider. At other times, each word might have many possibilities, and it would be better to tentatively establish a few letters before making use of the pattern word lists. We now continue our example, filling in the newly identified C, M, P, and Y.

```
H S MA ESTY S   TE ESTE  O LY    W  A  S, ALCH M STS,
GNX LBURXHW NX NYHRIRXHRD KYJW NY ANPBIDX, BJVGWLNXHX,
```

```
A  AL  TS A   THE L  E,  A      O E  E SE TO
MBFFBJNXHX BYD HGR JNMR, XSBINYO YK RCSRYXR HK QNYD
```

```
ALL    S O  T EAS  ES, LEA   SEC ETS A    SE
BJJ MNYDX KQ HIRBXTIRX, JRBIY XRVIRHX BYD TXR
```

```
SCA  ALO S WAYS O  HA M    H S E EM ES … HE ALSO HAS
XVBYDBJKTX ABWX KQ GBILNYO GNX RYRLNRX … GR BJXK GBX

A WHOLE L   A Y O  MA  C  OO S. HE ST   ES ALL THE
B AGKJR JNFIBIW KQ LBONV FKKMX. GR XHINERX BJJ HGR

T ME TO EL M  ATE  O  COMPLETELY SO THAT HE MAY
HNLR HK RJNLNYBHR OKD VKLSJRHRJW XK HGBH GR LBW NY

   T  E SE  E A     E E T MASTE .
QTHTIR XRIER B DNQQRIRYH LBXHRI.
```

With or without a pattern word list, we can see that H-S MA-ESTY must be HIS MAJESTY. We might have guessed HAS for H-S, but A was already used as the decipherment for B. Filling in I and J in all of the appropriate positions gives us

```
HIS MAJESTY IS I TE ESTE  O LY I  WI A  S, ALCH MISTS,
GNX LBURXHW NX NYHRIRXHRD KYJW NY ANPBIDX, BJVGWLNXHX,

 A  ALISTS A   THE LI E, S A I    O E E SE TO  I
MBFFBJNXHX BYD HGR JNMR, XSBINYO YK RCSRYXR HK QNYD

ALL  I  S O  T EAS  ES, LEA   SEC ETS A     SE
BJJ MNYDX KQ HIRBXTIRX, JRBIY XRVIRHX BYD TXR

SCA  ALO S WAYS O  HA MI   HIS E EMIES … HE ALSO HAS
XVBYDBJKTX ABWX KQ GBILNYO GNX RYRLNRX … GR BJXK GBX

A WHOLE LI  A Y O  MA IC  OO S. HE ST I ES ALL THE
B AGKJR JNFIBIW KQ LBONV FKKMX. GR XHINERX BJJ HGR

TIME TO ELIMI ATE  O  COMPLETELY SO THAT HE MAY I
HNLR HK RJNLNYBHR OKD VKLSJRHRJW XK HGBH GR LBW NY

   T  E SE  E A I  E E T MASTE .
QTHTIR XRIER B DNQQRIRYH LBXHRI.
```

The letter J appears only in MAJESTY, but I has many appearances and helps us recognize more words. If you haven't already, you may wish to take this opportunity to stop reading and try to complete the recovery of the message yourself. The solution appears at the end of this chapter.

With today's computers, attacks like those detailed above can be carried out much more rapidly. Indeed, more sophisticated approaches allow such ciphers to be broken in mere seconds, sometimes in just a fraction of a second. Thus, the fact that the underlying language is uncertain is not as great a handicap as might be expected. If a computer attack requires a few seconds, testing 100 languages takes only a few hundred seconds. Of course, depending on the exact nature of the computerized attack, frequencies of the letters and/or pattern word lists would be needed in each language tried.

Yet the pencil and paper methods detailed here would have worked well enough for any expert Rudolf tasked with recovering the secrets of the strange manuscript he purchased. The fact that its mysteries refused to yield suggests that the scheme used was not a simple MASC after all.

At this point, I invite you to take a break from reading and try to break the MASCs given below.

1. PUCM CZODM MUL NLE VSJL OJ HDFOVJ MUL NLYOKF? MUL
   ZSOXHCIUB ZB LWCKN NCBN UL UCF C "ILKYUCKM JOH
   ILHWLHNL HLVCMSOKN CKF LHOMSYC."[7]

2. JFZ ZSBZMIM INXZ MZSLMEZV, "A ENIU JFLJ A LS VZLV
   LNV VLSNZV; A LS L SLN BIHHZHHZV QP JFZ VZOAK."[8]

3. JAAHRMKVW GH EIWIVM, RJOOK EHLI HZ PRJWUI ARIJGIM
   J WHEIY HUG HZ AEJN GH MIZIVM GXJG AKGN'T CILT JG
   J GKYI LXIV RUMHEZ LJT KV PHLIR.

4. LDEUCAHE TUZHE'X TEEJ "EHW RZHCUWC XDSBK AUBDM-
   FWX ALADHXB BRW NABRWNABDMDAHX" PAX CWCDMABWC BE
   UZCEFI.[9]

5. TUG GOPGFQF AQLCUT QLT TUG ZGAT RI GWGFXTURIC
   UG EGARFGE. UG GWGI UME URA UQFQAJQPG JMAT ZX
   IQATFMEMOLA.[10]

Evidence against the MASC Hypothesis

I used average word length as evidence that a MASC was used, but this statistic oversimplifies the matter. In English, we have the one-letter words I and A and many two-letter words that see frequent use. These are largely responsible for keeping the average word length fairly small. Although our cipher hits about the same average word length, it has a scarcity of one- and two-letter words. It gets the low average by having few long words and a great many four-letter words. Unless we are looking at two hundred–plus pages of profanity, it would seem that the cipher is not MASCed English after all. In fact, the distribution of word lengths also fails to match any of the other European languages, including Latin. However, if we are willing to favor a more exotic origin (from the perspective of Prague) for the manuscript, it turns out that the distribution of word lengths in Chinese, Tibetan, and Vietnamese resemble what we have.[11]

Sravana Reddy and Kevin Knight made a comparison of the word length distribution in a portion of the manuscript (called "VMS B," for reasons to be made clear later) with English (both with and without vowels), Pinyin, and Arabic Buckwalter. It's reproduced in fig. 1.10.[12] As you can see, normal English (with vowels) is the worst fit. Removing the vowels improves the fit, but the best fit of all is provided by Arabic Buckwalter. Reddy and Knight noted, "This is an example of why comparison with a range of languages

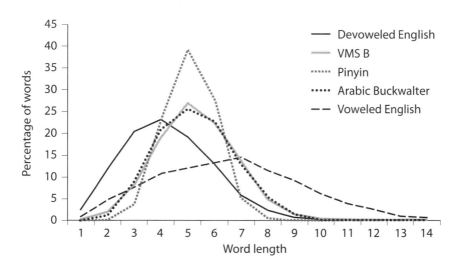

FIG. 1.10 A comparison of some word length distributions

is required before making conclusions about the language-like nature of a text."[13]

Fortunately, statistics is not the only weapon that can be used against a cipher. Context is often useful. In this case, it might help determine the language of the original message or the manner of encipherment. We have little to go on, but Rudolf II thought the cipher was created by Roger Bacon, so we'll begin there.

Roger Bacon

> The man is insane who writes a secret in any other way than one which will conceal it from the vulgar and make it intelligible only with difficulty even to scientific men and earnest students. On this point the entire body of scientific men have been agreed from the outset, and by many methods have concealed from the vulgar all secrets of science.
>
> —Roger Bacon, *De Secretis Operibus Artis et Naturae, et de Nullitate Magiae*, 1252[14]

Roger Bacon was a thirteenth-century English monk and philosopher. Over the centuries, his reputation has waxed and waned. At times, he was regarded as a pioneer of the experimental method, a man far ahead of his time, and at other times he was claimed to have been not so different from his philosophical contemporaries after all.

Bacon knew a wide range of languages. It's been claimed that he possessed Greek, Latin, Hebrew, Aramaic, and probably even a little Arabic, so presuming his authorship of the manuscript hardly narrowed down the plaintext possibilities for Rudolf II. However, this wouldn't be a setback that would easily deter Rudolf from trying to get at the plaintext, for Bacon wrote about all areas of science, including the occult realm. In *The Mirror of Alchemy*, Bacon told how to create an elixir that he claimed could not only transmute metals, but also prolong life. Was the true procedure concealed in the enciphered manuscript? Finding out would have been of the greatest importance to Rudolf.

The Inquisition began in the thirteenth century, so if Bacon wrote a heretical manuscript, he had good reason to hide the contents behind a MASC. But if he did take such precautions, he was still not careful enough. Plaintext comments made elsewhere, like "The whole clergy is given up to pride, luxury and avarice" brought him unwanted attention.[15] He spent fourteen years imprisoned for his daring statements, during which time he was not allowed to write. Finally released in 1292, he died later that same year.[16] Hundreds of years later, a historian claimed that all of Bacon's manuscripts that could be found at the monastery in Oxford were nailed to the walls by the friars and left to rot.

Much more can be said about Bacon's life, but what is most important to us is how involved he was with cryptology. The quotation that opened this section shows that he appreciated the importance of secret writing. But could he implement a system more complex than a MASC, as seems to be the case with the manuscript in Rudolf's possession?

In an essay contributed to a collection celebrating (roughly) seven centuries since the birth of Roger Bacon, Lieutenant Colonel H. W. L. Hime claimed, "... in the *De Secretis* he [Bacon] took extraordinary care to conceal, by cryptic methods and anagrams, the names and proportions of the ingredients which formed the explosive [gunpowder]."[17]

It's well known that the Chinese had gunpowder before England and Europe, but medieval science and alchemy historian Lynn Thorndike wouldn't even accept Bacon as having been the first in his part of the world to discover it. He dismissed Hime's claims, and the matter remains in dispute.[18] However, there are cryptologic writings by Bacon that are not in dispute.

In *De Secretis Operibus Artis et Naturae, et de Nullitate Magiae*, 1252, which was quoted at the beginning of this section, Bacon went on to describe seven techniques for hiding secrets:

1. Hide a message "under characters and symbols" (e.g., use code or jargon).
2. "In enigmatical and figurative expressions" (code or jargon). Bacon doesn't clearly distinguish these first two methods. He only gives one example, which is a code used by philosophers and alchemists.
3. "Writing with consonants only" (simply omit vowels to make reading harder).
4. "By commingling letters of various kinds" Some authors interpret this as the use of nulls—meaningless symbols inserted randomly to cause confusion—or the combined use of nulls and transposition, changing the order of the letters. However, Bacon wrote, "it is in this way that Ethicus the astronomer concealed his scientific knowledge

by writing it in Hebrew, Greek and Latin letters in the same written line,"[19] so I believe that their interpretation is wrong. I think Bacon meant writing a letter, like a, for example, variously as α (Greek), א (Hebrew), and a (Latin).

5. "By means of special letters, devised by their own ingenuity and will, and different from those anywhere in use" (substitution).

6. "Actual letters are not used but other geometric figures that function as letters according to the arrangement of points and marks" (substitution).

7. "Noting and writing down as briefly as we please" (shorthand).

Method 3 is especially weak. Bacon wrote, "In this way the Hebrews, Chaldeans, Syrians, and Arabs write their secrets," yet he also noted "Indeed, as a general thing, they write almost everything in this way."[20]

Bacon also wrote, "I have thought fit to touch upon these methods of concealment because I may perhaps, by reason of the importance of my secrets, employ some of these methods, and it is my desire to aid in this way, at least you, to the extent of my ability."[21]

So Bacon's interest in the subject went well beyond the casual. It's possible that he applied some combination of the techniques listed above to create the manuscript. But is there any evidence that the manuscript purchased by Rudolf II was old enough to have been penned by Bacon? And how could a manuscript created by an English monk end up in Prague?

There's been much speculation concerning the history of the manuscript before 1608. Wilfrid Voynich, a twentieth-century owner of the manuscript, came up with a plausible scenario of how the manuscript could have passed from the hands of Roger Bacon down to the court of Rudolf II. Of course, it was very much in the best interest of Mr. Voynich for Roger Bacon to be the author, instead of someone almost no one has heard of, as that would increase the value of the manuscript that he owned. Although the evidence isn't strong enough to confirm Bacon as the author, I relate it here for your consideration.

According to Voynich, Roger Bacon created the manuscript in the latter half of the thirteenth century, and it sat in some monastic library in England until about 1538. Not a very exciting beginning, but it gets us much closer to Rudolf II in a single sentence—see how easy this is? The next step occurs when John Dudley, Duke of Northumberland, pillages the manuscript as the monasteries are dissolving in the turmoil of the Reformation. He is known to have received many spoils from churches and other religious institutions, but that the manuscript was among them is pure speculation. If it was, then it's possible that in 1547 John Dee obtained it from Dudley.

Although we don't know that Dee possessed the manuscript in question, it is known that Dee had a large collection of Bacon's manuscripts; he bought every Bacon manuscript that he could find. In fact, Dee is credited with reviving interest in Bacon's largely forgotten work. Dee traveled throughout England in his quest for manuscripts and assembled a tremendous library. It became the largest private library in England—"in all neere 4,000 volumes." He saved money in assembling this collection by sometimes borrowing books and "forgetting" to return them.

The conjectured path from Bacon to Rudolf II ends sometime between 1584 and 1588 with Dee taking the manuscript to Prague, where Rudolf II bought it for six hundred gold ducats. We'll examine the evidence for this purchase later. Voynich speculated that Dee gave it to the emperor as a "present."

There are small bits of evidence both for and against this conjectured path:

1. Against: If the manuscript was created by Roger Bacon, then it predates Columbus, so how could it include an illustration of a sunflower (see the image of leaf 33v reproduced early in this chapter), which is indigenous to the Americas? Columbus only brought sunflower seeds back to Europe in 1493. One could argue that the illustration in question is not actually a sunflower (remember the weird roots?) or, with more difficulty, presume that the artist learned of the sunflower from some pre-Columbian visitors to the Americas (Vikings? Chinese?).

2. For?: Some experts, such as Professor Andrew G. Watson, coeditor of *John Dee's Library Catalogue*, say that the page numbers placed on the leaves of the manuscript match Dee's handwriting.[22] Opposing this identification, we have Rafał Prinke, who sees the 8 as clearly not that of Dee. He provides some subtler distinctions, as have others.[23] Dee also liked to put his name, or ownership symbols, in his books and make marginal notes, and we do not find these here.

3. For: Dee had a strong interest in cryptology and described another book on the subject, *Steganographia*, as "the most precious jewel" and worth more than "a Thousand Crownes." His interest was known to powerful men such as Sir Francis Walsingham, who visited Dee often to learn more about the subject.[24]

4. For?: John Dee's son, Arthur, spoke of his father having, and spending much time studying, a book "containing nothing but hieroglyphicks."[25] It might be taken as a subtle distinction, perhaps one Arthur did not care to make, but the manuscript we are interested in looks more like some sort of strange handwriting than hieroglyphics. Also, the cipher manuscript has some pages that are text only, but most bear large colorful illustrations. It's not accurate to describe it as "nothing but hieroglyphicks."

5. For: Dee did travel to Rudolf's court several times between 1584 and 1588, but we do not know anything concerning the sale of enciphered manuscripts during these trips.

As was stated before, the scenario posed by Mr. Voynich is plausible, but there is not nearly enough evidence for us to say if it is probable.

Rudolf the Gullible *or* Was It a Hoax?

Rudolf surrounded himself with alchemists, and there may have been some who believed in what they were doing. A few may have fooled themselves into thinking that they were making actual progress in acquiring the secret of transmutation. Even Sir Isaac Newton, generally considered the greatest scientist of all time, spent a tremendous amount of time in this area, more than on the scientific work that brought him fame. However, other alchemists were far from sincere in their efforts. They made use of tricks such as hollow ladles that contained gold, sealed at the end with wax. As they stirred their pots, the wax would melt and gold would appear, as if generated from the mixture being stirred. Another trick, favored by Edward Kelley, a friend and coconspirator of John Dee, involved a double-bottomed crucible in which the gold filings were concealed by wax, until the "transmutation" took place. More sophisticated methods existed, as well. For those wanting to believe, such demonstrations could be taken at face value.

The question is, if Rudolf II could be fooled out of some of his wealth by chemical fakery, could he also have been fooled by a meaningless manuscript? Was the strange manuscript he purchased simply random characters from a made-up alphabet? Could the amount of work that went into creating the manuscript be worthwhile? It was more than two hundred pages long ...

Well, as was mentioned earlier, on the chance that this manuscript could help him in his quest, Rudolf is said to have paid six hundred gold ducats for it. This is the equivalent of $100,000 today. So, the payoff did justify the effort necessary to create the document.

Such a con would not have been beneath John Dee, who could have created the manuscript himself (more on this later). Dee worked as a spy at a time when such work was considered dishonorable. It's only relatively recently that the occupation has become glamorized through characters such as James Bond (although we may trace the glamorization back to James Fenimore Cooper's *The Spy*, 1821). Dee's alchemical adventures were long remembered, and there's even a nineteenth-century painting of "John Dee performing an alchemical feet for Rudolf II." So, Dee was certainly willing to mislead people if it would profit him.

In any case, there was a lot of competition among the con artists Rudolf attracted. Dee himself was conned when Kelley convinced him that the angels he was in communication with were conveying God's demand that Dee and Kelley share wives.

Dee's lifestyle eventually caught up with him, and he had to flee from Rudolf II. While he was away, five hundred volumes were taken from his library. Kelley, who was also leading the doubly dangerous life of con man and spy, was much luckier at this time and was knighted.[26] However, matters eventually came to a head for Kelley as well. He was thrown into Rudolf's dungeon and tortured. He made an escape attempt, but it didn't go well and he received wounds that led to his death.

Other Possibilities

It should also be borne in mind that the manuscript doesn't have to be a hoax to be devoid of meaning. It's possible that the author was mentally ill, under the effects of drugs, suffering from migraines, or in religious ecstasy. Falling into one of these categories, he could have created the manuscript without the intent to deceive anyone. However, someone finding it later, not knowing the provenance, would expect a decipherment to exist. Brumbaugh wrote of the manuscript's illustrator, "His ability to see plants as animate verges on the mentally deranged."[27] Comparison with texts known to have been produced by people in a different sort of reality, for one reason or another, does not yield great similarities.

Or perhaps the manuscript was only intended to be taken as art. Such works do exist. One example is Luigi Serafini's 1981 book *Codex Seraphinianus*. The images indicate that this is an encyclopedialike book that covers an even wider range of topics than our strange manuscript. It can be appreciated as surrealist art, even if there's no meaning in it. However, the images are accompanied with what, at a glance, appears to be ciphertext rendered using a unique character set. It's not known whether the text is gibberish, but it has been determined that the pages are numbered in a sort of cipher.[28]

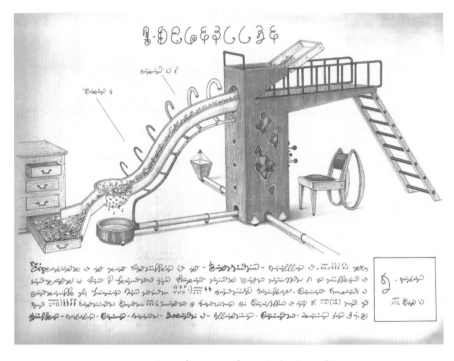

FIG. 1.11 A random page from *Codex Seraphinianus*

Rudolf's Efforts

There's no record of Rudolf II ever considering the hoax hypothesis. Indeed, if he came to this conclusion, we might well have another account of a con man like Kelley meeting a hard fate. Rudolf is not likely to have taken being

FIG. 1.12 Jacobus Horcicky de Tepenecz (1575–1622)

ripped off for six hundred gold ducats lightly! What we do have is a record of various experts close to Rudolf II trying to make sense of the manuscript for him.

We've speculated about the history of the manuscript before its arrival at the court of Emperor Rudolf II in Prague, but now some hard facts begin to appear. We begin this portion of the history with Jacobus Horcicky de Tepenecz, who was Rudolf's chief botanist and apothecary, and in charge of the imperial gardens. His greatest success was in finding a way to transfer the scent of flowers to a liquid, thereby inventing perfume. The "de Tepenecz" (of Tepenec) portion of his name was a patent of nobility that he earned in 1608 as a reward for saving Rudolf's life with a potion. With the large number of plants depicted in the manuscript, it would make sense for Jacobus to be among the first to try to decipher it.

Jacobus's connection with the manuscript was discovered accidentally when a later owner tried to have a photostatic reproduction made of the first page. The plate was underexposed, and this brought out a faded signature in the lower margin. It was then brought out more clearly by applying chemicals to the original. The signature seems to indicate that Jacobus owned the manuscript. For why would he write his name on it if Rudolf had simply loaned it to him for a decipherment attempt? You don't write your name in a book you borrowed, unless, like John Dee, you don't intend to return it!

Jacobus's signature included the title "de Tepenecz," so we know he possessed the manuscript sometime after 1608. Also, below his name, "Prag" is written, so it must have been before 1618, when Jacobus left the city.

Being in the good graces of Rudolf was not to be of lasting value. Rudolf fell from power and was replaced by his brother Matthias, although he officially retained the title of Holy Roman Emperor until his death in 1612. So the secrets of immortality, if present in the manuscript, were not set free in time to benefit Rudolf. It may well be that Jacobus took possession of the manuscript during Rudolf's last days or shortly after his death. In 1618, the Protestants had gained control of Prague, and Jacobus, a Catholic serving as district administrator in nearby Melnick, was thrown into prison. Events turned in Jacobus's favor again when Archduke Ferdinand's victory put the Catholics in charge of Prague. Jacobus's death followed from a riding accident in 1622.

Georg Baresch

There has, of course, been speculation on where the manuscript went next, but nothing is known for sure. The next confirmed owner was Georg Baresch (a.k.a. Georgius Barschius), although there could have been one or more owners between Jacobus and Georg. The identification of Georg as an owner came about only as the result of the discovery of a letter in the archives of the Pontificia Università Gregoriana in Rome.[29] It was first transcribed by M. J. Gorman, and then published by Smolka and Zandbergen in 2010. It's reproduced below. The lavish praise is due to Baresch's recognition of Kircher as a great linguist. Kircher wrote a book on Coptic, the language of early Christians in Egypt, and it was thought that he might be the one to unravel Egyptian hieroglyphs, which at the time were understood by no one.

The Letter of Georgius Barschius to Athanasius Kircher (1639)[30]

Most Revered Father

Without further ado, I pray for all happiness from the author of happiness to your Reverence.

Since a clerical individual is setting out for Italy and Rome itself I have seized the occasion to get him to take this letter with him. I hope it may reawaken your memory of a certain item of writing which I sent via the Reverend Father Moretus, priest of the Society of Jesus in Prague.

Your Reverence won worldwide acclaim by publishing the Prodromus Copticus. Now in it, among other things, you requested all those who might possess anything which might enrich the work for help with materials to increase the resources for bringing your work to fruition.

After that I do not doubt that people sent hordes of documents laden with such riches to the Universal City. Many must have turned up in person for the purpose of praising the author's unprecedented efforts for the republic of letters and his near superhuman labours on the very spot where they took place. When this most welcome news reached me my source also showed me a brief synopsis of this astounding work which was to see the light in our time (and I say the sooner the better). He also informed me of your unheard of skill in untying the riddles of some very obscure scripts.

Now since there was in my library, uselessly taking up space, a certain riddle of the Sphinx, a piece of writing in unknown characters, I thought it would not be out of place to send the puzzle to the Oedipus of Egypt to be solved. And so I ordered a certain old book to be transcribed in part, with the writing closely imitated (the bearer of this letter will inform you that he saw it with his own eyes). A year and a half ago I sent that writing to your Reverence. My hope was that (if your Reverence should see fit to assign some working time to investigating it and correlating those characters of unknown devising with known letters) the effort might (to the extent that the matters concealed in the book proved to be worthy of such first class work) benefit its Oedipus, and myself, and the common good.

It did not seem advisable to commit the book itself to a journey so long and full of perils. I could only suppose from the fact that I have had no word of the matter after all this time that the previous consignment did not reach Rome. I had therefore just decided to repeat the exercise when the above Father Moretus informed me to my great pleasure that it had ended up in the City. It would be an even greater pleasure if the said book could be opened by your Reverence's aid, and make all men of quality co-owners of whatever good it contains.

From the pictures of herbs, of which there are a great many in the codex, and of varied images, stars and other things bearing the appearance of chemical symbolism, it is my guess that the whole thing is medical, the most beneficial branch of learning for the human race apart from the salvation of souls. This task is not beneath the dignity of a powerful intellect. After all, this thing cannot be for the masses as may be judged from the precautions the author took in order to keep the uneducated ignorant of it. In fact it is easily conceivable that some man of quality went to oriental parts in quest of true medicine (he would have grasped that popular medicine here in Europe is of little value). He would have acquired the treasures of Egyptian medicine partly from the written literature and also from associating with experts in the art, brought them back with him and buried them in this book in the same script. This is all the more plausible because the volume contains pictures of exotic plants which have escaped observation here in Germany.

Your Reverence shines with ardour to bring the best things to light. I do hope you will not consider it beneath you to act for the common good and bring forth the good (if any there is) buried in unknown characters in this book. Indeed nobody here has the capability to take on the burden, seeing that it is of an obscurity requiring unique intelligence and mental dexterity or else certainly another method not easily grasped. I should be forever in your debt, not only for what the work contains but also for whatever it will make possible. I here append a line or two of the unknown script to revive your memory of it, having previously sent a whole file of similar characters.

With this I commend myself to your Reverence and I wish you a successful conclusion to your high labours. May God the Greatest and Best long preserve you for the republic of letters.

Prague AD 1639 27 April, on which day I once entered the Roman University of Wisdom to dedicate my work to medical wisdom.

Best wishes to your Reverence
Mr Georgius Baresch

FIG. 1.13 Johannes Marcus Marci (1595–1667)

The previous letter that Georg referred to is now lost, as are biographical details of its author. Nor is there any record of Kircher's response, or even an indication that he did respond. But we do know the next stop the manuscript made. When Baresch died in 1662, he left all of his books, including the enciphered manuscript, to Johannes Marcus Marci of Kronland.

Marci lived in a time when it was still possible to be highly respected in several different fields. He was a physician, but he was also highly regarded as a mathematician, physicist, and orientalist. He served as rector of the University of Prague.

It's not known how much time Marci spent trying to make sense of the manuscript. We do know that near the end of his life he mailed it to an even greater intellect, Athanasius Kircher in Rome, the same man Baresch had approached. A letter from Marci to Kircher that accompanied the manuscript was found with it centuries later. This letter is the source of much of the information we have concerning the manuscript's provenance. It's reproduced below.[31]

Reverend and Distinguished Sir;
Father In Christ:

This book, bequeathed to me by an intimate friend, I destined for you, my very dear Athanasius, as soon as it came into my possession, for I was convinced it could be read by no-one except yourself.

The former owner of this book once asked your opinion by letter, copying and sending you a portion of the book from which he believed you would be able to read the remainder, but he at that time refused to send the book itself. To its deciphering he devoted unflagging toil, as is apparent from attempts of his which I send you herewith, and he relinquished hope only with his life. But his toil was in vain, for such Sphinxes as these obey no-one but their master, Kircher. Accept now this token, such as it is, and long overdue though it be, of my affection for you, and burst through its bars, if there are any, with your wonted success.

Dr. Raphael,[32] tutor in the Bohemian language to Ferdinand III, then King of Bohemia, told me the said book had belonged to the Emperor Rudolph and that he presented the bearer who brought him the book 600 ducats. He believed the author was Roger Bacon, the Englishman. On this point I suspend judgment; it is your place to define for us what view we should take thereon, to whose favor and kindness I unreservedly commit myself and remain

> *At the command of your Reverence,*
> *Joannes Marcus Marci,*
> *of Cronland*
> *Prague, 19th August 1665*

Note: The date on the original could be 1666. The last digit of the year is difficult to read.

This letter posed a mystery for decades. Who was "The former owner of this book once asked your opinion by letter"? As was previously related, the correspondent's identity is now known to have been Georg Baresch.

FIG. 1.14 Athanasius Kircher (1602–1680)

There was very little that Kircher was not interested in. The frontispiece of his book *Ars Magna Sciendi Sive Combinatoria* (*The Great Art of Knowledge, or the Combinatorial Art*) has an inscription that reads "Nothing is more beautiful than to know all," and this seems to have been his goal. He made contributions to an incredible range of subjects, creating more than forty works that covered topics as diverse as Chinese calligraphy, baroque music, the cosmos, fossils, geology, magic lanterns, magnetism, medicine, pineapples (an exotic fruit that he said could devour iron nails), and the subterranean world. He risked death in seeking knowledge through direct experiences like being lowered into the boiling crater at Mount Vesuvius. It was bad enough at the top, as he explained.

> Methoughts I beheld the habitation of Hell; wherein nothing else seemed to be much wanting, besides the horrid fantasms and apparitions of Devils. There were pereceived horrible bellowings and roarings of the Mountain; An unexpressible stink; Smoaks mixt with darkish globes of Fires; which both the bottom and sides of the Mountain continually belch'd forth out of Eleven several places; and made me in like manner, ever and anon, belch, and as it were, vomit back again, at it.[33]

Kircher also compiled reports of others, such as Jesuit missionaries returning from the Far East. He even anticipated Darwin by suggesting that animal species evolve.

Such broad learning may have emboldened him to tackle a manuscript that, as the illustrations indicated, seemed to cover an impressive range as well. He would definitely have wanted to get at the knowledge concealed within the cipher manuscript, and now he owned it. As we saw, both Baresch and Marci thought Kircher stood the best chance at breaking the cipher. In addition to the impressive credentials described above, Kircher possessed special knowledge and skills that recommended him as a codebreaker.

First off, Kircher was employed as a professor of mathematics at the Roman College. It wasn't widely recognized until World War II, but mathematicians can make great cryptanalysts. Some of Kircher's mathematical work fell in the category of combinatorics, which at the time meant determining how many combinations were possible of items taken from some finite set. Ideally, the number is determined without counting directly by listing all possibilities. That is, shortcuts are desired. Kircher's attempts resulted in what may be the first sketches of complete bipartite graphs. These are graphs in which the points are divided into two groups and lines join every point in group 1 with every point in group 2. The total number of lines is the number of combinations with exactly one item from each group. Kircher also devised a system of logic.

Kircher addressed cryptography directly in 1663 in *Polygraphia Nova et Universalis ex Combinatoria Arte Detecta*, which had some original contributions, but mostly presented the work of others, including systems that are more sophisticated than the MASCs we've already examined. The most important part of the book was the portion in which he presented a system of *pasigraphy.* This is often called "universal writing." The aim was to create an artificial language that would be easy to learn, yet powerful enough to embrace complex ideas. Such writing could then serve all of humanity in its interactions. Kircher's version replaced the words we're used to with numbers, so it would only have been practical for written communication, not oral. Later systems, like Esperanto (1887), which was a universal *language*, sought to fill both needs.

In addition to this, as was mentioned earlier, Kircher was a great linguist, who was expected to unravel Egyptian hieroglyphs, hence, the references to Sphinxes in the letter from Marci. Kircher also studied Greek, Hebrew, and Syrian.

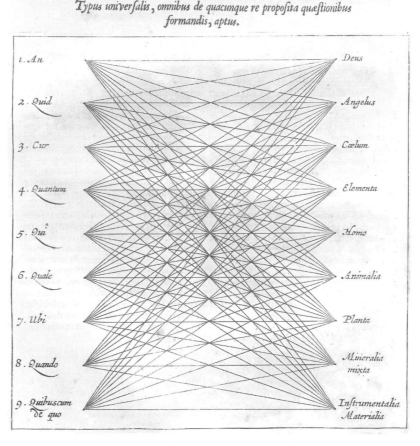

FIG. 1.15 A page from *Ars Magna Sciendi* showing
a complete bipartite graph

In some ways, Kircher mirrored Rudolf II. Like Rudolf, he was a collector. His museum at the Roman College, one of the first public museums in the world, contained "everything from stalactites to a stuffed crocodile."[34] He also had rooms filled with the latest technological innovations. Goldstone gave a list of examples: "magic lanterns (slide projectors), thermometers, megaphones (he linked these up from room to room to form an intercom system), spheres, *Smicroscopia* (Kircher's name for the microscope), magnets, trick mirrors. There were telescopes manned by Jesuits on the roof of the observatory tower, and as a practical joke, one of the megaphones opened up behind a Greek statue in the foyer, which Kircher nicknamed 'the Delphic Oracle.'"[35] He also built a magnetic clock and a musical fountain.

Again, like Rudolf, Kircher was interested in both science and the occult. This was normal for scientists up through Isaac Newton. After Newton, who has been called "the Last Sorcerer," the occult fell out of favor among scientific minds. Unlike Newton, Kircher rejected alchemy, but he did accept astrology, and he wrote about Atlantis for a few pages in *Mundus Subterraneus*. His mingling of scientific and pseudoscientific views may not have been a disadvantage when it came to trying to decipher the manuscript. For if it concealed ideas contrary to modern science, someone with one foot in each world might have an easier time understanding it.

Alas, Kircher did not succeed, or if he did, no trace of his work remains. Other than it having been mailed to him, there is absolutely no known connection between him and the document. No scrap of paper in which he even mentions it has been found, nor is there any listing for it in the catalogue of the Museo Kircheriano, detailing his collections. One might conclude that it got lost in the mail if the further history of the manuscript didn't indicate otherwise. Despite all of his qualifications, it's almost certain that Kircher failed to get at the manuscript's secrets. For if he did, his tremendous amount of writing would surely have included some mention of his result, and nothing survives addressing the subject.

As for his more public decipherment attempt, Kircher did publish several works on Egyptian hieroglyphics, and he was celebrated as having found the correct solution for some time, but later work would show that his "solutions" were completely wrong. That riddle would eventually be answered, more than 140 years after Kircher's death, by the Frenchman Jean-François Champollion (1790–1832). How long would the world have to wait for the solution to the mysterious enciphered manuscript?

Twentieth-Century Rediscovery
or The "Voynich" Manuscript

It's not known if anyone studied the cipher manuscript in the years immediately following Kircher's death, or even in the eighteenth or nineteenth centuries. Rudolf II was unable to make anything of it, nor were the intellectuals who were described in the preceding pages. Dormant for hundreds of years, the manuscript would arise again to challenge twentieth-century masters of the craft.

It was rare book dealer Wilfrid Voynich who rediscovered the manuscript in 1912. As a consequence, it is now typically referred to as "the Voynich manuscript." In a 1921 lecture delivered in Philadelphia, Voynich explained how he found it:

> In 1912, during one of my periodic visits to the Continent of Europe in quest of rare old books and manuscripts, I came across a most remarkable collection of precious illuminated manuscripts. For many decades these volumes had lain buried in the chests in which I found them in an ancient castle in Southern Europe where the collection had apparently been stored in consequence of the disturbed political condition of Europe during the early part of the nineteenth century. Most of these manuscripts must formerly have belonged to the private libraries of various ruling houses of Italy, now extinct, since many of them were embellished with the arms of such personages as the Dukes of Parma, Ferrara and Modena.[36]

Voynich then went on to describe the enciphered manuscript:

> It was such an ugly duckling compared with the other manuscripts, with their rich decorations in gold and colors, that my interest was aroused at once.[37]

Based on his great experience with old manuscripts, he estimated it as having originated in the "latter part of the thirteenth century."[38] This date, along with the wide range of subjects covered, led him to conclude that the author was either Albertus Magnus or Roger Bacon. But whereas Albertus had no need to use a cipher, Bacon had suffered persecution and may therefore have been motivated to do so. Thus, Voynich concluded that Bacon was the most likely author. Voynich's theory of how the manuscript could have passed from Bacon to Rudolf II and on to Kircher has already been recounted, but where it went after Kircher has not. Voynich addressed this, as well.

> My own impression is that Kircher left the manuscript to someone at the Court of Parma, where he had patrons and friends, and it probably remained in the possession of a member of the Farnese family until, with other manuscripts, it was removed to the collection in which I found it.[39]

Voynich concluded his lecture by repeating this conjectured path.

> ... we may suppose that it was presented by Kircher to a patron in one of the ruling houses of Italy, after which it remained buried until it was discovered by me in 1912.[40]

A footnote appearing earlier in the paper explains Voynich's reluctance to provide the exact location of the purchase:

> As I hope some day to be able to acquire the remaining manuscripts in the collection, I refrain from giving details about the locality of the castle.[41]

These details are important because, as will soon be shown, Voynich was lying.

When Voynich obtained the cipher manuscript in 1912, some of the leaves were missing. We know this because they were numbered and reveal that today we have 102 of an original 116. Recall that each leaf has a front (recto) and a back (verso). Hence, as Kahn and others state, there are 204 pages and 28 others are missing. Some sources refer to as many as 262 pages; such numbers are obtained by counting folding pages as more than one each. Were the missing pages gone, along with the cover, before Rudolf II purchased it, or were they removed later? What did the missing pages contain?

Biography of a Bookseller

The life of a bookseller may not sound exciting, but in Voynich's case, it is. He had adventures all over Europe. An employee of his said, "He spoke eighteen languages, all, so he told us, equally badly."[42] A look at his life raises many interesting suspicions. As the relevant details are presented, please consider what they might imply.

MICHAŁ WOJNICZ

FIG. 1.16 Wilfrid Voynich (1865–1930)

Wilfrid Voynich was born in Lithuania (his name was originally Michal Wo-
jnics). He earned a degree in chemistry from Moscow University and became
a pharmacist. But he did not pursue the safe, comfortable life such a career
might imply. Instead, he became involved with the movement to free Poland
from its Russian oppressors. Involved in a failed plan to free two other Polish
nationalists, who had been sentenced to death, Voynich ended up in prison
himself in 1885. Voynich didn't get a death sentence. He didn't even get a
trial. He simply spent a year in isolation in a tiny cell in the Warsaw Citadel.
Like a one-man A-team, he escaped from prison and continued fighting. Years
later, he pointed to three scars on his body saying, "Here I have sword, here I
have sword, here I have bullet."[43] Voynich was caught a second time and sent
to the salt mines in Eastern Siberia, from which he also escaped. This time he
didn't stick around, but rather made his way to England, arriving in 1890.

A woman later related some details of this trip:

Having fled from Siberia in 1890, he finally reached Hamburg with no means
of any sort. At that time there was a treaty between Germany and the Tsarist

government about the extradition of criminals. So Voynich hid in the docks, concealing himself in the stables. He fed himself what bits and scraps he could get, until some small vessel loaded with fruit would sail to London. Having sold all that he had, including his waistcoat and spectacles, he had barely enough money to buy a third-class ticket, a herring, and some bread.

After a long and stormy voyage, during which the vessel was carried off to the Scandinavian coast and lost its cargo, he finally arrived in the London docks without a penny, crawling with lice, half dressed and hungry.

This was in the evening of 5 October 1890.

Not knowing English, he walked along Merchant (Torgov) Street holding out a scrap of paper to passers-by with a single London address— Stepnyak's address [*Stepnyak had fled to England before Voynich and then contacted him with his address prior to Voynich's second escape*]. Finally, a Jewish student who worked in a tobacco factory in a slum where foreigners huddled, not far from the docks, came to him and asked him, "You have the look of a political man, are you from Siberia?" This student led him to Stepnyak's house. That very evening, Stepnyak was expecting another emigrant, and his wife, his wife's sister and me and (it seems) Felix Volkovsky. We were all there when the unknown man from the docks arrived.

"Here is yet another," said Stepnyak, explaining to us that this was not the man that we expected.

Later in the evening, when Voynich had washed himself and changed into clean and ill-fitting clothes belonging to someone else, he turned to me and asked in Russian: "Could I have met you before? Were you in Warsaw at Easter in 1887?" "Yes", I answered, "I was on my way to Petersburg." "You were standing in the square and looking at the citadel?"

When I again said "Yes," he told me that he was a prisoner in that same citadel and had seen me from there. Shortly afterwards, he was dispatched to Siberian exile.[44]

The woman relating the tale was Ethel Boole, daughter of the famous English logician George Boole. She was actively involved in the same cause as Voynich, and after his arrival in England, the two engaged in actions together, such as smuggling banned books (some by Marx and Engels) into Russia. Voynich continued his anti-tsar activities under the name Ivan Klecevski. He used the pseudonym because he feared the harm that could come to relatives in Europe if their connection to him became known. Ethel Boole used the name Bulochka. In 1897, the novel *Gadfly* appeared under the name Ethel

Voynich. It sold millions of copies in Russia, and later in China, but these were not nations from which royalties were received. Despite Ethel's most recent name change, the pair did not marry until 1902.

By this time, Voynich had made his reputation in England as an antiquarian book dealer by having somehow found and offered for sale many books believed lost, and in some cases not even known to have existed. E. Millicent Sowerby, an employee in Voynich's bookstore hired just after his purchase of the cipher manuscript, wrote of spies showing up at the bookshop and trying to keep tabs on Voynich in various ways.

Presumably, the spies weren't interested in the antiquarian book trade, but how Voynich made his spectacular finds was of interest to other dealers. He shared an example with a man whom he encouraged to join their ranks:

> Millions of books, shelves and shelves of the greatest rarities in the world. What I have discovered in Italy is altogether unbelievable! Just listen to this. I once went to a convent and the monks showed me their library. It was a mine of early printed books, codexes and illuminated manuscripts. I nearly fainted—I assure you I nearly fainted on the spot. But I managed to keep my head all the same, and told the monks they could have a most interesting and valuable collection of modern theological works to replace that dusty rubbish. I succeeded in persuading the Father Superior and in a month that whole library was in my hands, and I sent them a cartload of modern trash in exchange. Now take my advice: drop your present job and become a bookseller.[45]

Although this acquisition wasn't technically theft, a London Court established that there was a book in Voynich's possession that was stolen property from Lincoln Cathedral. Voynich described himself as "Catholic and agnostic" a few years before this court case.[46] Did his agnostic side consider religious institutions fair game? What did his Catholic side think of this?

Taratuta wrote

> He took an active part in the revolutionary movement. He was a member of the Polish Social Revolutionary Party, "Proletariat." He devoted himself to propaganda work, disseminating illegal literature. He got hold of a typewriter and a false passport and collected money for revolutionary work.[47]

So, Voynich had experience with fake documents.

Millicent indicates that Voynich was motivated by World War I to leave England in 1914 for America, and he did indeed open a shop in New York, but he also made trips to Europe, buying up collections at bargain war prices.

After the war, Voynich made serious attempts to get the manuscript deciphered. He made photocopies of some of the pages in 1919 and sent them to anyone he thought might be able to help. The experts included Professor A. G. Little (a Bacon scholar), H. Omort (a paleographer at Paris's Bibliothèque Nationale), Cardinal Gasquet (prefect of the Vatican Archives), U.S. Army cryptographers, and Professor Newbold (details in a moment!). In all, about twenty individuals were contacted.

In 1921, Voynich asked $160,000 for the manuscript. Why so much for a manuscript that couldn't be read, even if it was by Roger Bacon? Voynich felt that the price was justified by a partial decipherment offered by University of Pennsylvania Professor William Newbold.

FIG. 1.17 William Romaine Newbold (1865–1926)

Newbold was the Adam Seybert Professor of Intellectual and Moral Philosophy at the University of Pennsylvania. The first success he announced against the manuscript came in 1919. It indicated that Roger Bacon was indeed the author. Newbold's process of decipherment was tedious and time-consuming, but he eventually had enough recovered text to deliver a lecture on the secrets the manuscript had been hiding for all those centuries. He prepared the talk in the form of a paper that he presented to the College of Physicians and Surgeons on April 20, 1921, and to the American Philosophical Society at their meeting on April 21. His first talk followed a lecture by Voynich that gave some background on the manuscript and explained how it could have made its way from Roger Bacon to King Rudolf II. Following the talks, papers by Voynich and Newbold appeared back to back in *Transactions of the College of Physicians and Surgeons of Philadelphia*.

John Mathews Manly, a University of Chicago English professor and expert cryptanalyst, was in attendance and later reported that "rumors were rife" concerning what Newbold's decipherment would reveal. He pointed out that Bacon was suspected by some of his contemporaries of "practicing black magic." And he told how "Some believed that Professor Newbold had learned the secrets of this magic, and one poor woman came hundreds of miles to beseech him to cast out by means of Bacon's magic formulae the demons that had taken possession of her."[48] It was a lecture that Rudolf II would have eagerly anticipated, if it only could have been delivered more than three hundred years earlier.

Newbold began the lecture by talking about great discoveries that are sprinkled throughout history, such as the first creature to make use of a tool, or the first person to discover the lever, or to chip a stone and introduce the cutting edge, or to smelt copper, and how the people behind them are rarely thought of, if history has even bothered to record their names. He then went on to describe the great importance of the microscope and telescope and claim that his decipherment showed that Roger Bacon knew about and probably discovered both. He then went on at length, describing Bacon's life and contributions before getting to the information extracted from the cipher manuscript. Finally, the evidence for the telescope was presented. Newbold had identified one of the illustrations in the manuscript as the Great Nebula of Andromeda. It's not visible to the naked eye, so Bacon must have had a telescope. Newbold stated,

This conclusion is confirmed by the legend attached to Bacon's drawing of a spiral nebula in the Voynich manuscript, which, if it has been
correctly deciphered, states that the object in question was seen "in a
concave mirror," that is, a reflecting telescope. But this legend is so difficult to read that I would not adduce it as independent evidence of
Bacon's possession of a telescope until the reading has been revised and
verified.[49]

Notice how careful scholars can be to not overstate their conclusions! Newbold continued,

> That Bacon possessed a telescope I regard as an established fact, inde-
> pendently of the new evidence afforded by the Voynich manuscript. But for
> his possession of a compound microscope, or even of a simple microscope
> of sufficient power to enable him to make discoveries of real importance,
> there has been hitherto no evidence at all. At most one may say that, since
> he had lenses and was familiar with the idea of arranging lenses in such
> manner as to increase the size of the visual angle, there is no improbability
> in the hypothesis that he succeeded in so arranging them as to make the
> first compound microscope.[50]

> The doubt that has overhung the subject is now, in large part, dispelled by
> Mr. Voynich's discovery. That the author of the manuscript possessed both
> a telescope and a microscope, both of considerable power, is established
> by the drawings which it will be my privilege to show you. That the au-
> thor was Roger Bacon is established by the fact that the alphabets which I
> worked out from the Key on the last page of the manuscript when applied
> to the cipher elements interpolated into the Key spelled out the name
> *R Baconi.*[51]

Newbold went on to refer to spermatozoa and cells that Bacon saw with
his microscope and "so clearly depicted in the drawings" in the manuscript.[52]

> But even with the text unread the drawings alone throw a flood of light
> upon the achievements of Roger Bacon. They confirm to the full the infer-
> ence drawn by a few scholars from existing evidence, but denied by the ma-
> jority, that he possessed and was probably the discoverer of the telescope

and the microscope; they prove that he had seen anatomical and astronomical objects never seen by the human eye before, and not to be seen again for centuries, and show that he is here trying to weave them into and interpret them by a preconceived system of ideas drawn in the main from the Platonic tradition. Roger Bacon at last stands revealed as the true forerunner of modern science, as one of the greatest of the many men of genius born of the gifted English race.[53]

Newbold then showed dozens of slides of various manuscript pages and explained them. Some highlights follow:

Slide 29:

Upper corners schematized ovaries (nucleated ova); (Fallopian) tubes opened to show stream of ova descending into cavity (uterus), in which are seven souls or spirits (spermatozoa), three not yet awakened to consciousness, four expressing surprise and horror at their environment. Below, eight spermatozoa have discovered a "nest" of eight ova and view their destined dwelling places with expressions of surprise and curiosity, not unmixed with disgust.[54]

Slide 50:

This and several other of the disks have a marked general resemblance to magnified sections of a fertilized ovum, but the resemblance in detail is not close enough to warrant identifying any one of them with any particular stage of development as seen through modern microscopes.[55]

If you're wondering what Newbold made of the naked women standing in barrels, he thought they represented souls in their migration from and, after earthly life, back to stars. He said, "the 'barrels' represent the animal souls from which the spirits are not yet freed ... or the material bodies from which they are not yet entirely disengaged."[56]

Despite all of the great revelations in his talk, Newbold expected much more to be revealed by the manuscript, for he had only recovered about four percent of it thus far. He thought that the manuscript addressed the "secret of secrets," the prolongation of life. King Rudolf II would have been interested indeed!

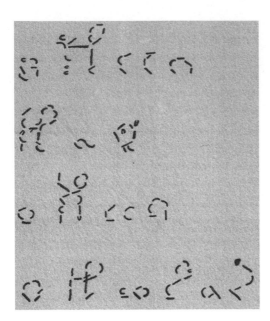

FIG. 1.18 A sampling of characters
within characters, as Newbold saw them

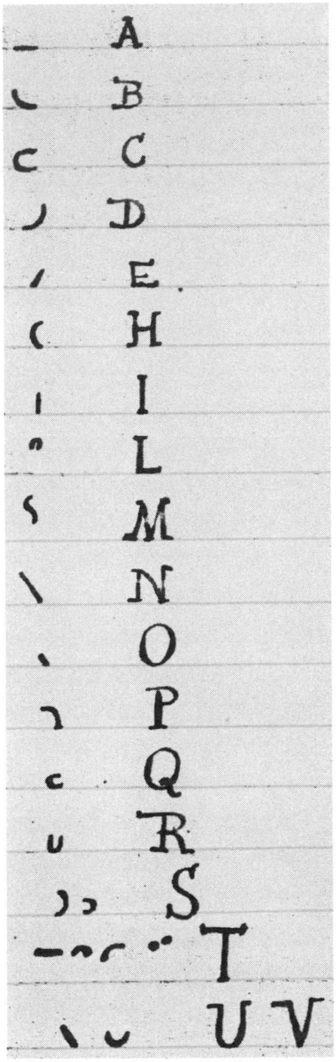

So, how did Newbold obtain his incredible decipherments?

Part of his solution was to look at the letters *very* closely. When he did so, he discerned that each character was really a composite made up of many smaller characters.

These tiny symbols each represented a letter. Newbold determined the appropriate substitutions, which are reproduced from his book in fig. 1.19.

According to Newbold, the symbols above, when fully deciphered, convey

FIG. 1.19 Newbold's substitutions
for the shorthand symbols

information about the comet of 1273. But the decipherment was no easy process. First off, Newbold noted that the minute characters are "to be read only with the aid of a microscope."[57] He added, "These letters are extremely difficult to see distinctly, and, as several closely resemble one another, mistakes in reading them are at present unavoidable."[58] Another problem was encountered by the fact that the letters obtained in this manner were not the plaintext message. The decipherment had barely begun! Newbold explained, "The text is buried under no less than six layers of cipher. Two of these can be eliminated, but there remain four operations which must be performed before the text can be read. Two of the four are merely mechanical, but two, the first and last, present difficulties which are as yet overcome only in part."[59] The first was reading the minute characters. The last is getting the letters of the message in the right order. Newbold revealed, "The letters of this final text, again, are always more or less displaced. Usually they are not displaced to such an extent as to make the words which they represent unrecognizable, but in many cases there is considerable doubt as to what word Bacon intended to be read."[60] This rearranging (also known as transposition or sometimes anagramming) could be done in a large number of ways. Newbold thought that Bacon rearranged in groups of 55 and 110 letters at a time. But there was no set pattern as to how this rearranging was to be done. It could be different for every group of letters.

The middle steps of the deciphering process won't be presented here in as much detail as the first and last steps, but we will take a look at what led Newbold to them, namely the "Key." Newbold claimed to have discovered this key on the last page of the manuscript. An image of the relevant lines appears below.

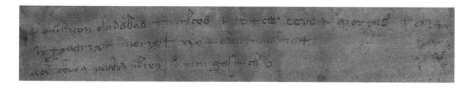

FIG. 1.20 The "Key" (leaf 116v)

Newbold read the beginning of this text as michiton oladabas multos te tccr cerc portas. He decided that ton ola te tccr cerc were "obviously cipher elements" and disregarded them to get michi dabas multos portas. He then changed multos to multas, fixing what he presumed was an error, to get a meaningful Latin sentence that translates to English as "To me thou gavest (or wast giving) many gates."[61]

Newbold then settled on pairing the letters in the Latin sentence with a particular twenty-two-letter alphabet (leaving out J, K, W, and X).

```
MICHIDABASMULTASPORTAS
ABCDEFGHILMNOPQRSTUVYZ
```

If these pairings are meant to define a MASC, we can interpret them in two ways:

1. The top sentence is the plaintext, and the ordered alphabet gives the cipher equivalents. That is, M is enciphered as A, I is enciphered as B, etc.
2. The ordered alphabet is the plaintext, and the top sentence gives the cipher equivalents. That is, A is enciphered as M, B is enciphered as I, etc.

Newbold made use of both interpretations, calling them the Primary Conversion Alphabet and the Primary Reversion Alphabet. The second is problematic because we have several instances of distinct letters becoming the same, when enciphered. For example, L, R, and Z are all enciphered as S. When it comes time to decipher, how is one supposed to know what to do with S? Should it be converted back to L, R, or Z? We'll come back to this problem later.

Newbold went on to use the alphabets above, and more of the hard-to-read writing on the "Key" leaf, to form four *biliteral* alphabets and eight auxiliary biliteral alphabets. The reasoning behind all of this seems to have been murky to everyone except Newbold. What he meant by "biliteral alphabet," though, is clear enough. It's just an alphabet of letters taken two at a time. Today's cryptologists call them *digraphs*. Newbold got the idea from the reference to gates in "To me thou gavest many gates." In Kabbalah (part of Judaism's esoteric tradition), gates can refer to such pairs of letters. For our alphabet, the digraphs are the following:

```
AA  AB  AC  AD  ...  AX  AY  AZ
BA  BB  BC  BD  ...  BX  BY  BZ
: :                       : :
ZA  ZB  ZC  ZD  ...  ZX  ZY  ZZ
```

The dots indicate pairs that were left out but that follow the pattern shown in the others.

The end result of all of Newbold's alphabets was that each single letter got paired with twenty-two digraphs.

In yet another step, Newbold believed that the encipherer took his twenty-two-letter alphabet and reduced it to eleven by making letters with similar sounds equivalent. For example, the letters B, F, and P were all written as P. The new alphabet was thus represented by the letters A, P, C, T, E, I, R, M, N, U, S. In undoing this process (i.e., trying to read the cipher), one would have to make frequent choices. Should the cipher letter P be taken as B, F, or P? It could be any of them, depending on what the message was at that point.

There's a lot more to Newbold's method, but already, we have several steps where there are choices for a decipherer to make—recognizing the shorthand characters, converting letters from an eleven-letter alphabet to a twenty-two-letter alphabet, and rearranging letters at the end. Whenever a decipherer has such flexibility, there's a risk of making wrong choices that lead to a meaningful message that was not intended by the encipherer.

Newbold summed things up:

> Therefore, until more progress has been made toward the removal of these difficulties no confidence is to be placed in the readings of the text except in the comparatively few cases in which they give facts unknown to the reader and afterwards confirmed, and, even in those cases, the confirmation attaches to the general idea only, not to the exact wording.[62]

The Critics (of Newbold) Weigh In

Voynich thought that Newbold would get a Nobel Prize for his work. Certainly Voynich would have *hoped* for this to be the case. If Newbold's decipherment were to be embraced in such a way, then the history of science would be rewritten, with Roger Bacon as perhaps the greatest scientist of them all, and the manuscript would become priceless. Some initial responses from experts were positive.

Scientific American began its coverage on May 7, 1921, with a small, more or less neutral piece announcing a longer feature to appear. When it appeared in June 1921, it endorsed Bacon as the author but was "not enthusiastic about Professor Newbold's translation."[63] The concern was the great amount of freedom provided by his substitutions and transpositions.

Somehow, a shorter *Scientific American* piece dated May 28, 1921, referred to and summarized the conclusion of the June piece and also addressed the illustrations and Newbold's interpretation of them. The author presented the opinion that "It really seems that at times Professor Newbold allows his wishes to run riot with his better judgment."[64] He also refers to the translation as "altogether unreasonable."[65] A June 25 letter from Lynn Thorndike that was printed mostly approved of this piece but questioned acceptance of Bacon as the author.

In a July 1921 piece for *Harper's Magazine*, John M. Manly seems to play both sides, giving evidence for and against Newbold's decipherments being accurate. Perhaps he was torn emotionally. Manly was a friend of Newbold's from the war years, so he might not have wanted to be too critical, but he also served as a cryptanalyst with MI8 (Military Intelligence—Section 8) and certainly knew a lot about the subject.

Newbold died in September 1926, and in the next few years opinions became stronger. In 1928, a book appeared with his byline explaining his decipherment technique at great length, although it remained murky, along with the amazing results it produced. It was not, as happens in some cases, completed before the author's death and released later. Rather, it was put together from Newbold's notes by his friend and fellow University of Pennsylvania professor Roland Grubb Kent.

In 1929, Lynn Thorndike chimed in again, in a review of the book. This time he was no longer simply questioning Bacon's authorship. He wrote, "There is hardly one chance in fifty that Roger Bacon had any connection with the production of the Voynich manuscript."[66]

Newbold's friend Manly, who was previously wavering, went on the record in a 1931 paper in *Speculum* with firm statements like this: "In my opinion, the Newbold claims are entirely baseless and should be definitely and absolutely rejected."[67] He might have kept quiet about it, but, as he explained it,

One of the most eminent philosophers of France, Professor Gilson, though bewildered by the method, has accepted the results; Professor Raoul Carton, the well-known Baconian specialist, in two long articles accepts both method and results with enthusiasm; and American chemists and biologists have been similarly impressed. The interests of scientific truth therefore demand a careful examination of the claims of the Newbold cipher.[68]

There was no issue of fraud or a hoax. Nobody questioned Newbold's integrity. Manly viewed the decipherments as being "the subconscious creation of Professor Newbold's own enthusiasm and ingenuity."[69]

> To me, the scattered patches of "shorthand signs" with which Professor Newbold operated seem merely the result of the action of time on the ink of the written characters. The vellum of the MS has a very rough surface, and the ink used was not a stain but a rather thick pigment. As the pigment dried out, the variations in sedimentary deposit and the cracking produced the phenomena which Professor Newbold has taken to be microscopic elements in the strokes. This view of the matter is shared by such experts in palaeography as Professors Beeson and Ullman of Chicago, Professor Fritz Saxl of Vienna, Mr Robert Steele of London, and Sir Frederic Kenyon and Mr Eric Millar of the British Museum.[70]

Manly added, "We can only hope that someone with equal courage and devotion but with a sounder method will be found to renew the attack upon the mysterious cipher of the Voynich manuscript."[71]

A Manuscript in Search of a Buyer

Voynich died in 1930, so he never saw Manly's devastating critique of Newbold's solution. Voynich's will specified that the famous manuscript was to be left in the joint possession of his widow and his secretary, Anne. M. Nill. He further stipulated that if a price of $100,000 could be obtained, it could be sold to a public institution, but a private individual could not be allowed to purchase it. In light of the Manly paper, who would pay so much? With no buyer in sight, Ethel Voynich placed it in a safe-deposit box, where it remained until her death in 1960.

Because Voynich refused to pinpoint where he got the manuscript, it's natural to speculate that he may have created it himself. After all, as an antiquarian book dealer he would have had no problem obtaining a sufficient number of blank vellum pages on which he could have placed the strange text and illustrations. Indeed, one of his London bookstore employees noted that Voynich sold blank fifteenth-century pages to the artist James McBey for a shilling apiece. He could also have faked the letter that provides some of the manuscript's provenance. Michael Barlow, in a paper published in 1986, conjectured that the letter was real but had no connection to the cipher manuscript. He thought the manuscript could have been created by Voynich, who then paired the genuine letter with it to lend it some authenticity.

But if Voynich created the manuscript, why would he leave behind instructions not to sell it to an individual? It's as if he wanted it to go somewhere scholars could have access to it, as a valuable document in the history of science.

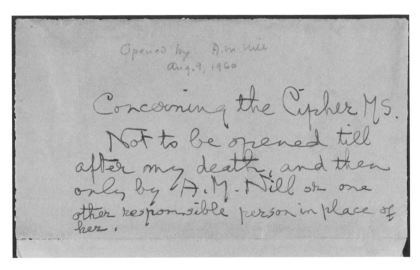

FIG. 1.21 Envelope containing a message about the cipher manuscript

When Ethel died, Anne Nill found herself in possession of a strange envelope that answered a different question.

The envelope Ethel left behind was marked "Concerning the Cipher Ms." and bore the instructions "Not to be opened till after my death, and then only by A. M. Nill or one other responsible person in place of her."

What was the secret of the manuscript that couldn't be revealed until after Ethel's death? The two-page letter inside, on her husband's stationery, read

The Cipher MS. was bought, with other MSS., by W. M. Voynich, in or about 1911. It was the property of the Vatican, and was (in a castle?) at Frascati. The intermediary through whom he approached the Vatican authorities was the English Jesuit Father Joseph[72] Strickland, who had, I believe, some connection with Malta. Father Strickland, who has since died, knew that the sale of certain MSS. had been decided upon, if a buyer could be found whose discretion could be trusted. Whether this secrecy was because of the strained relations with the Quirinal[73] I do not know. Father Strickland gave his personal assurance that W.M.V. could be trusted, and on that assurance he was allowed to buy, after giving a promise of secresy [sic]. He told me at the time, in confidence, feeling that someone should know, in case of his death. For the same reason I am leaving this statement in the safe, in case of my death.

E. L. Voynich.

July 19. 1930.

The location, "at Frascati," indicates Villa Mondragone, a former Jesuit College that happened to be strapped for cash. It was situated in Frascati, Italy, near Rome.

So, the letter provided a story quite different from Voynich's original claim of having found the manuscript in a collection belonging to the Dukes of Parma, Ferrara, and Modena. With this new information, the dots can be connected from Kircher to Voynich. When Kircher died, the many letters he had received over the years, and saved, went to the library of the first Jesuit university, Collegio Romano (Roman College), in Rome. Still in existence, it is known today as Pontifical Gregorian University. This is where the letters Kircher received from Baresch and Marci were found.

It has not yet been firmly documented, but it seems likely that the cipher manuscript also went to the library of Collegio Romano. Evidence for this is provided by the identity of a later owner, as we shall see. Rome, which was an independent state, was invaded in 1870, as part of the unification of Italy carried out by King Victor Emmanuel II. At this time, much church property, including the college's library, was confiscated by the new government. As a consequence of faculty members' personal libraries not being subject to confiscation, said personal libraries were rapidly improved at the expense of the college's library before the latter were seized. Among those whose home libraries grew was Pierre Jean Beckx (1795–1887),[74] rector of the college and twenty-second superior general of the Jesuit Order at Villa Mondragone.

Beckx took his "personal" library to Villa Mondragone for safekeeping. Knowing that Voynich obtained the cipher manuscript there, we can then turn to thirty other volumes that he bought at that location. Interestingly, almost all of these had notes attached to their covers identifying Petrus Beckx as the owner. Voynich left the notes on these other volumes, so if there was one attached to the mysterious manuscript, why would he have removed it?

This action, combined with Voynich's refusal to reveal where he found the manuscript, leads us to wonder if he stole it. Some of Voynich's biographical details were included in this chapter to allow you to decide whether the theft of this manuscript would have been out of character.

Whatever Voynich's true motivations were for his secrecy, with both him and Mrs. Voynich deceased, the decision as to what to do with the manuscript fell to Anne Nill. She was now the sole owner, and she decided to sell it.

FIG. 1.22 Anne M. Nill (1894–1961)

She found a buyer relatively quickly, selling it to Hans P. Kraus in 1961 for $24,500. Kraus, a Jew who had spent time in two Nazi concentration camps (Dachau and Buchenwald), came to America on Columbus Day with a copy of the rare Columbus letter of 1494 and had an incredible career as an antiquarian book dealer.

In his autobiography, Kraus wrote

In 1963 we were in Rome and I visited Monsignor José Ruysschaert at the Vatican Library. I knew that he had published the catalogue of the Mondragone Library and I hoped to get information about the Cipher Manuscript. To my great surprise he thought that the manuscript was still in the library.[75]

FIG. 1.23 Hans P. Kraus (1907–1988)

If Voynich had legitimately purchased the manuscript, why would Ruysschaert think it was still part of the collection?

Rather than selling the manuscript, Kraus presented it to the Beinecke Rare Book and Manuscript Library of Yale University in 1969, along with boxes of letters, pamphlets, notes, and other literature related to the manuscript. It remains there to this day as MS 408.

Other "Solutions"

Although Newbold's solution did not meet with much acceptance among experts, he did get a lot of attention from the media. No other proposed solution made as big a splash, but many have appeared over the years.[76] A few are indicated below, although space constraints prohibit examining them in detail. Follow the references if you want to learn more.

JOSEPH MARTIN FEELY, 1943

This Rochester, New York, lawyer hired a team of researchers to help him compile letter frequencies from the manuscript, as well as from Bacon's writings. He then concluded that the cipher was a MASC applied to abbreviated Latin, that was indeed penned by Bacon. The plaintext revealed a scientific journal with an emphasis on gynecology. Feely had previously written about ciphers in Shakespeare. Another of his efforts is in chapter 7 of this book.

DR. LEONELL CLARENCE STRONG, 1945

Strong was a Yale geneticist and cancer researcher who had his thoughts on the manuscript published in *Science*, a top academic journal.[77] He broke from tradition by claiming Anthony Askham as the author. According to Strong, the cipher made "peculiar use of a double system of arithmetical progressions of a multiple alphabet." He gave no further explanation but rather wrote, "Due, however, to present war conditions, it seems undesirable to publish, at this time, the details of the key." The portions he deciphered were in medieval English and dealt with "(1) the effects of plants on physiological processes in health and disease, especially the diseases of women, and (2) a conception of pre-Harveian generation and parturition." It included "several references to the use of antibiotics." A newspaper described the contents in layman's terms as "an extremely candid discussion of women's ailments and practical matters of the conjugal bed."[78]

DR. ROBERT S. BRUMBAUGH, 1974–1975

Like Newbold, Brumbaugh was a philosophy professor, but he was also a former U.S. Army intelligence officer and had the advantage over other researchers of working at Yale, where the cipher manuscript has resided since Kraus's donation.[79]

Brumbaugh announced his break in a brief paper in *Speculum*, promising that detailed results would follow in a future issue of *Yale Library Gazette*. The initial paper ended with a possibly new interpretation of the line Newbold interpreted as the key. Brumbaugh read it not as Newbold had, but rather as MICHI CON OLADABA. He then deciphered OLADA by shifting every letter forward by three (as Julius Caesar would do to encipher) to get RODGD and read it as ROGER. Taking other letters from the key, CON and BA, he rearranged them to get BACON. Thus, the complete result was MICHI ROGER BACON, or ME ROGER BACON. Brumbaugh thought previous research had shown that

Bacon was not the one who put the ink on the vellum, but he didn't rule out the possibility that the key and some of the text was copied from an older source. Or perhaps someone, maybe John Dee or Edward Kelley, manufactured the authorship cipher to make a potential buyer think Bacon had been the source.

In our image of the key, three lines of text can be seen. Brumbaugh looked at the third line, skipped over the first two words and read the rest as "valsch ubren so nim ga nicht o" which looks like broken, misspelled German for "the above is false so don't take it." This was further evidence that the so-called key should not be taken too seriously.

In the *Gazette* paper, Brumbaugh explained that the encipherment was carried out using the following substitutions:

```
A B C D E F G H I J K L M N O P Q R S T U V W X Y Z (-US)
1 2 3 4 5 6 7 8 9 1 2 3 4 5 6 7 8 2 4 6 8 1 3 5 7 9 (9)
```

Easy enough, but the decipherer has to deal with much uncertainty. Brumbaugh gave the following grid to be used to translate from ciphertext to plaintext.

```
1   2   3   4   5   6   7   8   9
A   B   C   D   E   F   G   H   I
J   K   L   M   N   O   P   Q
    R       S       T       U  -US
V       W  (X)  X       Y       Z(?)
```

Consider the cipher 2 8 2 4 6 4 5 2 1 3 6 5 6 5 9 6. To unravel this, we write beneath each number all possible letters that it could represent:

```
2 8 2 4 6 4 5 2 1 3 6 5 6 5  9   6
B H B D F D E B A C F E F E  I   F
K Q K M O M N K J L O N O N  -US O
R U R S T S X R V W T X T X  Z   T
    X   X
```

Now we try to select just one letter from each vertical column, so that when we read across the columns we get a meaningful message. But does the message begin with BURST or RUB? I'll leave it to you to try to determine the correct intended message.

In addition to this difficulty, the nine digits used for the ciphertext were disguised by being written in different ways. This is how we got the much larger Voynich alphabet.

A further wrinkle was provided by the text having been written in "an artificial language based on Latin, but not very firmly based there; its spelling is phonetically impressionistic; some passages seem *solely* repetitive padding."[80] Also, Brumbaugh believed that the key changed slightly every eight pages. He later wondered if there were more layers of encryption, for the results he got didn't make much sense—maybe each of his deciphered words represented a letter, or only certain words were meant to be read.

J. R. CHILD, 1976

This National Security Agency linguist, writing for *NSA Technical Journal*, a classified publication, claimed the manuscript "is a text in a hitherto unknown medieval North German dialect."[81] As the paper has since been declassified, you can examine his evidence for yourself. Child presented more evidence in a 2007 paper, in which he stated, a little less specifically than in 1976, "I am confident that the manuscript is a text in a Germanic language. ..."[82]

JOHN STOJKO, 1978

This Ukrainian amateur researcher discovered that the manuscript's decipherment was in Ukrainian. Before encipherment, all vowels had been removed, complicating the task of the cryptanalyst.

DR. LEO LEVITOV, 1987

The Cathars, a group with some similarities to the Knights Templar, and persecuted by the Catholic Church in the same way, were connected to the manuscript by Levitov. This New Jersey physician and WW II veteran described the writing not as a cipher but as "a highly polyglot form of medieval Flemish with a large number of Old French and Old High German loan words."[83]

JIM E. FINN, 2001

Finn claimed that the manuscript is based on a Hebrew plaintext. He asked if it could have been created by Nostradamus or was perhaps a book that the prognosticator used. A few quotes from the website on which Finn's work appeared give a sense of his solution. We find "The Voynich Manuscript is most likely part of an end times document given to mankind by Aliens, UFO's, in medieval times" and "The few pages he has deciphered appear to be referring to a cyclical terrestrial cataclysm brought about by a celestial 'EYE.'" When is the end coming? Finn seems to have made his decipherment just in time, for he warns, "We may live to see it."[84]

ZBIGNIEW BANASIK, 2003

The solution found by this Polish researcher may be the one that claims the greatest age for the manuscript. Banasik wrote, "The manuscript was written probably some centuries B. C." Banasik identified the language as pre-Manchu.[85]

BEATRICE GWYNN, 2004 OR EARLIER

This Dubliner, who worked at Bletchley Park in a noncryptologic capacity, concluded that the manuscript was in "mirrored Middle High German."[86] That is, the author made the writing as it would appear reflected in a mirror. Placing a mirror beside it would reveal the plaintext (in Middle High German). Leonardo da Vinci used mirrored writing, so the concept predates the sixteenth century, when Gwynn believes the manuscript was created as a hygiene manual.

URSULA PAPKE AND DIRK WEYDEMANN, 2005

In 2005, these two Germans issued a press release, in German, naturally, claiming to have translated the manuscript. A portion of the release translates as

> The book describes in a unique precision of words and images, the use of all pervasive universal energy by people. In China, this energy is called chi, in India, Kundalini or Prana, and by Christians, Light or Holy Spirit. A way is described as to how a human being can develop spiritually and ascend to higher levels. Starting from the core of the person, carried out as an internal promotion process, which closes a cycle of life. The heart plays a central role. Only after all of the transforming power of love can man reach the highest level.[87]

The press release also stated that they would post the entire manuscript in both German and English translation on the website www.ms408.com.

Attempting to go there now yields the message "This domain name is FOR SALE," but http://www.ms408.de/translation_d.htm gets one the translation page and links to more pages. As of this writing, thirteen leaves have been translated into German. No English is available yet.

Their translation was obtained not by treating the characters as encrypted text, but rather as representing concepts.

> We suspect, therefore, that MS408 or the eventual original was used as a textbook of an esoteric secret society for learning and for passing elevated states of consciousness.[88]

Another, completely different, solution from Germany was offered by Jonathan Dilas. It seems to defy terse summarization.

DR. EDITH SHERWOOD, 2008

Though not offering a decipherment, Sherwood does claim to have identified a young Leonardo da Vinci, around the age of 8–10, as the author. She believes that the cipher is "probably a series of Italian word anagrams written in a fancy embellished script."[89]

JODY MAAT, 2009

This researcher thought that the manuscript was written in "Old Dutch." He posted some translations online. Here's what he discovered.

> The Voynich manuscript was written by a man named de Rici. He got the order to sail towards the other side of the world, to the part where the sun does not shine. He uses the stars to navigate his ship towards this area and arriving here he discovers the sun does shine here and even that there is land. He makes drawings of what he sees and describes where he saw it. He also realizes this might be a shock for the homefront and they might not believe him. He therefore ends his book with the statement: IT IS AS IT IS THAT IS THE TRUTH.[90]

RICHARD ROGERS, 2009

Rogers, a U.S. Army/Navy computer programmer who was reported to have degrees in "ancient history, languages and computer science," gave a complicated explanation for the manuscript. Nick Pelling, a researcher, who has written a book on the manuscript himself, began a description of Rogers's work with the following:

> It took me ages to even begin to understand what he thinks he can see in the VMs—and I'm still miles off understanding why it might be so, as well as how he made the leap from (a) grasping that there is meaningful content, to (b) seeing how that meaningful content actually works. Which is why every time I tried to post about this, I've ended up giving up halfway through: but now he's gone public, I guess I'll have to complete the job as best I can ...[91]

If a solution cannot be communicated to other experts, it is almost certainly wrong.

VIEKKO LATVALA, 2011

This Finnish businessman obtained his decipherment with help from a "higher power." The self-described "prophet of God" described the manuscript as "sonic waves and vocal syllables" that contain a prophecy.[92] He claimed the predictions extend "for some decades and hundreds of years ahead from the time it was created."[93] Ari Ketola, a business associate of Latvala, refused to explain how Latvala deciphered it. Ketola did remark, "Mr Latvala said that no one 'normal human' can decode it, because there is no code or method to read this text, it's a channel language of prophecy. This type of persons are most rare to exist, yet they have always been on face of the Earth through millenniums up to today ... and Mr Veikko [*sic*] Latvala has had this gift of mercy last 20 years."[94]

A big year for solutions came in 2014. Three are described below.

STEPHEN BAX, 2014

This professor in applied linguistics at the University of Bedfordshire proposed solutions for "around *ten* of the words and some *fourteen* of the signs and clusters."[95] He thinks the manuscript is "probably a treatise on nature, perhaps in a Near Eastern or Asian language."[96] Much detail of his work can be found in a sixty-two-page paper available at his website, and in a YouTube video he made.

DR. ARTHUR TUCKER, 2014

In what appears to be a brand new theory, Tucker proposed that the manuscript was written in an extinct form of Nahuatl, which was the language of the Aztecs.

WAYNE J. PAUL, 2014

A direct quote from Paul's website is the best way to sum up his findings:

> It might never be 'understood' by Human beings exactly what the writing and text says, in the Voynich. Because it is written in AN ANGELIC LANGUAGE. The Voynich Manuscript is a school book for NEPHILIM SCHOOL CHILDREN. It is a 15th century copy of an ancient educational book that was used by the Nephilim to teach their half-human kids. Some miraculous copy from before the Great Flood survived into the early 1400's. It must have been in disastrous condition because it took a lot of work to REPLICATE it into the 'Voynich Manuscript.' It was 'recognized,' otherwise such effort and expense to replicate it would not have been spent.[97]

ОЛЕГ ЛАЙОНС, 2015

According to Лайонс, a merchant from northern Italy traveled to China sometime in the 1400s and stayed long enough to learn to speak, read, and write Chinese fluently. Upon his return, he found himself in need of money.

> He took one of the books that he owned (maybe, the only book that he owned) and translated it into Chinese. Then he 'Romanized' the translation (using a 'phonetic script') and re-wrote the book using the alphabet that he had invented (every character corresponded to a letter in a Latin alphabet). He added a few meaningless characters, some Latin comments; sophisticated pictures—and voila, he was in a possession of an instant "collector's item."[98]

Лайонс related that the Italian "received even more money than he hoped for, took care of his financial problems and went back to his merchant ways. In which he was highly successful."

Many more theories and "solutions" can be found at http://www.cipher mysteries.com/the-voynich-manuscript/voynich-theories.

What can we learn from all of these "solutions"? Some came from highly educated individuals, like Newbold, and warranted closer examination. Others can be dismissed very quickly. Yet, these have grabbed headlines as well. It seems that the lesson is, don't get too excited when you hear the next solution announced!

The Best of the Twentieth-Century Research

The nineteenth century is known for great decipherments. The irony is that the decipherments were of plaintext languages, the hieroglyphs used by any literate Egyptian of the ancient world, Old Persian, Akkadian, Sumerian, etc. Tremendous efforts resulted in the recovery of documents that were easily read by the texts' contemporaries.

But how would twentieth-century techniques fare against a manuscript that could presumably have been read by at least one individual (its creator) much more recently? The tremendous effort that was exerted in this direction in the twentieth century included efforts by top experts at the U.S. National Security Agency (NSA) and the corresponding organization in England, Government Communications Headquarters (GCHQ).

Transcription Alphabets

The first steps for all those concerned were to get a copy of the manuscript (obviously!) and then to transcribe the characters. That is, to represent the characters with letters and numbers, so that they can easily be typed and shared with other researchers. This format also makes it easier to compile statistics based on letter frequencies and combinations and allows tests to be run on computers. Any modern cryptanalyst would want to let his or her computer take a crack at it!

Many twentieth-century researchers used copies of copies. D'Imperio's copy of the cipher manuscript was four or five removes from Friedman's, even though both worked for NSA. The copies were sometimes out of focus around the edges or even partly cut off. Many copies also bore the markings made by earlier would-be decipherers. Joseph Feely, one of several men who claimed to have deciphered parts of the manuscript, didn't have a copy at all, but rather did all of his work from the illustrations in the book by Newbold and Kent! A notable exception was Brumbaugh, who had the advantage of being at Yale, where the manuscript resided.

Transcriptions are now available on the Internet for free and without legal restrictions, but some information is lost in the transcription process. For example, we cannot easily associate the text with the numerous pictures. Also, it's not even clear how many characters exist in the cipher alphabet, as can be seen in the transcription alphabets used by different researchers. All of the statistical tests that are applied must be applied with respect to a particular alphabet. If we use the wrong alphabet, our results may be meaningless.

Part of the problem in creating a transcription is that the characters sometimes blur into one another. For example are ♂, ♂, and ♂ different characters or not? What about ♃ and ♃ or ♃ and ♃? I believe the loop in the upper left hand corner to be a significant detail. There are no intermediate forms of this loop; it's always either clearly present or clearly not present. Perhaps this loop is used to make a distinction as we would in German between u and ü or o and ö.

As another difficulty, consider ⑂. Is this a single character or two? Before you decide, be aware that the manuscript also contains ⑂ and ⑂. If the initial slash is a character, then it's unlikely that we have a MASC because for such ciphers no letter would appear as a triplet.

Another symbol to consider is ⑂. The earlier symbols could represent the roman numerals that they resemble ⑂ = ix = 9, ⑂ = iix = 8, ⑂ = iiix = 7, but what would we then make of ⑂?

There's still more to consider before trying to come up with a transcription. Jacques Guy has noted that ℥ is always followed by a c with a little flat line at the top and this type of c always appears after ℥ or a regular c. From this, he concluded that c̃ and c̃ are single letters. He also noted that c c appears frequently and even c c c is not uncommon and concluded that c c is a single letter. Guy raised the possibility that ♋ and ♋ are single letters as well.

Reproduced in fig. 1.24 are transcription alphabets used by several researchers and two study groups. The study groups were formed at the National Security Agency, with the first running 1944–1946 and the second 1962–1963.

The alphabet attributed to "Kirscher" may look like a typo for Kircher, but as was mentioned earlier, we have no knowledge of Kircher's work on the manuscript. Kirscher is indeed a typo, but is intended to refer to Jeffrey *Krischer*, a fellow with a background in mathematics, computer science, medicine, and cryptology. While a graduate student at Harvard University, he carried out a computer analysis of the manuscript's text.

A more recent transcription alphabet, known as EVA, for European Voynich Alphabet, was made by René Zandbergen and Gabriel Landini in 1996 and 1997. There are several others.

We now examine a few of the statistical tests that can be carried out using a transcription of the cipher manuscript.

Tiltman		First Study Group		Second Study Group		Kirscher		Currier		DImperio	
⟨sym⟩	D	⟨sym⟩	P	⟨sym⟩	4	⟨sym⟩	4	⟨sym⟩	4	⟨sym⟩	A
⟨sym⟩	H	⟨sym⟩	F	⟨sym⟩	0	⟨sym⟩		⟨sym⟩	0	⟨sym⟩	B
⟨sym⟩	G	⟨sym⟩	FH	⟨sym⟩	9	⟨sym⟩		⟨sym⟩	8	⟨sym⟩	C
⟨sym⟩	8	⟨sym⟩	HD	⟨sym⟩	8	⟨sym⟩		⟨sym⟩	9	⟨sym⟩	D
⟨sym⟩	2	⟨sym⟩	G	⟨sym⟩	2	⟨sym⟩		⟨sym⟩	2	⟨sym⟩	E
⟨sym⟩	4	⟨sym⟩	AR	⟨sym⟩	B	⟨sym⟩		⟨sym⟩	E	⟨sym⟩	F
⟨sym⟩	0	⟨sym⟩	K2	⟨sym⟩	PV	⟨sym⟩		⟨sym⟩	R	⟨sym⟩	G
⟨sym⟩	A	⟨sym⟩	0	⟨sym⟩	F	⟨sym⟩		⟨sym⟩	S	⟨sym⟩	H
⟨sym⟩	C	⟨sym⟩	L	⟨sym⟩	#	⟨sym⟩		⟨sym⟩	Z	⟨sym⟩	I
⟨sym⟩	I	⟨sym⟩	NM	⟨sym⟩	$	⟨sym⟩		⟨sym⟩	P	⟨sym⟩	J
⟨sym⟩	T	⟨sym⟩	8	⟨sym⟩	%	⟨sym⟩		⟨sym⟩	B	⟨sym⟩	K
⟨sym⟩	S	⟨sym⟩	4	⟨sym⟩	@ S	⟨sym⟩		⟨sym⟩	F	⟨sym⟩	L
⟨sym⟩	L	⟨sym⟩	EC	⟨sym⟩	Z	⟨sym⟩		⟨sym⟩	V	⟨sym⟩	N
⟨sym⟩	R	⟨sym⟩	TS	⟨sym⟩	C	para start line		⟨sym⟩	Q	⟨sym⟩	O
⟨sym⟩	E	⟨sym⟩	I	⟨sym⟩	A	start line end		⟨sym⟩	W	⟨sym⟩	P
⟨sym⟩	DZ	⟨sym⟩	PZ	⟨sym⟩	E	space		⟨sym⟩	X	⟨sym⟩	Q
⟨sym⟩	HZ	⟨sym⟩	FZ	⟨sym⟩	I			⟨sym⟩	Y A	⟨sym⟩	R
		⟨sym⟩	HZ	⟨sym⟩	Y			⟨sym⟩	C	⟨sym⟩	S
		⟨sym⟩	DZ	⟨sym⟩	J			⟨sym⟩	I G	⟨sym⟩	T
		⟨sym⟩	V	⟨sym⟩	UK			⟨sym⟩	1 T	⟨sym⟩	W
		⟨sym⟩	Y	⟨sym⟩	G			⟨sym⟩	U 0	⟨sym⟩	X
			.	⟨sym⟩	Q			⟨sym⟩	D	⟨sym⟩	Y
		space para ?	0	⟨sym⟩	D			⟨sym⟩	N	⟨sym⟩	Z
				⟨sym⟩	N			⟨sym⟩	M 3	⟨sym⟩	1
				⟨sym⟩	M			⟨sym⟩	J	⟨sym⟩	3
				⟨sym⟩	W			⟨sym⟩	KL	⟨sym⟩	6
				⟨sym⟩	H			⟨sym⟩	5	⟨sym⟩	7
				⟨sym⟩	L			⟨sym⟩	6 7	⟨sym⟩	8
				⟨sym⟩	R			⟨sym⟩	/	⟨sym⟩	9
				⟨sym⟩	T			space line end para end	— =	⟨sym⟩	0
				⟨sym⟩	/					space line start line end para end ?	space ( ) / ?
				space line end ?	— *						

FIG. 1.24 Transcription alphabets of several researchers

Vowel Recognition Algorithms

There are several algorithms that can be applied to texts in order to determine which characters are most likely to be vowels. These tests work only for MASCs. As is clear from the descriptions, the algorithms do not give meaningful results for transposition ciphers or other more complex methods. As far as I can determine, vowel recognition algorithms began with Francois Viete (1540–1603), who looked at the frequencies of individual characters as well as digraphs and trigraphs.

The basic idea behind all of the algorithms is that vowels combine with more letters than consonants. Consider the vowel A. We can easily follow it with any letter of the alphabet. Observe: AARDVARK, ABOUT, ACT, ADD, AERO-NAUTICS, AFTER, AGAINST, AHA, AIR, AJAR, AKIMBO, ALE, etc. However, when we try to do this with a consonant, like B, we run into trouble frequently. We can start with BAR, but for BB we get stuck, unless we allow the B to be followed by B within a word, as in RUBBLE. We have the same problem with BC. Unable to think of a word that starts with BC, I resort to a word such as BOB-CAT, where BC appears in the middle. But this word was formed from two words pushed together, making the B and C touch in a sort of artificial way. As you try to continue your list with words containing BD, BE, BF, BG, BH, BI, BJ, BK, BL, etc., you'll find yourself resorting to compound words fairly often.

So, to determine if an unknown letter is a consonant or a vowel, simply count how many different letters it makes contact with. If you have a picky letter with only a small number of contacts, it's almost certainly a consonant, and a high percentage of its contacts are the vowels. On the other hand, if the letter contacts a large number of other letters, then it's a vowel. You could say that vowels are extremely social. Sometimes they're referred to as the sluts of the alphabet, because they'll go with anything.

The only potential problem is where to draw the line between consonants and vowels. What if a letter has a moderate number of contacts? Of course, if the language behind the MASC is known, then you know how many vowels to look for, so just take that many of the letters with the most contacts. There are, however, more sophisticated approaches that work a little better.

One simple algorithm by Helen Gaines is called the "consonant line method." This is described by Foster as follows:[99]

Let A(I) and B(I) denote the number of different letters that the character "I" appears after and before respectively. If the character "I" is a vowel, A(I) and B(I) should both be large and roughly equal. Foster found that there is little difference between maximizing A + B versus A × B.

The best method currently available is Sukhotin's algorithm. It consists of six steps.[100]

Step 1: Count the number of times each letter is in contact with each of the other letters and place these values in an *n* by *n* grid where *n* is the number of characters in the alphabet.[101]

Step 2: Make the entries along the main diagonal (from upper left to lower right) of the grid from Step 1 zeros.

Step 3: Assume all characters to be consonants. Sum the horizontal rows. Write the results to the right of each row.

Step 4: Find the presumed consonant with the highest positive row sum and assume that it is a vowel. If there is no "consonant" left with a positive row sum, terminate the algorithm.

Step 5: Subtract from the row sum of each consonant twice the number of times that it occurs next to the newfound vowel.

Step 6: Return to Step 4.

A much more detailed description of this algorithm, including an example, appears in my first book, *Secret History: The Story of Cryptology*.

Foster recommends using both the consonant line method and Sukhotin's algorithm. Having two tests gives us more confidence, and the run time is short for both algorithms. Foster also examined another method called "vowel distribution." This method was less accurate and took four hundred fifty times as long to run (almost fifteen minutes versus two seconds). Jacques Guy used his transcription (with a seventeen-letter alphabet) and the text of folios 79v and 80r in Sukhotin's algorithm. It resulted in six letters being identified as vowels.[102] These letters were ꜩ, ꜩꜩ, o, 9, ɑ, and ι.

Is it a coincidence that o, ɑ, and ι resemble our vowels o, a, and i?

Entropy and Redundancy Calculations

There's too much structure in the cipher manuscript for it to be a more complicated form of encryption than a MASC. Better forms of encryption leave the ciphertext looking more random, and a perfect cipher system generates results that appear *completely* random. In contrast to this, the Voynich manuscript has a great deal of order. Some combinations of characters appear frequently, and others never appear.

It turns out that there are simple calculations that can model how much order (redundancy) or disorder/randomness (entropy) a text has.

To see how this works, we'll look at some examples. Consider the incomplete message "I hate this job. I'm going to q . . . ." If I provide a little more of the sentence, such as telling you the next letter is a u, do you feel that more information has been conveyed? Probably not, as u almost always follows q in English. Actually, the next few letters won't provide much new information either. You are likely expecting the sentence to end with "quit." It's possible that it could wrap up differently. Perhaps the full message is "I hate this job. I'm going to query my boss about a raise." or "I hate this job. I'm going to qualify for unemployment when my boss finally fires me." But these completions seem much less likely than quit.

So, the main idea behind measuring the information content of a message is to evaluate how expected or unexpected each portion is. The expected (or high probability) portions, like a u following a q, contain little information, while unexpected (or low probability) portions carry more information.

We can evaluate this mathematically. The formula might look a little strange to some readers, but Claude Shannon proved, way back in the 1940s, that this is the only way to do it. His formula for information is $-\log_2(M)$, where M is the probability of the message. If the message naturally breaks down into n pieces, the total information conveyed is $-\sum_{i=1}^{n} \log_2(M_i)$. The Σ symbol is the Greek letter sigma. As used here, it stands for "sum" and means that we should add what follows for every value i takes. An example is given by a report card with grades in four classes. Here, i takes the values 1, 2, 3, and 4 and the grades could be thought of as four separate messages: M_1, M_2, M_3, and M_4. If you thought that there was an 80% chance of making an A in calculus, and you did, then the information content of that part of the report card would be $-\log_2(.8) \approx .32$. On the other hand, if you were 95% sure that you passed Street-Fighting Mathematics,[103] and your report card shows an F for it, then the information content of that portion is $-\log_2(.05) \approx 4.32$, which is significantly higher.

The negative log is a great way to measure information because it increases as the number being plugged in (the probability) decreases, and for messages that are guaranteed (probability = 1, no surprise factor whatsoever), the log returns 0 as the information content. Also, for two messages M_1 and M_2, we have $-\log_2(M_1 M_2) = -\log_2(M_1) + -\log_2(M_2)$, so the information content of two messages is the sum of the individual information contents. We want all of these things to be true, and we want the measure to change smoothly (be continuous) as the probabilities of the messages change. That is, the values should not make sudden jumps. Shannon proved that the negative log is the only function that does the trick. We can use another base (it doesn't have to be 2), but that just amounts to changing the units, or you could say scaling things differently. So, up to scaling, the negative logarithm solution is unique.

We are able to evaluate the information content not just of messages or strings of messages, but of entire languages. To do so, we take a weighted average of all possible messages. You can also think of this as an expected value, if you're familiar with that concept. We have

$$H = -\sum_{i=1}^{n} M_i \log_2(M_i)$$

Here, H is used to represent the entropy of the language and M_i indicates the probability of message i. For example, if we had only three messages, with probabilities M_1, M_2, and M_3, the calculation would be

$$H = -\sum_{i=1}^{3} M_i \log_2(M_i) = -(M_1 \log_2(M_1) + M_2 \log_2(M_2) + M_3 \log_2(M_3))$$

Realistically, though, we'd have a very large number of possible messages. Just think of all the different messages you could potentially send to friends! It would take way too long to come up with the probability of every possible message and then compute H. There are simply too many possible messages. More simply, we can just look at the probabilities of single letters. We call this first-order entropy and calculate it as $H_1 = -\sum_{i=1}^{n} p_i \log_2(p_i)$, where p_i is the probability of the ith letter and the sum extends over all of the letters of the language. It provides an estimate for H. Of course, this doesn't

consider effects over groups of letters. Better estimates can be made by de-
fining H_2, H_3, etc., where the probabilities in the summation are for digraphs,
trigraphs, etc. This allows us to capture structure in the language that happens over two or more characters, such as Q almost always being followed by
U, and rules like "I before E, except after C." As n grows, H_n becomes a better
and better approximation of H. We call H_2 and H_3 second- and third-order
entropy, respectively.

Redundancy

Because redundancy is essentially the opposite of disorder/entropy, it can
be calculated from the entropy value. The largest possible value for entropy is obtained when all letters are equally probable, as is the case for
completely random text. We call this value H_{max}. To get an idea of how
close an observed value for H is to H_{max}, we can look at H/H_{max}. This ratio is
equal to 1 for completely disordered text, but close to 0 for extremely redundant text. So, if we subtract this ratio from 1, the result is reversed, in
a sense, giving us a measure of the redundancy, which is typically denoted
by D. We have $D = 1 - H/H_{max}$. Because $H_{max} = -\sum_{i=1}^{n}(1/n)\log_2(1/n) = -(1/n)$

$$\sum_{i=1}^{n}\log_2\left(\frac{1}{n}\right) = -(1/n)(n)\log_2(1/n) = -\log_2(1/n) = \log_2(1/n)^{-1} = \log_2(n),$$ we

can write $D = 1 - H/\log_2(n)$.

Our measure of entropy is improved by looking at digraphs and trigraphs
to get H_2 and H_3, and we can, in turn, use these values to get the better, higher
order, redundancy measures D_2 and D_3.

Now that we have a reasonable way to model redundancy, let's look at
how the Voynich manuscript compares to various languages. But before
doing so, it should be made clear what we consider letters to be. In the
data that follow, the blank space is counted as a letter. Because it's so common (high probability), this gives a greater value for redundancy than a
measure that ignores it. If we simply take the twenty-six letters in the English alphabet, the first-order redundancy would come to about 0.7. This
smaller value is used in chapter 5 to help analyze the ciphers of a possible
killer.

Table 1.2 Redundancy Values for Various Languages

	Redundancy[a]	
	D_1	D_2
English	.119	.24
French	.131	.27
German	.098	.27
Italian (Military)	.127	.20
Spanish	.132	.26
Japanese (Romaji)	.127	.32
Voynich manuscript[b]	.167	.49

[a]These values were mostly taken from Fischer, Elliot, "Language Redundancy and Cryptanalysis," *Cryptologia* 3, no. 4 (October 1979).
[b]I calculated these two values using the corresponding H values from Bennett, W. R., *Scientific and Engineering Problem-Solving with the Computer* (Englewood Cliffs, NJ: Prentice-Hall, 1976), 193.

It's interesting to observe that natural languages have a range of values for redundancy. But the Voynich manuscript is much more redundant than any of the languages on our list above. In other words, it's too redundant to be MASCed English, French, German, Italian, Spanish, or Japanese. MASCing a text disguises it, but it doesn't change the redundancy.

Another way of saying that the redundancy is too high is to say that the entropy is too low. Remember that entropy, representing disorder, is the flip side of redundancy.

Jacques Guy quoted Bennett as having written that the "writing exhibits fantastically low values of entropy per character over that found in any normal text written in any of the possible source languages"[104] and went on to explain that Bennett's results were not meaningful since he mistakenly counted ⊄ and ⊄̂ as two letter pairs instead of the single letters that Guy believes them to be. Guy also remarked that the value for entropy obtained by Bennett is consistent with Hawaiian. Of course, there's no reason to suspect Hawaiian as the plaintext language. The values are given below.[105]

Table 1.3 Entropy Values for Various Languages

	H_1	H_2	H_3
English			
Contemporary	4.03	3.32	3.1
Chaucer	4.00	3.07	2.12
Shakespeare	4.106	3.308	2.55
Poe	4.100	3.337	2.62
Hemingway	4.055	3.198	2.39
Joyce	4.144	3.377	2.55
German	4.08	3.18	
French	4.00	3.14	
Italian	3.98	3.03	
Spanish	3.98	3.01	
Portuguese	3.91	3.11	
Latin	4.05	3.27	2.38
Greek	4.00	3.05	2.19
Japanese (Kawabata)	4.809	3.633	
Hawaiian	3.20	2.454	1.982
Voynich manuscript[a]	3.66	2.22	1.86

[a]These values were computed using the first ten pages of the manuscript and Bennett's twenty-one-letter transcription alphabet.

William Friedman was among the researchers who made comparisons, like those found above, showing a very low value for the entropy of the Voynich manuscript. It's almost too low for a real language, unless it's Hawaiian, or perhaps another Polynesian language. Friedman did not think it was a hoax, but he didn't think it was Polynesian either! We'll look at his theory at the end of this chapter. For now, we turn to work carried out by Captain Currier.

Currier's Findings

In January 1941, a four-man delegation departed from America to England on an extremely important mission. The American cryptanalysts had broken the Japanese diplomatic cipher codenamed Purple and built bootleg machines, known as purple analogs, to read intercepted messages. They carried one of these analogs with them as a gift for the British codebreakers at Bletchley Park. The hope was that their counterparts across the ocean would reciprocate by sharing the progress they'd made against Nazi ciphers. The U.S. delegation consisted of two representatives from the Army and two from the Navy. One of the naval men was Captain Prescott H. Currier.

FIG. 1.25 Captain Prescott H. Currier

Being entrusted with a share in such an important mission shows the high regard in which Currier was held. We'll return to this mission later, but for now let's jump past the war to a time when cryptanalysts could better afford to work on ciphers not composed by enemies.

Professor Albert H. Carter, a former technical historian for the Army Security Agency (ASA), claimed in 1946 that the handwriting is consistent throughout the enciphered manuscript. ASA was one of the precursor groups that evolved into the National Security Agency (NSA). In 1976, Currier was an NSA analyst, and he presented strong evidence that Carter was wrong.

Currier discovered that there are two distinct handwritings that represent two different "languages" in the Voynich manuscript. He started with the herbal section and called the first "language," which ran for 25 folios, A. The second half of the herbal section, another 25 folios, contained two hands, each paired with its own "language." Currier called the second language B. Earlier in this chapter, we looked at a graph showing the distribution of word lengths in the manuscript. It was actually just for language B. That's why it was labeled VMS B. After Currier's discovery, it made sense to study the statistically distinct portions of the manuscript in isolation from the other sections.

Languages A and B are both used elsewhere in the manuscript. *Languages* has been placed in quotes in its first appearances above because it is not meant to have its normal meaning in this context. Instead, it's used to mean that A and B are statistically different. Currier found no differences in the astrological section, and he found that the biological section was all in B. The middle of the pharmaceutical section is in a different language and hand from that of the beginning and the end. Currier actually found, in total, "an absolute minimum of four different hands in the pharmaceutical section." He suggested two languages with several scribes.

Currier gave the following statistical evidence to support his claim of a difference in languages in the herbal section:

1. Final 𝒹9 is very high in B, but almost nonexistent in A.

2. The groups cᴘocℍ and cᴘoↄ are very high in A, low in B.

3. cᴘaⁱℙ and cᴘaⁱⁱℙ rarely occur in B, but are of medium frequency in A.

4. Initial cᴘocℍℙ is high in A and rare in B.

5. Initial cℍℙᴘ is very high in A and very low in B.

6. Unattached finals are scattered throughout B texts in considerable profusion, but are generally much less noticeable in A.

A larger set of results is reproduced below from D'Imperio's book, *The Voynich Manuscript—An Elegant Enigma.*

The different languages identified by Currier could all represent the same natural language, with the variations arising from technical language specific

Voynich Symbol	Currier Language A (Herbal)	Currier Language B (Herbal)	Krischer (fo. 103–116)	D'Imperio (Herbal, Astronom.)
ᛌ	290	257	233	368
⊙	2249	1373	729	3389
𝟪	884	1250	406	1333
❾	1231	1529	464	1893
ↄ	205	151	41	425
ℜ	663	496	250	(all) 1005
ℜ	531	495	201	(all) 971
ᴀ	1315	752	376	1373
ᴀ	415	289	93	557
ᴪ	516	376	187	734
ᴪ	75	108	47	154
ᴪ	595	801	267	865
ᴪ	21	63	6	53
ᴪ	165	51	13	266
ᴪ	42	12	7	49
ᴪ	86	100	15	106
ᴪ	7	9	2	29
ᴀ	900	1085	546	1470
c	769	1390	730	1094
ℓ	16	8	2	ℓ 216
ıℜ	4	1	0	ꝟ 835
ııℜ	1	0	0	ꝭ 167
ıɑℜ	0	0	0	ℊ 23
ıɔ	22	45	35	ꭎ 689
ııɔ	8	24	11	ꝛ 12
ıᴙıɔ	3	2	1	ℴ 2
ꝯ	38	3	4	ℴ 0
ıꝯ	82	73	38	ꝯ 7
ııꝯ	455	286	153	ꝉ 3
ıııꝯ	18	22	0	ꝯ 36
ꝏ	78	99	23	ꝯ 13
ıꝏ	6	5	1	
ııꝏ	1	1	1	
ıııꝏ	0	0	0	
ꝺ	13	7	1	
ꝺ	5	5	11	
ıɪꝓ			2	
Totals	11709	11168	4896	18137

FIG. 1.26 More statistical evidence of two distinct languages

to certain portions of the manuscript, which, if the illustrations are to be believed, covers a lot of ground.

Eventually, twelve distinct hands were recognized in the manuscript.[106] Though this may seem to complicate matters, it does help us to eliminate one possible explanation for the manuscript. If many authors or scribes were involved, it would seem to be much less likely that it was created by a mentally deranged individual and has no meaning. Of course, it could still be an elaborate meaningless hoax created by several authors.

Currier also noted odd statistics that held throughout the manuscript. For example, it appears that the line is a "functional entity." Currier wrote "the frequency counts of the beginnings and endings of lines are markedly different from the counts of the same characters internally." Currier continued, "There are, for instance, some characters that may not occur initially in a line. There are others whose occurrence as the initial syllable of the first "word" of a line is about one hundredth of the expected."

The ends of lines often contain meaningless symbols, according to Currier, as if garbage was added to finish out the line. He describes these characters as "little groups of letters which don't occur anywhere else." One character that does occur elsewhere has 85% of its appearances at the end of a line. Our statistical tests, already weak from our uncertainty as to the size of the alphabet, are damaged further if the ends of lines do not represent real text. The possible alphabet characters may need to be reevaluated with emphasis placed on their use in the beginnings and middles of lines. However, a later researcher, Nick Pelling, concluded that one symbol (or symbol cluster, if you prefer) appearing far more frequently at the end of lines represents a hyphen used to split words over two lines. So maybe these odd end-of-line symbols are meaningful after all.

Other researchers have pointed out yet more statistical oddities, such as the lack of doubled letters in the manuscript. These claims ignore ⱳ and ⱳ, which are presumed to be single characters.

Currier's introduction in this section described how he was a member of an important intelligence sharing mission to England during World War II. We now turn to a cryptanalyst who was on the receiving side of America's offer.

John Tiltman had earned a reputation in England as being "without equal" in cracking hand ciphers. He showed his skill by breaking various Russian cipher systems in the 1920s, Japanese systems in the 1930s, and Nazi field ciphers in 1939 and 1940. In the summer of 1940, he played a key role in breaking the Railway Enigma system. Some of these messages revealed that the Germans were moving many troops to the east in 1941. This intelligence, and other indicators, motivated Winston Churchill to warn the Soviets that Hitler was going to attack. However, Stalin, who seemed to trust no one, except maybe Hitler, didn't believe it.

FIG. 1.27 Brigadier John Tiltman (1894–1982)

So, when the question of sharing intelligence with America arose, Tiltman's opinion counted for something. Concerns were aired that the Americans might unintentionally give away hard-won British progress on Enigma, through their own weak ciphers, but Tiltman argued against these objections and insisted that intelligence should be shared. A publication issued by NSA's Center for Cryptologic History noted, "His success laid the foundation for unparalleled cooperation that continues to this day."[107] When the British first reciprocated in the spring of 1942, it was Tiltman who was chosen to cross the ocean on an American warship with eight heavy mailbags full of reports and technical information for the American cryptanalysts.

Tiltman accumulated many more cryptanalytic success stories as World War II continued. He defeated another Japanese system in 1942 and made the first breaks into the Lorenz SZ40 cipher machine. Codenamed "Tunny" by the British, this was the machine Adolf Hitler used to exchange messages with his high command. Broken SS and police systems revealed the horrors of the Holocaust, horrors that would have gone on longer if the Allies hadn't had the benefit of the insight the cryptanalysts' work provided.

While details remain classified, Tiltman's successes continued through the 1950s, 1960s, and 1970s. At least some of the work concerned hand ciphers used in Eastern Europe and Third World countries.

Tiltman's accomplishments were recognized at the highest levels by both NSA and GCHQ. The American agency placed Tiltman in their Hall of Honor, and Sir Brian Tovey, director of GCHQ, said that Tiltman was one of the greatest cryptanalysts Great Britain had ever produced.

How would this master who had beaten the Russians, Japanese, and Nazis fare against the unknown author of a centuries-old cipher manuscript?

American cryptanalyst William Friedman brought the manuscript to Tiltman's attention around 1951. From 1958 to 1964, Tiltman actually worked at NSA and continued as a consultant thereafter. The years 1962–1964 coincide with the Second Study Group on the manuscript, run by Friedman.

Despite a massive amount of work having been carried out before he began his analysis, Tiltman turned up some new and important features of the manuscript. He noticed that the manuscript contains character groups that are stereotyped by occurring almost exclusively as beginnings, middles, or ends. That is, there were a decent number of prefixes, roots, and suffixes. Some of these are shown below. This could indicate that a synthetic language was used. The suffixes could indicate person and number as well as past, present, or future tense, or infinitive, when attached to the root of the verb. For example:

Table 1.4 Sample Meanings of Suffixes

	1st Person	*2nd Person*	*3rd Person*
Past	ɑⱳ	 ɑɯⱳ	 ɑɯⱳ
Present	ɑɾ	 ɑɯɾ	 ɑɯɾ
Future	ɑƴ	 ɑɯƴ	 ɑɯƴ

These assignments are intended to illustrate the idea and are not being proposed as correct. Other suffixes, such as �9 and ꝱ9, may have been used to denote the plural.

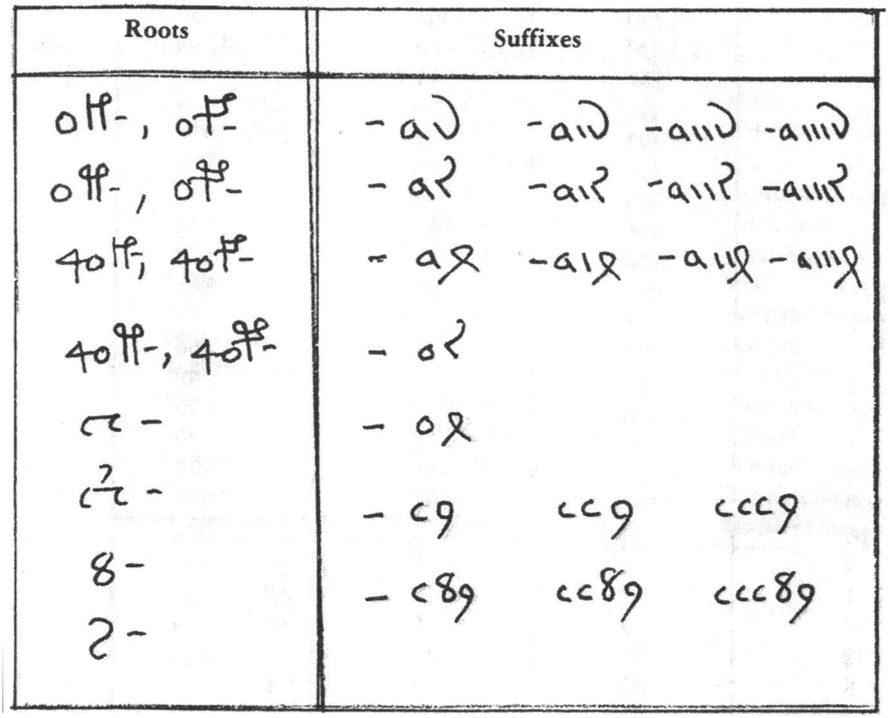

Roots	Suffixes			

FIG. 1.28 Common groups of symbols that appear at the start and end of words

Tiltman didn't list enough suffixes to provide for all the possibilities most languages offer, but perhaps other suffixes exist and were used too sparingly to stand out, or maybe they weren't used at all. For example, some authors avoid the word "I" and speak of themselves as "we" when writing. My editor at Princeton hates this!

In addition to finding prefixes and suffixes, Tiltman also made a list of common roots.

When I presented Tiltman's cryptanalytic résumé, I mentioned only systems he broke. It should be noted that he played the other side as well and created systems that stymied the Nazis. Some of the earlier British systems first converted the text to numbers, using a codebook, and then added other numbers to them, as a further disguise. These "other numbers" were referred to as additives, and it was important that they be as random as possible. If additives were reused, or predictable in some other way, there was a chance the enemy could break the system. Tiltman's improvement to the sheets of additives was to combine them with a grille. This was a flat piece of plastic with holes cut out in various places so that 100 four-digit additives would show through. The grille could be moved around on the additive sheet to show different sets of numbers. Still, the additive sheet was changed daily.

Tiltman's grille had an official name, the Stencil Subtractor Frame (a.k.a. S.S. Frame or SSF), and when it was introduced in December 1943, following a new codebook brought forth in June, the Germans could no longer read the communications of England's Navy.

But grilles can be used in another way. This was to form the foundation of a hoax theory put forth by British researcher Gordon Rugg in 2004.

Rugg's Theory

Gordon Rugg conjectured that a text with statistical properties matching those of the Voynich manuscript could have been created by starting with a grid of "syllables" like the one reproduced in fig. 1.29.

The next step would be to place a grille on top of the grid, so that only three syllables, from neighboring vertical columns, show through. These could then be copied down on another sheet of paper. Moving the grille reveals another set of three syllables, which could also be copied down. By using a variety of grilles, with holes cut out in different positions, and placing them on the grid in different positons, much text can be generated that has the sort of structure Tiltman and others observed in the Voynich manuscript.

Rugg noted that the grids could be of various dimensions, and the grilles needn't have just three slots cut out. His example was just for illustrative purposes.

The two languages observed in the manuscript by Currier, namely Voynich A and Voynich B, could have arisen from a different table being used for each of these sections of the manuscript.

So, who would have been behind such a hoax? Rugg pointed out that John Dee and Edward Kelley used a method something like this to generate Enochian, an "angelic" language.

Having a suspect, or pair of suspects, makes the theory even better. Yet more support was presented by Austrian physicist Andreas Schinner in 2007. His paper was more technical than Rugg's, but it led to the same conclusion from another direction. The abstract is reproduced here for those wanting a bit of technical detail:

9	ccett		ɤ	czc	anɔ	40	lł	anɔ
40	czc	89	40	lł	9	o	4ł	89
	4ł	89	o	2zc	o89	40	czc	9
40	lł	anɔ	40	lł	9		lł	89
ɤ	2zc	9		4ł	anɔ	40	2zcll	9
	lł	ax89	2	lł	89	2oɤ	lł	aɔ

FIG. 1.29 A grid that could be used to generate a pseudotext

In this article, I analyze the Voynich manuscript, using random walk mapping and token/syllable repetition statistics. The results significantly tighten the boundaries for possible interpretations; they suggest that the text has been generated by a stochastic process rather than by encoding or encryption of language. In particular, the so-called Chinese theory now appears less convincing.[108]

We've seen several different hoax theories. Do you favor any of them? Over the decades, top experts have been divided on the matter. NSA employee Mary E. D'Imperio, who wrote a great book on the manuscript, could accept a hoax theory, but Tiltman could not.[109]

I keep changing my mind. On some days, I believe that a method like Rugg proposed was used, and on other days I believe William F. Friedman. We'll now turn to Friedman's view on the matter.

Ronald W. Clark had a bestseller in 1972 with *Einstein: The Life and Times*. Probably wanting to duplicate that success, he looked for other geniuses to write biographies of. In 1977, he released a biography of William F. Friedman.[110]

Friedman was America's version of Tiltman, or you could say Tiltman was England's version of Friedman. Both excelled at cracking hand systems. NSA recognized Friedman's contributions by naming their main auditorium after him. They also placed him in their Hall of Honor. Most recently, NSA released 52,000 pages of material relating to his career.[111] Friedman managed the team that broke Purple, an important Japanese diplomatic cipher used during and before World War II. He also authored the textbooks that NSA used to train cryptologists for decades. He's even the one responsible for coining the term *cryptanalysis*. So, what were his conclusions in regard to the Voynich manuscript?

FIG. 1.30 William F. Friedman (1891–1969)

Well, he didn't form them casually. He began a study group on the manuscript at NSA on May 26, 1944 (while World War II was still ongoing!), with an initial meeting of sixteen people. Fortunately, these people met after normal work hours and were not attacking the manuscript at the cost of ignoring the German and Japanese communications that they were officially tasked with solving. The team began transcribing the manuscript characters and placing them on punch cards so that they could be analyzed statistically by the computing machines of that era. Although meetings were held in 1945 and 1946, the job was not completed by this study group.

Friedman began a second study group on September 25, 1962. Again, the goal was a complete transcription. But this group didn't start up where the previous group stopped. As the table of transcriptions reproduced earlier in this chapter shows, a different transcription alphabet was used. By the end of the summer of 1963, this second study group ceased its activities with the transcription once again incomplete. However, enough work was carried out by the first study group for Friedman to form an opinion, and the efforts of the 1960s did not change it.

Father Theodore C. Petersen, who supplied the groups with the photostats used to produce their incomplete transcriptions, completed a different daunting task. He made a concordance. Like a biblical concordance, for each word in the manuscript, it provided a list of every page on which that word appeared. It also provided, in each instance, the words that came immediately before and after the given word. One of the odd things this shows is that there are many instances of the same words appearing back to back.

But to get back to Friedman, this researcher published his theory of the Voynich manuscript in anagram form. That is, he rearranged the letters in the sentence explaining his viewpoint so that it formed another meaningful sentence. This is what saw print in 1959, between the years of activity for his two study groups:

> I put no trust in anagrammatic acrostic ciphers, for they are of little real value—a waste—and may prove nothing—Finis.[112]

Various solutions were turned in by readers:

> William F. Friedman in a feature article arranges to use cryptanalysis to prove he got at that Voynich manuscript. No?

> This is a trap, not a trot. Actually I can see no apt way of unravelling the rare Voynich manuscript. For me, defeat is grim.

> To arrive at a solution of the Voynich manuscript, try these general tactics: a song, a punt, a prayer. William F. Friedman.

This is why Newbold's random anagramming wasn't a good step! Recall that he used groups of 55 and 110 characters. Friedman's message had 96 characters and offered, as seen above, several different solutions.

The correct solution was revealed in 1970, after Friedman's death, as

> The Voynich MSS was an early attempt to construct an artificial or universal language of the a priori type.—Friedman.[113]

Such artificial schemes have been investigated for hundreds of years and known variously as synthetic, artificial, or universal languages. Earlier in this chapter, Kircher's attempt at creating a universal writing system was mentioned. However, he wasn't the only great intellect to take on this task. Gottfried von Leibniz, codiscoverer of calculus, had tried to create such a system, as had Sir Francis Bacon. Neither of these men was responsible for the cipher manuscript according to Friedman, for he thought it was written between 1480 and 1520, before they were born.

Tiltman was introduced to the manuscript by Friedman, who may well have been seeking a second opinion. As it turned out, Tiltman's conclusions matched Friedman's. When the titans of cryptology from England and America agree, it's an opinion that should be taken seriously!

Friedman was specific in the sort of artificial language used in the Voynich manuscript. He wrote that it was "of the a priori type." Such artificial languages divide everything into categories and then further divide these categories into subcategories. In this manner, we can specify a particular word by writing $x_1x_2x_3y$ where the xs are the categories and subcategories and the y is an ending used to denote the part of speech. This would be one explanation for the suffixes and prefixes Tiltman observed. Such languages are completely different in structure than ones that arise naturally, or even artificial languages patterned after real ones.

Although Friedman died in 1969, interest in the manuscript continued at NSA. Mary D'Imperio, an NSA analyst, chaired a Voynich seminar held in 1972. Her monograph, *The Voynich Manuscript—An Elegant Enigma*, appeared in 1976 and gave a great summary of work done up until that time. One later book on the manuscript questioned whether D'Imperio "was in fact a real person and not an NSA phantom."[114] Apparently, the authors' inability to learn anything about her other than the connection with the book led them to wonder if it was a pseudonym. However, she was indeed real. NSA historian David Hatch commented, "She never had occasion to publish in the open again—that was the era when NSA employees, like classic society ladies, were supposed to be mentioned publicly only thrice."[115]

Voynich Online, in Print, and in the Lab

In 1991, Jim Gillogly (a computer scientist and superb cryptanalyst who makes appearances in several chapters of this book) and Jim Reeds (at Bell Labs at the time) brought cipher manuscript research and collaboration online by starting a Voynich group mailing list. Reeds also prepared a massive bibliography for the manuscript that formed the basis, with permission, of this chapter's References and Further Reading list.

More excitement came in 2004 when a book-length history of the manuscript, titled *The Voynich Manuscript* (with different subtitles for the American and British editions) appeared. Written by Gerry Kennedy and Rob Churchill, this was the first book-length treatment since the volume edited by Brumbaugh in 1978, excepting "solutions." Then in 2005, *The Friar and the Cipher* by Lawrence and Nancy Goldstone appeared, offering another full-length treatment. Nick Pelling, who blogs on unsolved ciphers,[116] offered *The Curse of the Voynich* in 2006, although this book had a much smaller print run.

In 2011, almost a hundred years after Voynich went public with his rediscovery of the manuscript, carbon-14 dating revealed that the vellum on which the cipher was placed was from somewhere in the range 1404–1438.[117] The ink was added soon thereafter, according to the McCrone Research Institute.[118] However, Klaus Schmeh, in conversation with Greg Hodgins (the radiocarbon expert who made the datings), heard that nothing can be said about when the ink was added. The press report stating otherwise was wrong, according to Hodgins.

Hopes of Roger Bacon being the author were (mostly) dashed by the dating of the vellum alone. Although Bacon could certainly not have penned the work, diehards can still argue that it could be a later copy of a manuscript he created.

If the ink really was applied so early, we can definitively eliminate some hoax theories. Although the manuscript could still be a meaningless fake, it could not have been made with Rudolf II as the intended mark, not by John Dee or any other contemporary. Nor could it have been created by Wilfrid Voynich, as some suspected. If it was made to part a fool and his money, it must have been with a fifteenth-century dupe in mind.

Toward the Future

If a real solution is found, it will generate a tremendous amount of excitement. This could well happen, even after so many great minds have failed, because there's still hope that new information will arise that can aid in the decipherment. Have the missing pages survived the centuries? They might have been split off from the original manuscript by any of the owners before Voynich, and the information they contain could be exactly what's needed. Another natural question to ask is this: Did the unidentified creator of the manuscript write anything else? It would seem strange for a work of such length and complexity to be a first effort. Earlier works by the same hand(s) could shed light on the solution.

In 1967, David Kahn described the Voynich manuscript as "the longest, the best known, the most tantalizing, the most heavily attacked, the most resistant, and the most expensive of historical cryptograms."[119] Fifty years later, it still holds the lead position in all of these categories.

Solution to Example Cipher

His Majesty is interested only in wizards, alchymists, kabbalists and the like, sparing no expense to find all kinds of treasures, learn secrets and use scandalous ways of harming his enemies ... He also has a whole library of magic books. He strives all the time to eliminate God completely so that he may in future serve a different master.

The substitutions were as follows:

```
ABCDEFGHIJKLMNOPQRSTUVWXYZ  (Plaintext)
BFVDRQOGNUMJLYKSZIXHTEACWP  (Ciphertext)
```

2

Ancient Ciphers

Introduction

Carbon dating has established that the vellum used for the Voynich manuscript dates back to the fifteenth century. But even at an age of more than five hundred years, it's not the oldest unsolved cipher.

Almost every culture that develops writing quickly goes on to discover secret writing. Although we have unanimously accepted examples of ciphers from almost every literate culture that's existed, there are other possible examples that have been rejected by the experts on the ancient world. They look at these apparently random strings of symbols or letters and conclude that they must have been created by illiterate people imitating writing. They dismiss such writings as gibberish and often refer to them as nonsense inscriptions.

How can they tell the difference between a cipher, which looks like random letters, and the writing of an illiterate, which also looks like random letters? The individuals making these subtle distinctions usually do so casually. They aren't experts in cryptology and hardly ever even raise the possibility of cryptography having been behind the strange writings. Highly relevant is the fact that the unreadable inscriptions are in many cases not mere scribblings, but rather carefully constructed works of art, sometimes carved in stone and sometimes appearing in a serious religious context. Among the many examples are Viking runestones (the Vikings had a knowledge of cryptology on par with the ancient Greeks), painted vases from the Greeks themselves, and Egyptian sarcophagi.

This chapter details some of these little understood writings, providing many intriguing images. It's the first survey of such material that encompasses more than a single culture.

Languages and Scripts

We make the distinction between spoken *languages* and written *scripts*. Some languages, such as Japanese, can be written using more than one script; there's kanji, hiragana, katakana, and a version using the Latin script known as *Rōmaji*, after the Romans. Thus we can write "book" as 本 (in kanji), ほん (in hiragana), ブック (in katakana), or hon (in *Rōmaji*). Despite the different representations, all three would be pronounced identically.

In a similar manner, a single script can be used to represent various languages. The alphabets for English and German are almost identical (both are classified as Latin script), but they are used to represent distinct languages.

As a consequence of this fact, the script doesn't always indicate unambiguously what language is represented. When ancient writing is encountered that doesn't match anything previously known, and it cannot be read, we have a number of possibilities:

1. It can be a familiar script used to represent a language that it wasn't previously known to have been associated with.
2. It can represent a lost language.
3. It can be an encipherment of a language known or unknown.

No matter which option turns out to hold true, the tools of the cryptanalyst can be brought to bear. The techniques for recovering the meaning behind such writings have a lot in common with cryptanalysis, whether a cipher is present or not. But it can sometimes be hard to determine what category a piece of writing falls into. There's even historical precedent for a script representing a natural language being confused with a cipher.

In 1873, Joseph Halévy, a pioneer in the scientific study of languages, argued that Sumerian was nothing more than a form of secret writing used by priests that was never actually spoken. The debate lasted until 1900, when Halévy was finally proven wrong.

Although the "secret script" in this case turned out to be normal writing, once enough Sumerian documents were uncovered and understood, it was learned that they did indeed have a primitive form of cryptology. For example, the names of gods were sometimes replaced with code numbers. These are now known as Götterzahlen (German for "god numbers") in the literature. In general, cryptology follows close on the heels of the invention of writing. When few people are literate, writing itself is like a cipher, but as soon as there are literate people from whom secrets must be kept, secret writing is needed. It's also been heavily used in religious contexts. Even today, there are people who won't write out the word "God," for the paper it appears on could be destroyed. Instead, they write "G-d" if they have a need to use the word. Then, if something happens to the paper, it's not considered disrespectful.

In this chapter, I present examples from several ancient civilizations of what may well be unsolved ciphers. The experts on these civilizations typically dismiss these examples as gibberish or "nonsense inscriptions." However, if the experts of the past could mistake meaningful texts for "secret writing," then certainly today's experts could be mistaking secret writing for nonsense.

Ancient Egypt

The Sumerians represent the oldest civilization that we have some understanding of. Other sites, such as Göbekli Tepe in modern Turkey, indicate that the story of civilization goes back much farther, but we have not found writings at these locations, so our knowledge is minimal. Egyptian civilization came a bit later, and as with the Sumerians, their writings were incomprehensible to the people of the early nineteenth century. The story of the decipherment of Egyptian hieroglyphics is fascinating, but we must move on to another ancient Egyptian mystery that remains unsolved.

Our first example is a controversial sarcophagus in the Wellcome collection at Swansea University's Egypt Centre. The style of painting fits the Twenty-Sixth Dynasty, which dates to about 600 BCE. However, the sarcophagus is only a little more than twenty inches long and the hieroglyphs on it make no sense. Could it have been manufactured relatively recently to fill the demand created by scholars and collectors? Surely, someone creating a container for such a solemn funereal purpose would not place meaningless hieroglyphs upon it. An obvious line of investigation is to research the provenance of the sarcophagus. Sadly, the museum doesn't know how or where Sir Henry Solomon Wellcome (1853–1936) acquired this item (Accession number: W1013), only that it became part of the collection in 1971.

Those who favored it being a modern creation found themselves on much weaker ground in 2014, when CT scans were performed by Paola Griffiths at Swansea University's Clinical Imaging College of Medicine. The modern

FIG. 2.1 A controversial
sarcophagus. Is it ancient or modern?

technology revealed a fetus with a placental sac. It measured only about three inches long and was estimated by Egypt Centre curator Graves-Brown to have been between twelve and sixteen weeks along in development. The majority of the space inside the sarcophagus was filled with what is thought to be folded linen bandages. The CT scan also detected what may be an amulet and strings of beads or tassels, fitting Twenty-Sixth Dynasty customs.

Although it's hard to imagine anyone using a real fetus as material with which to fake an antiquity, similar perversions have been documented. Consider the following passage from a book on forgery in the world of art:

In 1485 on the Appian Way, which to a humanist in Renaissance times was virtually a time machine leading back into sacred antiquity, a tomb was found containing the perfectly preserved body of a beautiful young girl. A Latin inscription on the sarcophagus identified her as none other than Tulliola, Cicero's gorgeous only child, who had died at a young age. Cicero's words in the inscription were painfully poignant: "To Tulliola, his only daughter, who never erred except in dying, this monument was raised by her unhappy father Cicero. *Quae nunquam peccavit nisi quod mortua fuit.*"

The discovery was a sensation and accepted by everyone, even hardened skeptics. No one seemed to care that the quote belonged to a non-Ciceronian but wholly authentic ancient Latin source. Or that a bunch of other Tulliola tombs were found around Rome not long after and subsequently in Florence and Malta—each one with a well-preserved body of a young girl. As Anthony Grafton observes, "Tulliola's death scene continued to be a crowd-pleaser for a hundred years." Who these young girls were, how they died, or for precisely what reason, is almost too chilling to think about, but the perpetrators of the grim hoaxes have never been identified.[1]

Despite the unreadable inscription and the macabre possibility given above, the new evidence led experts to finally view the sarcophagus as genuine.

Let's now consider some extremes. If every Egyptian sarcophagus contained unreadable inscriptions, it would be reasonable to assume that they had some meaning to the ancients that we simply haven't been able to discern. On the other hand, if only one bore such markings, then perhaps it was an error of some sort, and is indeed devoid of meaning. What about values between these extremes? How many examples would there have to be to convince you that they have some sort of meaning?

Well, a lot more than one have been documented! Several examples of "mock hieroglyphic symbols," as they are sometimes called, were found at Saqqara on sarcophagi from the Twenty-First Dynasty. The official explanation is that they "acted as a magical aid in the afterlife."[2]

Is it convincing to have examples from just the Twenty-First and Twenty-Sixth Dynasties? What about the dynasties between these? Sir Flinders Petrie addressed the topic:

> The next period of importance at Illahun is from the twenty-second to the twenty-fifth dynasties. The hills near the pyramid had been much used for rock tombs and mastabas of the pyramid period; but these had been plundered and destroyed in early times, and the excavations were re-used during the later Bubastite and Ethiopian dynasties. These interments are generally rude, the coffins seldom having any intelligible inscription; but mostly sham copies of the usual formula, put on by a decorator who could not read.[3]

So during these dynasties, unreadable inscriptions were the norm! Could a tradition of truly meaningless inscriptions on the coffins of loved ones have endured through six dynasties? Because Sir Petrie thought the inscriptions were shams, he didn't spend much time on them. Another passage in which he gave them passing mention is reproduced here, emphasis added:

To conclude, I will describe the typical details of the burials of the XXIInd dynasty. The coffin thin, straight sloping sides with a slight shoulder, and round head; sides upright. Lid flat board with an edge to it, inscription down the middle, *usually nonsense*, or the personal name omitted at the end. Head and shoulders in relief, and sometimes the hands; the face a carved block of wood, the head-dress formed of stucco, or more usually of Nile mud; brilliantly painted with red, blue, yellow, black and white; the decoration a wig and vulture head-dress. Inside lies a cartonnage of linen and plaster, modelled to the body form, split down the back, where the mummy was slipped into it. The surface generally white with a band of inscription down the middle: a spread vulture at the top of it sometimes with a ram's head: the face carved in wood and inserted (hence come the multitude of such faces, as the cartonnages are often rotten or broken up) and the head painted with a wig. Sometimes the cartonnage is painted with scenes of offerings to gods all over it. The body inside seldom has any amulets and is usually mere black dust and bones.[4]

Alan R. Schulman, of Queens College, New York, was humble enough to put forth an unpopular view. It appeared in a paper he wrote about a scarab with a strange inscription, but seems intended to be taken much more generally. He wrote

> But an inscription should have a meaning. After all, the Egyptians were not that illogical as to write inscriptions which were meaningless gibberish. If a text does not make any sense to us, it is more likely that we do not understand it, rather than that the inscription itself is meaningless.[5]

Before any meaning can be found in such texts, a deeper understanding of ancient Egyptian cryptography is needed. How long a history did the subject have? What techniques were used? In a book based on his doctoral dissertation carried out at the University of Chicago, Yale University Assistant Professor of Egyptology John Coleman Darnell wrote,

> Beginning during the Old Kingdom and continuing through the First Intermediate Period and the Middle Kingdom, cryptography abounds in the New Kingdom, occurring in royal titularies, in inscriptions from private tombs, in private graffiti, and throughout the Netherworld Books preserved in the royal tombs. During the Late Period and throughout the Ptolemaic

and Roman eras, reaching a complicated height in the texts of the temple of Esna, a number of signs and sign values occurring earlier only in cryptographic texts became common in all hieroglyphic texts.[6]

In other words, crypto was used from the Old Kingdom right through to the end of hieroglyphs—the entire time! In later centuries, what was once considered cryptographic had gone so mainstream that it was simply one of the ways of writing. An example makes this a bit clearer.

One sort of substitution cipher that was used replaced hieroglyphs with others having a similar or related theme. For example, the hieroglyph ⬭ was replaced with ▷. Later, the symbol was replaced with ⬰. The common theme in this instance was mouths. The first was a mouth viewed from the front, the second from the side, and the third, the mouth of a crocodile.

In other instances, the substitutions were for symbols that had a similar appearance. Examples include replacing ⊏⊐ with ⎮⊔ and replacing ⊂ with ⟞⟝.

In addition to the substitutions described above, hieroglyphs could be replaced by others that didn't necessarily look the same, but did represent similar sounds.

In all of this, there was an escalation, for as alternate representations became more widely used and understood, those wishing to have their writing remain enigmatic were forced to come up with new forms. As this happened, Egyptian cryptography became more difficult. This is why older cryptographic inscriptions are, in general, more readily made sense of today.

The ancient Egyptians also had an encryption technique completely different from those described above. It involved keeping the hieroglyphs the same, but placing them in a different order. While mathematicians and computer scientists call it transposition, Egyptologists refer to it as perturbation.

Étienne Drioton, who carried out groundbreaking work on ancient Egyptian ciphers, wrote that this technique was mostly used for inscriptions on scarabs, and he provided a small example. Given the four hieroglyphs ⬤ ⟍ ↯ ⊥, he read them in a perturbed order, following the route 1 3 / 4 2 to get ⬰ ⊥ ⬟ ↯. The meaning becomes "Khonsu is (my) protection." Although he indicated that such perturbations are usually brief, he did point out an exception. A cup at the Guimet Museum in Paris has an inscription of twenty-one signs perturbed by this method.[7]

But there's a much flashier example. Perturbation was used on the second golden shrine of Tutankhamun, enclosing his sarcophagus, and many other places.

Darnell was quick to point out the gaps in our present understanding of ancient Egyptian cryptography.

> There has not been a detailed study of all pre-Late Period cryptography, the results of which one could with profit compare to an also needed but non-existent study of Late Period cryptography.[8]

These gaps exist, despite the efforts of a few people, because of the way the majority have regarded the matter. Alan R. Schulman noted,

> It has long been known among Egyptologists that the Egyptians frequently used cryptograms of various sorts to write certain of their inscriptions, but for one reason or another, probably due to the complexity of the subject, most scholars have preferred not to deal with it. The notable exception was the late Étienne Drioton whose brilliant researches in the realm of Egyptian cryptography, particularly in the field of cryptograms on scarabs, have established a solid footing for all future investigations and studies of Egyptian cryptography.[9]

Perhaps such studies will be carried out in the near future, bringing into the light texts now believed meaningless by many. What will the small sarcophagus have to tell us?

Sarcophagi inscriptions are not the only potential ciphers that ancient Egypt has to offer. The References and Further Reading section for this chapter includes leads on others.

Ancient Greece

We now turn to a culture whose use of cryptography we know more about, the ancient Greeks. The word "cryptography" in fact comes from the Greek κρυπτός (kryptos), meaning "hidden," and γραφία (graphia), meaning "writing."

The ancient Greeks are known to have had monoalphabetic substitution ciphers (MASCs) that turned letters into other letters. But they also had a method of turning letters into numbers. This is known as the Polybius cipher, and it uses a rectangular arrangement of the alphabet to facilitate the encryption. Using English letters, we have

```
  1 2 3 4 5
1 A B C D E
2 F G H I K
3 L M N O P
4 Q R S T U
5 V W X Y Z
```

The Greek version looked like this

```
  1 2 3 4 5
1 α ζ λ π φ
2 β η μ ρ χ
3 γ θ ν σ ψ
4 δ ι ξ τ ω
5 ε κ ο υ
```

Whether the alphabet is placed in the grid by rows or by columns, or in some random order, doesn't affect how the system works. The scheme was invented with long-distance signaling in mind. Polybius explained how it worked, crediting it to Cleoxenus and Democlitus, in his *Histories*.

In the example Polybius gave, the signaler wants to send a message that begins κ (kappa), ρ (rho), ... The Greek letter kappa would be represented by the signaler holding two torches on the left (for the second column) and five torches on the right (for the fifth row). Once the receiving end has been given sufficient time to note the number of torches in each hand of the sender, the sender would switch to show four torches on the left and two on the right, to represent rho. It's standard in books today to show the grid of letters as it was depicted above. However, Polybius described it a bit differently. For the ancient Greeks, each vertical column was a separate tablet. The senders and receivers would thus each have five tablets in front of them. So the number that's relayed on the left would represent which tablet is meant.

The Greek alphabet fit into the 5 × 5 grid with room to spare, but in English there's one letter too many. To solve this problem, the letter J was deleted (or you could say it was combined with I). We can use the grid for a written cipher by jotting down the numbers for the row and column in which each letter appears. For example, the Plato quotation "Only the dead have seen the end of the war" enciphers like so:

```
O   N   L   Y   T   H   E   D   E   A   D   H   A   V   E   S   E   E
34  33  31  54  44  23  15  14  15  11  14  23  11  51  15  43  15  15

N   T   H   E   E   N   D   O   F   T   H   E   W   A   R
33  44  23  15  15  33  14  34  21  44  23  15  52  11  42
```

It might seem that a cipher consisting entirely of letters could not be a Polybius cipher, but this is not the case. The rows and columns could be indexed by letters.

```
    A  D  F  G  X
A   A  B  C  D  E
D   F  G  H  I  K
F   L  M  N  O  P
G   Q  R  S  T  U
X   V  W  X  Y  Z
```

This was actually the first step in a cipher system used by the Germans during World War I, although they scrambled the order of the letters within the grid. They chose A, D, F, G, and X as the cipher letters because they were hard to confuse with each other when sent in Morse code. Later they expanded the grid to allow for encipherment of numbers, without having to spell them out, by adding the letter V to the row and column indexes.

So a cipher consisting solely of letters could represent a Polybius encipherment, but only if the number of letters is small!

Plutarch described a completely different form of encryption used by the Spartans in his *Life of Lysander*, which put forth the biography of a Spartan admiral active in the fifth century BCE. The relevant passage is reproduced here:

When the Ephors sent out any one as general or admiral of their forces, they used to prepare two round sticks of wood of exactly the same length and thickness, corresponding with one another at the ends. One of these they kept themselves, and the other they gave to the person sent out. These sticks they call skytales. Now when they desire to transmit some secret of importance to him, they wrap a long narrow strip of paper like a strap round the skytale which is in their possession, leaving no intervals, but completely covering the stick along its whole length with the paper. When this has been done they write upon the paper while it is upon the stick, and after writing they unwind the paper and send it to the general without the stick. When he receives it, it is entirely illegible, as the letters have no connection, but he winds it round the stick in his possession so that the folds correspond to one another, and then the whole message can be read. The paper is called skytale as well as the stick, as a thing measured is called by the name of the measure.[10]

The *skytale* (pronounced to rhyme with *Italy*) is the oldest known device for enciphering via transposition.

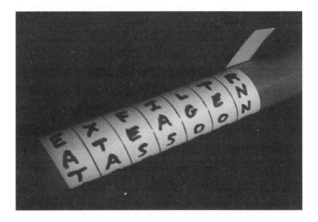

FIG. 2.2 A message written out using a skytale

It's been speculated that the origin of the commander's baton can be found in the skytale.

The Greek alphabet was used as a cipher alphabet by Julius Caesar, who would sometimes simply replace Latin letters with their Greek equivalents. For example, the message "ΓΟΛΔ" would be read as "GOLD." However, the Greeks didn't return the favor; they rarely used foreign alphabets for ciphers.

This summary of ancient Greek cryptography must, of necessity, be incomplete. It's likely that what scholars are aware of is only the tip of the iceberg.

Consider the situation with plays from antiquity. History has left us a record of Sophocles having written 123 plays, but only 7 have been preserved in complete form. A similar ratio applies to many other ancient writings. And these were writings for which distribution was encouraged. Surely, the scrolls that bore details of the secret art of cryptology were much rarer! Some hints of what we're missing managed to survive.

Albert C. Leighton, who had an interest in breaking historic ciphers, noted that in Aulius Gellius's *Attic Nights*, written in the second century CE, vague reference is made to ciphers more complex than those presently known to have been used in the ancient world. He also quoted a letter from the fourth century CE, in which Ausonius claimed that he knew "countless codes of the ancients for concealing and unlocking secret messages."[11] Leighton went on to speculate, "It may well be that ciphers in classical antiquity were more advanced than the literature would lead us to expect and that they influenced Renaissance cryptography."[12]

We'll soon turn to some strange writings that appear on ancient Greek vases. It's natural to investigate Greek writing through pottery, for the earliest extant Greek inscriptions, from the mid-eighth century BCE, are to be found there. Also, the surviving inscriptions on Attic vases (those from Athens) are more confidently dated and exist in greater numbers than any other inscriptions. And more are on the way; vases and vase fragments are still being excavated in large quantities today.

The words are sometimes read from left to right, which is more common, and sometimes from right to left. Thus, when looking at a potential cipher, two possibilities should be considered.

Many of the vases discussed below came from Athens, for her artists made inscriptions more often than those of any other ancient Greek city. This seems like a particularly bad location to fake literacy, as Athens was the intellectual center of ancient Greece.[13] Yet that is what many experts on Attic vases would have us believe happened. "Nonsense inscriptions" began appearing on the vases in the second quarter of the sixth century BCE. They are sometimes single "words" and sometimes series of words. Though there are some examples on Corinthian vases, it's believed that the practice began in Athens, because of the much larger number of inscribed Attic vases. This was a time at which the literacy rate was rising. As was mentioned before, writing itself can serve as a cipher if the literacy rate is low, but once the rate rises, more sophisticated methods are needed. That's when cryptography appears.

An example of a nonsense inscription is provided in fig. 2.3. This vase, now reduced to fragments, was made around 475–450 BCE. An Amazon is shown along with the meaningless Greek letters ΓΥΓΑΜΙΣ (Gugamis).

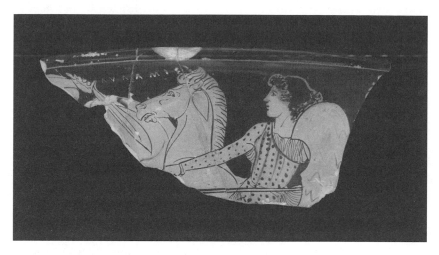

FIG. 2.3 An Amazon and the nonsense inscription ΓΥΓΑΜΙΣ (Gugamis) on an Attic vase fragment © Musée du Louvre, Dist. RMN-Grand Palais / Hervé Lewandowski / Art Resource, NY

Another fragment of this vase bears the inscription ΟΙΓΜΕ (Oigme).

Sign Language?

Upon hearing about nonsense inscriptions, the first thought that crossed my mind was that the "nonsense" might represent encrypted signatures. I imagined that the artists who made these vases were not allowed to sign them for some reason, but got around the rule in some cases by signing their names in an encrypted form. Looking into the matter, I quickly learned that signatures were allowed, although they often failed to appear. In some cases, two signatures were present, one for the sculptor/potter and one for the painter.

After the potter's name, the Greek word ἐποίεσεν (*epoiesen*, meaning "made it") followed, although it may have appeared after the name of the owner of the workshop or the designer instead. The name of the painter was followed by ἔγραπσεν (meaning "painted it"). Oddly, on some vases ἐποίεσεν (*epoiesen*) appears but is not paired with a name. When both potter and painter are named, the potter's name appears first.

There were artists who both sculpted and painted vases. In these instances, the signature, if present, was often of the form ἔγραπσεν κἀποίεσεν, which meant "--------- made and painted it."

For artists whose signatures appear, it is done inconsistently, and not necessarily on their best works. Stylistic comparisons provide the identification of about 900 artists, but we know the names of only about 40. Others have been given pseudonyms by researchers, for convenience.

Although we would expect only a fraction of the vases produced during this time period to have survived, we still have a great many to study. For example, we have 250 from an artist named Douris alone.

There are also fake signatures, although the vases are otherwise legitimate. That is, the names were placed on the vases dishonestly by ancient artists. Altogether, the vases with signatures, real or not, represent less than one percent of the total.

In any case, if signatures were on occasion encrypted, it was not out of necessity. Although the initial signature idea fails quickly as a general explanation, it's possible that signatures are for some reason represented in this manner for vases that do not bear them elsewhere. One way to test this would be by looking at the patterns formed by the letters in the nonsense inscriptions. Do any of these match the patterns in the names of known artists, or common Greek names of the time period? Is the artistic style consistent with the artist (if any) thus identified, or with other vases bearing the same nonsense inscription?

One popular theory is that the nonsense inscriptions were added by illiterate artists in order to make the vases more attractive to potential purchasers. But this would limit the market. A customer who was literate would likely turn his nose up at such fakery. And what would an illiterate purchaser think when a literate friend told him the words on his vase were gibberish? He certainly wouldn't buy another from the same seller. Therefore, it doesn't seem like a wise marketing ploy to allow nonsense to appear on vases.

A New Approach

Not only do we have the same artists producing meaningful and nonsensical inscriptions, we also have this happening on the very same vases in some instances. And it's not just vases. The Memnon pieta (ca. 490–480 BCE), which resides at the Louvre in France, also features this odd combination. An image of it is reproduced in fig. 2.4.

FIG. 2.4 In the Memnon pieta, Eos, goddess of the dawn, picks up
Memnon, her dead son © Musée du Louvre, Dist. RMN-Grand Palais /
Hervé Lewandowski / Art Resource, NY

Memnon died defending Troy, but the story gets a lot better after the sad
scene depicted above. Eos's tears moved Zeus to make Memnon immortal.
The inscriptions on the right give the name Memnon and also indicate who
made the plate and who painted it. The inscriptions on the left are more interesting. The image below zooms in on a rotated version of the image above to
better show these inscriptions.

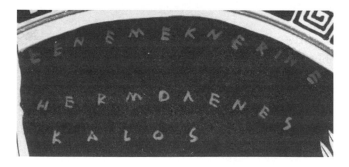

FIG. 2.5 A closer look at some of the Memnon pieta's inscriptions
© Musée du Louvre, Dist. RMN-Grand Palais / Hervé Lewandowski /
Art Resource, NY

The last two lines, HERMOΓENES KALOS can be translated as "Hermogenes is beautiful," but nothing has been made of the top line, EENEMEKNERINE.

The experts offered a variety of unlikely explanations as to why sense and nonsense inscription sometimes appear together. Henry R. Immerwahr, Professor of Greek at the University of North Carolina from 1957 to 1977, wrote,

> It seems clear, from the vases which have both sense and nonsense inscriptions obviously written by the same painter, that in the majority of cases these inscriptions are written by literate painters who felt that the scenes should have inscriptions even where no precise information was to be transmitted.[14]

How in the world does that "seem clear"? It sounds unlikely to me. Artists are creative people. I have trouble believing that they frequently found themselves in the situation where they couldn't think of anything to write and therefore had to fall back on gibberish. I think this paragraph should be longer, but I don't have any precise information to transmit, so ehujf ki fengoku wfscph!

See how silly this is?

Sir John Boardman had this to say on the matter:

> There are examples too of scenes where sense and nonsense stand side by side suggesting a mixture of copying and imagination.[15]

So an illiterate painter carefully copies a meaningful inscription from somewhere and then ruins the possibility of any literate customer buying the product by scrawling nonsense on it in another spot.

For someone who suspects that the "nonsense" might often mean something, the vases with both sense and nonsense provide the best examples for deeper study. This was the path taken by Adrienne Mayor of Stanford University's Department of Classics. She looked at a small set of vases that depicted Amazons and/or Scythians and contained both sense and nonsense inscriptions.

> "It all started from a hunch," she says. "What if these illiterate gibberish scribbles on ancient Greek vases depicting Amazons and Scythians meant something?"[16]

FIG. 2.6 Adrienne Mayor

Mayor thought that some of the inscriptions might represent foreign names spelled out phonetically using Greek letters. The reason they made no sense in Greek was because they were never intended to represent Greek names.

To test her hunch, she and David Saunders, an expert on Greek vase painting and iconography, selected a small group of vases that fit her criteria and sent the nonsense portions, with minimal context, to John Colarusso, a comparative historical linguist. With a specialization in Northwest Caucasian (Circassian, Abkhazian, and Ubykh), Ossetic, Old Georgian, Ancient Greek, Iranian, and other languages, Colarusso stood a good chance of recognizing the languages that would have been used by the foreigners on the Greek vases, if they were real.

Making Fun of Foreigners?

But then, maybe the inscriptions on the vases were just making fun of the foreigners. Greeks were known to do this. To Greeks, some foreign languages sounded like bar-bar-bar-bar .... They called the people who talked this way βάρβαρος (barbaros). This is where our English word barbarian originated.

A modern version of this sort of humor appears in Umberto Eco's novel *Foucault's Pendulum*, where a computer program, Abu, creates a fake foreign paragraph.

Abu, do another thing now: Belbo orders Abu to change all words, make each "a" become "akka" and each "o" become "ulla," for a paragraph to look almost Finnish.

Akkabu, dulla akkanullather thing nullaw: Belbulla ullarders Akkabu tulla chakkange akkall wullards, makkake eakkach "akka" becullame "akkakk-akka" akkand eakkach "ulla" becullame "ullakka," fullar akka pakkarakka-grakkaph tulla lullaullak akkalmullast Finnish.[17]

Eco was a professor of semiotics (symbology), which makes him the closest thing to a real-life version of novelist Dan Brown's hero Robert Langdon.

Some examples of nonsense inscriptions on the Greek vases contain four *kh* sounds in a single word. This certainly sounds like exaggerated imitation of someone's language or accent made with comical intent. Or maybe the painter knew someone who stuttered, and had Edward Elgar's sense of humor (this is a forward reference to chapter 3). The Greeks rendered the *kh* sound by using the letter X (chi), and repeated use of it was made in Greek comedies ridiculing Scythians.

Another possibility is that some nonsense inscriptions represent sound effects. Such a representation would be confusing to someone who isn't a native speaker of the language and who would thus be unfamiliar with such conventions. This problem was mentioned in the television sitcom *Big Bang Theory*, Season 4, Episode 9, "The Boyfriend Complexity," when Sheldon asked Raj,

Are you sure you have enough comics? You're going to be monitoring the telescope for 12 hours, and by my estimate, you've only selected seven hours of reading material. That's even factoring in your difficulty in parsing American comic book idioms like Bamf and Snikt.[18]

It could well be that the inscriptions had no deeper meaning than ridicule or sound effects. Among experts, these seemed to be the only acceptable alternatives to the inscriptions being pure gibberish. But Colarusso wasn't bothered by the naysayers, for he had no idea that he was working on something that had already been rejected repeatedly by experts as meaningless.

As Mayor was hoping to get foreign names back as the meanings of the nonsense words, a look at how Scythians were (and still are) named is appropriate here. The Scythians actually received two names, one private and one public. The public name was based on a personality trait or an episode from the person's childhood. Mayor et al. provided examples:

> [O]ne family might have the public name "God-let (you)-live" (= "Thank You") because a founder was known for habitual politeness. Examples of modern Caucasian nicknames for individuals include "Corpse-cause-die," meaning something like the English phrase "beating a dead horse," denoting extreme stubbornness as a character trait; "Dog-shit," commemorating a childhood accident in which one stepped in dog feces; and "Stuck in the Oven," for someone who as a child crawled into a cold oven and was trapped.[19]

So much for high hopes of getting a cool nickname! But seriously, some people did get cool names. And now that you know the range of possibilities that might arise, it's time to look at the results of Colarusso's analysis.

Case 1 (510–500 BCE): The name ΧΥΧΟΣΠΙ (Khukhospi) appeared next to a Scythian archer. Interpreted as an ancient form of Abkhazian, this translated to "Enthusiastic Shouter" or "Battle-Cry." Much cooler than dog-shit!

Case 2 (525–500 BCE): The Amazon label ΠΚΠΥΠΗΣ (Pkpupes) was interpreted as an attempt to approximate ancient Circassian; it yielded "Worthy of Armor."

Case 3 (510–500 BCE): The three Amazons preparing for battle are named Andromache, Hyphopyle, and Antiopea, with the last also paired with ΧΕΥΧΕ. This is one of those nonsense inscriptions with a repeated chi that could be used to make fun of foreigners. But *kheukhe* matched the Circassian for "One of the Heroes/Heroines."

Case 4 (525–500 BCE): An Amazon or a Scythian (the sex is uncertain) is labeled ΣΕΡΑΓΥΕ (Serague). In Circassian, this means "Wearing (Armed with) Dagger or Sword."

Case 5 (510–500 BCE): Herakles is shown with an Amazon named ΒΑΡΚΙΔΑ (Barkida), meaning "Princess/Noble Kinswoman" in Circassian.

Case 6 (475–450 BCE): An Amazon labeled ΓΥΓΑΜΙΣ (Gugamis) suggested "iron" in Circassian and another Amazon labeled ΟΙΓΜΕ (Oigme) resembled Ubykh forms for something like "Don't Fail!"

Case 7 (525–510 BCE): This vase showed two Amazons and a dog. The two inscriptions, OHE(Y)N and KE(Y)N, were interpreted by Colarusso, who didn't know the vase showed a dog, as ancient forms of Abkhazian meaning "They/She were/was over there," or "We are helping each other," for the first, and, *Keun,* "Set the dog loose," for the second.

Case 8 (550–530 BCE): On this vase, Herakles fights Amazons, who are labeled ΓΟΓΟΙΟΙΓΙ and ΓΟΓΙΓΙΚΙ. Colarusso thought *gogoioigi* and *gogiwiki* suggested the Georgian word for "maiden."

Case 9 (ca. 550 BCE): On this vase, Herakles fights the Amazons again, but actually has help from Telamon and other Greek warriors this time. Four of the Amazons have Greek names, but the fifth, labeled ΚΕΠΕΣ (Kepes), sounded like an ancient form of Circassian that translated to "Enthusiastic Sex." I'm starting to understand why other warriors showed up to help Herakles for once!

FIG. 2.7 "Enthusiastic Sex" in action © Musée du Louvre, Dist. RMN-Grand Palais / Hervé Lewandowski / Art Resource, NY

Case 10 (550–500 BCE): Herakles fighting Amazons yet again. One has a Greek name, but the other is labeled ΟΑΣ ΟΑΣ (*oas oas*). This sounded like an Ossetian word for "sacred" or "spirit." So, Herakles fights alone this time.

Case 11 (ca. 510–500 BCE): This vase shows a Greek citizen-soldier with a handful of entrails. It wasn't a scene following a battle, but rather an attempt to read an omen before departure. A boy holds more entrails, while a Scythian archer looks over his shoulder. A woman looks on as well, and a dog is present. The non-Greek inscriptions have meaning in Circassian. They are TΛΕΤΥ (*tletu*) by the boy meaning "Jumper"; ΙΤΕΙΣ (*iteis*) by the citizen-soldier indicating position as in "here stands the warrior"; ΕΙΟΣ (*eios*) over the dog, also indicating position, meaning roughly "the dog is beside him"; and by the woman ΙΣΛΕΙ (*islei*), meaning "I am dressing him (in armor/garments)."

Case 12 (510–500 BCE): Greek names for three gods were on one side. The other side had a Greek warrior, a Scythian archer, a dog, and an old man, plus five nonsense inscriptions. Taken one at a time: ΚΙΣΙ (*kisi*) appears behind the Scythian's head and in Circassian means roughly "here is his friend." To the right of Scythian's head appears ΓΕΧΓΟΓΧ (*gekhgogkh*), which means "Brave Adversary" in Abkhazian. Above the dog, the inscription is missing a letter. It looks like ΧΛΕ[.]ΣΙ (*khle[.]si*), and Colarusso couldn't make anything of it. Along the old man's back, the label ΧΛΕΙΟΠΧΙΟ (*khleiopkhio*) seemed to reveal him as the descendant of "the daughter of a big man" in Circassian.

I saved the weirdest inscription, in Case 12, for last: the nonsense word ΧΕΧΓΙΟΧΕΧΟΓΕ (*khekhgio-khekhoge*). Here we have X (chi) appearing four times in one word. That is, this is one of the nonsense words with four *kh* sounds. Surely, this can't be meaningful! Could it serve to ridicule a foreigner, as happened in Greek plays? Well, it isn't likely, as the word doesn't go with the Scythian, but rather appears along the Greek warrior's back. Also, single words with four *kh* sounds do exist in Circassian, which has seven distinct sounds that the Greeks would all approximate in this way, being unable to distinguish them in their own language. It turns out that *khekhgio-khekhoge* means "One Chosen from among the Brave" in Circassian.

Note: Circassian dialects get by with only two or three vowels but have between forty-eight and eighty-one consonants. So there are, in general, far more sounds than the Greek alphabet can represent. With some Greek letters having to represent multiple distinct Circassian sounds, the Greek versions often end up having a repetition of letters that didn't occur in the original Circassian.

All of this compression raises the question of whether there might be too many choices (i.e., too much flexibility or leeway) in converting back from Greek letters to Circassian words. With multiple choices for some of the Greek letters, maybe it's always possible to find Circassian words. Anticipating this concern (or maybe even *objection*), Mayor ran another test.

A Controlled Experiment

To test whether or not Colarusso was seeing words that were not intended, a controlled experiment was carried out using nonsense inscriptions from vases that did not feature Amazons or Scythians and thus were not expected to have any connections with the language groups within Colarusso's area of expertise.

The results were satisfying. Of the ten examples provided, his responses were characterized by[20]

1. "junk"
2. "*xargekexs* might have potential, but nothing obvious here"
3. "possibly dialect Greek?"
4. "could be Paleo-Balkan words? Illyrian, Moesian, Macedonian, Thracian?"
5. "nothing recognizable"
6. "looks like gibberish"
7. "incomprehensible, nothing familiar, some look vaguely Greek with odd endings"
8. "cannot make any sense of this"
9. "a three-letter sequence might be Abkhazian for 'we, us,' but nothing looks promising"
10. "overall, this looks like glossolalia, unintelligible, unless drinking and dancing are depicted"

So mostly he saw nothing when it was expected that there was nothing represented in the languages he was familiar with. It's intriguing, though, that the tenth inscription was indeed on a vase that depicted a drinking party.

The combination of reasonable results for the inscriptions that seemed likely to yield something and almost complete failure for the control inscriptions makes for a convincing case that the results are valid. Saunders, the vase specialist on Mayor's team, commented, "It certainly has made me a lot more careful about what I call nonsense."[21]

Amazons?

Not long ago, scholars considered Amazons, first mentioned in the *Iliad*, to be entirely mythical. But they also rejected the possibility of meaning in the so-called nonsense inscriptions on vases. Now that inscriptions paired with Amazons and Scythians have been shown to have meaning, a natural question arises. Mayor both raised and answered it in her book *The Amazons: Lives and Legends of Warrior Women across the Ancient World.*

> But were Amazons real? Though they were long believed to be purely imaginary, overwhelming evidence now shows that the Amazon traditions of the Greeks and other ancient societies derived in large part from historical facts. Among the nomad horse-riding peoples of the steppes known to the Greeks as "Scythians," women lived the same rugged outdoor life as the men. ... Archaeology reveals that about one out of three or four nomad women of the steppes was an active warrior buried with her weapons.[22]

Still Left to Solve

Mayor, Colarusso, and Saunders did great work, but other scholars needn't despair that there's nothing more to say on the topic. We still have the Memnon pieta, for example, in need of a meaningful interpretation for its EENEMEKNERINE inscription.

Another example is provided by a cup found near the village of Koropi in the Attic Midlands in the 1930s. A paper on the cup published by Eugene Vanderpool in 1945 said that it was broken into many pieces when found, but since mended, with missing pieces restored with plaster. In a private collection at the time, the cup is pictured in figs. 2.8 and 2.9.

The inscription by the male, ONƎNOIƎN, is in retrograde and was dismissed by Vanderpool as meaningless, as was KNIONI, which is associated with the female. Another inscription near the woman, KAVITINE is interpreted as possibly representing a female name such as Καλλιτίμη. Can anyone do better than this? This cup might have a tiny story to tell, if only someone

FIG. 2.8 An Attic cup with nonsense inscriptions

can figure it out. Also, ONƎNOIƎN, might be identified by its unusual pattern of letters. What Greek words fit this pattern? Is there a foreign language that could reasonably have been represented on this cup, for which the word has meaning? Pattern words and their value in cryptanalysis were discussed in chapter 1. They could also prove valuable here.

Some painters repeated nonsense inscriptions, letter for letter, on many vases. These include the Sappho Painter (*lilislis, loloslos, etotot*), the Guglielmi

FIG. 2.9 A detail of the cup, showing
a pair of nonsense inscriptions

Painter, and the Pointed-Nose Painter. Why bother to make the inscriptions consistent, if they're just nonsense? Why not just make up a random inscription each time?

Reflections and Thoughts on Future Work

It's interesting to reflect on how progress was made by Mayor's team and how they think that further progress can be made. Although the names were not exactly enciphered, but rather written as plainly as it was possible for a Greek to do with his limited alphabet, there was a cryptologic connection in a number of ways. First off, Adrienne Mayor explained,

> Part of my interest in deciphering the apparent "gibberish" on vases stems from my first marriage during the Vietnam War to an ASA [Army Security Agency] cryptanalyst stationed at RAF Menwith Hill, near Harrogate, Yorkshire, NSA's intel gathering site and then NSA Ft Meade near DC.[23]

Secondly, as you may have noticed, the method is similar to Caesar's use of the Greek alphabet to encipher his own Latin. In general, recovering lost writing, even writing that was never meant to be secret, uses many of the same techniques as cryptanalysis.

Yet another connection is provided by a change made in the paper before publication. In Version 2.0 from 2012, we have the following:

> This paper considers only a small fraction of potentially interesting "nonsense" inscriptions. One can envision designing statistical studies, perhaps undertaken by a team of archaeologist/vase experts and mathematicians, to determine whether the distribution of letters in "nonsense" inscriptions are randomly chosen from the Greek alphabet, or whether the selection of letters indicate certain sounds and consonant clusters distinctive of Caucasian or other languages. ... Another statistical project would be to compare the distribution of phonemes in foreign context "nonsense" inscriptions with those in, say, mythological scenes to see if there is a statistically significant variance.[24]

But no mention of mathematicians possibly being able to help is made in the final published version from 2014. Mayor explained why the change was made.

I believe that we revised the paragraph suggesting that archaeologists/vase experts collaborate with mathematicians to study nonsense inscriptions because of the comments of Anonymous Reviewer # 4. I quote the comment in full because it might be of value: "At the end, where the authors propose a possible statistical analysis, I would amend this slightly in a couple of ways. First, I think that computational linguists rather than mathematicians would be the more appropriate collaborators since they are more used to dealing with the statistics of text than are mathematicians. And second, while I realize that this was only intended as a tentative proposal, I would a priori be a little skeptical that such analysis would yield much. The texts are all short, and the whole corpus of such texts is limited. We do, of course, have a large amount of Greek text from the relevant period (though the vast majority of that is in modern redacted form), but we have no text from the other languages of interest from that period, and we have to rely on their modern descendants. I would, therefore, be surprised if one could do very much. However, it might be worth a try, given the appropriate collaborators."[25]

Mayor commented, "I would stand by our original suggestion about mathematicians in Version 2.0, even though we revised the Hesperia version."[26]

The phrase, from Reviewer # 4, "they are more used to dealing with the statistics of text than are mathematicians" is a bit ironic in that cryptanalysts pioneered the statistical handling of text. Also, the National Security Agency, whose mission is cryptologic in nature and deeply involves the statistics of text, is the largest employer of mathematicians in the world.

The final version of the paper did still contain ideas for future work that obviously involve mathematics:

If "meaningless" inscriptions were random gobbledygook, one might have expected the painters to use all the letters available in the Greek alphabet.[27]

Another change that was made between Version 2.0 and the published version of this groundbreaking paper is a decrease in the estimated number of nonsense inscriptions.

Version 2.0 stated, "More than 2,000 'nonsense' inscriptions (meaningless strings of Greek letters) appear on ancient Greek vases," whereas the published account claimed "more than 1,500." This was simply caused by discrepancies in how such inscriptions are to be counted. Because there are borderline cases, a smaller value was used, to be on the safe side. But even the smaller number leaves a tremendous amount of room for future work!

So, far from having had the last word on the topic, it seems that Mayor, Colarusso, and Saunders have rather opened the door to a potential flood of new scholarship, if others can simply open their minds.

And don't fear that Mayor's team already solved all of the "easiest" ones. There are about a hundred forty Attic vases with both meaningful and nonsense inscriptions.[28] Expecting that some readers would want to make their own investigations, I wanted to provide references to Attic vase listings, but I quickly learned that they are numerous and often far from complete. So I asked David Saunders which were best. He replied thus:[29]

Dear Craig,

The best resource, at least for searching, is Rudolf Wachter's Attic Vase Inscriptions (AVI) project: www.avi.unibas.ch.

This builds on Henry Immerwahr's *Corpus of Attic Vase Inscriptions,* which is incorporated into the Beazley Archive Database (http://www .beazley.ox.ac.uk/databases/inscriptions.htm). In terms of publications, those by Immerwahr cited in our article offer a good starting point.

Immerwahr, H. 1990. *Attic Script: A Survey,* Oxford.
___. 2006. "Nonsense Inscriptions and Literacy," *Kadmos* 45, pp. 136–172.
___. 2007. "Aspects of Literacy in the Athenian Ceramicus," *Kadmos* 46, pp. 153–198.

Hope this helps—let me know if there's anything else,
David

The Vikings

Although the popular conception of the Vikings is that of a violent, unintellectual people, their cryptography was actually as advanced as that of the Greeks. To see this, consider the Swedish Rotbrunna stone, from 1000 CE, pictured in fig. 2.10.

FIG. 2.10 Swedish Rotbrunna stone

Notice the long and short lines near the end of the snake's tail. These are re-drawn here, with the lines closest to the end of the tail placed on the right:

‖••••‖••‖‖••••‖••‖‖••••••‖‖••••

Notice that the lines don't seem to be random. They appear in groups of long lines followed by groups of short lines. Counting the number of lines within each group, we have

    2, 4, 2, 3, 3, 5, 2, 3, 3, 6, 3, 5

These numbers can then be paired to give

    24, 23, 35, 23, 36, 35

Does this look a bit like the result of encipherment via the Polybius cipher? It should, for that's exactly what the Vikings had done! Over the centuries, they had at least two different arrangements of their runes in the form of a grid that allowed for easy encipherment.

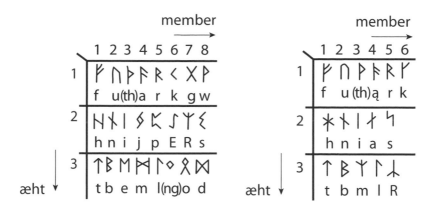

FIG. 2.11 Pre-Viking form with 24 letters and Viking form with 16 letters

The runes are sometimes referred to as an alphabet, but this isn't the best term to use. "Alphabet" derives from alpha and beta, the first two letters in the Greek alphabet. The Latin alphabet may also be referred to as such, for it begins with Λ and B, letters corresponding to alpha and beta. However, runes appear in a completely different order, so it's preferred to refer to it as a rune-row or fuþark (futhark) (of which there are two varieties, "older" and "younger"), after the order in which the initial letters actually appear. There are variants, so researchers also refer to fuþorc (futhorc) (in the "Anglo-Frisian" or "medieval" variety). The single rune þ is transcribed as the two-letter pair *th*, for that is the sound it represents.

To complete the decipherment above, we need to use a modified version of the second grid above. The correct grid labels the *æhts* (the rows) in an order different from how the runes appear in the futhark.

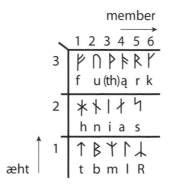

FIG. 2.12 A variant of the Viking form

Using this, we find the decipherment is

```
24,      23,      35,      23,      36,      35
 a        i        r        i        k        r
```
= airikr or Erik, a man's name.

Another example of Viking cryptography is provided by the Rök stone of Sweden (fig. 2.13).

FIG. 2.13 The 13-foot-tall Rök stone of Sweden

This stone makes use of various forms of encryption. For part of it, the key is provided in fig. 2.14.

III	II	I
ᚠᚢᚦᚭᚱᚲ	ᚺᚾᛁᛅᛋ	ᛏᛒᛘᛚᛦ
f uþạr k	hn i a s	t bm l R
1 2 3 4 5 6	1 2 3 4 5	1 2 3 4 5

FIG. 2.14 Key for part of the Rök stone
in Östergötland, Sweden

Once again, the futhark is split into three groups of characters, or three *æhts*. To decipher, we move each rune forward one position within its group. Thus, the runes that correspond to

```
a i r f b f r b n h n   become   s a k u m u k m i n i
```

= sakum ukmini = I say to the youth / I tell the young

or, if interpreted a bit differently (repeating the first m)

= sakum mukmini = I tell the great memory

It might seem that I cheated a bit by inserting an extra m, but this is completely valid. The Vikings wouldn't write the first rune of a new word if it duplicated the last rune of the previous word. The reader had to decide from context whether a rune was omitted.

The Vikings had variants of the cipher described above. For example, they are known to sometimes have shifted back by one or two characters. Julius

FIG. 2.15 More views of the Rök stone

Caesar is famous for enciphering by shifting each letter forward by three. So, although the shift may seem simple to us, a great military figure is not known to have progressed farther than the Vikings. Nowadays, any cipher that shifts letters by some fixed amount is known as a Caesar shift cipher.

To see how the Vikings went beyond what is known to have been done by the older civilizations of Greece and Rome, consider another portion of the Rök stone, for the Vikings didn't just inscribe the fronts and back of the giant stones; they also often made use of the tops and sides (fig. 2.15).

There's been some damage to the runes, but they're complete enough to allow experts to fill in the missing parts. One side is represented in fig. 2.16.

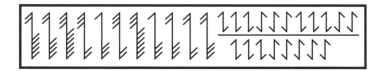

FIG. 2.16 Another Rök stone cipher

Once the viewer is ready to recognize the Polybius cipher in a disguised form, mere "ornamentation" translates to numbers, which can then be converted to meaningful text:

25 24 36 32 13 32 36 13 23 22 23 – 33 32 35 =

  s   a   k   u   m   u   k   m   i   n   i   th   u   r

We end up with a message like the one in the previous example.

The vertical lines, with various numbers of shorter lines protruding at an angle from each side, were sometimes crossed to represent two letters at once. This can be seen on the top of fig. 2.15 on the left, as well as at the top of the first image of the Rök stone (fig. 2.13).

FIG. 2.17 Enciphered letter pairs redrawn for clarity

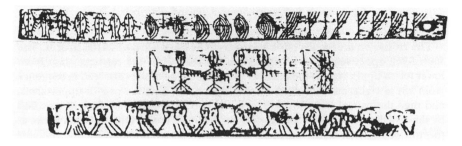

FIG. 2.18 Carvings on wooden sticks made by Vikings

Other variations on this idea were used by the Vikings (fig. 2.18). A viewer unfamiliar with the Polybius cipher would mistake these messages for illustrations of fish, designs, people, and bearded men, having no other meaning. In fact, the illustrations could be masking whatever message the carver desired.

The top wooden stick deciphers to "God give us ..." The short middle stick masks "Klim," which is the beginning of the runic representation of St. Clement. The bottom stick yields "as one's life."[30]

When, as in the case of the ciphers above, the very existence of a message is disguised, we have what is called *steganography*. Over the centuries, it appeared in forms as various as invisible inks, microdots, and messages embedded in digital images (see Cicada 3301 in chapter 9 for this last form).

Now that we have an idea of the range and subtlety of Viking encryption, we turn to stones for which the inscriptions are typically dismissed as nonsense.

Stig Eliasson wrote, "The Scandinavian Viking-Age/early medieval runic tradition includes a small number of inscriptions, seemingly written in a normal natural language, which cannot be interpreted. Modern runologists almost unanimously dismiss these inscriptions—along with other, more deviant ones that may constitute writing exercises, etc.—as being linguistically meaningless."[31]

However, he believes that Denmark's Sørup runestone does indeed have meaning. The runic inscription certainly doesn't look like the work of an illiterate hack. Eliasson described the stone as having a "carefully and elegantly fashioned Danish inscription."[32] He went on to present evidence that it represents Basque. If he's correct, his solution is of the same sort as that of Mayor et al.

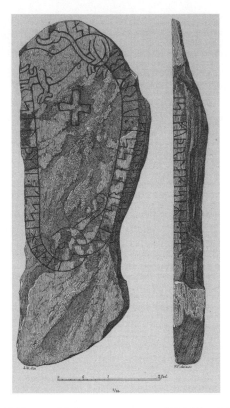

FIG. 2.19 Denmark's Sørup runestone, DR 187, c. 1050–1250

Speaking of the care which the carver obviously took in making his inscription, it should be noted that this was the norm. There are some runic inscriptions in which the carver made a mistake, but these are extremely rare.

So if we have only the extremes of complete nonsense and almost perfect inscriptions, it seems that all carvers were either completely illiterate or fully literate. This is a strange situation. Why were there no semiliterate carvers? In some cases, experts have ways of identifying which stones were made by a particular carver, even if no signature is present. Yet we don't have any examples of a carver who made a nonsense inscription, then later learned to read and write and went on to make literate inscriptions.

Ten more images of stones with "nonsense inscriptions" are reproduced in figs. 2.20 to 2.29.

Michael P. Barnes commented on the inscription on U 811 (seen in fig. 2.24):

FIG. 2.20 U 370   FIG. 2.21 U 466   FIG. 2.22 U 468

FIG. 2.23 U 483   FIG. 2.24 U 811

Some rune-stones carry texts that do not seem to be cryptic and are yet uninterpretable, or partly so. The eleventh-century Hjälsta inscription, Uppland, Sweden, provides an example. It runs: **fast...R:þuliak:oaRtþiol:ati-urai:fasatiR:þaloi:oaRfsai**, a sequence that hardly contains a recognizable Scandinavian word. Attempts have been made to find a hidden message here, not least because the inscription is otherwise competently executed, but none has proved successful. In this and similar cases we are likely to be dealing with a carver who knew runes but had little or no understanding of how to use them to write language (and was unable to seek help, or disinclined to do so). Such "nonsense" inscriptions also seem to presuppose a commissioner who was illiterate and thus unable to check the wording.[33]

But an illiterate commissioner could easily check the wording by asking someone who is literate.

FIG. 2.25 U 835   FIG. 2.26 U 983   FIG. 2.27 U 1061

FIG. 2.28 U 1125 Side A   FIG. 2.29 U 1125 Side B

To see what a relatively small area the ten stones pictured above were sprinkled over, consider the map of Sweden reproduced below, with Uppland indicated.

I expect that a survey of all such stones would fill a book by itself. None presently exists. In fact, surveys of runic inscriptions are missing for other

FIG. 2.30 Sweden, with Uppland indicated

categories. Michael P. Barnes pointed out, "There currently exist no corpus editions of the Anglo-Saxon, Frisian, Greenlandic or Manx runic inscriptions."[34]

Runes were used to represent a range of languages, mostly Germanic (e.g., Old English and Old Norse), but representations of Latin, Greek, and Hebrew are known as well. Perhaps, whether Eliasson is right about Denmark's Sørup runestone or not, it will be learned that some of the runic inscriptions regarded as nonsense actually represent texts in unexpected languages.

Unintentionally offering possible support for some meaningless inscriptions being ciphers, Michael P. Barnes noted "⌐ rarely occurs in meaningful inscriptions."[35] Why the qualifier "meaningful"? If ⌐ simply rarely occurred, he could have written "⌐ rarely occurs in inscriptions." The way he actually phrased it suggests that it's more common in the allegedly meaningless inscriptions.

All ciphers (other than pure transposition systems) change the frequencies of the letters. An enciphered English text can expect to have E less than 12.7% of the time and Z much more than 0.07% of the time. That is, substitution ciphers result in frequent letters becoming rarer and infrequent letters becoming more common. The latter is suggested by Barnes's comment above.

A statistical study needs to be conducted on groups of related stones, such as those from Uppland, Sweden, with currently unreadable runic inscriptions to see if they might have been enciphered in the same manner. If the frequencies are found to differ from readable runic inscriptions, but are consistent with each other, it would provide strong evidence of some meaning being present after all.

3

Dorabella

As we saw in the last chapter, ancient ciphers, arising from simple systems, may elude decipherment by simply not being recognized as ciphers. Once recognized, the language barrier affords a small margin of safety against some modern-day codebreakers. But could such a simple cipher remain unsolved if it were modern, and in English? A British composer may have given us an example.

FIG. 3.1 Edward Elgar (1857–1934)

Edward Elgar is best known for *Pomp and Circumstance*,[1] which has become an obligatory piece for graduation ceremonies, but he had other hits including *Enigma Variations*. In this composition, fourteen of his friends and family members are represented musically with a "variation" for each. There was also an overarching hidden theme that Elgar never revealed. The fun was for people who knew some of the individuals represented to try to guess for whom each variation stood and what exactly it was meant to signify.

For example, Variation I (*L'istesso tempo*) "C.A.E." stood for Caroline Alice Elgar, who was represented musically by four notes that Elgar would whistle to her when he arrived home.

Other variations are much more loosely connected with Elgar's acquaintances. Elgar explained the significance of Variation XI (*Allegro di molto*) "G.R.S.," for Hereford Cathedral organist George Robertson Sinclair.

> The variation, however, has nothing to do with organs or cathedrals, or, except remotely, with G.R.S. The first few bars were suggested by his great bulldog, Dan (a well-known character) falling down the steep bank into the River Wye (bar 1); his paddling up stream to find a landing place (bars 2 and 3); and his rejoicing bark on landing (second half of bar 5). G.R.S. said, "Set that to music." I did; here it is.[2]

Another variation was for the daughter of a close friend of his wife. At the time, her (maiden) name was Dora Penny, but Elgar called her Dorabella, a name he took from Mozart's opera *Così fan tutte*. Many people who knew her found her variation to be very pretty.

FIG. 3.2 Dora Penny (1874–1964)

It was in a letter to Dora that Elgar penned a cipher that remains unsolved. Not even Dora could uncover its meaning. The letter was dated July 14, 1897, and the cipher portion appears in fig. 3.3.

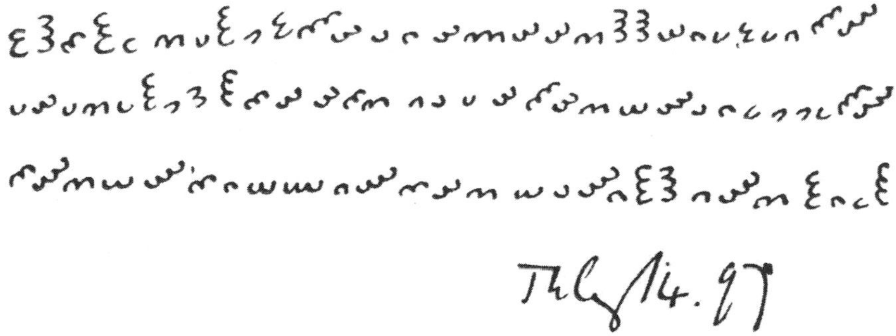

FIG. 3.3 The "Dorabella Cipher," so-called after Elgar's nickname for recipient Dora Penny

Where did Elgar get the idea to use squiggles to represent letters? A look at how Elgar signed his initials suggests an answer.

FIG. 3.4 Elgar's initials

Was it while writing his initials that Elgar got the idea of writing entirely in such squiggles? Do the letters that look like E and W actually stand for E and W? Elgar never provided an explanation as to the origins of his cipher symbols, so all we can do is speculate.

Although squiggles are used instead of letters, the cipher looks pretty simple, as if it might be a monoalphabetic substitution cipher (MASC) with word spacing removed. But none of the techniques used to solve MASCs yield anything reasonable for this puzzle. Some attempts by various individuals are shown below. None of these have met with general acceptance.

Attempted Solutions

In 1970, an attempt was made by Eric Sams, a man who admitted to experience as a wartime codebreaker but didn't often say much more. A few pieces can be put together, though. Sams had won a scholarship to Corpus Christi College, Cambridge, at age sixteen but didn't enroll. Instead, he volunteered for the intelligence corps at age seventeen. In later years, he felt free to talk about the interview process:

> [T]he interviewer had certain questions to ask, and one of them was, "Can you play chess?" as you might imagine and the second was, "Do you like crosswords, and are you good at them?" and so on, but the third was, "Do you like music and can you read a musical score?" and it seemed to me then and it seemed to me ever since that a clear correlation between interest in music and interest in cryptology. One point being, for what it's worth, that generations of composers, whether they're interested in cipher, or not, have thought of music as a kind of communication and, in particular, a secret language which is accessible only to the initiated. It's often thought of like that exactly and that I think clear structural resemblances also exist between music and cipher.[3]

Author Ron Rosenbaum got a few details from Sams about his actual work.

> Sams told me he'd been a World War II cryptologist stationed at the legendary Bletchley Park code-breaking station, home of the Enigma machine, where he said he worked with the team that cracked the Japanese code called MAGIC.[4]

Another account has him decrypting both Japanese and German codes.

Sams was interested in cryptology before becoming a professional, and this interest stayed with him after he left the army in 1947. He described his passion in a later interview.

> I began my working career as a cryptographer in the English Army Intelligence Corps in wartime and that interest has never left me. I mean it was a passionate boyhood interest about codes and ciphers generally and I find that sort of thing in music, and I find it also, actually, in the archives all over

the world if you consider the world as a kind of "collection of archives" a great deal of it is still hidden in darkness, shorthand people can't read, codes and ciphers that people can't read. The English Civil War, the American Civil War, the American War of Independence, and all manner of shorthand diaries like Pepys and Cromwell's secretary in Scotland, William Clark, and so on. A great many of these things are unread and virtually unreadable.[5]

Sams left the Army in 1947 and earned bachelor's degrees in French and German in 1950 from Cambridge University. Like many of the people who attacked the Voynich manuscript, Sams seemed overqualified for the task at hand. Yet his "solution" to the Dorabella cipher was

STARTS: LARKS! IT'S CHAOTIC, BUT A CLOAK OBSCURES MY NEW LETTERS, A, B [alpha, beta, i.e. Greek letters or alphabet] BELOW: I OWN THE DARK MAKES E. E. SIGH WHEN YOU ARE TOO LONG GONE.

While Elgar was famous as a composer of music, not literature, it's hard to imagine he would pen something so disjointed.

Another solution was offered by Jean Palmer author of *The Agony Column Codes & Ciphers*. This book contains more than a thousand ciphers that were published in the personal columns of British newspapers, mostly during the Victorian era, by people who thought that this was a safe way to communicate secretly! Solutions appear in the back of the book, but not for every message. Many remain unsolved. Space limitations prevent me from discussing them here, despite their being very intriguing.

Palmer's solution, obtained in 2007[6] after "warming up" with more than a thousand newspaper ciphers, initially looked like

BLTACEIARWUNISNFNNELLHSYWYDUO
INIEYARQATNNTEDMINUNEHOMSYRRYUO
TOEHO'TSHGDOTNEHMOSALDOEADYA

This is disappointing! It's not possible to turn this into readable English by simply inserting spaces to separate words. Palmer explained the next step as "Reversing the 'backslang.' " Doing this and *then* inserting spaces, Palmer came up with

```
B  LCAT  IE  AWR  USIN  NF  NN  ESHLL  WY  YOUD
INTAQRAYEIN  NET  DMINUNEHO  M  SRRY  YOU
THEO  O  'THS  GOD  THEN  M  SO  LA  OD  E  ADYA
```

Next, came the process of "changing the phoneticised spellings and correcting a few minor errors (the orientation of a few symbols being unclear)," which yielded

B (Bella) hellcat i.e. war using ?? hens shells is why your
antiquarian net diminishes hem sorry you
theo oh 'tis God then me so la do E (Elgar) Adieu

Again, the result hardly seems natural. Another sort of cipher is offered by the person behind this result. I read that "Jean Palmer" was not a real person, but rather a pseudonym used by Tony Gaffney. When I wrote to Gaffney to confirm this fact, he replied,

Yes—it comes from my sense of humour—it is an anagram—had I solved them all [the 1,000+ Agony Columns ciphers] my pseudonym would have been an anagram of Sherlock Holmes.[7]

Recall that anagrams are rearrangements of the original letters in a name, word, or set of words that are also meaningful. Some are comical, like DORMI-TORY = DIRTY ROOM. Here are a few challenges for you:

1. Can you rearrange the letters in ROAST MULES to get a single English word? This anagram was included by Ira Levin in his novel *Son of Rosemary.*
2. Can you turn MR. MOJO RISIN into the name of a famous musician?
3. Can you make three other words out of the letters in THE MORSE CODE, in a way that connects with Morse code?
4. Can you convert PUNCH EVILDOERS to a completely different pair of words?
5. Can you figure out what name JEAN PALMER was an anagram of?
6. Can you rearrange the letters in I CARRY NO LAST END HOUR to get the name of a famous author?[8]

The website http://www.ssynth.co.uk/~gay/anagram.html is useful for making your own anagrams!

In 2007, in celebration of the one hundred fiftieth anniversary of Elgar's birth, the Elgar Society offered a prize of £1,500 to whoever could find the real solution. They received submissions, including Tony Gaffney's, presented above, but none were found acceptable.

Yet another attempt was made in 2009 by Tim S. Roberts, a lecturer at CQUniversity of Australia.[9] His solution was

```
P.S. Now droop beige weeds set in it—pure idiocy—one
entire bed! Luigi Ccibunud luv'ngly tuned liuto studo
two.
```

Rough as this plaintext is, it represents an improvement from the substitutions Roberts used to obtain it. He originally got, for example, "drocp" and presumed that the c was a typo. Thus, "drocp" became "droop," and "pure" was originally "bure."

Roberts offered an explanation of what this corrected decipherment means, but real solutions rarely require explanations. "Layers of deciphering" usually indicate that the solution is incorrect.

If these are the best solutions that can be found by assuming that the cipher is a MASC, then maybe it's something more complicated. To get an idea of what other system Elgar might have used here, we examine all of his known encounters with cryptology.

Elgar's First Known Cipher

Rolling the clock back to about a decade before the Dorabella cipher, we have the first known sample of an Elgar cipher. It's reproduced in fig. 3.5.

As you can see, this cipher was written by Elgar in the margin of a concert program. Because the music was composed by Liszt, this page is typically referred to as "the Liszt fragment." A close-up view of just the cipher appears in fig. 3.6.

A solution to this cipher, found by Anthony Thorley, was presented by Elgar expert Jerrold Northrup Moore.

On 10 April 1886 he went to the Crystal Palace to attend a performance in honour of the seventy-five-year-old Franz Liszt. Liszt was visiting England

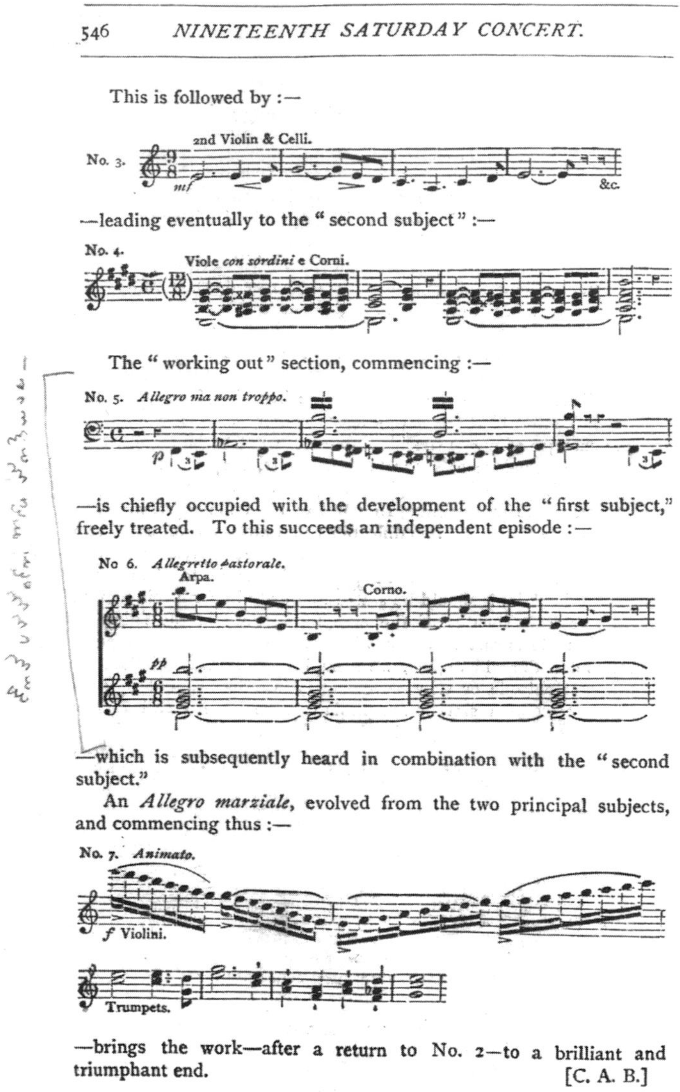

546   *NINETEENTH SATURDAY CONCERT.*

This is followed by :—

No. 3.   2nd Violin & Celli.

—leading eventually to the " second subject " :—

No. 4.   Viole *con sordini* e Corni.

The " working out " section, commencing :—

No. 5.   *Allegro ma non troppo.*

—is chiefly occupied with the development of the " first subject," freely treated. To this succeeds an independent episode :—

No 6.   *Allegretto pastorale.*
Arpa.   Corno.

—which is subsequently heard in combination with the " second subject."

An *Allegro marziale*, evolved from the two principal subjects, and commencing thus :—

No. 7.   *Animato.*
f Violini.   Trumpets.

—brings the work—after a return to No. 2—to a brilliant and triumphant end.   [C. A. B.]

FIG. 3.5 An earlier cipher created by Elgar

FIG. 3.6 A close-up view of the Liszt fragment

for the first time in many years, and Manns's orchestra surpassed itself. The old Abbe shook hands with Manns and bowed many times to a cheering audience. Edward recorded his own Judgement on this music with a cipher pencilled into his programme. It was an early instance of his interest in this arcane art, more secret than a pun yet with its own show of cleverness. Many years later the cipher on Liszt's music was decoded to read: "GETS YOU TO JOY, AND HYSTERIOUS." (footnote: communicated by Anthony Thorley, 1977) "Hysterious" was a portmanteau invention: it married "hysteria" with "mysterious" throu a "tear."[10]

But this doesn't fit! As you can see below, not only do repeated letters fail to line up with the same squiggles each time, but we don't even have the right number of squiggles. The last six letters of the proposed decipherment have nothing to line up with.

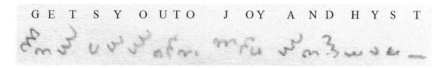

FIG. 3.7 Checking a proposed solution

Other solutions exist that have the right number of letters, but they don't make much sense. So, it seems that this one must also be considered unsolved, and therefore not as helpful as we had hoped! Jumping way ahead, Elgar's last known encounter with cryptology is examined next.

Elgar's Last Known Cipher

Another use of the strange squiggles came much later than the Dorabella cipher, in a pair of pages from a 1920s notebook of Elgar. Because the key is given in the notebook, you can pause and decipher the squiggles yourself. If you keep reading beyond the image, you'll see part of the message revealed.

The substitution alphabet is provided at the top of the left-hand page. Notice that I and J are enciphered using the same squiggle, as are U and V. These substitutions work for the ciphers shown in the notebook, but yield gibberish in the Dorabella cipher and in the Liszt fragment.

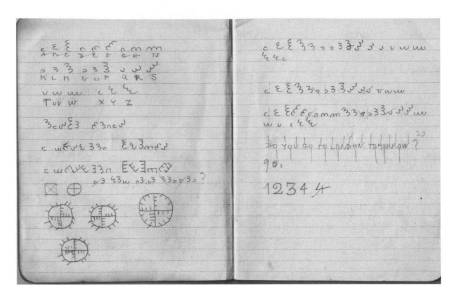

FIG. 3.8 Some more of Elgar's cryptologic writings

If you paused to apply the key to the ciphertext in the notebook, you found the name "Marco Elgar." This was not a relative of Edward's, but rather his pet spaniel. Elgar loved dogs. At least, he loved *his* dogs.

So, the Liszt fragment (unsolved) didn't help, and the notebook ciphers (solved) didn't help either. There's one more document left to consider that bears Elgar's squiggles. It dates back to well before the notebook, and it arose in the context of a cipher challenge issued by someone else.

Elgar Accepts a Challenge

John Holt Schooling wrote a series of four monthly articles on cryptology for *The Pall Mall Magazine*, beginning with the January 1896 issue. These articles surveyed the history of the subject chronologically, and each ended with a cipher to challenge the readers. In the fourth installment, Schooling introduced what is sometimes called the Nihilist cipher, after its use, and possibly invention, by the nineteenth century anarchistic foes of the czars. It's an easy cipher to use, but it requires a few steps. Schooling provided the grid reproduced in fig. 3.9 for converting the original message to numbers, in the manner of the Polybius cipher discussed in chapter 2. This is the first step.

	1	2	3	4	5
1	A	B	C	D	E
2	F	G	H	I	K
3	L	M	N	O	P
4	Q	R	S	T	X
5	V	W	X	Y	Z

FIG. 3.9 Grid for converting letters to numbers

Notice that the letter X appeared twice in Schooling's table. This was a typo—the first X should be a U. The letter J was left out intentionally, as a five by five grid only allows room for twenty-five letters. Something has to give!

Given any letter, you simply find it in the grid and note the horizontal row and vertical column it appears in. For example, M appears in row 3 and column 2. Thus, M is replaced by 32. Schooling took the message "MEET ME IN PARIS ON MONDAY" and noted all of the corresponding row and column numbers. You can see this in lines 1 and 2 in the image below, which is reproduced from Schooling's article. Line 2 is a MASC. But to make the encryption harder to break, we write some keyword under the numbers in line 2. In Schooling's example, the key was TYRANT. Because the key, in this instance, is much shorter than the message, it needs to be written repeatedly.

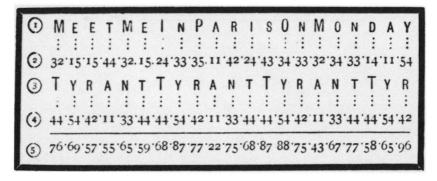

FIG. 3.10 A sample encipherment using the keyword TYRANT

Beneath the key, the numerical values, according to the grid again, for the key letters are written out. This is line 4 in the image. In the last step of the enciphering process, the numbers in lines 2 and 4 are added together to get the final ciphertext, which is shown in line 5.

Notice that although E appears three times in the original message, it is enciphered as three different values, 69, 57, and 59. This is because each E lined up with a different letter of TYRANT (the keyword). Sometimes repeated letters line up with the same letter in the key, as happens here with the letter I, but we have the *possibility* of different encipherments. Such systems are called "polyalphabetic" because they don't always use the same (single) substitution alphabet as a MASC does. People seeing polyalphabetic ciphers for the first time are often impressed with their apparent resistance to attack. None of the attacks described against MASCs work here. Certainly Schooling was impressed with this cipher.

As with the first three installments of his series, Schooling ended this final piece with a challenge for his readers. It's reproduced in fig. 3.11.

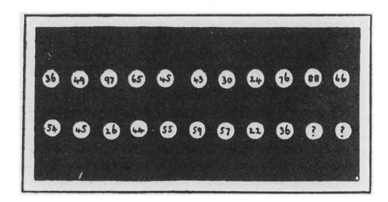

FIG. 3.11 Schooling's final challenge cipher

Schooling warned that it was like the example just described

but the keyword TYRANT there used has not been used for this concluding sentence; therefore the meaning of the cipher which now follows will never be solved by any one. The "box," which for us has been unlocked to let out the cipher secrets of past centuries, has now closed and firmly shut its fastenings for one hundred years of future time.[11]

Despite these strongly worded declarations, Elgar was not intimidated. He went ahead and solved Schooling's challenge cipher. A set of cards in which he showed exactly how he did it are part of the collection at the Elgar Birthplace Museum in Worcester, U.K. We examine them now to understand how Elgar found the solution. Perhaps it will be relevant to his unsolved Dorabella cipher, which he created soon thereafter.

The first card simply presented the cipher.

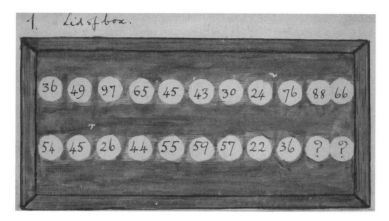

FIG. 3.12 Card 1: Elgar presented the cipher, labeling it "Lid of box"

The second card showed the grid used to convert the message and key letters to pairs of numbers. On it, Elgar corrected Schooling's error of having the letter X appear twice.

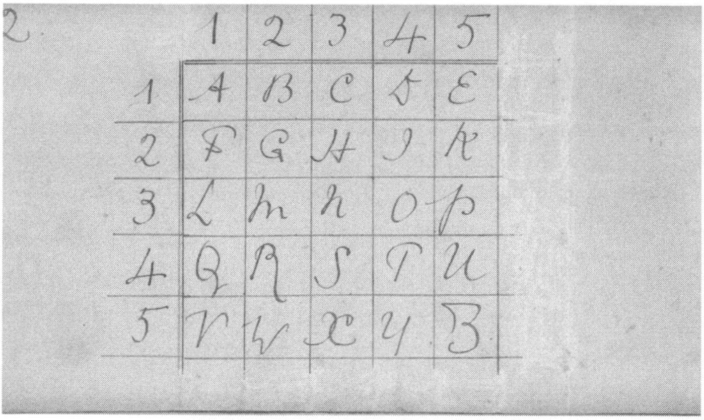

FIG. 3.13 Card 2: Letter to number conversion scheme

The third card provided an example of the enciphering process. It drew special attention to the fact that repeated letters sometimes encipher to different values. Elgar noted, "Any word in any language may be the keyword." Without a computer, Elgar didn't have the option of simply trying all of these possible keywords!

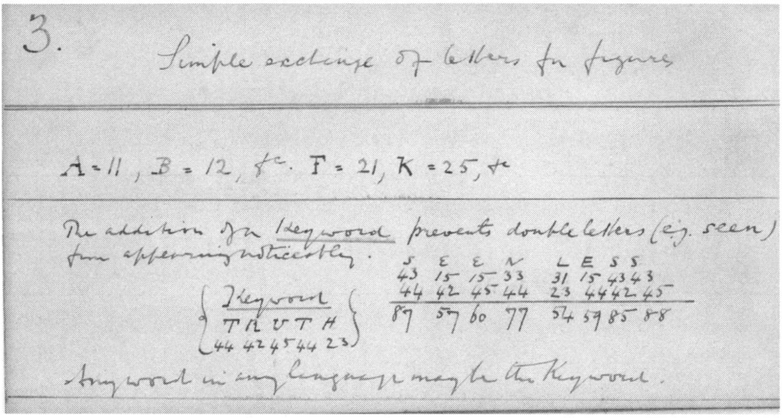

FIG. 3.14 Card 3: An example of how the cipher works

On the fourth card, Elgar provided another example. He pointed out that after the first step (converting the message to numbers), the result is "easily decipherable." He showed some sample frequencies (15 = (10 times) E) to indicate this. It's the step in which the key is added that makes the system stronger than a MASC.

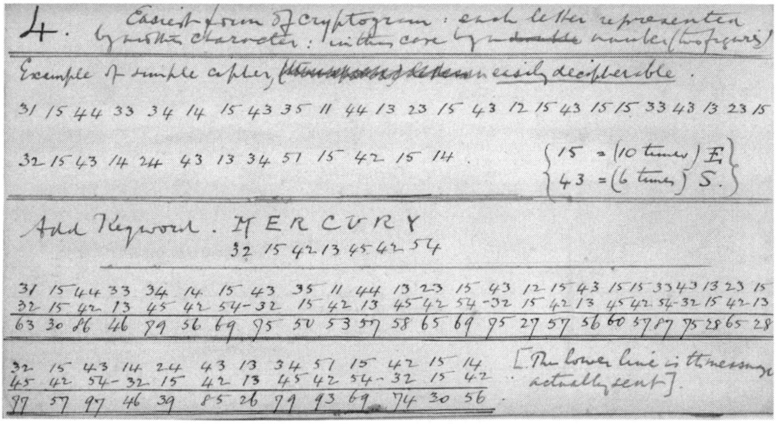

FIG. 3.15 Card 4: The Nihilist cipher is contrasted with an easy-to-solve MASC

The fifth card explained the manner in which the intended recipient, who knows the key, can get the original message back again.

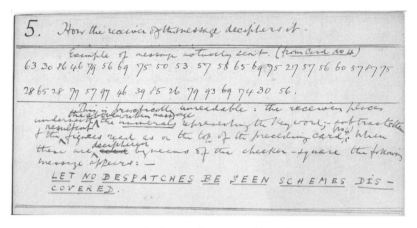

FIG. 3.16 Card 5: Explanation of decipherment,
by intended recipient, using the key

On card six, Elgar finally got down to business and began to crack Schooling's challenge cipher. He noticed that the second to last number in the cipher was 22. The only way this can arise is when the message letter and the key letter (that is added to it) are both 11. That is, both the message and the key letter must be A for their sum to be 22. So Elgar quickly deduced a letter

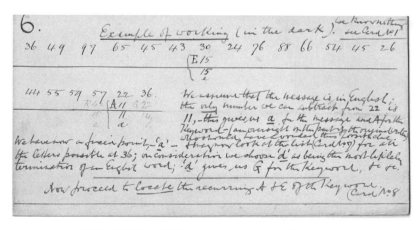

FIG. 3.17 Card 6: Elgar begins to crack the message

7. Table of possible letters. (under each number)

1. 36		2. 49		3. 97		4. 65			5. 45	
11 a		44 d		42 r			41 q		11 a	12 b
12 b		15 e		43 s		11 a	42 r		13 c	14 d
13 c		24 i		44 t		12 b	43 s		21 f	22 g
14 d		25 h		45 u		13 c	44 t		23 h	24 i
15 e		34 o		52 n		14 d	51 v		31 l	32 m
21 f		35 p		53 x		21 f	52 w		33 n	34 o
22 g				54 y		22 g	53 x			
23 h				55 z		23 h	54 y			
24 i						24 i				
25 k						31 l				
						32 m				
						33 n				
						34 o				

6. 43	7. 30	8. 24	9. 96		10. 88	
11 a	15 e	11 a	21 f	41 q	33 n	
12 b		12 b	22 g	42 r	34 o	
21 f	E.	13 c	23 h	43 s	35 p	
22 g			24 i	44 t	43 s	
31 l			25 h	45 u	44 t	
32 m			31 l	51 v	45 u	
			32 m	52 w	53 x	
			33 n	53 x	54 y	
			34 o	54 y	55 z	
			35 p	55 z		

11. 66	12. 54		13. 45		14. 26	15. 44	
all	11 a	31 l	11 a	31 l	11 a	11 a	31 l
letters	12 b	32 m	12 b	32 m	12 b	12 b	32 m
are	13 c	33 n	13 c	33 n	13 c	13 c	33 n
possible	21 f	41 q	14 d	34 o	14 d	21 f	
	22 g	42 r	21 f		15 e	22 g	
	23 h	43 s	22 g			23 h	
			23 h				
			24 i				

16. 55		17. 59	18. 57		19. 22	20. 36	
11 a	31 l	14 d	12 b	32 m	11	11 a	
12 b	32 m	15 e	13 c	33 n		12 b	
13 c	33 n	24 i	14 d	34 o		13 c	
14 d	34 o	25 h	15 e	35 p	A	14 d	
21 f	41 q	34 o	22 g	42 r		15 e	
22 g	42 r	35 p	23 h	43 s		21 f	
23 h	43 s	44 t	24 i	44 t		22 g	
24 i	44 t	45 u	25 h	45 u		23 h	
						24 i	
						25 k	

FIG. 3.18 Card 7: Elgar brute-forces the remaining possibilities

in the message and a letter in the key! He then noticed the value 30 in the cipher. This could also only have arisen in one way: 15 + 15 = 30. That is, there must have been an E in the message and an E in the keyword. Mathematically, there are other possibilities, like 14 + 16 = 30, but 16 is not a value in the grid that converts letters to numbers, so it isn't an option here. Elgar filled in the known values.

The other numbers in the cipher offered more possibilities. Elgar listed all of the possibilities for each number on card seven. For example, the second cipher number, 49, could have arisen as 14 + 35, 15 + 34, or 24 + 25. Of course, any of these pairings could be reversed. The first could be written as 35 + 14, for example. But Elgar wasn't concerned with which letter in each of these pairings was from the message and which was from the key. He was only concerned with which pairs of letters could have been involved in some way. Elgar placed the numbers that could lead to 49 in numerical order and wrote out their alphabetical equivalents, D, E, I, K, O, and P. Some of the positions gave a lot more possibilities, but that was okay; Elgar had another trick up his sleeve.

On card eight, Elgar noted all of the positions in which the key could possibly be A or E. He chose these two letters because he already had a definite position in which each of them must appear (see card six), and he knew from Schooling's examples that the keyword repeats. If he could use all of these data to determine the length of the keyword, he would be that much closer to knowing the keyword itself. Cryptanalysts often have to be content with progress in small increments like this. They are patient, persistent people who keep poking small holes in ciphers until the answers are finally completely revealed.

Elgar didn't squeeze much detail about how he found the keyword length into the space left on card eight, but he might have tried various possibilities until he found one that was consistent with what he already knew. The process could have begun with Elgar trying length 2. If the keyword was of length 2, then having an A as the key letter in position 19 would mean that all odd positions would have to have A as the key letter. But an A is not possible in position 3, so this keyword length must be wrong. Next, length 3 can be investigated. If the keyword was of length 3, then the E in position 7 means that we'd have to have an E in positions 1, 4, 7, 10, etc. Position 4 cannot be an E, so this is also wrong. Okay, time to try keyword length 4. The E in position 7 would then imply we also have E in position 3 (and elsewhere), but position 3 cannot be E, so this length is wrong too. On to keyword length 5! With this in mind, the E in position 7 implies an E in positions 2, 7, 12, and 17. Position 12 cannot be E, so keyword length 5 is ruled out. For keyword length 6, the E in position 7 implies that we also have E in positions 1, 13, and 19. Position 13 cannot be E, and position 19 is already known to be A, so the keyword is not of length 6. Remember what I said about patience and persistence? Finally, keyword length 7 looks good. The E in position 7 implies that we also have an E in position 14, which is possible, and the A in position 19 implies an A in positions 5 and 12, which is also indicated as being possible. Elgar placed the newly determined values in positions 14, 5, and 12, and circled them.

With the key length looking like seven, Elgar wrote out what he knew of the key, using an x to represent each unknown letter. He had xxxxAxE.

Now, looking back at card 7, Elgar had various possibilities for positions 18 and 20. (He knew position 19 was A). Showing the possibilities for each of these positions as columns of letters, we have

Position 18	Position 19	Position 20
B	A	A
C		B
D		C
E		D
G		E
H		F
I		G
K		H
M		I
N		K
O		
P		
R		
S		
T		
U		

With sixteen choices for position 18, one choice for position 19, and ten choices for position 20, there are $16 \times 1 \times 10 = 160$ possibilities altogether. But these three letters represent the end of the message, so many combinations seem unlikely. For example, do you really think the message ends UAI? Elgar wrote on card 8, "assuming the last three letters of the text to be 'E A D', . . ." This was a good guess! But maybe he tried other possibilities first and only wrote down the one that worked out. With just 160 choices, a patient

8. possible positions of A and E

1/36 e a	2/49 e	3/97 X (neither)	4/65 a	5/45 a (A)	6/43 a	7/30 e (E)
8/24 a	9/76 X (neither)	10/88 X (neither)	11/66 a e	12/54 a (A)	13/45 a	14/26 a e (E)
15/41 a	16/55 a	17/59 e	18/57 e	19/22 a (A)	20/36 a e	

Keyword proved to be a word of seven letters by counting (backward) from (22) noting the recurrence, impossible squares, of A and E. Experiment points to the following – x x x x A x E; assuming the last three letters of the text to be 'sad', this gives AGE for the end of the Keyword: further experiment suggests the word 'Courage'. PTO

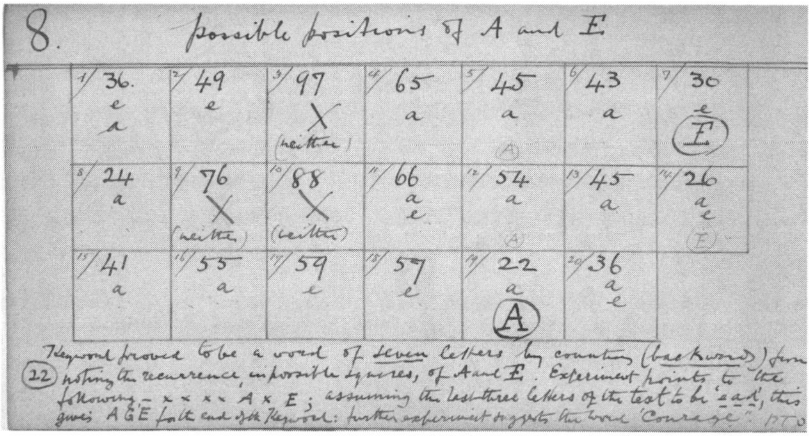

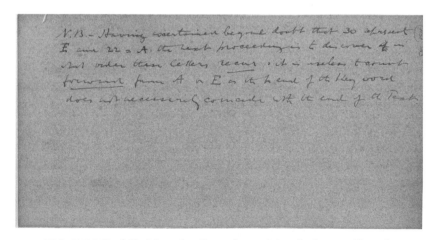

FIG. 3.19 Card 8: Elgar determines the keyword

individual could try all of them. In any case, the D (which equals 14, according to our letter to number grid) in position 20 of the message implies that the key letter in that position must be 22 (= G). Otherwise, we don't get the desired sum of 36. From this, Elgar concluded that the key was xxxxAGE. If he had also looked at the implication of having an E in the message at position 18, he could have expressed the key as xxxRAGE. Elgar greatly enjoyed crossword puzzles, and the skill he developed by pursuing this passion probably helped him recognize the rest of the keyword. With "further experiment" as he put it, he determined that it was "COURAGE."

N.B. – Having ascertained beyond doubt that 30 represents E and 22 = A: the next proceeding is to discover in what order these letters recur: it is useless to count forward from A or E as the end of the Key word does not necessarily coincide with the end of the Text

FIG. 3.20 Card 8b: More details on determining the keyword length

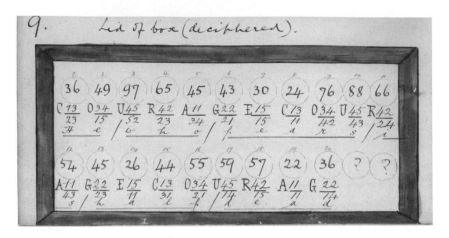

FIG. 3.21 Card 9: Elgar gives the solution

Elgar provided more information on how he determined the keyword length on the back of card eight. However, it's almost easier to figure it out ourselves than to try to decipher his handwriting.

With the keyword established, Elgar went on to subtract the key from the ciphertext, use the grid to convert the numbers back to letters, and place the final decipherment on card 9.

Card nine is a little crowded, so the calculations are shown again (neatly) below.

```
  36 49 97 65 45 43 30 24 76 88 66 54 45 26 44 55 59 57 22 36
-  C  O  U  R  A  G  E  C  O  U  R  A  G  E  C  O  U  R  A  G
  13 34 45 42 11 22 15 13 34 45 42 11 22 15 13 34 45 42 11 22
  _____

= 23 15 52 23 34 21 15 11 42 43 24 43 23 11 31 21 14 15 11 14
   H  e  w  h  o  f  e  a  r  s  i  s  h  a  l  f  d  e  a  d
```

The original message is now recovered as "He who fears is half dead."

According to Nick Pelling, "Elgar took such great delight in cracking it that he later had it painted on the floor."[12] Almost a decade later, the cryptanalytic feat still made Elgar's highlight reel. He recounted it for his first biographer. The book appeared in 1905, with the remark, "During railway journeys amuses himself with cryptograms; solved one by John Holt Schooling who defied the world to unravel his mystery."[13]

While the solution is interesting, and I did promise some *solved* ciphers in this book, there's one more card to present that connects this solution with Elgar's unsolved cipher. It appears in fig. 3.22.

FIG. 3.22 The "Cryptogram Card"

Notice that the main line of squiggle letters is made up of 10 characters, as is the word "Cryptogram," which also appears on the card. At first, I thought that the squiggles represented an encipherment of "`Cryptogram`," but I quickly realized that "Cryptogram" includes the letter R twice, whereas there is no repeated character in the squiggle cipher.

Looking more closely at the squiggles on this card, it looks like the first one starts to lean, then falls over and flips over, as if doing a forward roll. Eventually, when the roll is about to return to the starting position again, the number of squiggles decreases.

This could be interpreted as a doodle, but we know that such squiggles had meaning for Elgar both before and after he solved Schooling's challenge, so it seems reasonable to assume that they had meaning here. With the pattern of the roll, it's doubtful that they represent an enciphered word in this instance. But having ten symbols could be significant in light of the context. Did Elgar have the idea of using these squiggles to represent the ten digits, zero through nine? With these, he could express a Nihilist cipher in terms of squiggles. Christian Schridde has speculated about this, but I doubt that this approach will help us solve the Dorabella cipher, as it has more than these ten squiggles. However, it is possible that some numbers could have been paired with more than one squiggle. That is, when it came time for Elgar, as the encipherer, to jot down a particular number, he would have had a few possibilities to choose from. The card doesn't provide any indication that he had something like this in mind, though. Another argument against a cipher like Schooling's somehow being behind the Dorabella cipher is that Elgar *broke* the Schooling cipher. Why would he base his own system on something he knew wasn't secure?

Yet Another Cipher

Elgar used a different sort of cipher in a letter to his daughter Carice on August 20, 1917. It's given, in context, below.[14]

> I was glad to get your letter because I feared the North End road has para pary parri p (I will) pare paryli parali paralyside pare paralysed your writing thumb—I had not heard for deax—I am reading GK phuriosli —hence

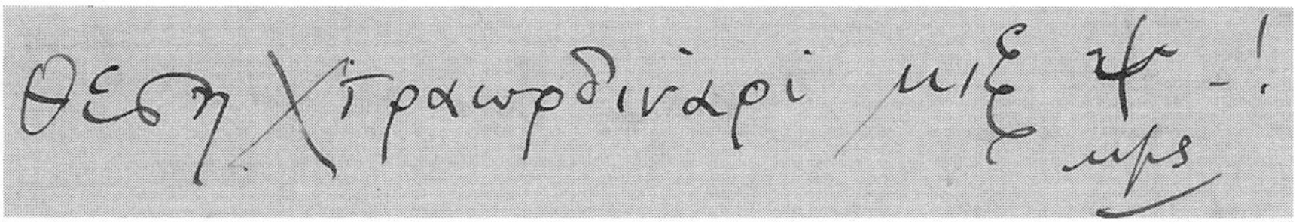

FIG. 3.23 A letter with another Elgar cipher

The sort of cipher shown in fig. 2.23 was used by Julius Caesar, as mentioned in chapter 2. A reader who doesn't know any Greek might try looking up the words in a Greek dictionary, to no avail, but someone who can replace the letters with their equivalents in the Latin alphabet will get the phrase "these extraordinary mix-ups!"

Elgar loved such word games. In fact, his daughter's name represents a sort of wordplay. Before she was born, Elgar compressed the name of his wife, Caroline Alice, to "Carice" for a song's dedication. Wordplay was also to credit for "Craeg Lea," the name Elgar gave to one of his homes. He came up with this by rearranging the initials of his name and that of his wife, along with the rest of his last name, that is, the underlined letters in <u>A</u>lice <u>C</u>arice <u>E</u>dward <u>Elgar</u>.

But such games didn't turn any letters into squiggles.

A Cruel Hoax?

It would be easy to conclude that the Dorabella cipher is a hoax, just a random jumble of symbols with no meaning. But, in this instance, what motive could there be? The letter was not created for resale value (as the Voynich manuscript of chapter 1 ended up having) or to lead to treasure (see the Beale ciphers in chapter 9). The only motive for a hoax in this instance would seem to be cruelty—to tease Dora and waste her time. So the question arises, was Elgar a cruel man? Rather than rely on the testimony of others, let's first take a look at a letter written by Elgar himself.[15]

Battenhall Manor, WORCESTER
27 MAR 1928
Dgck[16]

First of all—we went on Sunday afternoon to S-on-A (all very nice) & on the way back 'killed' a dog: it ran out: I got out & the man was very nice—said it belonged to a <u>crippled</u> son, a boy, & was his pet & slept with him: sd. the boy had been told <u>not</u> to let him out—not our fault at all— they carried the poor beast into their garden—the cripple fled in hopeless tears & it was all too sad—then, oh, joy! This morning (Tuesday) I recd. the enclosed.

You will have recd. the puzzle & you have gone one better than me, e.g. 29→ 29↓ I evolve thus; water-ice=SORBET—extract the ore & you have SBT—it seems feeble though.

~~I return Walker, Barnards letter. If you have not answered it you might write.~~

No, I don't—I am sending them a note to say that you made enquiries for me; see?

All well here & Mary better: she is walking to do the shopping today which may be a good thing.

I called 3 times on Frank & he's hopeless.

Love
EE.

Well, Elgar did love crossword puzzles! To relieve your tension, I'll let it be known that the dog made a full recovery. The odd thing about this anecdote is that, despite Elgar's apparent lack of concern, he was very fond of his own dogs. The next anecdote represents how he sometimes interacted with family.

On Sunday E.E. and Mr. Jaeger shut themselves into the study all morning. When I came back from church, having called for Carice at her school on the way, I found them still busy. The Ninepin came to luncheon and it was all most amusing. When they came into the dining-room, E.E. saw Carice, standing by her chair, waiting.

"Hullo, Fishface! Quite well?"

"Yes thank you, Father"

"Yes thank you, father," imitated E.E. in a high sort of squeak; after which the unfortunate child was expected to say Grace![17]

So, Elgar could be cruel, but was he cruel to Dora in particular? Sadly, yes, both physically and emotionally. An example was provided by Dora in a book she wrote on her relationship with Elgar.

We went in and had our dinner. He never spoke. When he was not look-ing at his plate he looked straight in front of him with a rather tense ex-pression. He was very pale and looked tired and drawn. Half-way through dessert he pushed his chair back, hit my hand, which happened to be on the table, quite sharply, and left the room. He banged the study door and turned the key. For an instant I thought, "That's to keep me out!" I looked at the Lady enquiringly.

"He always locks himself in now that the study is downstairs," she said, "he feels safer!"

As a matter of fact I never, in all the many times I stayed with them, went into the study unless I was asked, or sent, or unless the door was open. "Oh, *dear* Dora, look at your *poor* hand! That *was* naughty of Edward, really!"

"It looks much worse than it is," I said rubbing it, "I expect it will be all right soon."[18]

The way the story is recounted, we cannot be certain if the injury was ac-cidental or not, although Mrs. Elgar clearly believed it was not. However, all doubt was later removed by Elgar himself, when he returned in a more talk-ative mood.

"At first I hoped you wouldn't and then, as dinner went on, I hoped you would. Finally I went away; you'd won, and that was why I hit your hand so hard. Did it hurt? I meant it to!" He picked up my hand and inspected it.[19]

Again, Dora tried to downplay the injury. She took this approach, years later, to a realization concerning her *Enigma* variation, which must have caused some emotional pain.

What I thought of this composition when I first heard it on the piano and also when I first heard it on the orchestra will be found earlier in this book. I had no idea what it really meant. It was not until many years afterwards that it dawned on me that I had been as much the victim of E.E.'s impish humour as had R.B.T.[20] I stammered rather badly at times when I was young and as is the case with so many of those so afflicted it was a terrible trial to me. Now, I am thankful to say that this is to a certain extent a thing of the past. Elgar exploited his humour at my expense with such marvelous delicacy that no one could help laughing with him—if they understood it. Anyhow, I have the satisfaction of knowing that nothing like the "Intermezzo" has ever been written before or since.[21]

Elgar biographer Ian Parrott noted, in regard to Dora, "She was inclined to over-emphasize the happy side of Elgar."[22]

Morse Code?

Dora had an interest in Morse code and even used the Morse alphabet to decorate the flyleaf of her 1899 diary. Elgar toyed with Morse as well. Kevin Jones, writing for *New Scientist* in 2004, noted

Morse code may also crop up in a cryptic letter that Elgar sent to Dora in 1901. Within the message he inserts short, distinctive motifs from his Enigma Variations, in particular a fragment from the Dorabella variation and the opening of the initial theme. The segment reads: "Whether you are as nice as", three short notes, three short notes, "or only as unideal as", two short notes, two long notes. Interpreted as Morse code, these mysterious notes become SS and IM, inviting the interpretation: "Whether you are as nice as sugar and spice or only as unideal as I am".[23]

An image of the motifs is reproduced in fig. 3.24.[24]

Whether you are as nice as

or only as unideal as

Eh? No. Perhaps??

FIG. 3.24 Musical Morse

The squiggles making up the Dorabella cipher don't look like Morse code, but then neither do the motifs! There would be many ways to split the symbols in the Dorabella cipher into two groups, and label one dots and the other dashes. But for each of these, many different interpretations would be possible, for Morse code requires another symbol. Without a pause, we don't know where one letter ends and another begins. Morse operators actually have yet one more symbol they use, a longer pause, used to separate not letters, but words.

In any case, anyone attempting to use Morse code, in some manner, to break the cipher should be aware that there have been three versions, American, Continental, and International.

Elgar never hinted to Dora that the squiggles represented Morse code, but in general, he didn't seem to like to have puzzles or mysteries he created solved. A few pages back Elgar's naming a home "Craeg Lea" was used as an example of his love of wordplay, but an anecdote about it now serves to show that he preferred his mysteries to be impenetrable:

> Edward called the place Craeg Lea and challenged me to guess how he had found the name. But by some stroke of luck, I realized that the key lay in the unusual spelling of "Craeg" and immediately saw that the thing had been built up anagrammatically from (A)lice (C)larice (E)dward ELGAR. I think he was a little annoyed that this mystification had fallen flat.[25]

Dora married Richard Powell, a man who made a convincing case for the hidden theme in Elgar's *Enigma Variations* being *Auld Lang Syne*. The last time Dora spoke with Elgar, he told her that her husband's theory was wrong. However, Dora later revealed that Elgar was a bad liar and that she could tell from his behavior that Richard had guessed correctly. This was an emotional confession for her to make, for she hated the idea that Elgar had lied to her, and she made the man she revealed it to promise to keep it a secret. He did, until years after Dora's death.

Another Theory

If Elgar's cipher were just a MASC, with word spacing removed, automated programs would be able to break it quickly. But they cannot! I think a look at a letter Elgar received from his daughter Carice will show why.[26]

The Mount
Malvern
Jeudi [2 Oct. 1902]

My dearest Mother & Faser
I hope zu are bof enjoying yoursouses vesy much.
    I'se dood & happy & cean & well, & busy & polite.
    Drefful many sanks for the beauful card.
    I had to put Jeudi on my letter, because I have been talking French & cannot get it out of my head.
    I hope "Gerontius" will be beautiful to-day.
    Our class & the other are going to Mr. Montagnon for Grammar this evening. I am looking forward to it.
    I am going to have a lovely music-lesson "KEKKY", & I shall show Miss Reynolds "<u>the</u>" postcard.
    Drefful much love from your loving little ~~fille~~ daughter
Carice

This letter contains not only misspelled words, but also a made up word (KEKKY). Some words that are spelled correctly here are intentionally misspelled in other letters. Such is the strange manner in which Elgar often corresponded with family members!

Elgar also frequently interchanged first letters or syllables in consecutive words, creating phrases such as Histish englory, Uther or Fancle, exwrite the skusing, busted a sculp, Bung yirds, Kig-spin, clise wub, I'm waiting for my dimple sinner & am hungfully paingry, It's raining dots & cags, oracular jerkations, Philharmonic Qreisler in the Kueen's Hall, I smowed & biled, chiserable meque, a box of pam, jickles fished meat & mated pot!, preieping wedderations, hates and dours, dimes and tays, and tace and plimes.[27]

I suspect that the reason the automated programs fail to break the Dorabella cipher is because much of it consists of misspelled words and words invented by Elgar and his family. If I'm right, culling these "words" from his collected letters and adding them to a standard dictionary file would create a database that would allow the automated programs to succeed.

FIG. 3.25 Elgar honored on a British twenty-pound note

The note pictured in fig. 3.25 was introduced in England in June 1999 and remained legal tender through June 30, 2010. Elgar wasn't being honored for his cryptographic skills, but he's not the only man pictured on currency who also appears in the history of our subject. Another will be mentioned briefly in chapter 8. In the meanwhile, can you guess his identity?

4

Zodiac

A First Date

The story of the Zodiac Killer appears to have begun on the night of Friday, December 20, 1968. Betty Lou Jensen, just sixteen years old, headed out for her first date. She was with David Arthur Faraday, a seventeen-year-old, who was a top student and a varsity wrestler. They ended up parked just off Lake Herman Road, a lovers' lane in Vallejo, California.

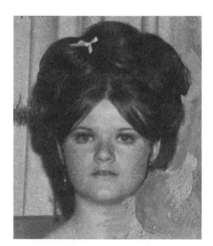

FIG. **4.1** Betty Lou Jensen
© Bettmann / Getty Images

FIG. **4.2** David Arthur Faraday
© Bettmann / Getty Images

Sometime between 11:10 and 11:15 p.m., they were attacked. The killer put his gun to the boy's head and pulled the trigger, killing him while he was still in the car. Betty exited the vehicle, but the killer shot her in the back five times as she tried to run away. Both teens died. Police later found ten expended .22 casings.

Another Attack

Minutes after midnight, as July 4, 1969, turned into July 5, the crimes continued with an attack on a man and woman at another lovers' lane in Vallejo—
Blue Rock Springs Golf Course. This time, the man survived and offered a
description of his attacker. There were also strange details surrounding the
episode that may provide further clues as to the identity of the killer. The victims were Darlene Ferrin, age twenty-two, and Michael Mageau (pronounced
May-hew), age nineteen. Darlene was married to Dean Ferrin (her second
husband) and had a baby girl, so her presence at a lovers' lane with Mageau
requires some explanation.

Investigators learned that shortly before midnight, Darlene told her
babysitters that she needed to buy fireworks for a party she was having that
night. She asked the babysitters if they could stay until 12:15 a.m., so whatever she was up to, she didn't expect it to take long. Indeed, fireworks were
readily available, sold at booths all over town. She picked up Michael Mageau,
who noticed that they were being followed when Darlene pulled away from
his house. Darlene drove faster, and a chase ensued.

FIG. 4.3 Darlene Ferrin   FIG. 4.4 Michael Mageau
© Bettmann / Getty Images

The chase ended when Darlene hit a log at the golf course, causing her
car to stall. Whoever was pursuing them parked behind and to the left of

Darlene's car but then quickly drove off again. Any sense of relief was temporary, as the car returned five minutes later, parked the way it had before, and left the lights on. The driver shined a bright light in the faces of Darlene and Michael, got out of the car, and approached them. The light went out. Michael thought it was a policeman. The two went for their wallets to get their IDs out. The driver came to the passenger side, shined his flashlight in Michael's face again and then began shooting, hitting Michael, but doing far more damage to Darlene, who was hit nine times. When the shooter returned to his car, Michael got a good look at him, but he also saw the man returning. He fired another two shots at each of the victims.

The course caretaker heard the gunshots around midnight, but the police didn't take his report seriously, thinking he heard only Fourth of July firecrackers. Three teenagers found the victims soon thereafter. When one of the teens called the police to report the shooting, the police realized they had made a serious mistake in not responding to the previous call as quickly as possible. They arrived too late to catch the killer, but they were in time to save Michael's life.

Although it was a hot July, Michael was wearing "3 pairs of pants, 3 sweaters, a long-sleeved button shirt, and a T-shirt."[1] Some authors make this into a minimystery, but there's a simple explanation. Michael was incredibly self-conscious about being six feet, two inches tall and very thin. He habitually wore extra layers to make his skinny frame look more filled out.

Michael's survival is the reason that many more details can be related for this second attack than for the first, in which both Betty and David died. Michael's story of the car chase can be questioned, though. It would not be unexpected for someone found on a lovers' lane with a married woman to manufacture some explanation as to how he got there. Indeed, multiple inconsistencies have arisen in his retellings of events that night. However, the speed with which Darlene got from her house to Michael's house, and on to the scene of the attack is consistent with a car chase. Also, when police got to Darlene's vehicle, the ignition was turned on, the car was in low gear, and the hand brake wasn't set, which is consistent with Michael's claim of the car stalling at the end of the chase.

Another contradiction needs to be addressed, though. Darlene told the babysitters that she was leaving to buy fireworks, but her wallet contained only thirteen cents at the time of her death. The simplest explanation is that she picked Michael up, hoping that he would pay for her purchase.

Investigators learned that Darlene went on dates with quite a few men, including policemen from the sheriff's office, despite being married. Thus, jealousy issues may have produced a long list of suspects, but a phone call made things more complex, in a way. The call was received by the Vallejo Police Department switchboard at 12:40 a.m., just two minutes after Darlene was declared dead, and consisted of a man saying this:

> "I want to report a double murder. If you will go one mile east on Columbus Parkway to the public park, you will find kids in a brown car. They were shot with a 9-millimeter Luger. I also killed those kids last year. Good-bye."[2]

Although the call was recorded, it was lost sometime before 1986, so only memories of the voice from those who heard it before then could serve as evidence now.

The call was traced to a pay phone in front of the Vallejo Sheriff's Office. Given that the call was received before anyone other than a few police knew of the murders, and a nine-millimeter gun was used as claimed, it was deemed genuine.

Within an hour of the murder, someone made "heavy breathing" phone calls to three of Darlene's relatives. This suggests that her killer knew her. A friend of Darlene later quoted her as saying that she knew the first two victims. Darlene was also quoted as saying that she saw a guy, who she was obviously afraid of, murder someone! Various stories told by people who knew Darlene seem to indicate that she got in over her head in some sort of criminal matter, but the exact nature of the trouble hasn't been satisfactorily pinned down. After her death, Darlene's husband found a yellow photo envelope with strange writing on it in Darlene's hand. He saw the words "hacked," "stuck," "testified," "seen," and the partial words "acrqu," "acci," "calc," and "icio." The words "on," "by," and "at" appeared in circles. The word "highly" was present, but scratched out. On the back was the phone number to Mr. Ed's Restaurant and Drive-In.

The First Ciphers

The next communication from the killer came in the form of a July 31, 1969, letter sent to three area newspapers, the *San Francisco Chronicle*, the *San Francisco Examiner*, and the *Vallejo Times-Herald*. A transcription of the letter to the *Chronicle* follows. As with the phone call, the killer gave details not publicly known to verify his identity. All spelling errors in the transcription appeared in the original.

Dear Editor
This is the murderer of the 2 teenagers last Christmass at Lake Herman + the girl on the 4th of July near the golf course in Vallejo To prove I killed them I shall state some facts which only I + the police know.
Christmass
1 Brand name of ammo   Super X
2 10 shots were fired
3 the boy was on his back with his feet to the car
4 the girl was on her right side feet to the west.
4th July
1 girl was wearing paterned slacks
2 The boy was also shot in the knee.
3 Brand name of ammo was western
Over

The next page continued

Here is part of a cipher the other 2 parts of this cipher are being mailed to the editors of the Vallejo Times and SF Examiner. I want you to print this cipher on the front page of your paper. In this cipher is my idenity. If you do not print this cipher by the afternoon of Fry.1st of Aug 69, I will go on a kill ram-Page Fry. night. I will cruse around all weekend killing lone people in the night then move on to kill again, until I end up with a dozen people over the weekend.

In lieu of a signature, the killer drew the following symbol.

Investigators, both professional and amateur, have found a great many interpretations of what this symbol represents. What does it make you think of? A few possibilities are discussed later in this chapter.

The other two newspapers received similar letters. Each of the three letters was accompanied by a distinct cipher. They're all shown in figs. 4.5–4.7.[3]

FIG. 4.5 Cipher sent to the *Vallejo Times-Herald*

FIG. 4.6 Cipher sent to the *San Francisco Chronicle*

FIG. 4.7 Cipher sent to the *San Francisco Examiner*

Without a great deal of work, it cannot be determined if the ciphers offer three separate messages or are meant to be combined to form a single message. What is clear is that we're not dealing with a MASC. Simply counting the fifty-four distinct symbols makes that clear. If we had thirty-six symbols, they might represent the twenty-six letters and ten digits (0 through 9), but with fifty-four, we still have an extra eighteen symbols to account for. Punctuation can get a few more, i.e., period, question mark, comma, apostrophe, etc. There are other punctuation symbols, but getting to fifty-four still seems like an unreasonable stretch.

So, we have too many symbols for a MASC, but we have far too few for the symbols to represent pairs of letters, or syllables, or words. It appears that the killer took one of the most basic steps in improving on a MASC, using multiple symbols to represent the most common letters. We could have, for example, R, M, T, and + all representing the letter E. Although this is a simple way to improve on a MASC, it does makes breaking the cipher significantly harder. The multiple symbols for each letter are called *homophones*, and the entire scheme is a homophonic cipher. Such systems go back to at least 1412, so the killer wasn't treading new ground cryptographically.

Frequency counts and pattern words, two of our most useful tools, are worthless against homophonic ciphers. Also, it seems likely that the deciphered message will have misspelled words, just like the "Dear Editor" portion of the letter, providing another complication.

On another note, it's interesting that we have every letter of the alphabet except for C. Several letters appear in more than one form (i.e., backward or upside down). We do have a backward C, but no normal C. Why is this?

With every communication a killer makes, he or she risks unwittingly revealing a clue that could lead to his or her identification and capture. Knowing this, Vallejo Police Chief Jack E. Stiltz made an open request to the writer to send more facts to prove that he was the killer. The killer accepted this invitation.

August 7, 1969, Letter

The killer's next letter was three pages long. It was the first in which he referred to himself as "the Zodiac." It's reproduced below.

Dear Editor
This is the Zodiac speaking.

   In answer to your asking for more details about the good times I have had in Vallejo, I shall be very happy to supply even more material. By the

way, are the police haveing a good time with the code? If not, tell them to
cheer up; when they do crack it they will have me.

On the 4th of July:

I did not open the car door, The window was rolled down all ready. The
boy was origionaly sitting in the front seat when I began fireing. When I fired
the first shot at his head, he leaped backwards at the same time thus spoiling
my aim. He ended up on the back seat then the floor in back thashing out
very violently with his legs; thats how I shot him in the knee. I did not leave
the cene of the killing with squealling tires + raceing engine as described in
the Vallejo paper,. I drove away quite slowly so as not to draw attention to my
car. The man who told the police that my car was brown was a negro about
40-45 rather shabbly dressed. I was at this phone booth haveing some fun
with the Vallejo cops when he was walking by. When I hung the phone up the
dam X@ thing began to ring & that drew his attention to me + my car.

Last Christmass

In that epasode the police were wondering as to how I could shoot + hit my
victoms in the dark. They did not openly state this, but implied this by saying
it was a well lit night + I could see the silowets on the horizon. Bullshit that
area is srounded by high hills + trees. What I did was tape a small pencel flash
light to the barrel of my gun. If you notice, in the center of the beam of light if
you aim it at a wall or celling you will see a black or darck spot in the center of
the circle of light about 3 to 6 inches across. When taped to a gun barrel, the
bullet will strike exactly in the center of the black dot in the light. All I had to
do was spray them as if it was a water hose; there was no need to use the gun
sights. I was not happy to see that I did not get front page coverage.

NO ADDRESS

When the killer wrote, "By the way, are the police haveing a good time with
the code? If not, tell them to cheer up; when they do crack it they will have
me," he had no idea that his cipher had already been broken.

A Solution to the Cipher

The ciphertexts were made available to the NSA, CIA, and FBI, but it was a pair of
amateurs, who saw the ciphers in a newspaper article, who broke them. Donald

Harden, a high school history and economics teacher, began working on the ciphers and was later joined by his wife, Bettye Harden, who had no previous experience with cryptology. Their method of decipherment is detailed below, but you may wish to stop reading at the end of this paragraph and try cracking the cipher yourself. Be warned that it is difficult! However, if you're clever enough, you may be able to beat the husband and wife decipherment time of twenty hours.

FIG. 4.8 Donald Harden
© Bettmann / Getty Images

The Decipherment

Donald had been interested in cryptology as a boy and had books on the subject to refer to, but Bettye had two characteristics that are tremendously valuable in breaking ciphers—persistence and creativity. When Donald was ready to give up, Bettye wasn't. She dragged him back into it. Finally, she had the key insight of applying the psychological method. The killer obviously wanted attention, or he wouldn't have written the letters. Someone so self-centered would likely begin his message with "I." Also, she suspected that the word "KILL" would appear in the message. She even went as far as to suggest that the message might begin with a phrase like "I LIKE KILLING."

Trying to get inside the head of the person who created the cipher and guess words or phrases he or she might use is a technique that has been successfully applied numerous times. It even helped break a great many Nazi Enigma ciphers during World War II, when combined with some more sophisticated mathematical techniques and specially designed machines (the Bombes).

Cryptanalysts refer to "probable words" or sometimes a "probable word search" when they are trying to locate a word such as "KILL" in a particular cipher. The word is also called a "crib" and the attempt to locate it "cribbing." Cribbing paid off for the Hardens.

"KILL" ends with LL, so in a MASC, we would simply look for a doubled ciphertext letter. However, in the cipher in question, L might be represented by more than one symbol, so we have to have some flexibility. By a lucky break, some doubled cipher letters do appear. The pair ◼◼ appears twice (in separate cipher pieces) and ◼◼ appears once. We also have one appearance of **R R**. Some letters, like Q, never appear doubled in normal English. The letter L is, in fact, the one that's doubled the most often. So, it's reasonable to assume that the pair ◼◼ represents LL. But again, it must be recalled that there may well be other symbols in this cipher that represent L. In any case, assuming that ◼◼ is LL, and that this is part of the word KILL, we can make the appropriate substitutions above the ciphertext symbols of one of the appearances of ◼◼ to get $\overset{K}{/}\overset{I}{\Delta}\overset{L}{◼}\overset{L}{◼}$. That is, **/** is tentatively identified as K, and **△** is tentatively identified as I.

For the record, we could do the same for the other appearance of ◼◼, but that will lead us in the wrong direction. When trying to break a cipher, one encounters many false leads. It eventually becomes clear that a mistake was made, and the attacker then backtracks. Describing such false leads and backtracking would drastically increase the length of the discussion, so authors typically only show the approaches that worked.

Now that we believe we know some of the symbols that replaced K, I, and L, these may be identified throughout the cipher. When this is done, the first line of one piece becomes $\overset{I}{\Delta}\overset{L}{◪} P \overset{K}{/} Z \overset{K}{/} U \overset{L}{B} ◪ \lambda O R \kappa 9 X \kappa B$.

Imagine the excitement when the Harden's saw this and realized that the message apparently did begin with "I LIKE KILLING," just as Bettye had guessed! Not only did this reveal which portion of the cipher came first, but it also identified **P, U**, and **λ**, as the letter I, **Z** as the letter E, **B** as the letter L, **O** as N, and **R** as G.

As often happens with ciphers, once a major break is made, the rest unravels much more rapidly. Feel free to stop reading at the end of this paragraph and try to complete the decipherment yourself. But remember, it's likely that there will be some misspelled words. I present the complete solution below.

The decipherment hinged on finding letters that represented LL. Our initial choices were ■ ■, ■ ■, and R R. The first seemed like the best choice, because it appeared twice, while the other pairs occurred just once each. However, as the complete decipherment reveals, both ■ ■ and ■ ■ represent LL. The pair R R ends up deciphering as GG.

The Hardens' work was complicated, as expected, by the presence of several errors. They are all shown here. The message deciphered to

```
I LIKE KILLING PEOPLE BECAUSE IT IS SO MUCH FUN IT
IS MORE FUN THAN KILLING WILD GAME IN THE FORREST
BECAUSE MAN IS THE MOST DANGEROUE ANAMAL OF ALL TO
KILL SOMETHING GIVES ME THE MOST THRILLING EXPERENCE
IT IS EVEN BETTER THAN GETTING YOUR ROCKS OFF WITH A
GIRL THE BEST PART OF IT IS THAE WHEN I DIE I WILL
BE REBORN IN PARADICE AND ALL THE I HAVE KILLED WILL
BECOME MY SLAVES I WILL NOT GIVE YOU MY NAME BECAUSE
YOU WILL TRY TO SLOI DOWN OR ATOP MY COLLECTIOG OF
SLAVES FOR MY AFTERLIFE. EBEORIETEMETHHPITI
```

Newspapers published solutions in various forms—correcting spelling errors, introducing new errors, etc. The deciphered message above doesn't contain the letters J, Q, or Z. This could explain why the symbol C didn't appear in the cipher. Maybe C wasn't needed because it stood for either J, Q, or Z.

Many attempts have been made to make a name out of the letters EBEORIETEMETHHPITI that appear at the end, as if it could reveal the killer's identity.[4]

It seems unlikely that any could be correct, as the portion already deciphered stated "I WILL NOT GIVE YOU MY NAME." Donald Harden himself doubted that the letters served any purpose other than to fill out the last rectangle of ciphertext.

Although the Hardens broke the message on August 4, 1969, they had trouble convincing anyone that they had really done it, and the decipherment wasn't published until August 12, 1969.

It was only after this decipherment that Robert Graysmith, a political cartoonist for the *San Francisco Chronicle* who had taken a strong interest in the case, realized that the envelopes provided a clue as to the order of the three-part cipher message. The envelopes for parts 1, 2, and 3, bore 2, 3, and 4 postage stamps, respectively.

Oddly, before the Harden's solution was published, an anonymous writer mailed the cipher key to Vallejo Detective Sergeant John Lynch. Was this a communication from the killer or a scared "Concerned Citizen," as the letter was signed, who didn't want to attract the killer's attention?

Every confirmed communication from the killer has potential value. As you should suspect, the case would not be included here if there didn't remain an unsolved cipher. Will one of the letters reproduced here give some reader the psychological insight that helps break the unsolved cipher soon to be presented?

Lake Berryessa, September 27, 1969

Another attack came on Saturday, September 27, 1969. This time it occurred in daylight, but still in a secluded area.

Bryan Hartnell, twenty, arrived by car with Cecilia Sheperd, twenty-two, at the man-made Lake Berryessa, thirty-six miles north of Vallejo, at 4:00 p.m. They enjoyed a picnic on a peninsula that jutted into the lake.

FIG. 4.9 Cecilia Sheperd   FIG. 4.10 Bryan Hartnell
© Associated Press   © Associated Press

As the killer approached the couple, he was armed with a gun, as in his past attacks, but this time he also had a large knife in a wooden sheath on his belt, some lengths of rope, and a costume. The costume included a black hood that made him somewhat resemble an executioner. A reconstruction for a magazine cover is reproduced below. As you can see, the hood extended over the chest and

FIG. 4.11 Lake Berryessa © Bettmann / Getty Images

bore the symbol the killer used as a signature in his letters. The reconstruction is not completely accurate, though. The killer's hood bore a slit for his mouth, and he also wore clip-on sunglasses over the hood, so his eyes were not visible.

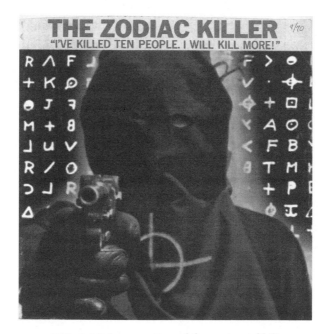

FIG. 4.12 A re-creation of the costumed killer

Holding out a semiautomatic pistol, he said, "I want your money and your car keys. I want your car to go to Mexico."[5] Bryan handed over all the money he had on him, less than a dollar, and his car keys. He was cooperative, believing that he was only being robbed, albeit it by a deranged individual. The killer pocketed the items and holstered his gun. He told the couple, "I'm an escaped convict from Deer Lodge, Montana. I've killed a prison guard there. I have a stolen car and nothing to lose. I'm flat broke." There was more conversation, and then the killer took his sections of rope in hand. It was hollow-core plastic clothesline cut into three-foot lengths. He said, "Lie face down on the ground. I'm going to have to tie you up."[6] Bryan argued a bit, but eventually complied. The killer ordered Cecilia to do the tying, which she did loosely. He then tied her up, and redid the ropes on Bryan very tightly. More calm conversation ensued as the area gradually darkened. Finally, the killer said, "I'm going to have to stab you people."[7]

He began stabbing Bryan first, in the back. He moved on to Cecilia and did the same, but she rolled on to her back in her struggles and was then stabbed repeatedly on her front side as well. Bryan pretended to be dead. The killer left, but not before tossing the change and keys on the blanket, proving that the encounter had nothing to do with money or transportation, as he originally claimed. On his way out, he paused to write a message in black felt-tip pen on Bryan's car door. It began with the symbol that had appeared on his letters, a circle with a pair of perpendicular lines intersecting at the circle's center, and was followed by words and dates. A picture of the door and the message left on it appears below.

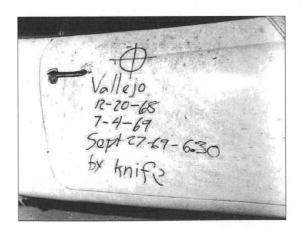

FIG. 4.13 A message left by the killer on a car door

This message provided the first link to previous Zodiac crimes. So many elements of the attack were different (the time of day, the use of a knife instead of just a gun) from previous Zodiac crimes that it's not certain the incidents would have been connected otherwise.

Amazingly, both Bryan and Cecilia were still alive when the killer departed. They even managed to get out of the ropes used to bind them, but they were too weak to get to the car. A pair of fishermen heard their moans and went to get help.

The car door was not the only link provided to previous Zodiac crimes. Following this attack, Zodiac called the Napa Police Department at 7:40 p.m. and said,

"I want to report a murder—no, a double murder. They are two miles north of Park Headquarters. They were in a white Volkswagen Karmann Ghia. I'm the one that did it."[8]

Police traced the call and got a palm print off the public phone the killer used. The attacker didn't know that help had arrived for his victims. Both made it to the hospital, but only Bryan survived. Cecilia had been stabbed twenty-four times and died at 3:45 p.m. the day after the attack.

October 11, 1969

To recap, Zodiac began his string of murders by killing a teenage man and woman, but then in his next two attacks, although he thought he murdered couples, he only actually killed the women. It seems clear that the men survived because they weren't attacked with quite the same intensity as the women. Perhaps because of this, he felt that he had to prove something and therefore targeted a male. In any case, he struck again only two weeks after the Lake Berryessa attack.

On Saturday, October 11, 1969, the killer was in downtown San Francisco, where he hailed a cab and instructed the driver, Paul Stine, to head to Washington Street and Maple, in the residential district of Presidio Heights. Upon arriving at the specified location, the cab's headlights revealed a man walking his dog. The killer, seeing this, instructed the driver to "Go another block."[9] At the corner of Washington Street and Cherry, the killer put a nine-millimeter gun to Paul's head while wrapping his free arm around Paul's neck. He pulled the trigger, exited the rear of the cab, and then opened the front passenger door and got back in the cab. While holding Paul's destroyed head in his lap, he took his wallet.

FIG. 4.14 Paul Stine
© Associated Press

Paul was a twenty-nine-year-old San Francisco State University English Ph.D. student. All previous victims were also students, at one level or another. This may well be a consequence of the killer stalking lovers' lanes, where he would be likely to encounter younger individuals. Like Darlene, Stine was married.

In the minutes after this murder, the killer came close to being caught. Although he had Paul drive an extra block, presumably to avoid a potential witness, he was still seen. A fourteen-year-old girl saw him in the front of the cab with Paul's head in his lap as he robbed him and proceeded to wipe down the cab's interior. She called to two brothers, who also became witnesses. After the killer exited the cab, the teens saw him wipe down parts of the cab's exterior. He then closed the door and walked away.

The witnesses phoned the police at 9:58 p.m. and gave a description of a white man wearing a dark jacket, but this was mangled on the receiving end and turned into "NMA—Negro Male Adult."[10] Within minutes, patrolmen Donald Fouke and Eric Zelms arrived and spotted a white man walking in the direction the killer was seen heading. They shouted to him, asking him if he'd seen anything, and he responded, claiming that he saw a man waving a gun running east on Washington. The police then followed the false lead, and the white man was not seen by the police again. Given the combination of their distance from the man, his dark clothes, the night, and the shadows, they did not see what must have been substantial blood stains. Later that night, a corrected description of the killer was put out. Also, a police sketch was made based on the teenage witnesses' descriptions.

Fouke and Zelms became aware of the error and eventually revealed to their superiors that they believed the man they talked to was likely the killer. It took courage for these men to do the right thing and speak up, adding what they could recall to his description. Just a few months later, on New Year's Day 1970, Zelms was killed in the line of duty, at the age of twenty-two.

Less courageous was the manner in which the new information brought forth by Fouke and Zelms was handled by their superiors. A new composite sketch of the killer was made, but the police never publicly explained why, and they officially denied that any police officers had seen the killer.

Looking back, we see that the killer lied when he claimed that his three-part cipher would reveal his identity; police lied when they said that they hadn't stopped the killer following the Stine attack; and witness Michael Mageau's inconsistent accounts may have included some lies. Are there other lies out there relating to the case that have passed undiscovered? When everything can be brought into question, it makes getting at the truth all the more difficult.

The killer was spotted again, after his encounter with the police, by other witnesses, who saw him go into "the dense undergrowth of the Presidio."[11] This park was searched, with dogs, but perhaps too late. Nobody was found.

But the police had another clue from this murder—bloody prints of someone's right hand left on the cab. They decided to keep this lead confidential. A check showed that the prints did not belong to Stine.

October 13, 1969, Letter

As with the other murders, the killer followed up with a letter to the *San Francisco Chronicle* for which the authenticity cannot be disputed. It was accompanied by "a three-by-five-inch piece of gray and white cloth, neatly torn, not cut, and spattered with blood."[12] It was a fit for a piece missing from Paul Stine's shirt. The letter read

This is the Zodiac speaking.

I am the murderer of the taxi driver over by Washington St + Maple St last night, to prove this here is a blood stained piece of his shirt. I am the same man who did in the people in the north bay area.

The S.F. Police could have caught me last night if they had searched the park properly instead of holding road races with their motorcicles seeing who could make the most noise. The car drivers should have just parked their cars and sat there quietly waiting for me to come out of cover.

School children make nice targets, I think I shall wipe out a school bus some morning. Just shoot out the front tire + then pick off the kiddies as they come bouncing out.

As was noted, the killer didn't balk at lying in his letters, but the risk of dismissing his threat on the school bus as a bluff, if it wasn't, was a risk too great to take. The response from area school officials was, "How can you overreact to a threat like that?"[13]

Napa Valley Unified School District placed an extra man on every bus as a lookout for a sniper attack. Seventy police units also rode guard on the buses, heavily armed. Others followed the buses in forestry department and ranger station pickup trucks. Plainclothes cops watched the buses in San Francisco from unmarked cars, and more than a hundred police vehicles were alerted. There were even air patrols.

Whether one or more of these measures deterred the killer from acting on his threat or if he never intended to carry it out in the first place may never be known. What we do know is that the children were kept safe. No attack on them was attempted, but letters from the killer continued to arrive.

November 8, 1969, Letter

The killer's first letter in November was written in a greeting card. His contribution was

This is the Zodiac speaking I though you would need a good laugh before you get the bad news you won't get the news for a while yet

PS could you print this new cipher on your frunt page? I get awfully lonely
when I am ignored, so lonely I could do my Thing!!!!!!   ⊕

Des July Aug Sept Oct = 7

By this time, the killer had attacked seven people, two of whom survived. Was
the seven meant to indicate seven attacked or seven killed? If the latter, we
don't know who the next two victims were. The only unsolved murders from
August were a pair of teenage girls, but someone unconnected with established Zodiac murders was tied to these murders almost two years later.

If he wasn't claiming some previously unconnected attack, why include
August? Maybe he read about the murder of the teenage girls and thought
he could "get credit" in this way. But, if it was a months-old murder, why
would he write "you won't get the news for a while yet"? Absolutely no evidence or inside information was provided this time, unlike in previous communications. A convincing case was later made by Michael D. Kelleher and
David van Nuys[14] in their book *"This Is the Zodiac Speaking": Into the Mind of
a Serial Killer* that the killer was severely rattled by his close call with police
following the Stine murder and that he, in fact, committed no more attacks
after this. He did continue to write and claim victims, though.

His card was accompanied by a new cipher. It's often called the 340 cipher,
because it contains three hundred forty characters. It's reproduced in fig. 4.15.

This appears to be another homophonic cipher, like the one that came
in three parts and was solved by the Hardens. However, they didn't turn in
a solution for this one. Nor did anyone else. The substitutions that worked
before give gibberish this time. Could this be what the killer wrote? Was
he playing a joke on the police by sending them a cipher that didn't really contain a message? According to Graysmith, "The NSA said that there
was definitely a message in the cipher."[15] Besides this argument from authority, Graysmith provided another convincing reason why there should
be some meaning to the cipher, namely, the crossed out and corrected
character in the sixth row, twelfth column. If the killer was simply writing
random characters, why change one of them? What possible difference
could it make?

Graysmith claimed a decipherment in his book *Zodiac*, but it has not met
with wide acceptance. It's reproduced below, so you can judge its merits for
yourself.[16]

FIG. 4.15 The 340 Cipher

HERB CAEN:
I GIVE THEM HELL TOO.
BLAST THESE LIES. SLUETH
SHOELD [SHOULD] SEE A NAME
BELOW KILLEERS FILM. A PILLS
GAME. PARDON ME AGCEPT TO
BLAST NE [ME]. BULLSHIT.
THESE FOOLS SHALL MEET
KILLER. PLEAS ASK LUNBLAD.

SOEL [SOUL] AT H LSD UL
CLEAR LAKE. SO STARE I
EAT A PILL. ASSHOLE. I
PLANT MR. A. H. PHONE LAKE B.
ALL SSLAVES BECAUSE LSD
WILL STOLEN EITHER SLAVE
SHALL I HELL SLASH TOSCHI?
THE PIG STALLS DEALS OC [OF]
EIGHTH SOEL [SOUL] SLAIN.

In my opinion, the phrasing is too awkward for this to be the correct solution. And Graysmith had to use some flexibility to get it even this readable. He not only made substitutions, but he also reordered some of the letters. That is, he performed some transpositions, at the level of words (letters could be exchanged within what he perceived as a word, but not swapped for letters farther away than that). In his solution, ⊃ doesn't stand for anything. That is, it's a null, and ⋉ stood for both K and S.

The police sought help with this cipher from the American Cryptogram Association and Mensa. One Mensa member, Gareth Penn, was the first to notice that the cipher ends with a distortion of "Zodiac," followed by two other symbols. Is this a coincidence or does it indicate that some letters are enciphered as themselves and that transposition does indeed play some role in the cipher?

The killer followed up on this still unsolved cipher (as I see it), with a long letter.

November 9, 1969, Letter

This letter was sent to the *San Francisco Chronicle*. In it, he makes it clear that he's claiming seven murders, not just *attacks* on seven people.

This is the Zodiac speaking up to the end of Oct I have killed 7 people. I have grown rather angry with the police for their telling lies about me. So I shall change the way the collecting of slaves. I shall no longer announce to anyone. When I committ my murders, they shall look like routine robberies, killings of anger, + a few fake accidents, etc.

The police shall never catch me, because I have been too clever for them.

1 I look like the description passed out only when I do my thing, the rest of the time I look entirle different. I shall not tell you what my descise consists of when I kill

2 As of yet I have left no fingerprints behind me contrary to what the police sayin
   my killings I wear transparent finger tip guards. All it is is 2 coats of airplane
   cement coated on my fingertips -- quite unnoticible + very effective

3 my killing tools have been boughten through the mail order outfits before
   the ban went into efect. Except one & it was bought out of the state.
   So as you can see the police don't have much to work on. If you wonder
   why I was wipeing the cab down I was leaving fake clews for the police to
   run all over town with, as one might say, I gave the cops som bussy work to
   do to keep them happy. I enjoy needling the blue pigs. Hey blue pig I was
   in the park -- you were useing fire trucks to mask the sound of your cruze-
   ing prowl cars. The dogs never came with in 2 blocks of me + they were to
   the west + there was only 2 groups of parking about 10 min apart then the
   motor cicles went by about 150 ft away going from south to north west.

p.s. 2 cops pulled a goof abot 3 min after I left the cab. I was walking down
the hill to the park when this cop car pulled up + one of them called me
over + asked if I saw anyone acting suspicious or strange in the last 5 to 10
min + I said yes there was this man who was runnig by waveing a gun & the
cops peeled rubber + went around the corner as I directed them + I disappeared into the park a block + a half away never to be seen again.

Hey pig doesnt it rile you up to have your noze rubed in your booboos?

If you cops think I'm going to take on a bus the way I stated I was, you de-
serve to have holes in your heads.

Take one bag of ammonium nitrate fertilizer + 1 gal of stove oil & dump a
few bags of gravel on top + then set the shit off + will positivily ventalate
any thing that should be in the way of the blast.

The death machine is all ready made. I would have sent you pictures but
you would be nasty enough to trace them back to developer + then to me,
so I shall describe my masterpiece to you. The nice part of it is all the parts
can be bought on the open market with no questions asked.

1 bat. Pow clock -- will run for aprox 1 year

1 photoelectric switch

2 copper leaf springs

2 6V car bat

1 flash light bulb + reflector

1 mirror

2 18" cardboard tubes black with shoe polish inside + oute

[A full page diagram of the "death machine" appears at this point.]

the system checks out from one end to the other in my tests. What you do not know is whether the death machine is at the sight or whether it is being stored in my basement for future use. I think you do not have the manpower to stop this one by continually searching the road sides looking for this thing. + it wont do to re roat + re schedule the busses because the bomb can be adapted to new conditions.

Have fun!! By the way it could be rather messy if you try to bluff me.

The letter then presented a diagram. It's reproduced in fig. 4.16, along with the accompanying text. What do you make of it?

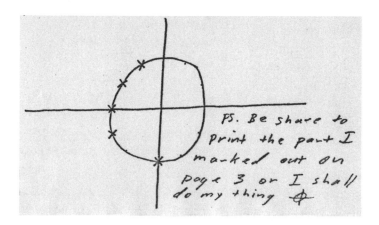

FIG. 4.16 An odd diagram from the killer

Police believed the diagram was a sort of calendar with the months of the killings marked on it. The letter finally ended with a postscript.

PS. Be shure to print the part I marked out on page 3 or I shall do my thing. To prove that I am the Zodiac, Ask the Vallejo cop about my electric gun sight which I used to start my collecting of slaves.

Although he gave details of the Stine murder, and his getaway, the killer provided no evidence of a connection to any unsuspected crimes. The claim "I have killed 7 people" is doubtful.

December 20, 1969, Letter

Melvin Belli, a high-profile lawyer, received the next Zodiac letter. It's reproduced below.

Dear Melvin
This is the Zodiac speaking I
wish you a happy Christmass.
The one thing I ask of you is
this, please help me. I cannot
reach out for help because of
this thing in me won't let me.
I am finding it extreamly dif-
icult to hold it in check I am
afraid I will loose control
again and take my nineth &
posibly tenth victim. Please
help me I am drownding. At
the moment the children are
safe from the bomb because
it is so massive to dig in & the
triger mech requires much work
to get it adjusted right. But
if I hold back too long from
no nine I will loose ~~complet~~ all
controol of my self & set the
bomb up. Please help me I can
not remain in control for much
longer.

Much has been made of the words "happy Christmass." In America, people say "Merry Christmas." Spelling aside, Zodiac's phrasing is a better match for England and Canada. Was the killer British?

Previously, on October 22, 1969, someone claiming to be Zodiac had phoned the Oakland Police Department saying that he would talk to F. Lee

Bailey or Melvin Belli, if he would appear on the Channel 7 talk show. Belli did, and he spent a great deal of time talking to someone claiming to be Zodiac, but the call was eventually traced to a mental institution.

The strange thing is that the letter reproduced above contained another piece of Paul Stine's bloody shirt. Is it a coincidence that an institutionalized man and the real killer both reached out to Melvin, or did the killer hear the call-in show and get the idea to contact the lawyer from someone having no connection to himself?

Graysmith speculated on the identity of the possible eighth murder victim, but one (Leona Larell Roberts) doesn't fit Zodiac's style at all, and the other (Elaine Davis) was never found, so we cannot make a comparison. I side with Kelleher and Van Nuys in that I believe Stine was the last victim. The theory of the killer being too scared to act on his impulses is reinforced by this letter. His fear may have been what motivated him to think of a lawyer in the first place. Someone not worried about being caught wouldn't anticipate needing a lawyer's help. If he really wanted help to stop killing, he might contact someone in the medical, psychiatric, or religious community, but probably not a lawyer.

Was the killer's repeatedly expressed fear of losing control genuine? This a question to ponder when considering an incident that followed almost three months later.

March 17, 1970, 92 Miles East of San Francisco

Kathleen Johns, who was seven months pregnant and also had her ten-month-old daughter with her, was driving close to midnight on 132 west, heading away from Modesto, California, when a man in a car behind her began flashing his lights and honking his horn. Having caught her attention, the man then sped up, pulled alongside her, and shouted that her left rear wheel was wobbling. Kathleen, though scared, eventually pulled over. The man who had yelled to her parked behind her. He had a tire iron and said that he would tighten the lug nuts holding her wheel on. After he indicated that he was done, he went back to his car and drove off.

However, when Kathleen tried to drive off again, her left rear wheel almost immediately fell off. It seems that instead of tightening the nuts, the man had removed them. The man hadn't driven far, and he backed up to where Kathleen's car was now stranded, and said, "Oh, no, the trouble's worse than

I thought!" That is, he was trying to indicate that the problem was not caused by him. He then offered her a ride to a service station.

Kathleen could see the station about a quarter mile down the road, but she accepted the ride, rather than walking there. Once she was in the man's car, he drove right past the station. He got on deserted farm roads and repeatedly began to pull off to the roadside, only to get back on the road again without stopping. He kept telling Kathleen, "You know I'm going to kill you" and "You know you're going to die."[17] The ride along the back roads went on for two or three hours.

When the man made a mistake by driving onto a freeway off ramp, Kathleen realized that she finally had a chance to escape. As he stopped, she fled the vehicle with her baby daughter in her arms. With nobody in the area to appeal to, she hid in an irrigation ditch. The man searched for her with a flashlight, but before he could find her, a trucker trying to exit the freeway saw the car parked there and slammed on his brakes. When he got out of the semi and yelled, "What the hell is going on?" Kathleen's abductor got in his car and sped off. Not such a brave hunter after all!

By this time, Kathleen was too scared to ride with another man, but the trucker stayed at the scene until a woman came down the road. Kathleen rode with the woman, who dropped her off at a police station. As she reported her abduction, she noticed a wanted poster featuring a composite sketch of Zodiac and immediately recognized him as the man who said he was going to kill her. Police later found her car. Someone had put the tire back on, moved it to another location, and set it on fire.

There are some big unanswered questions here. Why did the ride last for hours? Why did the killer repeatedly pull to the side of the road, only to get back on again? Was he enjoying Kathleen's agony as she wondered for hours if the next minute might bring her death? Was he dragging things out for his own perverted pleasure? Or was he fighting to control his evil urges, as he indicated in his letter to Melvin Belli? Perhaps he was just too afraid of getting caught to follow through on his murderous plan.

On many occasions, his behavior was seen to be cowardly. He fled and hid from the police after the Stine murder, and he fled the scene on this night as soon as the trucker got out of his semi. The hunter of the "most dangerous game" preferred unarmed, unsuspecting targets.

A little more than a month later, he got his nerve up enough to return to the safer activity of letter writing.

April 20, 1970, Letter

The letter mailed to the *Chronicle* began

> This is the Zodiac speaking
> By the way have you cracked the last cipher I sent you?
> My name is ——
>
> A E N ⊕ ⊛ K ⊙ M ⦿ ↲ N A M
>
> I am mildly cerous as to how much money you have on my head now. I hope you do not think that I was the one who wiped out that blue meannie with a bomb at the cop station. Even though I talked about killing school children with one. It just wouldn't doo to move in on someone else's teritory. But there is more glory in killing a cop than a cid because a cop can shoot back. I have killed ten people to date. It would have been a lot more except that my bus bomb was a dud. I was swamped out by the rain we had a while back.

The letter continued on another page with "The new bomb is set up like this" and a diagram that fills most of the page. Text on the diagram labels items and explains how the whole thing is supposed to work. The letter ends with a taunt and the score thus far, according to Zodiac.

> PS I hope you have fun trying
> to figgure out who I killed

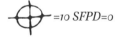 =10 *SFPD*=0

He now claimed ten kills, but again, unlike early letters, there was absolutely no evidence provided that he took lives. Why would he suddenly stop providing inside information and evidence, if he was still at it? In searching for victims nine and ten, Graysmith came up with Kathleen Johns (who, as was mentioned earlier, was taken for a frightening ride, but was not killed) and Marie Antoinette Anstey, who was found dead off a country road in Lake County on March 21, 1970.

The letter writer also paints a false portrait of himself when he states, "But there is more glory in killing a cop than a cid because a cop can shoot back."

He never risked taking on a cop. He fled the scene of every murder to *avoid* the cops. He hid from them following the Stine murder. There was no glory in any of his attacks, only defenseless unsuspecting targets.

Although the "My name is ..." cipher is considered unsolved by some, a claimed solution has been presented, and I believe it's correct.[18] Recall that the killer previously promised to reveal his identity, but the deciphered message didn't deliver (unless the last few letters were not used simply to fill out the rectangle). It seems to me that he was, once again, teasing the reader. The solution I accept is

A L F R E D E N E U M A N
A E N ✦ ☻ K ❺ M ● ↲ N A M

It isn't quite perfect—N deciphers to F in position 3, but deciphers to M in position 11. Still it's damn close, and the killer did make frequent mistakes in his writings.

For those who missed this aspect of pop culture, Alfred E. Neuman has served as a sort of mascot for the humor periodical *Mad Magazine* since the mid-1950s.

FIG. 4.17 Alfred E. Neuman

The killer seemed to still be worried about being caught eight days later, for when he struck again, it was with another greeting card.

April 28, 1970, Letter

The message inside the card received by the *Chronicle* read

I hope you enjoy your selves when I have my Blast.

The postscript on the back was much longer.

> If you don't want me to have this blast you must do two things. 1 Tell everyone about the bus bomb with all the details. 2 I would like to see some nice Zodiac butons wandering about town. Every one else has these buttons like, 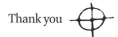, black power, melvin eats bluber, etc. Well it would cheer me up considerably if I saw a lot of people wearing my buton. Please no nasty ones like melvin's

Thank you

Graysmith wrote, "No buttons were ever made."[19] Punk rockers did eventually make some, but they were meant to be worn ironically.

Almost two months elapsed before the letter writer was heard from again.

June 26, 1970, Letter

The letter that arrived at the *Chronicle* near the end of June is of such interest that I'm reproducing the full image instead of simply transcribing the text.

There was a murder like the one described in the letter, but police already had an arrest warrant for a man who was not the Zodiac. This combined with the complete lack of any details not publicly available makes a convincing case that Zodiac was indeed not connected with this crime.

As for the cipher, no one was able to break it. But this letter was different from any that came before, for it was paired with a map, shown in fig. 4.19.

The connection between the cipher and the map was not immediately clear to anyone. The map had a symbol drawn on it similar to the one the killer used to sign his letters. It was centered at Mt. Diablo, in particular, at the location of a Naval Radio Station on the south peak, although this detail isn't shown on the map the letter writer modified.

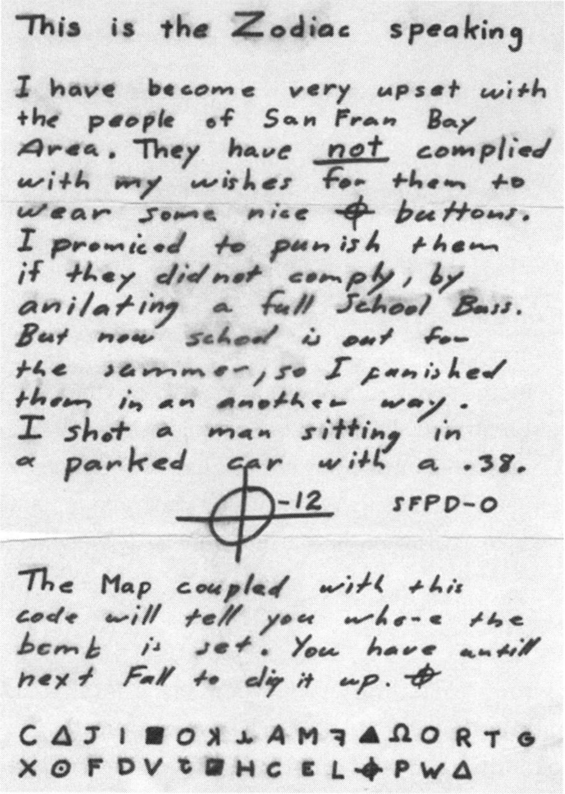

This is the Zodiac speaking

I have become very upset with
the people of San Fran Bay
Area. They have **not** complied
with my wishes for them to
wear some nice ⊕ buttons.
I promised to punish them
if they did not comply, by
anilating a full School Bass.
But now schol is out for
the summer, so I punished
them in an another way.
I shot a man sitting in
a parked car with a .38.

⊕-12   SFPD-0

The Map coupled with this
code will tell you where the
bomb is set. You have antill
next Fall to dig it up. ⊕

C ⊿ J I ■ O λ ⅃ A M ⊐ ⊿ Ω O R T G
X ⊙ F D V ⵀ ⬚ H C ε L ⊕ P W ⊿

FIG. 4.18 June 26, 1970, letter (came with a cipher and a map)

The cipher that ended the letter is unsolved, but I have an idea that's worth considering. Could it be that the Zodiac tried to throw off would-be decipherers by making a big change in how he made his substitutions? In his solved three-part cipher, no letter represents itself. So, in this new cipher, I think he may have made several letters represent themselves, figuring that those attacking the cipher wouldn't consider such a possibility. I believe the message could begin with "CALIFORNIA." This method would pair the letters C, I, O, and A with themselves. One letter of plaintext is missing between the O and the A, but we could attribute this to his lousy spelling. Some speculation on what the rest of the cipher letters might represent is provided after considering a clue from a later letter. But another letter without a cipher or any clues came first.

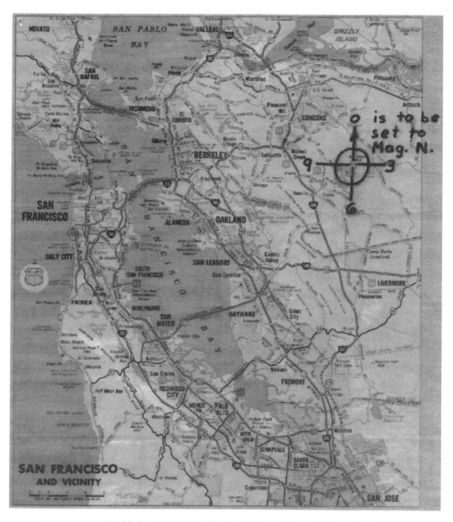

FIG. 4.19 Could this map provide a key to the cipher and the killings?

July 24, 1970, Letter

Near the end of July, the *Chronicle*'s correspondent felt the need to talk about his proposed buttons once again.

This is the Zodiac speaking
I am rather unhappy because you people will not wear some nice ⊕ buttons. So I now have a little list, starting with the woeman + her

baby that I gave a rather intersting ride for a couple howers one evening a
few months back that ended in my burning her car where I found them.

This letter seems to confirm that the man identified by Kathleen Johns was
indeed Zodiac. If so, it raises an interesting psychological question because
the killer failed to act as he did in the past. Again, the question can be posed,
was it an attempt at self-control or did fear of being caught prevent him from
carrying through with the attack? Perhaps he simply couldn't get his nerve up.

The letter that might shed the most light on the crimes arrived at the
Chronicle two days later.

July 26, 1970, Letter

Zodiac's next letter was mostly plagiarized (with many errors, perhaps intentional, perhaps not) from "A More Humane Mikado" and "As Some Day It May
Happen," songs in the Gilbert and Sullivan comic opera *The Mikado*, but it
contained a very interesting postscript about the killer's most recent cipher.
So, be patient—the important material is at the end!

This is the Zodiac speaking
Being that you will not wear some nice ⊕ buttons, how about wearing
some nasty ⊕ buttons. Or any type of ⊕ buttons that you can think
up. If you do not wear any type of ⊕ buttons, I shall (on top of everything else) torture all 13 of my slaves that I have wateing for me in Paradice.
Some I shall tie over ant hills and watch them scream + twich and squirm.
Others shall have pine splinters driven under their nails + then burned.
Others shall be placed in cages + fed salt beef untill they are gorged then
I shall listen to their pleass for water and I shall laugh at them. Others will
hang by their thumbs + burn in the sun then I will rub them down with
deep heat to warm them up. Others I shall skin them alive + let them run
around screaming. And all billiard players I shall have them play in a darkened dungen cell with crooked cues + Twisted Shoes. Yes I shall have great
fun inflicting the most delicious of pain to my slaves

SFPD=0 ⊕ =13

As some day it may hapen that a victom must be found. I've got a little list. I've got a little list, of society offenders who might well be underground who would never be missed who would never be missed. There is the pestulentual nucences who whrite for autographs, all people who have flabby hands and irritating laughs. All children who are up in dates and implore you with im platt. All people who are shakeing hands shake hands like that. And all third persons who with unspoiling take thoes who insist. They'd none of them be missed. They'd none of them be missed. There's a banjo seranader and the others of his race and the piano orginast I got him on the list. All people who eat pepermint and phomphit in your face, they would never be missed They would never be missed And the Idiout who phraises with inthusastic tone of centuries but this and every country but his own. And the lady from the provences who dress like a guy who doesn't cry and the singurly abnormily the girl who never kissed. I don't think she would be missed Im shure she wouldn't be missed. And that nice impriest that is rather rife the judicial hummerest I've got him on the list. All funny fellows, commic men and clowns of private life. They'd none of them be missed. They'd none of them be missed. And uncompromising kind such as wachmacallit, thingmebob, and like wise, well - - nevermind, and tut tut tut tut, and whashisname, and you know who, but the task of filling up the blanks I rather leave up to you. But it really doesn't matter whom you place upon the list, for none of them be missed, none of them be missed.

PS. The Mt. Diablo Code concerns Radians & # inches along the radians

The postscript bears a closer look. The original for this portion is reproduced below.

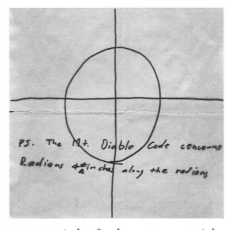

FIG. 4.20 A clue for the most recent cipher

So, in addition to confirming that the killer was not making a living as an author, this letter reveals that he knew what a radian is. For those who are unfamiliar with the term, it's a way of measuring angles. Most people first learn to measure angles in degrees. A full circle is three hundred sixty degrees, but this is artificial. The number three hundred sixty is completely arbitrary. It may have something to do with there being roughly three hundred sixty days in the year, or the fact that the Sumerians used a base sixty number system (which in turn may have been influenced by the number of days in the year). In any case, there is no good reason to divide a circle up into three hundred sixty degrees. Any other number could have been used.

On the other hand, radian measure is natural. Consider the angle drawn below.

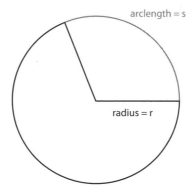

FIG. 4.21 Angle measure in radians

We say that its radian measure is $\frac{s}{r}$ radians, where s is the distance along the curve of the circle (an arclength) and r is the radius. This is lovely because it doesn't force us to make up some number of degrees to go full circle. Rather it *gives* us values. A circle has circumference $2\pi r$, so the radian measure of a full circle is $\frac{2\pi r}{r} = 2\pi \approx 6.28$.

To get the radian measure to be exactly 1, we need to have the arclength match the radius. This happens when the angle is about $57.296°$.

So why is a serial killer talking about radians? Did he have a little deeper background in mathematics than most people?

Gareth Penn found an explanation that connects radians with the map the killer provided along with his fourth cipher. I'll let him explain his theory in his own words.

> I was shocked and horrified to the point of nausea to discover that a radian whose apex rests on the peak of Mount Diablo, and one of whose legs passes through the scene of the murder at Blue Rock Springs, has another leg which passes directly through Presidio Heights in San Francisco, where the Zodiac murdered the cabby. All of a sudden, there *was* an m.o., a mathematical one. The Zodiac had been constructing a gigantic geometric design on the map of the San Francisco Bay Area, using the lives of his victims as markers.[20]

A map illustrates this observation (fig. 4.22).

Penn's discovery led him to reevaluate the intelligence level of both the killer and the police investigating the case. He wrote

> It was shocking to discover that so many lives had been wasted in the furtherance of such a vain objective and that a coldly calculating, highly intelligent mind was hiding behind the guise of an idiot. I was almost as shocked to discover that in the more than 10 years that had elapsed since the receipt of this letter, the police had never tried the radian exercise. In fact, in more than a decade of having been confronted with this problem, they had not even taken the trouble to look up the word "radian" in a dictionary to find out what it meant. Once I pointed out to them why the information was significant, they immediately understood why the Zodiac had murdered the cabby, which had been troubling them for years. What I got out of this was to be promoted, for about two weeks, to the status of chief suspect. When it was discovered that the most I could be charged with was possession of a dictionary and protractor, the police lost interest in me and turned their attention to more pressing matters.[21]

Penn further endeared himself to law enforcement by writing, "My other research has established that the Bay Area police are of the learned opinion that pi is 3.1, and that binary numbers larger than 12 cannot be written."[22]

This raises an interesting point. Very few people find themselves in professions that require knowledge of radian measure for angles. Those who

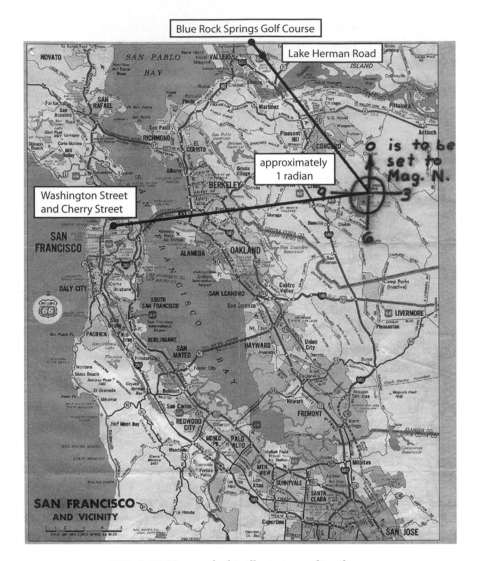

FIG. 4.22 Map marked to illustrate a radian theory

happened to see it in some high school or college math class will certainly not store it in long-term memory if they don't anticipate ever needing it again. So, why did Zodiac not only know the meaning of radian, but also choose to make use of it? Did he use it in his career? This is one of many pieces of evidence that indicates he was smarter than his poorly spelled letters indicate. For that matter, his misspellings were likely intentional.

Though he misspelled simple words, he sometimes wrote more complicated ones correctly.

In light of the radian theory, the Stine murder may have had two motivating factors. One, a cab driver could be led anywhere the killer desired. Thus, Stine's occupation made him an easy victim for someone intent on committing a murder at a specific location to create a pair of deaths separated by a radian subtended from Mt. Diablo. Two, in his previous attacks, the killer had unintentionally allowed the male victim to survive. Psychologically, he may have felt a need to prove that he could kill men as well. Stine, being male, conveniently fit this need.

Some researchers claim that the angle formed by these killings is not 1 radian, but a bit larger, about 60 degrees, or 1.047 radians.[23] But should we really fret about an error of 0.047 radians coming from a guy as deranged as Zodiac? I accept Penn's radian theory as being what the killer intended. On the other hand, I disagree strongly with much of the other content of Penn's writings, including his claimed identification of the killer. I believe that Penn has created another victim by publicly accusing an innocent man of the crimes.

There are a great many names bandied about online, in books, and in Kindle titles, as to the identity of Zodiac. Although it's possible that he spent time in prison for other crimes, there were never any charges brought for the attacks described here. Thus, a great deal of room remains for speculation. Several people have claimed that their fathers were the Zodiac. Others claim that it was an ex-husband. Without a smoking gun, I'm not comfortable throwing names around in print. It's highly unlikely that more than one of the suspects had anything to do with the crimes, so the rest are all victims. You will not have a hard time finding suspects. Indeed, some of the references provided at the end of this chapter discuss them, but you won't find any names in this book.

Now, what more can we do with the radian clue? The killer's hint includes "# inches." So, I would expect the portion of the cipher that follows what I conjectured to be "CALIFORNIA" to contain some numbers. It might also contain the word "INCH" or "INCHES." A decimal point could be indicated by the word "POINT." These are all cribs that could possibly be exploited, but there are still many possibilities. The tentative cipher letter identifications provided by CALIFORNIA don't help tremendously, as we only see three of those letters reappear later in the cipher. Perhaps the best clue is that one of these, the triangle presumed to represent A (which it does actually looks like), ends the message. Not a lot of English words end with A. According to one study, it's only about 24.7 words per thousand, when looking at meaningful text.[24] That is, 2.47%. Such text, of course, contains many appearances of the one-letter word A. Because we don't expect a message to end with this word, the 2.47% is an inflated measure from our perspective. Anyway, maybe someone can make something of this. Or maybe I'm just pointing in the wrong direction! You be the judge.

Not Quite the End

There are more Zodiac communications that are deemed to be genuine, but there are no more ciphers, so the rest of the letters are ignored here. It's worth noting, though, that a long gap between letters suggests that the killer may have been incarcerated on lesser charges. Nothing was received from the killer between his letters of March 22, 1971, and January 29, 1974. The body count, according to the killer's letters grows to thirty-seven, but I believe that this is pure fiction.[25] None of the letters provide any evidence of Zodiac killing anyone after Stine. As was detailed, Zodiac came close to being caught following this murder. So, bravado aside, it seems that he was afraid to commit another murder. Some of the victims he claimed were really killed, but he failed to provide any evidence that he had anything to do with the murders. He had no knowledge of these killings that couldn't be found in newspaper accounts. He was eager to provide insider knowledge, and even physical evidence, for other claimed victims, so one would have expected him to do the same for later crimes, if they were indeed his.

For a contrasting view, the reader may see Robert Graysmith's excellent book *Zodiac*, which provides a list of forty-nine "Zodiac Attack Victims and Possible Victims."[26]

Cheri Jo Bates—The First Victim?

Just as some authors try to link Zodiac to more killings after Stine, there are also attempts to link him to an early murder, before the attack on Betty Lou Jensen and David Arthur Faraday with which this chapter began. The victim was Cheri Jo Bates, a freshman at Riverside City College. She was attacked on October 30, 1966, in the parking lot by the college library and died from knife wounds, which included a severed jugular. Despite some details in common with other Zodiac activities (the killer followed the attack with a letter to the Riverside *Press-Enterprise*, and others), I believe that attributing Bates's death to Zodiac is incorrect.[27] On March 13, 1971, four months after a newspaper article identified Bates as a possible Zodiac victim, the *Los Angeles Times* received a letter from Zodiac in which he took credit for the murder. The letter, however, provided no evidence to support Zodiac's claim.

Graysmith lists Bates as a "Definite Zodiac victim."[28]

Pop Culture

Probably in large part because of his having eluded justice, the Zodiac killer, like Jack the Ripper, has attracted a lot of attention in popular culture. The first movie based on his crimes, titled *The Zodiac Killer*, hit theatres in April 1971. In December of the same year, the Clint Eastwood film *Dirty Harry* came out. It was loosely based on Zodiac as well and set in San Francisco, where Stine was killed. It even has the murderer calling himself Scorpio, in place of Zodiac. Typical of Hollywood exaggeration, though, in the film, the killer actually hijacks a school bus filled with children, instead of just *threatening* to attack one, as the real Zodiac had. Another major difference was that the film had the ending that people in the real world wanted to see in the Zodiac case. Ulli Lommel made an extremely low budget film, *Zodiac Killer*,[29] in 2005, a year that also saw the release of *The Zodiac*.[30] A far superior film, *Zodiac*,[31] based on Robert Graysmith's book of the same title, appeared in 2007.

William Peter Blatty, best known for his 1971 novel *The Exorcist* and the hit film based on it, also wrote a sequel, *Legion*, in which the murderer, based on Zodiac, calls himself the Gemini killer. Some other books inspired by the case can really be labeled only as "fan fiction," although they managed to see print. I'm not going to get into these. Nor am I going to survey the volumes in which an author identifies a relative as the killer. A few of the latter appear in the References and Further Reading section, though.

As with the Voynich manuscript, the Zodiac's unsolved ciphers, especially the 340 cipher, have generated many claimed solutions. Graysmith's was worth mentioning because of his great knowledge of the case, but the others are without merit, in my opinion.

But the art and science that is cryptanalysis is constantly improving, and computers are becoming ever more powerful. Although I don't expect the unsolved Zodiac ciphers to reveal the killer's identity directly, perhaps their eventual solutions will provide that one extra clue that helps unravel the case.

5

More Killer Ciphers

Like Batman's rogues gallery, there's been a real-life series of criminals who purposefully left clues to their crimes in taunting letters. Some of these, like Zodiac, included ciphers that have yet to be solved. This chapter covers several such examples. We begin with a case from 1882–1883.

Henry Debosnys

On May 1, 1882, a man using the name Henry Deletnack Debosnys arrived in Essex, New York. He spoke six languages, was a competent artist, and could be very charming. This may have been part of what attracted Elizabeth "Betsey" Wells to him. He told Betsey that he was from Paris. She too was an immigrant, having been born in Ireland in 1840. She moved to America with her father after her mother's death. Tragedy struck again when her first husband died in a mining accident in 1870. At the time, she was pregnant with the couple's fourth daughter.

Betsey didn't seem to be in a hurry to remarry, as she remained a widow for a dozen years, but she wed Henry on June 8, 1882, only five weeks after they met. How well could she have known him? It seems unlikely that he revealed that he was married twice before, that both of these wives were dead, or that he was a suspect for the murder of the second, Celestine, who died less than two months before he arrived in Essex.

Betsey didn't need a new husband for financial support. She was doing well on her own. In fact, she had received a settlement after her first husband's death and had placed about $300 in a hiding place (a compartment her husband made under the windowsill), for safekeeping and easy access when she needed it. She also owned a property of 15 acres.

From the moment she married Henry, he begged her to turn the deed to the property over to him. She refused, and she also refused to tell him where her settlement money was hidden. The two argued constantly about this. Nobody described them as a loving couple.

On August 1, 1882, after less than two months of being married to Henry, Betsey was found dead. It was definitely not an accidental or natural death. She had been shot twice in the head, and her throat was sliced to the spine. As this chapter covers several cases, it's not possible to go into as much detail as was done with the Zodiac killer in chapter 4, but the interested reader can see Farnsworth's book[1] for much more detail. Suffice it to say that there was great deal of compelling evidence against Henry presented in a two-day trial and that the jury only needed nine minutes of deliberation before returning the verdict—guilty of first-degree murder.

For Henry, the long wait had been to get to the trial. Although he was arrested in August 1882, the trial didn't begin until March 6, 1883. During these months of confinement, Henry spent a tremendous amount of time writing. Much of it was gloomy poetry and prose, but some ciphers were included as well, and none of them have ever been solved. We can't be sure that the ciphers are in English, for his other writings included French, Greek, Latin, Portuguese, and Spanish, in addition to English.

Despite the low quality of his output, many women were attracted to him, perhaps because of his worldliness, and visited him in prison. You may judge his physical attractiveness (and artistic ability) by the self-portrait reproduced in fig 5.1.

FIG. 5.1 Henry: self-portrait of a serial killer (with cipher)?
From the Collection of the Adirondack History Museum/
Essex County Historical Society, with permission

Henry's self-portrait doesn't show the two-inch scar on the left side of his forehead that he claimed was made by a rebel saber in 1862 (which also cut across the knuckles of his right hand) or the tattoo on the back of his left wrist of "two tri-colored French flags with crossed staffs, and an anchor with a coiled cable."[2]

In addition to the mystery of the ciphers, we have the mystery of Henry's real identity. He admitted that the name he used in Essex was an alias, but it's the same name that he'd been using in Philadelphia before moving to Essex. How far back did this alias stretch? What motivated him to use it in the first place? These are questions that he didn't answer. Yet another mystery arose in the form of three trunks that arrived at the jail for Henry. The sender wasn't identified, but the contents indicated a desire to help Henry. Two of the trunks contained guns, knives, and powder horns. Fortunately, these were found and confiscated before reaching Henry. He was only allowed to take possession of the other trunk, which contained personal papers, ciphers, poems, and sketches he had made. While in prison, he created more.

Who sent Henry these items? He was thought to have come to Essex alone and to have become close to only Betsey. Certainly Betsey's daughters would not have tried to aid or comfort him. Who else would have access to his papers?

More of Henry's ciphers are reproduced on the following pages.

The length of the ciphers indicates a fair amount of information being concealed. Henry said that his real name was included therein, but he didn't pinpoint a particular paper. He may have even meant that it was concealed in autobiographical details he provided in plain English. In any case, context is often extremely useful in breaking ciphers, so it's worthwhile to examine what Henry revealed about himself in his writings and interviews. Bear in mind, though, that it could all have been fabricated. What follows is the account Henry gave to *Post & Gazette*.

I was born May 16, 1836, at Belem, Portugal, on my uncle's plantation a few miles from Lisbon, on the shore of the Tagus River; removed to France with my parents in November 1836; removed to Havre de Grace in April 1838 and remained there until January 1839, when I went with my father to Giromaunt; attended the communal school of Giromaunt from 1843 to 1845. Removed from there to Compiegne College in June 1845 and remained until September 1847. Was with the North Pole Expedition under McClure from February 1848 to October 1850; at Paris from January 1852 to 1853; admitted to the College la Grande May 1853 to 1854; volunteered

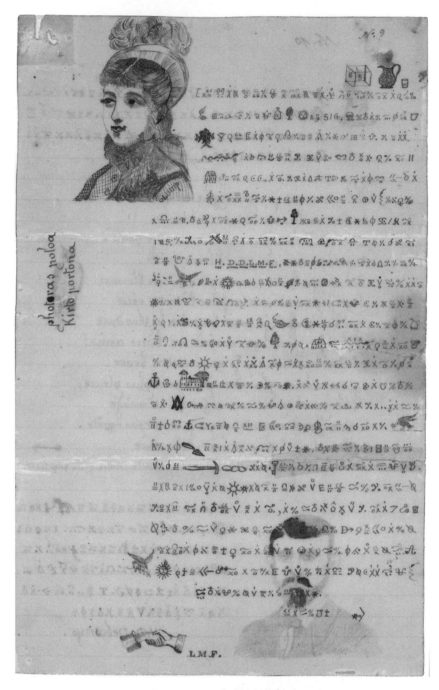

FIG. 5.2 A Debosnys cipher.
From the Collection of the Adirondack History Museum/
Essex County Historical Society, with permission

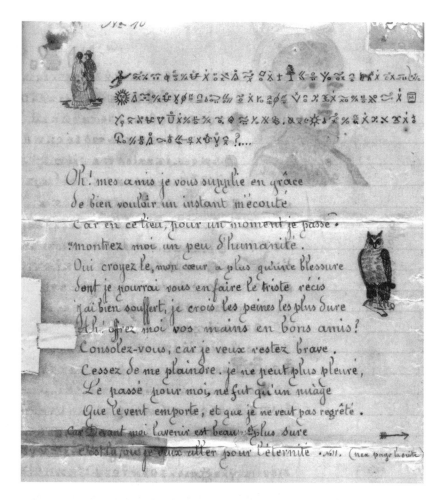

FIG. 5.3 A cipher, bad poetry, and illustrations—another
illustration on the flip side also shows through.
From the Collection ofthe Adirondack History Museum/
Essex County Historical Society, with permission

for the Crimean War with my father and one brother from June 1854 to
1856; admitted to the seminary of St. Brieuc (Cotes-du-Nord) July 1856
and remained until January 1858 [training to become a Catholic priest!]; at
Normal Superior School at Paris from February to December 1858.[3]

So, according to this account, Henry was still in Paris at age 22. This contradicts a briefer biography that he previously gave the police:

I was born in Portugal forty-six years ago and was educated for the Catholic priesthood in Paris. I stayed at college two years and then left for America, as I decided to become an Episcopalian. I was then about seventeen years old. I came to Essex from Philadelphia four months ago and sailed my own yacht all the way. I came via New York, the Hudson River, and the Champlain canal and lake. I came to work at my trade—that of a painter. I married the deceased June 8. I am innocent of the crime charged against me.[4]

Was this version closer to the truth? The tale provided to the newspaper continued with more, possibly fictitious, details of travel and adventure:

Volunteered for Italian War in 1859 under MacMahon; embarked at Genoa for China under the Count of Palikao August 1859. In Pekin [Beijing] in 1860. In Mexico under Bazaine August 1861. Removed to Mexican side after Bazaine's treason. Captain of Guerillas under Lopez; wounded at Delpass Amidjo, June 7, 1862. In America February 1863, volunteered 4th Pennsylvania under McIntosh; at Gettysburg July 1863, wounded in head and left hand with sword. Back to France in March 1864. Married in May 1864 to Miss Judith Desmarais. In the Arctic expedition from July 1864 to February 1867 under Captain Frank. Back to Rome in February 1867. Went to Paris in March 1868 and to New York in June 1868. Moved to the Indian Territory with the Osage tribe to Canville Creek on the Canadian river, and went back to New York June 1869. Volunteered for Franco-Prussian War in 1870 with 600 men from America.[5]

The papers Henry used to dictate these biographic details included marginal comments not printed in the newspaper articles. One of these indicated that he changed his name in October 1870 and was then sent to the army of the Rhone. The published account moves through the month of October as follows.

In France under General Bourbaki in October and General de Busseroles from October to November. I was Colonel under General de Busseroles engaged in skirmishing in December at Autin; drove the Prussians out of Nuit Sousbone 15th December, 3 o'clock A.M., and took the place, capturing 55 prisoners, arms and baggage. December 16th, Battle of Autin and capture of the road and stone bridge; 11 men and 2 horses killed; retreated to woods and lost 4 men. December 19th attacked advance guard on Autin Road, 4 Prussians killed and 7 prisoners; 8 Frenchmen killed. December 28 pursuit of Prussian detachment of Hussars "of the death," and capture of 5 horses, 1 man and 1 wagon of provisions; January 1, 1871, St. Claude skirmish along the road; 4th and 5th January, Battle of Vergam; 6th and 7th January, Great March; 8th January, Battle of Maisondon by the whole army under General Bourbaki at Chateau with the 4th Company of Chasseur; Prussian retreat with heavy loss, 1800 prisoners; whole army pursued enemy to St. Marie. January 12 assault of Plateau de Montpellier, and capture of 7 pieces of artillery and 50 prisoners, 21 men killed, 48 wounded, 3 cars burned. Retreat of General Bourbaki to Switzerland by the Pontavlier Road and Verriere le Suisse.[6]

Return of Colonel Henry to Marseilles in March. Colonel Henry stayed at Marseilles until after the bombardment of the Prefecture of the Police by the fort St. Nicholas and Notre-Dame de la Garde, and the surrender of Gaston Cremieux, the chief of the Commune of Marseilles; then returned to Paris with 400 of his men—some of them volunteered in the regular army at Versailles, some at St. Claude. Lived with his father in Paris during the Commune and had 400 guns in his care which he refused to give to the members of the Commune. He was [threatened] to be shot by the generals Cluseret and Rosselle, in company of a man named Delescluze, for refusing to take part in the Commune and deliver his guns. Colonel Henry made some wood boxes and packed the guns and put them away in the cellar of his house where the regular army of Versailles found them upon their entrance into Paris. Colonel Henry was arrested and sent to Versailles; was tried and sentenced to be shot the same day of the sentence, but he made his escape through a platoon of 30 bayonets and went back to Paris to his father; took 300 francs and then took the cars for Brussels, Belgium, where he remained until all the errors of the sentence were corrected. He then returned to Paris and went to Havre de Grace where he embarked on board the Cimbria for New York in June. His wife, Mrs. Judith Debosnys, died in July and her body was sent to France to be buried in her family vault; the father of the wife accompanied the body to France. She had given all her property to her children, except a little money which was placed on interest.[7]

I find it odd that Henry didn't mention the cause of death for his wife Judith. By contrast, he finds the disposition of her property and money important enough to detail. Notice that nothing went to Henry. A newspaper reported that Judith was found drowned in a river in New York City and that Henry was a suspect. His autobiography continued with no mention of this.

> Henry Debosnys went back to France every spring following and stayed a month or two with his family. He has two children, a daughter and a son; the daughter was 17 years of age in February 1883, and the son was 14 in October 1882; they live in England with a sister of Henry Debosnys; they are educated in English, French, Spanish, and German. The father and children of Henry Debosnys do not know of the downfall of today; his death will remain a secret to them. He leaves a house which, after the death of his father, goes to his two children. In 1871 Debosnys lost $9,000 worth of property, destroyed by fire on Grand Street, New York, in September; in August 1872 he went to Canada; in December he married Mrs. Celestine, his second wife, and in March 1873 they moved to Keeseville. In April came to Essex village and remained until July 1873; then to Burlington, Vermont, until August 27; in N.Y. from September to November 1873; in Philadelphia from December 1873 to November 1875; Omaha, December 1875 to May 1878. Left his farm in his brother's care and went with his wife to California in May 1876, and in September went to the Black Hills; there until October 1877; in May 1878 sold his farm and returned to Philadelphia, June 1879; at Wilmington, Delaware, October 1879 to May 1881. Death of Mrs. Debosnys (Celestine) March 5, 1882, in Philadelphia.[8]

Again, Henry can't find room in his narrative to indicate the cause of death of a wife. This time it appeared to be due to starvation. How could a man allow his wife to starve to death, while he eats enough to survive? It certainly seems suspicious, but Henry avoided arrest, just as he did when his first wife Judith drowned.

Also left out of Henry's self-serving autobiography is the fact that following Celestine's death, Henry quickly hired a housekeeper, Elizabeth Brown, to cook for him. That's right, although his wife starved to death, Henry had enough money to hire a cook!

Henry's narrative then turned to his third wife, and a death for which he would finally be held responsible, even though he never accepted the responsibility.

> Sold his place in Greenwich Point and came back to Essex village May 1, 1882. Married Elizabeth Wells June 8, 1882, and arrested August 1 for the

murder of his wife and sentenced March 7, 1883, to be hanged by the neck on the 27th of April 1883.[9]

My history since 1871 until this day can be recorded without difficulty. I have all the references required to prove where I have stopped in each place mentioned in the index of my life; the references are in my trunk with some family papers (if not stolen), I have traveled more than I can write in this little recital. I keep some passages of my life so as to complete my family story. I could only give to you here a few words of my traveling and my occupation in this country.[10]

The trunk Henry referred to can't be the one he had with him in prison, for in that case he would know if it had been stolen or not. Instead, he probably meant another trunk that was at his deceased wife's house. If so, it's far too late to view the papers in an attempt to confirm details of Henry's story. Betsey's family didn't steal the contents as Henry feared; rather, they destroyed them, along with all of Henry's other possessions, after Betsey's murder. So we can move ahead only with the account Henry gave the newspaper.

Am a good ornamental painter by trade and have worked in New York, Philadelphia, Chicago, New Orleans, Buffalo, and Albany; but I was a better trapper and lover of sport, and followed this last in preference to any other occupation. I went to the Indian Territory as a trader with my uncle, and I have made a very good and independent living. But some accident, sickness and mortality pursued me, and I became poor. I had no friends because I never stayed long enough in the same place. I would have had plenty of friends if I was a man that loved the saloon and would pay a drink to any of the men that worked with me. As long as your money lasts, you have good friends, but when that is gone, your good friends are gone, too; such kind friends never were mine. I never had any bad company and never anticipated the property of others; if I be poor, I be poor by sickness, I don't say that I was the very best kind of man; no, but I never did harm to anybody; more false reports against me than anything else.[11]

Henry Debosnys has a good education, conversant with six languages, reads and writes five fluently. He draws very well and passes for a fair artist; but he is not so fair as it is believed. He says he knows the value of such compliments, and he thanks the people of Essex County for their good judgment in this matter and for their good history they have built against him since he has been arrested. But that doesn't hurt his name, nor his own family—they never had any opportunity to find his family name. And he says he is ready to go to his grave; that death does not scare him at all.[12]

For the present, he keeps a menagerie of living animals – he has got a little baby that cries, a donkey that brays and kicks, a dog that barks, a goat that bleats, a crow that caws, and a rooster that crows, all of which can be seen free of charge at Mr. Jenkin's Hotel [the jail].[13]

The secret of his life will be printed after his death.[14]

The newspaper account wrapped up with this:

The foregoing is copied from manuscripts furnished us by Debosnys. He says he will give us more of his history for our next issue. In explanation of his menagerie: he is a perfect mimic and afternoons, when locked in his cell, he opens his menagerie, mimicking the several animals and birds very perfectly.[15]

One of Henry's drawings included a tombstone that read "To Henry D. Debosnys—Colonel of the 2[d] Legions of Satory—International Society 1882."[16] But oddly, the "2[d] Legions of Satory" was not mentioned anywhere in his autobiography.

In another account, Henry provided further details that, if true, could help identify him.

[Regarding the children from the marriage to Judith] They have a good education in English, French, German and Spanish, Both have a revenue of 600 dollars yearly, interest at 4 per cent from their mother, and a property in Paris France. Henry Debosnys daughter and son do not know nothing of the sad case and sentence of their father, and will never hear from, by the members of their family. One brother E. A. Deletnack Debosnys leave in this country some were near Nebraska, or Texas. Henry had left to his care some property, when he was in the state of Nebraska, and went to California, January to May 1875; and September 1876 went to the Black Hills until

October 1877. A farm of 160 acres (property of Henry D. Debosnys) was sold at Omaha for the sum of $600 cash, and H. Debosnys and wife came back to Philadelphia. Henry bought a fishing boat at Newburg, and hired the boat to a party from Atlantic City, N.J. for $20 monthly. At Wilmington, Deleware, he bought the sloop of Mr. Martinez Coollie for the sum of $157 in cash. This boat was hired by H. C. Hamlen of Camden. Those boats were for H. D. Debosnys' son; they gave a profit $50 a month during the fishing season; in the summer season they were used for the transportation of vegetables to the market of different places [like] Philadelphia, Rochester, and Camden, N.Y.[17]

I could only give you here a few words of my traveling and my occupation in this country, on account of my family's name and their present situation. Not because my life past has been dark. No. But it is only for my family and my friends. It is enough for me to suffer this degradation without leaving the dishonor upon my family. My downfall will never reach a member of my family in this world! Many curious and ignorant peasant farmers of Essex County spoke evil about me and built every kind of story against me, without any knowledge of my name or my relation. Such persons are worse than the criminal by their false reports from one place to another. "No timebo millia populi circumdantis me." I am not afraid of ten thousand such people that set themselves against me. In pace in idipsum dormian, et requiscam.[18]

In the first paragraph of Henry's autobiography, he mentioned living and attending school in Giromaunt, a village in Oise, France. Researching this location, Farnsworth found that the spelling has since been changed to Giraumont. To confuse matters, Henry also spelled it "Giromond." He described his time there as follows:

The present number represents a view of the residence of the Debosnys family in 1850, the residence or castle was built and furnished by Mr. Midole D. Debosnys in 1740. In its dimensions and ornaments, it is such a one as presents the characters and fortune of the family. It stands upon an elliptic plain, formed by cutting down the apex of a mountain; and on the east and north way, big hills; and on the south and west side, it commands a view of the prairies for a hundred miles and brings under the eye one of the boldest and most beautiful horizons in the world; while on the south it presents an extent of prospect, bounded only by the spherical form of the earth in which nature seems to sleep in eternal repose, as if to form one of her finest contrasts with the rude and rolling grandeur on the west. In the wide prospect, and scattered to the north and south, are several detached mountains which contribute to animate and diversify this enchanting landscape; and, among them, to the south Clairoy Mountain which is so interestingly depicted in his notes. From this point, the philosopher was wont to enjoy that spectacle, among the sublimest of nature's operations, the looming of the distant mountains; and to watch the motions of the plants and the greater revolution of the celestial sphere from this summit, too, the patriot could look down, with uninterrupted vision, upon the wide expanse of the world around for which he considered himself born; and upward, to the open and vaulted heavens to which he seemed to approach, as if to keep him continually in mind of his responsibility.[19]

Approaching the tower on the southwest, the visitor instinctively pauses to cast around one thrilling glance at this magnificent panorama; and then passing to the hall where, if he had not been previously informed, he would immediately perceive that he was entering the house of no common man. In the spacious and lofty hall which opens before him, he marks no tawdry and unmeaning ornaments: but before, on the right, in the left, all around, the eye is struck and gratified with objects of science and taste, so classed and arranged as to produce their finest effect. On one side, specimens of sculpture set out in such order as to exhibit at a coup d'oeil the historical progress of that art. From the hall he was ushered into a noble salon from which the glorious landscape of the west side again burst upon his view; and which within is hung thick around with the finest productions of the pencil—historical paintings of the most striking subjects from all sections of the country, and all ages; the portraits of distinguished men and patriots, both of Europe and America, and medallions and engravings in endless profusion. Few homes were more attractive and imposing than the castle where Mr. D. Debosnys and family were born. It possessed a very fine garden from which the most enchanting views were obtained of mountain, river, and valley. Part of it was arranged and most carefully kept as a bowling green. This lay on one side of the house and ran parallel with the high road from which it was separated by a hedge.[20]

The best apartment of the entire building was appropriated to my father, mother, and sisters; and I was accommodated in a small apartment next to the kitchen, on the ground floor, with a window opening upon the bowling green.[21]

One night it was unusually hot and close. We had retired early, according to my father's wont; but my room was so stuffy that I could not sleep or even rest, and after tossing about most uncomfortably for a long period, I got up. And putting on a few clothes, threw open the window and stepped out onto the bowling green. The night was exquisite; the full moon was shining in all her glorious splendor—it was, in fact, nearly as light as day. After walking about the garden, I returned to the bowling green and sat down in a pretty arbor covered with creeping plants. The air was soft and deliciously cool, and everything seemed to induce to calm enjoyment which was enhanced by the profound stillness that reigned around, broken only by the murmur of a distant waterfall. While thoroughly enjoying this beautiful scene, the village clock struck one, and I fancied I heard the sound of wheels and horses' feet approaching. In a short time I saw a vehicle come in sight and pass slowly along the high road; and, as my arbor was on the opposite side of the green, I could readily observe in the bright clear moonlight that it was a large family coach, such as country squires often drove drawn by two tall, fat horses and attended by coachmen and footmen in liveries and cocked hats. It turned the corner before mentioned to proceed through the village, as I supposed. Not so, however, for it stopped immediately, and I heard the door open and the steps let down and the sound of feet approaching the castle. "Belated travelers," thought I. "It's little use, your trying the King Head, for we certainly can't take you in." But this was not the intention of the party. It was not the castle but the big garden which they required; for they all stopped at the end of the bowling green furthest away from the castle, where the hedge happened to be very loose and thin. One of the party instantly pushed himself through, and walking a few steps into the green, stood still and looked carefully round. From my having been brought up entirely among soldiers, all military uniforms were perfectly familiar to me; and I therefore instantly recognized the huge gold-laced cocked hats, blue and green cloak, jack boots and spurs, and a heavy ...[22] [if there was another page, it is now missing]

In addition to all of the above, Henry also left an account of his uncle, Captain Maurice Debosnys, whose son died in a duel. With all of these details, why hasn't anyone been able to determine who he really was? Perhaps a reader in France will be able to do so!

Henry's glorious account of his former circumstances, and his life in general, contrasts strongly with a description printed in a Philadelphia newspaper, following his arrest, of what his life had been like in that city.

> Debosnys is spoken of as a ne'er do well who loafed about the vicinity until forced by hunger to perform manual labor, which he could have had at any time in the oil mill. The French Society keeps a record of all moneys given to the needy French. In December 1878 appears an entry to Henri De Boisnys and wife, $2, and from that date on to March 9, 1882, when his wife Celestine died, De Boisny's name constantly appears on the books as a recipient of charity.[23]

Henry's offenses were much worse than simply loafing around and avoiding work. He was a murderer and, if his first two wives were also his victims, he qualifies as a serial killer. But there may have been even more murders. While in prison, Henry revealed that he had lived in Essex for a short time nine years earlier, and he spoke of a man, Henry Lemaire, who disappeared during this time. From his cell, he provided the police with a suspect for Lemaire's murder, but it would be a hell of a coincidence if Henry, one of very few men to kill someone in Essex, just happened to be able to help the police solve another murder. It may well be that Henry had a male victim, in addition to the woman or women he killed. Perhaps he brought up the older murder as a way of teasing the police and entertaining himself while awaiting his fate.

It should be repeated once more, we don't know how much of Henry's autobiography was fact and how much was fiction! However, we do know how his story ends. Henry was hanged for the murder of Betsey on April 27, 1883. He claimed to see his brother in the crowd of people gathered to watch his demise.

After Henry was executed, it was reported that his whole body was covered with tattoos, but further details were not given. The tattoos were apparently "too explicit to describe in the local press in those days."[24]

It was mentioned previously that many women were attracted to Henry and visited him in prison. Before his execution, Henry gave most of his writings and illustrations to one of these ladies. She kept them to herself for thirty years. After her death in 1917, her granddaughter, Nellie Turner, found the papers but did nothing with them other than keep them safe. She eventually loaned them to her daughter, but upon seeing a newspaper article about the case in 1957, she asked for them back. Turner then wrote to the paper, sharing stories her grandmother had told her of the executed man. She also sought out a museum that would be interested in the documents. She found the Adirondack History Museum in Elizabethtown, where the papers can be found today. Others who had papers of Henry's passed down to them also placed them with the museum. Some, from a Mrs. Taylor, were added to the museum's collection as recently as 1991.

If decipherments ever shed more light on this mysterious man, the breakthrough is likely to arise from the enciphered poem reproduced in fig. 5.4.

Notice that every pair of lines ends with the same symbol(s). This is what we would expect of a simple poem. However, we don't usually get rhymes by having just the *last* letter in a pair of words match. We typically need two or more letters to match. But we don't have that very often in this example. Notable exceptions are 𝄞 𝄢 (lines 15 and 16) and % 〰 (lines 17 and 18).

Thus, we have strong evidence of a substitution cipher and that some of the symbols represent more than one letter. For example, ẋ (lines 3 and 4) could be some letter represented by a dot, followed by some other letter represented as x. And the symbols + and o probably represent distinct letters that can be shown together as ♂ (lines 13 and 14).

The fact that this last symbol also appears upside down indicates that it's a pair of letters that can occur in either order. One common example in English is ER and RE. By contrast, we can have QU in English, but not UQ.

This sort of conjectured solution gains support if we can find similar ciphers that saw use. Indeed, Henry claimed that his cipher was widely used in Europe. Although I don't have an example of an identical cipher system to present, there are some with common features that are worth examining.

The first is known as the Copiale cipher. It's believed to date to somewhere in the range 1740–1780. It was found in an East German library in 1970, but wasn't broken until April 2011. The cryptanalysts in this instance were Kevin Knight, a computer scientist in California,[25] Beáta Megyesi, a computational linguist in Sweden, and Christiane Schaefer, another linguist in Sweden. The cipher they attacked can be referred to as a book, as it has 75,000 characters spread over 105 pages. A sample page is reproduced in fig. 5.5.

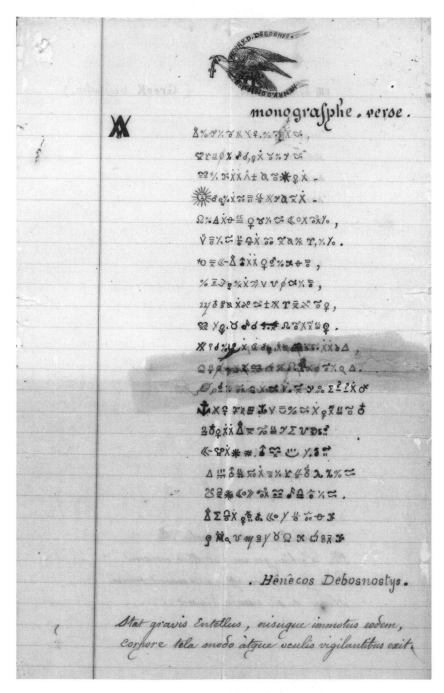

FIG. 5.4 An enciphered poem.
From the Collection of the Adirondack History Museum/
Essex County Historical Society, with permission

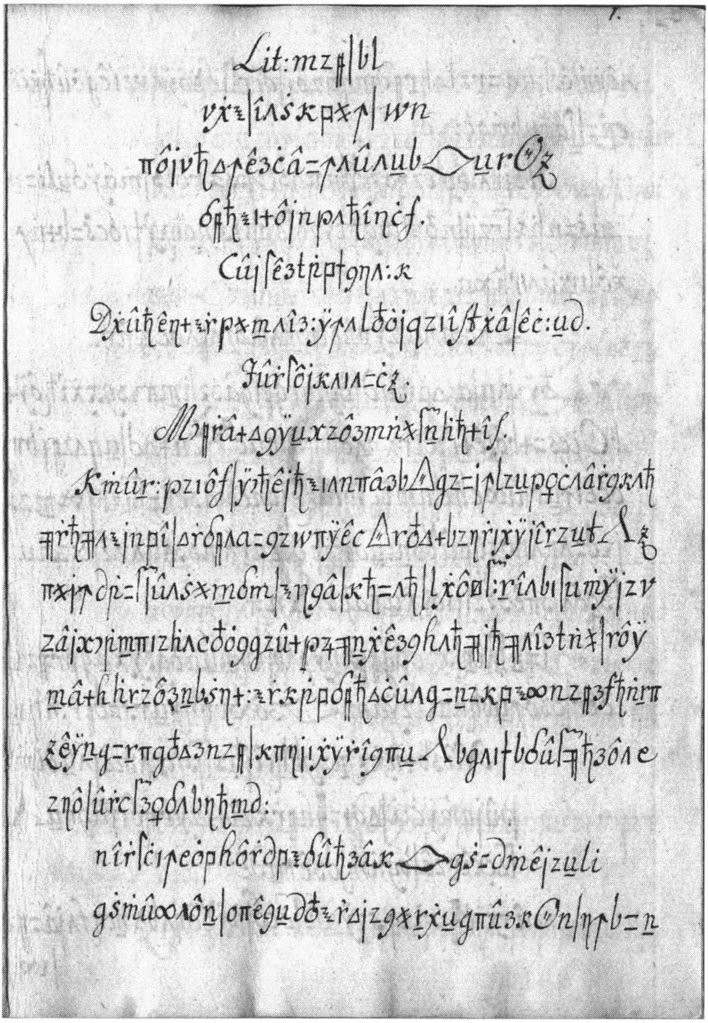

FIG. 5.5 The first page of the Copiale cipher

There are 90 distinct characters used in this book. Many of them, along with their plaintext equivalents, are shown in fig. 5.6. As each character consistently represents the same plaintext letter, this is a homophonic substitution cipher, like that used by California's Zodiac killer (chapter 4).

Once this book was deciphered, it was revealed to be an eighteenth century German Masonic ritual. Interestingly, none of the three people who deciphered it knew German.

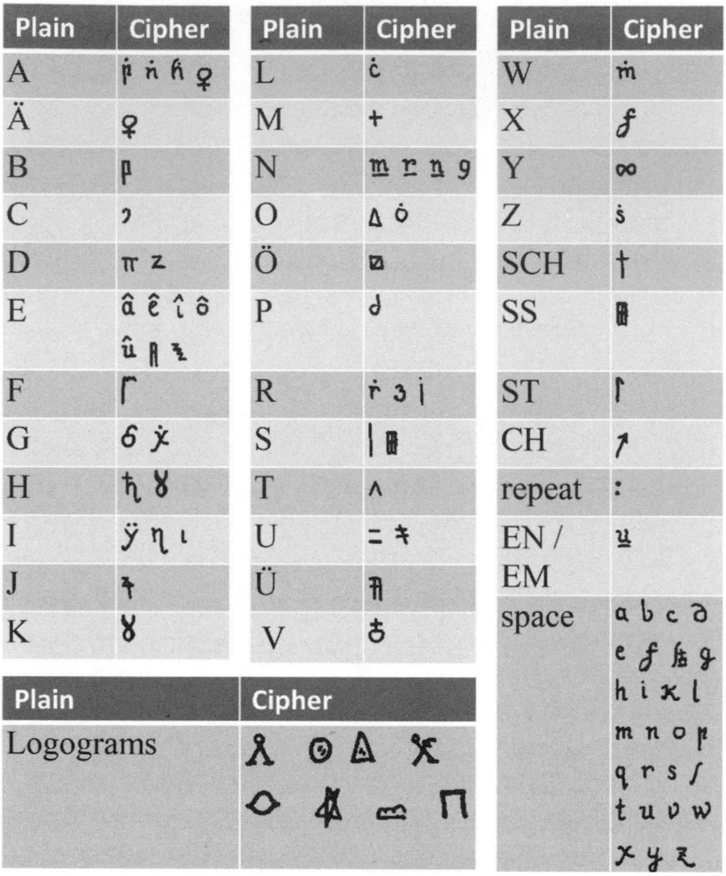

Plain	Cipher	Plain	Cipher	Plain	Cipher
A	ṗ ñ ĥ ǫ	L	ċ	W	ṁ
Ä	ǫ	M	+	X	ƒ
B	ρ	N	m̱ ɾ ṉ g	Y	∞
C	?	O	Δ ȯ	Z	ṡ
D	π z	Ö	◪	SCH	†
E	â ê î ô û ⱥ ⱬ	P	ɗ	SS	▦
F	Γ	R	ṙ ɜ i	ST	ſ
G	6 ẋ	S	| ▦	CH	↗
H	ℏ ỿ	T	Λ	repeat	:
I	ӱ η ι	U	= ⱬ	EN / EM	u̱
J	ⱬ	Ü	Ꞁ		
K	Ɣ	V	ȧ	space	a b c ð e ƒ ẞ g h i x l m n o p q r s ʃ t u v w x y z

Plain	Cipher
Logograms	⅄ ☉ Δ ✗ ◇ ⚔ ⌒ Π

FIG. 5.6 Key to the Copiale cipher

We have another example of a similar cipher in what is known as *The Folger Manuscript*. A page from this intriguing book appears in fig. 5.7. Brent Morris presented this cipher to NSA employees in a paper in the classified journal *Cryptologic Spectrum*. Donald H. Bennett, also at NSA, then solved it and published his findings in the same classified journal. However, both papers later saw print in *Cryptologia*. Morris also went on to write a book on *The Folger Manuscript*, which is highly recommended for anyone wanting a thorough description of how it can be read, as well as the decipherment itself. For this book, a quick explanation must suffice.

Letters and some common words were replaced with various shapes, but the end result was not a simple MASC. The individual shapes were then

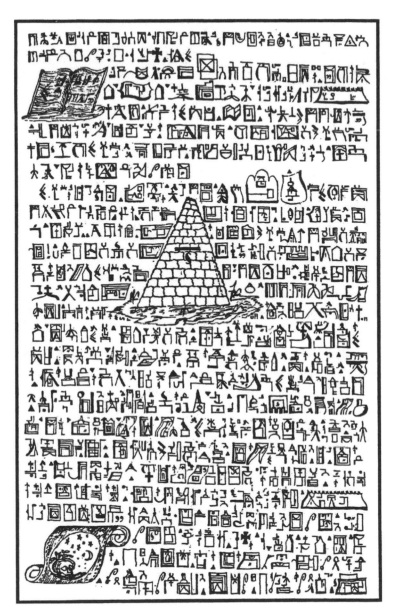

FIG. 5.7 Page 21 of *The Folger Manuscript*

combined in various ways, so that several possible representations existed for the final encipherment of any word with more than a pair of letters. The illustration reproduced in fig. 5.8 demonstrates this.

To check my conjecture on Debosnys work, I contacted the best authority possible, Brent Morris. His qualifications include not only a paper on Masonic ciphers, and the book on *The Folger Manuscript* mentioned here, but also a career as an NSA cryptanalyst with much work that he'll likely never be allowed to discuss in public! In addition to all of this, he's a thirty-third degree Mason and the editor in chief of *The Scottish Rite Journal,* the largest circulation Masonic magazine in the world. He echoed my thoughts in an e-mail, writing "The Debosynys cipher looks like a typical Masonic cipher in that it's composed of discrete, mysterious-looking symbols."[26]

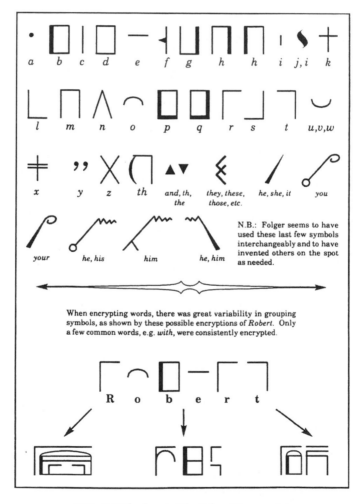

FIG. 5.8 The key to *The Folger Manuscript*

But Morris didn't stop there. He pointed out another Masonic cipher that was unsolved. It's shared later in this book. If you can't wait, flip ahead to the final ciphers of chapter 9.

If we believe that Henry's cipher is Masonic, the natural question to ask is, did Henry in fact have connections to Masonry? It seems that he did, but something went wrong. As evidence in this direction, I point out that Debosnys always drew the famous square and compass masonic symbol upside down. We saw an example of this in the top left margin of his enciphered poem:

FIG. 5.9 The Mason's square and compass—inverted.
From the Collection of the Adirondack History Museum/
Essex County Historical Society, with permission

Farnsworth noticed this, and more. She wrote,

Beyond the symbols drawn in his cryptograms, Henry made other references to Freemasonry. For example, in his poem, "Poor Henru—My Last Voyage and My Adieu," he wrote, "For my son, whom I leave in this moment like a shadow of my life, will not be initiated into the first degree." The first degree of what? My guess is the first degree initiation into Freemasonry. At the bottom of one cryptogram, he drew the secret handshake used during the initiation of the First Degree (or Entered Apprentice) that is known as the "Boaz," or "grip of an Entered Apprentice Mason."[27]

Perhaps in an attempt to make things a bit easier for would-be decipherers, Henry placed a translation of his enciphered poem on the back of the same piece of paper.

Matching up the Greek words with their cipher equivalents should help crack the system. However we cannot necessarily pair them up symbol by symbol. The cipher side might represent English, French, Latin, Spanish, or Portuguese, the other languages Henry knew. If this is the case, a solver will need to consider the various ways that the Greek lines can be translated into the appropriate language before matching up symbols. Still, I think this will be the first cipher from this book to be moved to the solved category by a reader!

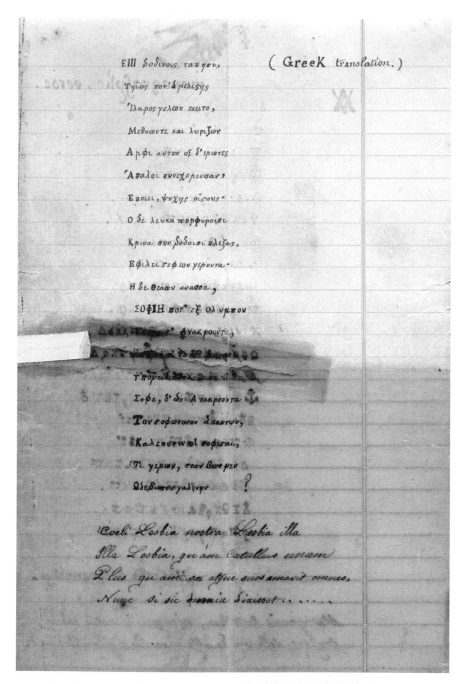

FIG. 5.10 Greek translation of the poem cipher.
From the Collection of the Adirondack History Museum/
Essex County Historical Society, with permission

Henry's drawings included pictures of several pretty women. It's known that he sketched Betsey while in prison, but the other ladies have not yet been identified. Will they turn out to be his previous wives? Will his ciphers reveal anything about their deaths? Will they provide Henry's real identity? I think we'll soon know.

Some killers, like Henry, use ciphers to record thoughts or facts they want to keep hidden from *everyone*, but other killers, like those belonging to terrorist organizations, use ciphers to *share* information and plans, although with only a specific set of people! In the twenty-first century, the ciphers used by terrorists vary widely in sophistication, but back in the 1920s things were simpler.

IRA Ciphers (1920s)

In 2001, Tom Mahon, a medical doctor and historian originally from Dublin, was carrying out research at the University College, Dublin (UCD) Archives, when he came upon a number of enciphered documents. They were composed in the 1920s by the terrorist group the Irish Republican Army (IRA). As his research continued, he found more and more of them. The papers had been donated to UCD by the family of Moss Twomey, IRA Chief of Staff, 1926–1936.

Finally, in 2004, Mahon sought help. After many fruitless inquiries, he turned to the National Security Agency. They pointed him to the ACA. The ACA is not another three-letter government agency, but rather the American Cryptogram Association, described on their website as "a non-profit organization dedicated to promoting the hobby and art of cryptanalysis."[28] Mahon sent six of the ciphers to the group, and James J. Gillogly responded with solutions in just two hours. The collaboration had begun!

One of the ciphers they encountered is reproduced here.[29] It begins with some unenciphered English from "Mr. Jones," who was based in New York and spying there on behalf of the Russians to help raise money for the IRA.

My job is getting very hard I was asked to concentrate and give results. I have, but it means:—

```
OEOCH HOGNW TIWAT CEROS TDHOT NOLOO UGLAI AHUEI AMGVW
NGTEE NOENI EREYI RLEWH NUAIO NLGDG FIHDT OTIST WEEIM
PARDB KAROP HNGNA HEHDH HPMDT EDFTG OLNNO FENOP ARIYU
NEESA TPRLB NAPKS WTMRW EIADA IIHV (164).
```

A quick check of the frequencies of the letters in the cipher reveals that they are very close to those of normal English. This means that the letters have not been replaced by substitutes, but rather rearranged to disguise the meaning. Such systems are known as transposition ciphers. We saw examples of these used in ancient Egypt and Greece in chapter 2. The one above is a bit more sophisticated.

The (164) that ends the message tells how many characters should be present in the cipher. If there are fewer, then the recipient knows that one or more letters were somehow missed. He might then ask for the message to be re-sent.

To get the letters back in the original order, so that the message can be read, the intended recipient would make use of a key. In this instance, the key was the sequence of numbers

        2 8 5 11 12 3 10 6 1 13 4 9 7

The recipient would write these numbers out across the top of a piece of paper and then begin filling in the letters he received below the numbers, in the indicated order. So, his first step would give the result shown here:

2	8	5	11	12	3	10	6	1	13	4	9	7
								O				
								E				
								O				
								C				
								H				
								H				
								O				
								G				
								N				
								W				
								T				
								I				

After the next step (filling in the column headed by 2), he would have

2	8	5	11	12	3	10	6	1	13	4	9	7
W								O				
A								E				
T								O				
C								C				
E								H				
R								H				
O								O				
S								G				
T								N				
D								W				
H								T				
O								I				
T												

After filling in eleven more columns, the result would be

2	8	5	11	12	3	10	6	1	13	4	9	7
W	I	N	E	A	N	D	W	O	M	E	N	I
A	M	O	N	T	O	T	H	E	R	I	G	H
T	P	E	O	P	L	E	N	O	W	A	N	D
C	A	N	P	R	O	D	U	C	E	M	A	T
E	R	I	A	L	O	F	A	H	I	G	H	O
R	D	E	R	B	U	T	I	H	A	V	E	T
O	B	R	I	N	G	G	O	O	D	W	H	I
S	K	E	Y	A	L	O	N	G	A	N	D	S
T	A	Y	U	P	A	L	L	N	I	G	H	T
D	R	I	N	K	I	N	G	W	I	T	H	W
H	O	R	E	S	A	N	D	T	H	E	P	E
O	P	L	E	W	H	O	G	I	V	E	M	E
T	H	E	S	T	U	F	F					

The original message might then be read out along the rows of the grid. With word spacing inserted, it reads

WINE AND WOMEN. I AM ONTO THE RIGHT PEOPLE NOW AND
CAN PRODUCE MATERIAL OF A HIGH ORDER BUT I HAVE TO
BRING GOOD WHISKEY ALONG AND STAY UP ALL NIGHT DRINK-
ING WITH WHORES AND THE PEOPLE WHO GIVE ME THE STUFF.

You may have noticed that when we were filling in the columns, some contained twelve letters and others thirteen. The recipient knows the length of
the message. In this instance, it was 164 letters. He also knows that it is to be
split over 13 columns, When we divide 164 by 13, it goes in evenly 12 times,
but there's a remainder of 8. This means that some columns will have 12 letters, but 8 of them must have one more. That is, the leftmost 8 columns will
have 13 letters and the rightmost 5 will have 12. The recipient makes this little calculation before beginning the deciphering process and carefully counts
how many letters he places in each column. If no mistake is made, the message will be easily read at the end.

The person who created the cipher in the first place didn't have to worry
about any of this. All he had to do was write his message out in the rows of
the grid, as one normally would, and then copy out the columns in the order
imposed by the key. The result is the ciphertext.

Of course, Gillogly was not the intended recipient, so he didn't know the
sequence of numbers that provided the key. He didn't even know how many
numbers should be in the key. There's no immediate way to tell if one should
form a grid with 13 columns or just 5, or perhaps as many as 20. After recovering many messages, Gillogly learned that the IRA varied the number of
columns in their grids from 6 to 15. Not knowing the right number for any
particular cipher presented an extra challenge to Gillogly. But this was a challenge he was up for! Mahon and Gillogly reported

In the end we were very successful. We broke nearly all of the transposition
and substitution ciphers, and we were able to read more of them than the
original correspondents had been able to manage because of mistakes in
key selection or encryption. A few messages remain undeciphered, including a transatlantic cable using a code system that cannot be solved without
much more material and a munitions' list using a substitution system for
the digits that we have been unable to break from the context. However, we
can now read the vast majority of the encrypted material, and it has given
us a rare look into the inner workings of the IRA.[30]

There was only one transposition cipher that they couldn't break. It's shown in fig. 5.11.

(52) GTHOO KCSNM EOTDE TAEDI NRAIE EBFNS INSGD AILLA YTTSE AOITDE.

FIG. 5.11 An unsolved IRA transposition cipher
from November 16, 1926

It's likely that whoever made this cipher botched it in some way. Indeed, the (52) at the start of the message indicated that a fifty-two-letter message should follow, but only fifty-one letters appeared. Gillogly wrote,

> I tried a number of approaches, including assuming the missing letter was in each of the fifty-two possible positions in turn (or in none of them, leaving fifty-one letters as shown), but none of my attacks succeeded. If you crack this one, please let us know.[31]

If you take this cipher on, be aware that some of the IRA transposition ciphers included an extra little wrinkle, a column of nulls. A small example shows how this works.

Message: I have other interests. I'm a magician.

If our key is 5 3 6 1 4 7 2 and the column under 6 is understood to consist of nulls, then to encipher, we begin filling in our grid like so

5	3	6	1	4	7	2
I	H		A	V	E	O
T	H		E	R	I	N
T	E		R	E	S	T
S	I		M	A	M	A
G	I		C	I	A	N

The 6 column is then filled in with letters that have no bearing on the message, but with frequencies roughly approximating those of normal English. These are the nulls. One possibility is

5	3	6	1	4	7	2
I	H	N	A	V	E	O
T	H	O	E	R	I	N
T	E	E	R	E	S	T
S	I	I	M	A	M	A
G	I	T	C	I	A	N

The ciphertext is then read out by columns, in the order dictated by the key. It is AERMC ONTAN HHEII VREAI ITTSG NOEIT EISMA. Someone attempting to unscramble this might be thrown off the right track because IHNA doesn't look like the beginning of a meaningful message. The intended recipient, however, knows to ignore the third column from the left.

Recovering a message that eluded Gillogly would be a great feather in your cap, for he conquered 312 documents. Many contained multiple ciphers. There were about 1,300 recovered cryptograms in all.

For the record, the IRA also had some codebreaking success. Mahon and Gillogly related

> During the Anglo-Irish War the IRA in Cork and Kerry was familiar with the use of cipher by the police. And in 1920, after the IRA in Cork city obtained the keywords for the police's ciphers, it decrypted a police dispatch, leading to the capture of a British spy by the name of Quinlisk, who was shot and his body dumped in a ditch.[32]

We can take on old unsolved ciphers just for fun, but when these ciphers were fresh, it was no game!

Also not a joking matter were the anonymous ciphers received by John Walsh within the context of threatening letters.

The Scorpion Letters (1991)

Few readers are unfamiliar with the work of John Walsh. This is because when the foulest of crimes occurred and his six-year-old son Adam was murdered, Walsh fought back—broadly. He began a crusade against crime. His life before his terrible loss didn't prepare him to wage a war on crime. He worked as a hotel marketing executive. But he has done far more to carry out justice than many people with formal training.

This success occurred, in part, because of his innovative approach. As Walsh put it, "I figured out how to catch fugitives without a gun."[33] By profiling wanted criminals on television in a way that attracted a massive audience, Walsh generated tips that, over the course of more than a thousand episodes, enabled *America's Most Wanted* to announce a thousandth capture on the show's official website on May 2, 2008.[34] A criminal who wanted to avoid capture would do well to avoid capturing the attention of the show's star, John Walsh. Yet, one lunatic actually sought him out, as the document reproduced in fig. 5.12 relates.

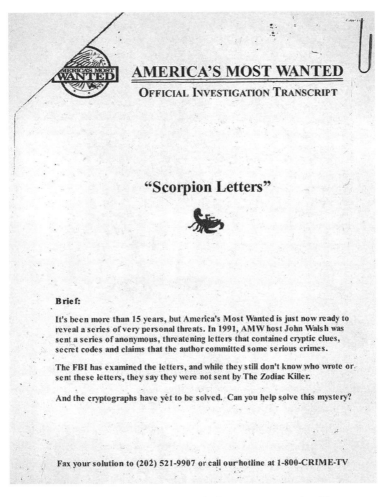

FIG. 5.12 *America's Most Wanted* takes on a personal cold case

Text and cipher portions of the letters are reproduced in figs. 5.13 to 5.17, with some analysis interspersed between them. There are more pages, but those shown here are all that have been released thus far:

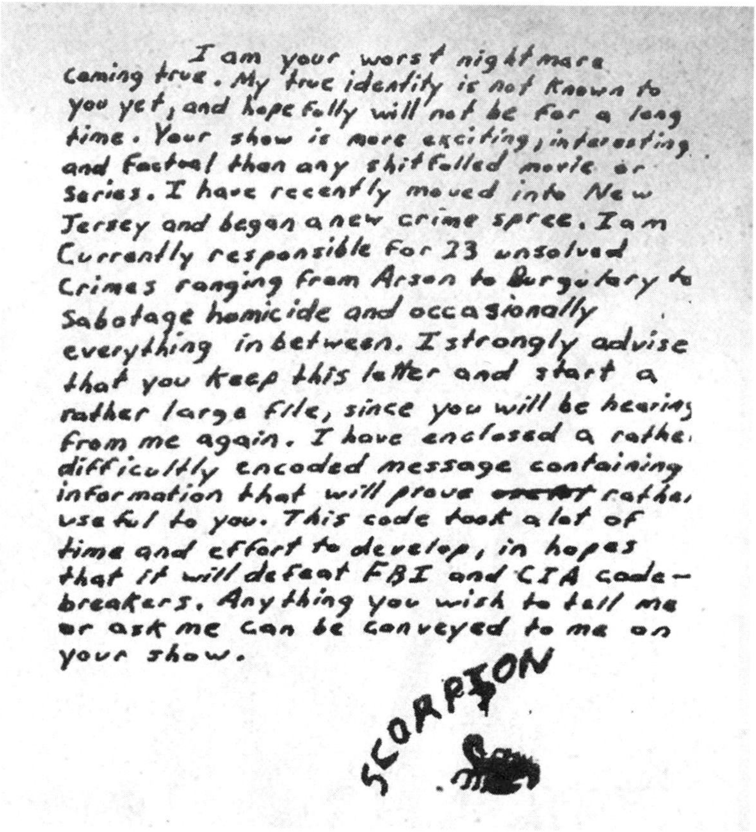

I am your worst nightmare coming true. My true identity is not known to you yet, and hopefully will not be for a long time. Your show is more exciting, interesting, and factual than any shit filled movie or series. I have recently moved into New Jersey and began a new crime spree. I am currently responsible for 23 unsolved crimes ranging from Arson to Burgulary to Sabotage homicide and occasionally everything in between. I strongly advise that you keep this letter and start a rather large file, since you will be hearing from me again. I have enclosed a rather difficultly encoded message containing information that will prove rather useful to you. This code took a lot of time and effort to develop, in hopes that it will defeat FBI and CIA code-breakers. Anything you wish to tell me or ask me can be conveyed to me on your show.

SCORPION

FIG. 5.13 A page of text from a Scorpion letter

Although the FBI ruled out Zodiac as the composer of these letters, there is a resemblance, as you can see. I think it's likely that the Zodiac letters influenced or inspired Scorpion.

Here we have a cipher that's 180 characters long, so I'll refer to it as "Scorpion 180." The problem is that 155 of those characters appear only once. Without even trying, we get the sense (a gut reaction) that we could find multiple meaningful messages that fit such a weak requirement. Unable to decide which is correct, we'd have to declare the message unbroken. But we

FIG. 5.14 Scorpion 180

don't have to go by such gut reactions. Nor do we have to find such distinct messages to show it's possible. This problem has, in general, already been addressed mathematically.

In chapter 1, we briefly looked at Claude Shannon's concepts of entropy and redundancy in a message, but these are topics that go far deeper. Entire volumes have been dedicated to them and their implications. We'll just investigate a single topic closely related to what we've already covered, namely *unicity* points. These vary from system to system, but unicity points always indicate the length at which multiple possible meaningful solutions give way to a unique decipherment.

To consider how the number of possible keys factors into all of this, consider the ciphertext PNXX RJ. If this is the result of a MASC, then there are many possible decipherments, because the ciphertext is so short. The possibilities include CALL ME, CALL US, KILL ME, and KILL US.

However, if the encipherment resulted from a shift cipher, then we have a much smaller set of keys to consider, and it turns out that only one (shifting forward by five to encipher and back by five to decipher) yields anything meaningful. In this case, we know that the message is KISS ME.

This example suggests that the unicity point increases with the size of the keyspace. The question remains, at what rate does it increase? It turns out that it's at a pretty slow rate—logarithmically.

But there's another factor that governs the unicity point—how redundant the underlying language of the message is. In a highly redundant language, letters are more highly determined by previous letters. On the other hand, for a language of extremely low redundancy, where almost anything goes, previous letters have little influence on what follows. The loss of freedom in a highly redundant language means that there are fewer possible messages of a given length. With fewer possibilities, it becomes less likely that a meaningful message will arise from an incorrect key. So, a higher redundancy means a lower value for the unicity point.

Now that we understand the variables that influence the value of the unicity point, it's time to look at the formula. If we represent the unicity point by U, we have $U = \log_2(K)/D$, where K is the number of possible keys and D is the redundancy per letter of the message.

To apply this to Scorpion 180, all we need are the values of K and D. The value for D is not a problem; it was given in chapter 1 as about 0.7 for English (if we don't include the blank space as a character). But how do we calculate the keyspace for the cipher when we know so little of its nature? We observed 155 distinct symbols. If we assume that any symbol can represent any plaintext letter, then we have 26 choices for each of them, for a total of $26^{155} \approx 2.1 \times 10^{219}$ possible keys. But this number includes ridiculous possibilities, like every symbol representing Z. If we assume that the homophones are evenly distributed, and that we've seen all of them in Scorpion 180, then we have almost enough for 6 distinct representations for each plaintext letter. So the number of possible keys is approximately the number of ways we can assign 6 symbols to each of the 26 letters of the alphabet.[35] For the first symbol to be assigned to plaintext letter A, we have 155 choices. For the second, we have 154 choices (one has already been used). For the third, we have 153 choices. And so on, down to 150 choices for the sixth. When we get to plaintext letter B, we'll have 149 choices, then 148, and so on. So the total number of choices by the time we get through plaintext letter Z is the product of all of the whole numbers from 155 down to 1. This product may be written tersely as 155!, where the exclamation point is pronounced "factorial." However, we actually counted too many possibilities. Think about the first plaintext letter A. Is there any difference between assigning it the homophones @, #, $, %, ^, & or the homophones &, ^, %, $, #, @? If I change only the order of the possible substitutions, there's no real difference. Because there are 6! = 6 × 5 × 4 × 3 × 2 × 1 ways to order the 6 homophones, we've overcounted the possibilities for A by a factor of 6!. In fact, we overcounted every letter's possibilities by this same factor. To correct this error, we need to divide our previous result, 155!, by 6! twenty-six times. That is, the number of keys is really this:

$$\frac{155!}{(6!)^{26}} \approx 2.45 \times 10^{199}$$

This is smaller than our first estimate, but it's still a ridiculously large number. It can be argued that we still have some bad assumptions. It's unlikely that every letter of the alphabet is represented in Scorpion 180, much less that each is represented 6 times. It's more likely that the homophones were not evenly distributed. Scorpion may have had 12 representations for E and only 1 for Z, for example. We can keep investigating and find smaller estimates of the keyspace, but they will still be incredibly large, and we won't know if they're really any more accurate. So, let's just go with the number found above.

Plugging into our formula for the unicity point, we get $U = \log_2(2.45 \times 10^{199})$ /0.7 ≈ 946. That is, with fewer than 946 letters, this system can be expected to have multiple possible solutions. At 180 letters, we are *way* below the unicity point. So, we can expect a very large number of possible meaningful solutions. With no way of knowing which was intended, we might give up hope of ever solving this cipher.

However, many of the symbols in this cipher have a geometric nature. Could letters have been assigned to these symbols in some sort of systematic way? If, for example, we find a meaningful solution in which $\boxed{\cdot}$ = A, $\boxed{\cdot\cdot}$ = E, $\boxed{\cdot\cdot\cdot}$ = O (i.e., *n* dots indicates the *n*th vowel), and many other symbols match up with letters in a logical manner, then we would be very confident that the solution is correct. I'm not suggesting that I've identified the vowels—these assignments were simply for illustrative purposes. Any kind of *logical* way of converting from symbols back to letters that yields a meaningful message is likely to be correct. This is because the number of keys that connect symbols to letters in a logical way would be small, and therefore lead to a value for the unicity point below the length of the cipher in question. We'll see reasonable possibilities for some of the letters when we get to the Masonic ciphers that appear in chapter 9.

Nick Pelling used his Cipher Mysteries blog to relate an interesting pattern someone else found:

> In May 2007, user "Teddy" on the OPORD Analytical forum pointed out that if you transpose S5 [what I call Scorpion 180] from a 12-column arrangement to a 16-column layout, shape repeats *only ever occur within a single vertical column*. In fact, every single 16-way column except one (column #5) includes one or more repeated shapes.[36]

Pelling goes on to suggest that each of the sixteen columns might have its own cipher alphabet. If this is correct, we have another avenue of attack.

We also have another Scorpion cipher to consider.

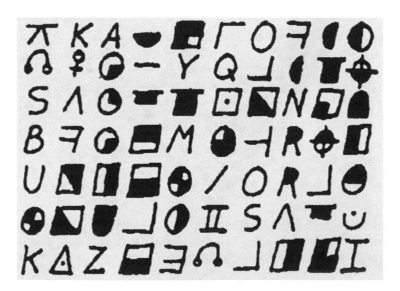

FIG. 5.15 Scorpion 70

This Scorpion cipher is shorter than the other. Because it's 70 characters long, I'll call it "Scorpion 70." We have the same problem that we encountered with the longer message. Because 53 of the 70 characters appear only once, this system apparently has enough homophones to place the message below the unicity point. But again, a pattern may emerge.

Also, there's still hope that these ciphers can identify the criminal in another way—not by being solved, but rather by being *recognized*. Perhaps a reader will know someone who used these same symbols for other messages. Did the criminal have enough self-discipline and control to only use these symbols in letters to John Walsh? If he used them anywhere else, or allowed anyone else to see them, we could have a lead.

This may sound like a very slim chance, but there is a precedent of a sort. The Unabomber was captured based on a tip from his brother, who recognized the style of the Unabomber's manifesto. Bearing this in mind, it seems like less of a stretch that someone might recognize the method of encryption that Scorpion used. But in all likelihood, the text portions will be more useful in matching Scorpion to a known person. The rest of the text portions that have been made public are in figs. 5.16 and 5.17.

Again we may wonder, how strong was Scorpion's self-discipline? How controlled was he in using odd phrases like "mindracking experimentation" and "See you in hell amigo!" here and nowhere else? Is there a reader out there who knows someone who used phrases like this?

Scorpion's threatening letters were not made public until 2007. There's been much less time for the general populace to see what they might have to offer than on colder cases, like Zodiac. Also, with the letters having been composed much more recently than the Zodiac letters, there's a much greater chance that this writer is still alive. So, despite the discouraging insight provided by looking at the unicity point, there is some hope for progress on this case.

The FBI may have leads of which we are unaware. There are other letters and ciphers that they have not yet released. John Walsh may come forward with more information at some point.

Walsh's campaign has brought him celebrity status, but he stays focused on his mission. He's not one to show up on *Dancing with the Stars*. In fact, he said, "I didn't want to be on television . . . didn't want to hunt men down, but you know what, as my wife always said to me, 'Let's make sure Adam didn't die in vain.' "[37]

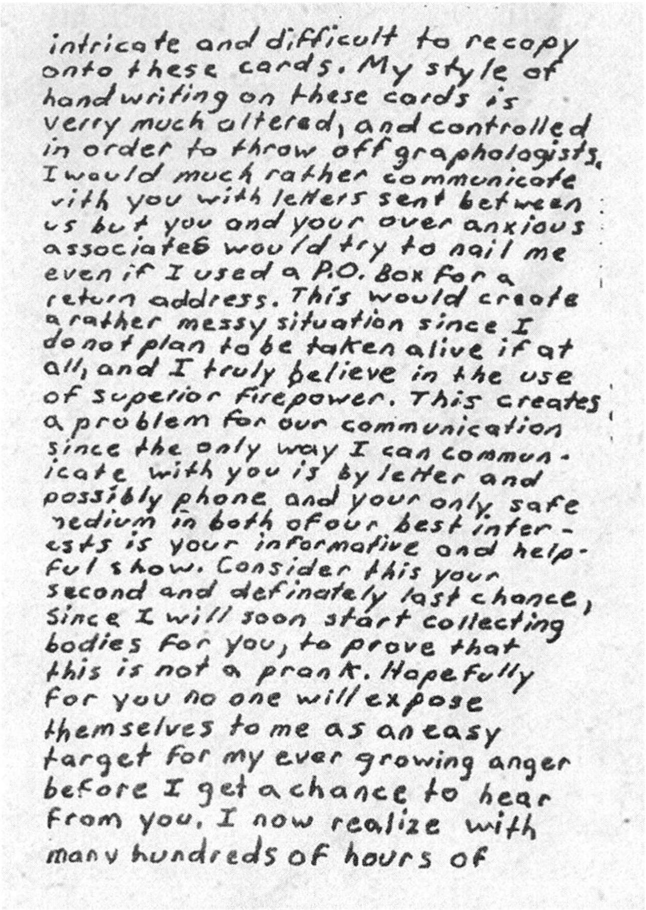

intricate and difficult to recopy onto these cards. My style of handwriting on these cards is verry much altered, and controlled in order to throw off graphologists. I would much rather communicate vith you with letters sent between us but you and your over anxious associates would try to nail me even if I used a P.O. Box for a return address. This would create a rather messy situation since I do not plan to be taken alive if at all, and I truly believe in the use of superior firepower. This creates a problem for our communication since the only way I can communicate with you is by letter and possibly phone and your only safe medium in both of our best interests is your informative and helpful show. Consider this your second and definately last chance, since I will soon start collecting bodies for you, to prove that this is not a prank. Hopefully for you no one will expose themselves to me as an easy target for my ever growing anger before I get a chance to hear from you. I now realize with many hundreds of hours of

FIG. 5.16 Do you recognize the style of this letter?

Even though he never wanted it, Walsh is a celebrity, and celebrity status brings its share of crank mail. I presume, like most celebrities, he ignores most of it, turning over anything of a threatening nature to the police. But why, in the instance of the letters reproduced here, did he go public after such a long wait? It would seem that if there was no reason to believe the claims made in the letters were legitimate, then they might have simply been forgotten after so many years. We can only speculate, as the pages appearing here are all we have to go by. It seems to me that there must be some concealed reason for treating this matter seriously. That's why I've included it in this book.

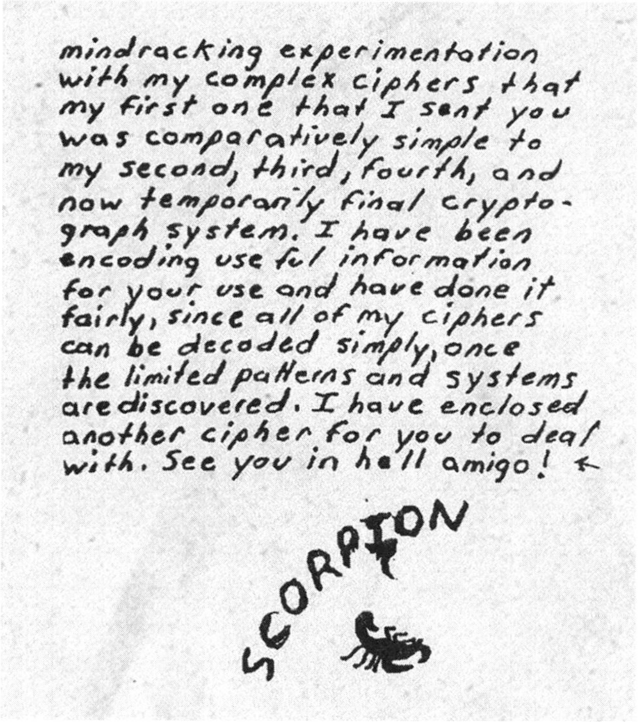

mindracking experimentation
with my complex ciphers that
my first one that I sent you
was comparatively simple to
my second, third, fourth, and
now temporarily final crypto-
graph system. I have been
encoding use ful information
for your use and have done it
fairly, since all of my ciphers
can be decoded simply, once
the limited patterns and systems
are discovered. I have enclosed
another cipher for you to deal
with. See you in hell amigo! ←

SCORPION

FIG. 5.17 Why do these guys all want nicknames?

The letter writer claimed to have disguised his handwriting, but, as I asked before, did he disguise his word choices? Again, I ask you to look closely and see if you recognize Scorpion's style. As Walsh would say,

And remember, you can make a difference.[38]

Sometimes there's absolutely no doubt as to the identity of the killer, but a cipher might shed some light on other aspects of the crime. Such is the case in our next example.

A Cipher or Passwords? (2004)

The National Security Agency has a retiree organization, the Phoenix Society, which, among other activities, puts out a small black-and-white magazine called *The Phoenician*. The summer 2006 issue presented an unsolved cipher associated with a double murder and suicide from two years before. The article is reproduced below. Like the viewers of *America's Most Wanted*, NSA retirees are willing to use their special skills to help where they can.

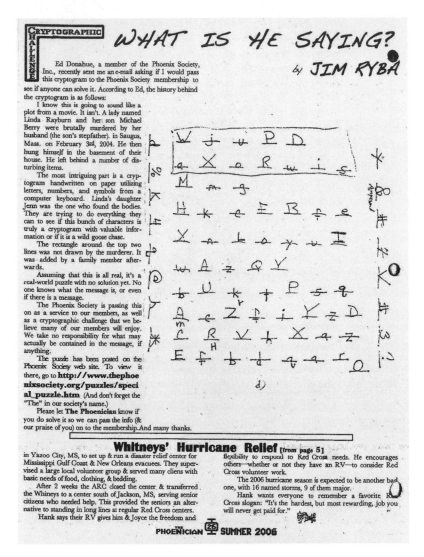

FIG. 5.18 Getting the word out
—an unsolved cipher sees print in *The Phoenician*

Here's a clearer image of just the cipher:

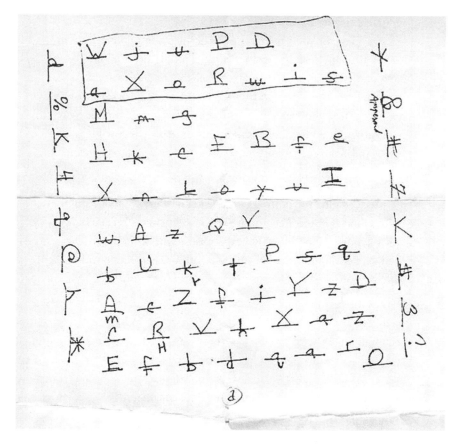

FIG. 5.19 A strange cipher associated with three deaths

Earlier in 2006, Bruce Schneier mentioned this cipher in his Cryptogram newsletter/blog, quoted below, with permission.

January 30, 2006
Handwritten Real-World Cryptogram

I get e-mail, occasionally weird e-mail. Every once in a while I get an e-mail like this:

> I know this is going to sound like a plot from a movie. It isn't. A very good friend of mine Linda Rayburn and her son Michael Berry were brutally murdered by her husband ... the son's stepfather.

They were murdered on February 3rd, 2004.[39] He then hung himself in the basement of their house. He left behind a number of disturbing items.

However, the most intriguing is a cryptogram handwritten on paper utilizing letters, numbers and symbols from a computer keyboard. Linda's daughter Jenn was the one who found the bodies. Jenn is a very good friend of mine and I told her I would do everything within my power to see if this cryptogram is truly a cryptogram with valuable information or if it is a wild goose chase to keep us occupied and wondering forever what it means.

I have no idea if any of this is true, but here's a news blip from 2004:

Feb. 2: Linda Rayburn, 44, and Michael Berry, 23, of Saugus, both killed at home. According to police, Rayburn's husband, David Rayburn, killed his wife and stepson with a hammer. Their bodies were found in adjacent bedrooms. David Rayburn left a suicide note, went to the basement, and hanged himself.

And here is the cryptogram:
[image deleted]
The rectangle drawn over the top two lines was not done by the murderer. It was done by a family member afterwards.
Assuming this is all real, it's a real-world puzzle with no solution. No one knows what the message is, or even if there is a message.
If anyone figures it out, please let me know.
Posted on January 30, 2006 at 10:15 AM · 411 Comments

Cryptologists tend to be suspicious, some would even say paranoid. That's why both Ryba and Schneier questioned whether this cipher was even real. But news reports show that the murders and suicide, at least, did indeed happen. I spoke on the phone with Lieutenant Ronald Giorgetti in August 2015. He was at the crime scene and did not see anything resembling the cipher shown above. But he might not have been aware of everything in the midst of such a terrible scene. He referred me to the district attorney's office as another possible source. My query in that direction is being processed as of this writing.

Newspaper articles reveal that the authorities were baffled as to why Rayburn committed the murders (with a sledgehammer, according to *Boston*

Herald sources) and took his own life. He did not seem psychotic, was not seen fighting with his family, and had even thrown a Super Bowl party twenty-four hours earlier. Police did find a suicide note on the kitchen table, but it provided no motive or explanation for Rayburn's actions. *The Boston Herald* had a source who described the note as "very brief." He said, "It basically said he was sorry and some other things but not why he did it."[40]

There were indications of financial trouble. In a town survey filled out less than a month before her death, Linda gave her husband's occupation as "Internet technician," but he was in fact unemployed.[41] According to *The Boston Herald*, he was about to be tossed out of his house due to financial difficulties, but Saugus Police Chief James. J. MacKay said that the suicide note didn't mention financial troubles.

Could the cipher shed some light where the suicide note did not? Schneier's blog generated 411 user posts in the years that followed. They dealt primarily with the cipher, but other aspects of the case were discussed.

ChrisW was the first to point out something that may be highly relevant. He wrote, "No idea if it's a hoax or not, but it's notable that only lowercase characters are intersected by underscores, and non-letter characters only occur on the sides."[42]

Perhaps the person who wrote the characters (presumably Rayburn, but who knows?) didn't want any ambiguity between letters like C and c. Putting a line through every lowercase letter is overdoing it, but it does clearly indicate case throughout. It could well be that the need for case sensitivity was motivated by these being passwords. This conclusion (without the argument given here) was posted by Davi Ottenheimer.[43]

Looking into the context for what it might reveal of the cipher, another poster, Cher, quoted a portion of the original e-mail Schneier received ("He left behind a number of disturbing items.") and commented, "Would be good to have some idea of what those disturbing items were."[44]

A post responding to Cher's comment is reproduced below, in full.

Ken · February 1, 2006 8:17 AM
Hello—
I am the one who e-mailed the cryptogram to Bruce. Thank you for posting it Bruce.
To answer the most common questions, the cryptogram was found in an open briefcase next to where he hung himself. There was a bunch of child

pornography CD's in there and a bunch of new sex devices like velvet covered handcuffs.

The suicide note that the story refers to was on the kitchen table. It says
"forgive me, it had to be this way". The other family members names are Pat,
Jenn, Kristen. There was also Pat's husband Charlie.

Jenn, Kristen and Michael's last name is Berry. Not Rayburn. They also had
2 cats. Ozzie and Luke.

This [cipher] is a copy of the original. I whited out the areas where friends
were trying to decipher it. I didn't want to make it more confusing. The
folds in the paper are also mine. The murderer was diagnosed Bi-Polar
many years ago. My friend Linda was not told by him or his parents. He was
a tech. junkie and loved Star Wars and music. There wasn't much of a inves-
tigaion because as far as the police were concerned ... it was an "open and
shut case". They had the murderer and the murder weapon. I understand
that some may think I sent this as a hoax. I didn't. I was just hoping that
someone may recognize this and say "I seen this type of cryptogram before".
The password scenerio may be something. There was also a laptop near the
briefcase. As I said, the police didn't investigate all that much. They couldn't
(or wouldn't) even narrow down the time-of-death to less than a 24 hour
period.

Bruce is the only person I sent this to. I recently saw the cryptogram in my
drawer and decided to pursue it a little bit. I appreciate everyone's time and
effort.

Thank you,

Ken

I asked Bruce if this poster really was the one who sent him the cipher in
the first place, and if he could put me in touch with him, but he responded,
"No, I cannot. I have no idea who he is."[45] He also gave a negative answer in regard to whether he had any insights into the case beyond what he had posted.

Lieutenant Ronald Giorgetti told me that he did not see any child pornography or handcuffs at the scene, but again, I was referred to the DA's office for
a more definitive answer.

After Ken's post, things quickly became ugly. Ken, who claimed to be trying only to help and seemed sincere, saw his friendship with Jenn destroyed,
when she found the posts. A big part of the problem was the obnoxious posts
that seem to be a part of any much-responded-to article. Beyond that, posts

appeared attacking a victim, and a feud ensued between people claiming to be (truthfully, in my opinion) friends and relatives of the victim and, on the other side, of the killer. I only quote people who were trying to help in what follows.

Ganzfeld suggested that the "cipher" was simply a list of file and/or directory names, with the vertical lines being the passwords. He continued

> The style of the directory names is very similar to the random (8 letter) directory names you get in the 'userdata' directory. That none of the lines are more than 8 characters supports this. It's highly unlikely to be a coincidence that most of the lines are 7 or 8 characters. This suggests part of a file system ... and NOT encrypted words.

> I would suggest that the 3 and 5 line items are directories, and the 7 and 8 line items are their respective sub-directories, or file names within those directories.

> So the whole thing would be a list of directories on a server somewhere, and the passwords to them.[46]

Ganzfeld posted again the next morning, shooting down the idea that all the page contained was passwords. In particular, the line "Mmg" would have made a ridiculously weak password. He also offered a possible explanation for another aspect of the note.

> There's also a major clue in the 'floating' letters. I think the one below the 'k' is actually the letter 'L'......it follows the same style of the other letter 'L' of having the downward stroke extend too far. This would give the floating letters as L, M, and H. And to my mind, that is Low, Medium, and High. Could be a reference to file resolution.[47]

It might not be possible to ever determine if the cipher represents file directories, passwords, or a message, but the query I made to the DA's office should at least resolve whether it was real or a hoax.

Our next case is almost like a long, drawn-out version of what's just been covered.

Susan Powell (2009)

You know, everybody uses this word [closure] and banters it around.
... I don't have any closure and most parents of murdered children or
crime victims don't really have closure because your life is changed
forever by that event.   ———John Walsh[48]

Susan Cox, like Betsey Wells, was an innocent woman who apparently became
a victim by marrying the wrong man, Josh Powell, in April 2001 and becoming
Susan Powell. Her first difficulties, however, were with her new in-laws. Josh's
brother Johnny had a history of mental problems, but Susan's father-in-law,
Steve, became a much greater problem. He wanted his daughter-in-law to act
as a wife to both his son and himself. This caused the couple to move out of
Steve's house in the state of Washington, which provided a shelter when the
couple was weak financially, to Utah. After the move, Steve, apparently unable to take a hint, mailed naked pictures of himself to Susan.

Steve's fixation led him to write dozens of love songs about Susan and to
fill seventeen spiral notebooks, more than 2,330 pages, with writings about
her. Although Susan was apparently unaware of much of it, while she was still
living in Steve's house he had taken hundreds of surreptitious pictures of her
doing everything from eating cereal to using a tampon.

Though she thought things would be better in Utah, her expectations were
not met. As their family grew with the addition of two sons, Josh changed,
becoming extremely controlling and a financial burden.

When Susan went missing on December 7, 2009, there was a large amount
of circumstantial evidence implicating her husband. Only forty-eight hours
later, he rented a car and drove 807 miles in two days. He never explained what
he was doing, nor did his father account for his whereabouts during the same
two days, which he had taken off from work. Susan's father and the police suspected that Josh and Steve moved Susan's body during this time. Police declared
Josh a person of interest and described his "unusual lack of cooperation."[49]

If the police thought that they could continue gathering evidence and
eventually have an airtight case against Josh and one or more of his family
members, they were badly mistaken. The case became the largest missing
person's investigation in the nation's history, but despite this the body was
never found and nobody was ever arrested for murdering Susan, or for being
involved to a lesser extent.

On September 22, 2011, police did manage to get Steve on other charges. He was arrested for possession of child pornography and fourteen counts of voyeurism. The latter was not connected with Susan, but rather the eight- and nine-year-old daughters of a neighbor. Investigators eventually concluded that Steve Powell was not directly involved with Susan's disappearance but may have learned details of it later.

Police discovered that Josh's preference was for cartoon porn, which though depicting underage individuals in many instances, was in a gray area legally because it didn't portray real people.

Susan's parents, Chuck and Judy Cox, believed that Josh was responsible for their daughter's disappearance and fought him in the courts for custody of their two grandsons. The toxic environment at Steve's house and Josh's own odd pornography, as well as his being a person of interest in his wife's disappearance, all contributed to Chuck and Judy eventually prevailing.

Josh was allowed to have his sons, now five and seven years old, over only for supervised visits. They ran up to his house for a visit on February 5, 2012, and, after Josh opened the door to let them in, he closed and locked it again before the supervisor could enter. Earlier, Josh had spread gasoline around the house; when the children entered, he ignited it. The house exploded.

Investigators later found that both of Josh's sons had portions of their skulls missing from being attacked with a hatchet, yet their deaths were attributed to smoke inhalation. They also learned that Josh had planned the killings well in advance; he had given away his sons' toys the day before.

A little more than a year later, on February 11, 2013, Josh's brother Mike committed suicide. The investigation had begun to focus on his possible role in the murder and/or cover-up. It was only after his death, though, that police publicly revealed that they thought Mike was "heavily involved" in getting rid of Susan's body.[50]

Just weeks after Susan's disappearance, Mike had abandoned a car at a junkyard in Oregon. A detective learned of this by chance. He was at an imaging company in Boulder, Colorado, when Mike called to purchase a high-resolution satellite photo of the junkyard to see if his car had been cut up and sold yet. Police raced to the vehicle and found that their dogs indicated an odor of decomposition inside.

Circumstantial evidence for Mike's having helped his brother is also provided by changes Josh made to his life insurance beneficiaries approximately two months before his suicide. He revised it to leave 93% of the 2.5 million dollar payout to Mike.

Before any of the deaths, the police had phone taps on all of the Powells, but Josh and Mike didn't trust the phones. That was no problem; the police also monitored both Josh and Mike's computers from August 2011 until October 2011. Unfortunately, the brothers used encryption that the police, the FBI, and the software manufacturer were all unable to break.

The exact type of encryption was never publicly revealed. There are many unsolved ciphers of this kind. A frustrating aspect of such cases is that, in my opinion, NSA can probably break most of these ciphers. Yet, if this is true, it's also true that they have good reason not to! Consider the following separate scenarios.

1. A group of terrorists is planning another attack on the scale of 9/11. To maintain operational security, they rely on encryption. They pick a system that they think NSA can't break, but they don't really know because the agency does a good job of keeping many of its capabilities secret. The result is that they choose the wrong product and the intelligence garnered from their broken ciphers allows the U.S. intelligence community to take action and prevent the attack.

2. A group of terrorists is planning another attack on the scale of 9/11. To maintain operational security, they rely on encryption. They've closely followed many U.S. court cases and seen NSA employees testify as to the accuracy of decipherments from various cipher systems. Such testimony aided the prosecutors in gaining convictions in a wide range of criminal cases. However, the terrorists observed that there's a particular system that never seems to be discussed in court. Presumably, NSA can't break that one. This is the system they use. Some of the terrorists' communications are likely intercepted, but none are broken. The U.S. intelligence community fails to get advance warning of the attack, and it goes through. Thousands die.

Everyone in law enforcement and the intelligence community would have preferred that the Powell case end differently, but breaking the encrypted e-mails and using them as evidence in a trial could have had negative consequences elsewhere.

In an unusual twist, two victims of Steven Powell's voyeurism purchased his house and allowed Susan's family to go through it. A possible find was reported.

> The Cox family told HuffPost that among the items of interest they found is a box that contains computer-coding files. The family said they hope the files might contain information that could help the FBI get at files on a heavily encrypted computer previously owned by Josh Powell, as investigation agents have been struggling to access those for several years.[51]

What exactly is meant by "computer-coding files" is unclear.

Huge advances in both cryptanalysis and computing power have been made in recent decades, and such progress is expected to continue. If the Powell e-mails and computer drives have been retained, there's hope that they will become vulnerable, even without help from the government, in the years to come. Eventual breaks won't bring closure for Susan's parents, but they might end the mystery of what exactly happened and possibly allow Susan's body to be recovered and given a proper burial.

The Powell case mainly involved encrypted e-mail, but in many other cases, it's a hard drive that's enciphered. An example from Germany shows how problematic this can be.

Germany's Masked Man (2011)

In 1992, German police were seeking a man matching the sketch reproduced in fig. 5.20. The press called him the "Masked Man" and the "Black Man" after the disguise he wore on nights when he sexually assaulted children at school camps and youth centers. In 1994, he began breaking into homes to find victims.

There were also murders. Investigators observed the pattern of a killing every three years, beginning in 1992 with thirteen-year-old Stefan Jahr. In 1995, eight-year-old Dennis Rostel was abducted from Germany, but his body was found in Denmark. The next murder that fit the profile occurred in the Netherlands. The victim was eleven-year-old Nicky Verstappen. The 2001 murder victim was nine-year-old Dennis Klein. He was both abducted and found in Germany, but the next victim, eleven-year-old Jonathan Coulom, went missing in France, and his body was found there as well.

FIG. 5.20 Sketch of the Masked Man

Because of the international aspect of the crimes, special commissions in Germany, the Netherlands, and France all worked together closely. A great deal of media attention was also brought to bear.

Finally, a tip from a 1995 sexual assault victim led police to arrest Martin Ney in April 2011. After the first interrogation, Ney confessed to the three murders committed in Germany and about forty incidents of sexual assault against children. He denied killing Nicky Verstappen and Jonathan Coulom, and no proof could be found in those cases, but Ney remained a suspect.

For the murders he confessed to, Ney is now serving a life sentence in a German prison. Whether or not he can be tied to other murders will make no difference in his life, but it would make a difference to relatives of his possible victims. It would also determine if the killer of Verstappen and Coulom is off the streets or not. If he was not connected with these killings, the responsible individual needs to be found, as the threat remains. Can codebreakers help find these answers? Maybe. And the chances are constantly increasing.

After Ney's arrest, police seized his computer, and it was found to have an encrypted hard drive. Additional storage devices Ney had hidden in his Hamburg apartment were placed so well that police missed them. These were found only months later by a new tenant. If the contents could be accessed, they might help close more cases. But the encryption was strong, and Ney refused to reveal his password.

German cryptologist Klaus Schmeh suggested that the encrypted contents could be made available online. Then codebreakers the world over could try to crack it. However, I thought that this was unlikely ever to happen. The contents could be horribly obscene, and depending on who breaks the encryption, there would be the possibility of the contents spreading, further exploiting victims. Certainly police would not release the encrypted material to the general public for this reason.

Yet, there might be some hope for the sort of approach Schmeh proposes. He pointed out

> It is not necessary to reveal a part of the ciphertext. A file encryption program usually checks if a key provided for decryption is correct (decrypting with a wrong key will produce nonsense). This function is usually achieved by using some meaningless non-secret information that is encrypted beforehand. If the police make this piece of non-secret information (both encrypted and in the clear) public, everybody could try his luck without the necessity of revealing the actual data.[52]

Neverthless, not only are the authorities refusing to let the larger cryptologic community have a try at the encryption; they won't even reveal what scheme was used. There are a tremendous number of encryption products currently on the market, and the police wisely don't want to specify which ones stymie them. But progress in cryptanalysis is made on a constant basis, and computers are becoming ever faster. Ney's encryption might fall later this year, or next year. Only time will tell, but with Ney currently in his 40s, I'll be surprised if he doesn't live to see his encryption fail.

For experts who may have something to contribute, a paragraph on the case is reproduced that contains a point of contact and some details.

> Martin Ney's computer and hard drives are currently in the Forensic Science Institute of the Lower Saxony Police Department, which also employs IT forensics specialists. According to Kai Thomas Breas, two computer science professors were also consulted. With their help it was possible to raise the tempo of the decoding software from testing 90 passwords per second to 125,000.[53]

Although there are a large number of cases with enciphered e-mail and hard drives that remain unsolved, they've been left out intentionally. Part of the reason is that they are not easy to reproduce for readers who wish to attempt breaking them. They could be posted on this book's website, but law enforcement would be unlikely to share them for the same reasons given in this last case.

If you really want to help, learn as much mathematics and computer science as you possibly can and get a job or contract work with law enforcement!

6

From the Victims

In the last chapter, we looked at unsolved ciphers associated with murderers, but these aren't the only challenges connected with lost lives. Sometimes ciphers are left behind by the victims. Could cracking one of these reveal the identity of the killer? Anthony Boucher imagined such a situation in his short story "QL696.C9."[1] In this tale, a librarian, who had been shot, sat at a typewriter as her life drained away and typed something quite strange, the Library of Congress card catalog number that gave the story its title. Her killer would have destroyed the paper, if she had typed something like "I was killed by …", but might simply have been amused by her adding a catalog number to the list of books to purchase that she had been working on before being attacked. Was she so obsessed with her job that her last efforts were exerted in the direction of completing a work-related task? The story's investigators narrowed the list of suspects to junior librarian Stella Swift, children's librarian Cora Jarvis, library patron James Stickney, and high school teacher Norbert Utter, but that list and the strange message the librarian left behind were essentially all they had to work with. Can you determine who the murderer was? The answer is revealed at the end of this chapter. In the meantime, three real-world cases are examined. Though the ciphers associated with these deaths may not directly implicate a killer, they may provide valuable leads in otherwise cold cases. Such a lead could ultimately result in a resolution for one of them.

Case 1: Somerton Man

ON THE BEACH

Few Americans would imagine December 1 as a nice day to spend at the beach, but in Australia, that date marks the first day of summer, and early on that date in 1948, a man was found dead at Somerton Beach in Adelaide, South Australia.

FIG. 6.1 Map of Australia showing Adelaide and the capital, Canberra

The man's identity could not be determined, and even today he is usually referred to simply as "Somerton Man," or within the context of the case, as "the unknown man."

FIG. 6.2 The body of the unidentified man was found at the spot marked by X in the picture above

The man appeared to be in his 40s but in very good physical condition, with broad shoulders. In particular, his muscular calves were noted. It was conjectured that these were developed by wearing high heel riding boots, but other reasonable possibilities include that he was a hiker, or fencer, or mountain climber, or danced ballet.

His hair was described as "fair" and "blonde to reddish" and "gingery" with some gray. His eye color was also tricky to pin down and was described variously as gray or hazel. The latter deserves a bit of explanation:

> Hazel eyes are one of the least understood eye colors. What color are hazel eyes? This color is semi-rare and is a combination of several other colors including green and brown. Hazel eyes have less melanin than brown eyes, but more than blue eyes. This eye color can be difficult to define since there is often substantial variation in this eye color. Hazel eyes often appear to shift in color from brown to green. People with this eye color often have a multicolored iris with one color being found close to the pupil and a different color found around the edges. One study indicates that 74% of hazel eyes have a brown ring around the pupil.
>
> ... Eye color is currently divided in to four main categories: brown, blue, green and hazel. However, to accurately define eye color there would need to be many more classifications.[2]

He was 5 feet 11 inches (180 cm) tall and weighed in at about 165 to 176 pounds (75–80 kg). His clothes were of decent quality. He was dressed too formally for a day at the beach, if we are to judge by today's standards, but his attire wasn't unusual for 1948. The image in fig. 6.3, from the cover of Gerry Feltus's comprehensive book on the case, shows how he appeared.

FIG. 6.3 A re-creation of how the Unknown Man appeared when found

The man's coat was made in America. His pants were made in Australia but not sold in South Australia. Clothing was more expensive in the 1940s, so it was normal to write one's name on the labels, but this man's clothes bore no labels at all—they had been torn out. Was he murdered by someone who wanted him to go unidentified? The man appeared to care about his appearance and had recently polished his shoes, which were a good fit. They were not high heeled. Basically, he seemed like someone reasonably well off both physically and financially.

His only unhealthy habit was indicated by heavy nicotine stains on his hands.[3] Indeed, a cigarette still hung from his mouth. Of course, in 1948 smoking was extremely popular, and it wasn't at all unusual for athletic types to smoke. Dental care back then was also less informed, so the fact that he was missing eight upper and eight lower teeth fails to tell us much about the man. The missing teeth were all in the back, with two exceptions, as was typical for the time.

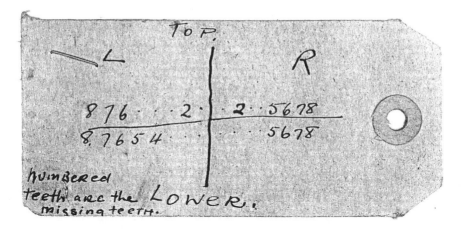

FIG. 6.4 Tooth chart for the Unknown Man

Although the unidentified man was found dead at 6:30 a.m. on December 1, witnesses came forth who recalled seeing him the evening before. A married couple saw him at 7:00 p.m. and noted that he moved his arm, and was thus still alive at that point. Another couple saw him at 8:00 p.m. at the same location but didn't observe any movement.

Another eyewitness account came out more than a decade later on December 5, 1959, and seemed to contradict initial reports. The new witness, who wished to remain anonymous, spoke with Detective Don O'Doherty. The detective reported,

> He stated that at about 10 pm on the evening prior to the body being found on Somerton beach the following day, he saw a man carrying another man on his shoulders along the foreshore. At the time he was single and in company with another man, and two girls. He cannot remember the name of the man but he did nominate the name of the girls who were two sisters. He believed the two sisters later went interstate. They were walking along Somerton Beach towards or near a location known as the 'dugouts'. They saw a man carrying another on his shoulder, walking south, near the water's edge. He could not describe the man.[4]

Why did this witness wait so long to tell his story? He claimed that he discussed reporting what he saw with the sisters, and they all decided that someone else would have already reported it. Such a diminished sense of responsibility when others are present is sadly a common phenomenon. But the question remains, if the man was placed on the beach around 10:00 p.m., how was he seen there at 7:00 p.m., and again at 8:00 p.m.? One can construct scenarios where all of the witness reports are accurate, but it is easier to assume that this last witness was somehow confused. After all, he was recalling an event more than a decade later. Could he have seen what he reported on another night, and therefore simply conflated an unrelated event with our mystery man? Feltus managed to identify this witness and spoke with him in 2003. The witness stuck to his story, including the detail that the event took place the night before the body was found.

Dr. John Barkley Bennett made a quick examination of the body in the police ambulance at the hospital. He estimated the time of death at 2:00 a.m. and thought that the man "appeared to be just like a person who had a coronary seizure."[5] In particular, there were no signs of violence.

As far as possible signs of violence from the past, the unidentified man did possess some scars. Three small ones were present on the inside of his left wrist, another 1-inch (2.5-cm) long curved scar was located inside his left elbow, and finally there was a scar between these on his upper left forearm, although this last one was only a half inch and may have been a boil mark. Author Kerry Greenwood offered a possible explanation for the wrist scars:

> Someone who wears an oilskin, standing in salt sea-spray, gets the sleeve of his non-dominant hand wet and the sleeve then scrapes across his inside wrist, where the skin is thinner. Salt is a powerful abrasive. It produces scrapes, then sores and then scars. A cargo master on a ship, giving orders about stowage in heavy weather, might easily have such scars. Somerton Man was probably right handed because it is his left wrist that has what the fishermen call 'gurry sores'.[6]

Despite the wrist scarring occasioned by being a cargo master, the hands and nails could be kept in nice condition, as were those of the unknown man. The job was a skilled position and wouldn't entail manual labor.

Greenwood's explanation seems possible; I'm not aware of anything that contradicts it, but there isn't, at present, enough supporting evidence to make it convincing. More peculiar features of the deceased are discussed toward the end of this section. For now, we'll make a closer examination of his clothes.

With all of the clothes the man had on, there were many pockets to be checked. We would expect the objects therein to tell us more about him. His bodily possessions were

A handkerchief
A railway ticket to Henley Beach
A bus ticket (number 88708)
A metal comb
A partial packet of Juicy Fruit chewing gum
Two combs
A partial box of Bryant and Mays matches
Cigarettes (the box was Army Club, but it contained seven Kensitas cigarettes)
A pair of underwear
A singlet

Notably absent was any form of identification, as well as money. One would expect such a well-dressed man to have cash or a checkbook in his pockets, but there was nothing, not even pocket change. Not having ID would've been problematic for the unknown man when alive. Without it, he would have had difficulty finding a job, and he might've even run into trouble with the law. Police in Adelaide would ask for ID from individuals for a multitude of reasons, including simply being unknown to them.

So the contents revealed nothing as to his identity. Another unanswered question was this: Did this seemingly healthy man pick the spot on the beach to commit suicide, or was he murdered? Suicides may leave notes, but it cannot be expected. Estimates range from only one in six leaving a note, to as high as fifty percent for some demographics. In any case, if it was a suicide, the unknown man was not someone who felt he needed to leave an explanation for his actions.

A THEORY FROM THE AUTOPSY AND THE SEARCH FOR AN IDENTITY

John Matthew Dwyer performed the postmortem on December 2, 1948, and believed the man to be a victim of poisoning, although he found no traces of poison in the body. That was a detail left for a chemist to verify. To that end, Dwyer collected urine, blood, liver, muscle, and stomach content specimens. We'll get to the results in the section on the inquest.

In the meantime, in an attempt to learn the man's identity, the police released pictures taken the day after the autopsy to the media.

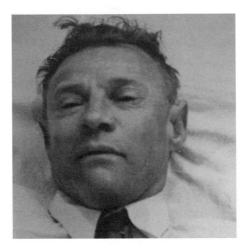

FIG. 6.5 Do you know this man? Two views of Somerton Man

Police are often confronted with false confessions, but in this case, they had to contend with false identifications. Early on, it was reported that thirty-five people viewed the body and eight were able to positively identify him. The only problem is that they provided a total of four different names! As the various "identifications" failed to pan out, the police broadened their search. On December 21, 1948, they reached out to all English-speaking countries, but the result was ultimately disappointing. None of the leads held, although they kept coming in. By November 1953, the list of names provided to police had grown to 251. The police ruled out all of them.

AN UNCLAIMED SUITCASE

The police suspected that the unknown man might have left his possessions checked somewhere when he went out on the beach. It was just a hunch, for no deposit ticket was found on the body. They appealed on January 12, 1949, to anyone with unclaimed items that might belong to the man to come forth. Although inquiries into his identity had failed, this one succeeded quickly.

His possessions were found on January 14. They were in a suitcase that had been deposited at the Central Railway Station in Adelaide sometime after 11:00 a.m. on November 30, the day before he was found dead.

Whatever motivated someone to make the man hard to identify seems to have also been applied to his luggage. Today, tags are attached to luggage at the handle with little strings or pieces of elastic material, but in 1948 they were glued on and difficult to remove. Yet, the ones on the unknown man's suitcase were gone. So, not only was his identity obscured, but also where he came from.

The contents of the suitcase were as strange as the other aspects of the case:

1 red checked dressing gown and cord

1 laundry bag (with "Keane" on it)

1 pair of scissors in a sheath

1 shortened table knife in a sheath

1 stencil brush

4 singlets (1 with the name torn off and 1 with "Kean")

4 pairs of underwear (2 jockey and 2 regular)

2 ties (1 with "T Keane" on it)

1 pair of red felt slippers (size 8)

1 pair of pants with dry cleaning or laundry marks, 6 pence in coins in a pocket, and sand in the cuffs (indicating that that pair had also been worn on a beach)

1 sports coat

1 coat shirt

1 pair of pajamas

4 pairs of socks

1 yellow coat shirt

1 shirt missing the name tag

6 handkerchiefs

1 scarf

1 piece of light cord

1 cigarette lighter

8 large envelopes and 1 small envelope

2 coat hangers

1 razor and razor strop

1 shaving brush

1 small screwdriver

6 pencils, 3 of them H type (drafting pencils)

1 toothbrush and paste

1 glass dish

1 soap dish with 1 hairpin, 3 safety pins, 1 front and 1 back stud (used to attach a collar to a collarless shirt), 1 brown button, 1 teaspoon, 1 pair of broken scissors, and 1 card of tan (or orange) thread

1 tin of tan (or brown)[7] boot polish

2 air mail stickers

1 rubber eraser

Conspicuously absent were his missing ID, money (beyond 6 pence) or a checkbook, and, for 1948, a hat. None of these things were found on the unknown man or in his suitcase. The contents were pretty generic and could have, allowing for the missing items, belonged to almost any man. The depositor's ticket was never found. So, how could the police even be sure that the suitcase was his? There were several indicators:

1. The coat in the suitcase and the coat the unknown man was wearing were of similar size; the one in the suitcase was just a bit smaller, and both had name tabs (tags) removed.
2. The shirt in the suitcase was similar in size to the one worn by the deceased.
3. The Jockey underwear was a similar brand to those worn by the deceased.
4. There were similar handkerchiefs in the suitcase and on the body.
5. The pants in the suitcase and on the deceased were the same size/length.
6. Most impressively, the thread found in the suitcase provided a strong link. Professor John Cleland, University of Adelaide, testified (a bit redundantly) at the inquest, "In the suitcase was an orange coloured linen thread. I found a similar thread in the clothing on the body and in the clothing in the suitcase. In the suitcase was some orange coloured thread. This was examined microscopically and corresponded in colour and size of fibres to similar thread used to sew up a trouser pocket in the suitcase, buttons on the trousers taken off the deceased, and to repair where the coat collar of the

deceased had given way. The colour was a warm sepia colour, which is an unusual colour."[8]

The first five indicators could simply be coincidences. The unknown man wasn't extremely small or large, so the odds of someone else's clothes fitting him wouldn't be too long. However, combined with indicator six, the case for a match is greatly strengthened. Also, the police found no other candidate suitcases to possibly pair with the unknown man. If they had found, for example, a hundred unclaimed suitcases, we would be less impressed, as a "match" would likely arise by chance, at least in regard to clothing sizes.

On the whole, the contents of the suitcase tell us little about the unknown man. They are mostly items any male would be expected to have while traveling in 1948. Greenwood interpreted the presence of the stencil brush, knife, screwdriver, pencils, and scissors as support for her theory of the unknown man as a cargo master. These are tools that a man so employed would be expected to have, but many other possibilities exist, as the equipment is hardly superspecialized.

The itemization of the suitcase contents included a note on dry cleaning or laundry marks on a pair of pants. This was a sort of safety feature at the cleaners, in case the tag was lost. The marks provided a code that would allow identification. In this case, the marks said 1171/1, 4393/7, and 3053/7.[9] Laundry workers thought that the marks were made in England, but they couldn't be narrowed down beyond that. A few people have commented online that they thought the numbers were actually some sort of cipher. If so, it is a particularly bizarre way to convey a message. There are enough stories of clandestine communication carried out by both amateurs and professionals during this time period that I think I would already have heard about a "laundry mark cipher system" if one had seen use.

Fans of today's forensics shows are familiar with how seemingly trivial items can crack cases wide open, so it should not surprise them that the potential lead provided by the dry cleaning or laundry marks was vigorously pursued. The marks were even reproduced in a specialized periodical, *Dry Cleaner's Journal*, with the hope that a reader could identify who handled the unknown man's pants, but no answer came forth.

The most obvious lead provided by the suitcase is the names T. Keane, Keane, and Kean that appeared on clothing items. However, it needs to be noted that names only appeared in garments where they could not be

removed without damaging them. Thus, they might not relate to the unknown man's identity. Greenwood, a native Australian, spoke from her own
experience, as well as that of older relatives, when she wrote that used clothes
often retained labels when offered for resale. These labels saw heavy use because clothes were much more expensive back in the 1940s. New owners
would normally replace the old labels with their own.

Despite doubts that the names belonged to the deceased, the police
checked into many possible identities based on T. Keane, Keane, and Kean.
They even considered variations such as Keanic, Kane, and Tommy Reade,
and the possibility that the first initial T was really an Arabic J.

A LATE DISCOVERY—"TAMAM SHUD"

Detective Sergeant Leane turned over the clothing from the suitcase, and the
clothes the deceased was wearing, to Sir John Burton Cleland, M.D., professor
emeritus in the University of Adelaide's Department of Pathology. It is no longer known on what date he did this, but on April 19, 1949, he learned that he
had given the task to the right man, for Cleland informed him on that day of
an important discovery that was missed at the postmortem.

On June 21, 1949, at the coronial inquest, Cleland went on record with his
discovery from the pants the unknown man was wearing when he was found.
It is this find that would eventually lead to a cipher. He explained thus:

> In examining the clothes, in a fob pocket which was rather difficult to find,
> just on the right of the fly, I found a piece of paper. After I found it and put
> the paper back, it took me a good deal of time to find it the second time as
> it was a pocket which could be easily missed.[10]

Some media reports made a big deal of this secret pocket, as if it was something only a spy would have. Actually, such pockets were common at the
time. They were made to hold fob watches, not spy communiques. In this
case, though, the latter may be close to the truth. The piece of paper was
rolled up, but otherwise in good condition, and bore the words "Tamam
Shud." These words were printed, not written or typed, and so indicated that
the piece of paper had been torn from a book.[11] The good condition of the
paper suggested that it had been torn out of the book shortly before the unknown man's death. If he had been carrying it in his fob pocket for months,
it would have been more heavily worn. In some media accounts, the phrase

was presented as "Taman Shud" with an *n* ending the first word, instead of an *m*. This is an error that has propagated widely; you can find many references to the "Taman Shud Case."

Before detailing how this odd piece of paper became a major lead and brought a cipher into the case, we take a look at other testimony presented at the inquest.

THE INQUEST—IN SEARCH OF A POISON

An inquest was held from June 17 to June 21, 1949. It failed to confirm the cause of death, despite testimony from several experts. The consensus was that the unknown man's death was caused by poison, but this could not be proven and was not supported unanimously. Indeed, Deputy Government Analyst Robert James Cowan thought natural causes were more likely! Those favoring poisoning had difficulty narrowing it down to anything specific.

John Matthew Dwyer, who had performed the postmortem, noted, "The spleen was strikingly large and firm about 3 times normal size."[12] He also reported, "The blood in the stomach suggested to me some irritant poison, but on the other hand there was nothing detectible in the food to my naked eye to make a finding, so I sent specimens of the stomach and contents, blood and urine for analysis."[13]

However, his suspicion of a poison was not to be confirmed. He explained, "When I sent in my report, the poison I suggested was a barbitur[at]e or a soluble hypnotic, and I think that is still consistent with the finding. Assuming Dr. Cowan found no barbiturate or any common poison, I was astounded that he found nothing, as I thought he would. I know he is a chemist of considerable experience, and if he did not find any I accept his finding."[14]

Despite this statement, Dwyer had trouble believing an amount sufficient to kill the unknown man could fail to appear in tests. After some more discussion, he concluded with the contradictory, "On the whole, I think it is probably correct that barbiturate is not the cause of death, except that as I said earlier it is a possible explanation. It is my opinion that in view of the chemist's findings it is unlikely that barbiturates are responsible for death. On the other hand, being driven as far as one can possibly go, I find that the cause might be the cause which I originally suspected."[15]

Professor Cleland was similarly stuck on barbiturates, and testified, "Apart from the special case of barbiturates, there is no case of poison known to the average person which would not be discernible on analysis."[16]

Dr. Cowan then put the results of his analysis of various specimens (stomach and contents, liver and muscle, urine, blood) on record, stating, "I tested for common poisons. Cyanides, alkaloids, barbiturates, carbolic acid, are the most common poisons. If any of the poisons for which I tested were the cause of the death, they would not be absent from the body after death if they were taken by mouth."[17] Previous testimony revealed that no needle holes were apparent on the deceased. Cowan became the lone dissenter to the poison theory. He never claimed that his tests ruled out rare poisons, but he expressed his view that a natural death was more likely than an application of a poison that is rarely used in suicide or homicide.

In contrast, Cleland concluded that "death was almost certainly not natural, and in all probability that some poison had been taken, with suicidal intent."[18] Yet he was uncomfortable with the absence of vomit, noting "Most of the common poisons would give vomiting or evidence of convulsions, something which would have drawn attention to the deceased."[19]

With the next case in this chapter in mind, I draw attention to a particular poison that Cleland essentially ruled out at the inquest: "Cyanide would be very quick, and no bottle was found, nor was there any smell of cyanide."[20] Echoing Dwyer, Cleland stated, "Barbiturates are the things which could have caused death, if only they could have been found."[21] One senses in his testimony the frustration of a man with a square peg, but no square hole, just a round one, repeatedly trying to make it fit.

Sir Cedric Stanton Hicks provided the most interesting testimony, and a possible solution. He had a poison in mind that was consistent with the evidence, but took a roundabout path to naming it. He first teased with an apparent solution, stating, "In cases where death has been said to be due to barbiturates, and in which barbiturates have not been found, in cases mentioned, the poison is sulphonal."[22] But he then went on to explain why that wasn't the case this time!

Finally, he began to present his answer, but in a coy way. He said, "Therefore I incline to conclude that a member of a group of drugs causing the heart to stop systole might have been used. The first word on the exhibit is the name of the group, and the other words are members of the group."[23] So those in attendance had to look at the exhibit; Hicks wouldn't name the poisons. Instead, he had written them down and called it an exhibit! The reason for his being so indirect soon becomes apparent.

His testimony continued, "Of the members of the group, I would say that there are several variants of number 1, and I had in mind more particularly number 2, which would be extremely toxic in relatively small dose, I mean even in an oral dose, and would be completely missed by any of the tests applied and would in fact be extremely difficult if not impossible to identify even if it had been suspected in the first instance. I mean it would not be identifiable to ordinary chemical tests. Such a substance would be quite easily procurable by the ordinary individual, I do not think even special circumstances would be required."[24]

Although Hicks's testimony comes off as the most confident, he did express some discomfort with his conclusion, saying "The only missing fact which would have made me confident is the absence of signs of vomiting, but there is sufficient variation between individuals to account for it or he may have vomited before he took up his position by the seawall, but I confess that I would have been more confident in drawing a frank conclusion had there been signs of vomit somewhere about him."[25]

The reason for Hicks's coyness was obviously concern about a nefarious individual reading his testimony in the inquest document and using it to carry out an experiment in consequence-free murder. Thomas Cleland, charged with writing an introductory summary of the inquest acted similarly at first, writing, "The only poison which Sir Stanton Hicks can think of, and which is consistent with the postmortem findings, is one of the group he mentioned. But here again there are difficulties. There was no vomit, although there was some evidence of convulsion."[26] But he continued with "I would be prepared to find that he died from poison, that the poison was probably a glucoside and that it was not accidentally administered; but I cannot say whether it was administered by the deceased himself or by some other person."[27] Glucosides was the group that Hicks declined to utter aloud! Cleland did not go on to name the two members of the group that Hicks included as part of his exhibit, but they were 1. Digitalin and 2. Strophantin.

Almost a decade later, another, final, inquisition concluded on March 14, 1958, with statements such as "I am unable to say how he died or what was the cause of death."[28]

Although urine, blood, liver, muscle, and stomach content specimens were taken and tested, they are believed to have been lost when the morgue closed and its activities were moved elsewhere. So, there is little hope of applying more modern tests to the samples. The fingerprints fortunately survive, but it is unlikely that they will yield his identity at this late date, when they couldn't be found in any of the databases of the time, and it is hard to imagine a way in which they could shed light, even indirectly, on his cause of death.

Even if poisoning is accepted, the evidence did not allow those testifying to conclusively decide between murder and suicide. If it was murder, how was the poison administered? Was his last meal, determined by analysis of stomach contents to be a pastie, poisoned? Greenwood considered the few possibilities as to where he could have obtained this—the Glenelg railway station pie cart, a private home, or even at the Crippled Children's Home. But the speculation didn't lead anywhere, and Greenwood admitted that she found it difficult to imagine the Home "as a haven for murderers or secret spies."[29] Perhaps the meal wasn't his downfall. Were some of his cigarettes poisoned? Why did he have two kinds in the same pack? The cigarettes were apparently never closely examined, as this means of poisoning had not occurred to the investigators.

Sadly, it is too late to run a test on the cigarettes now. They, like the specimens from the body, are long lost. Indeed, much of the evidence no longer exists. Some items, like the suitcase and its contents, were destroyed, and others were simply lost.[30] Even the police homicide file for the case went missing. A small portion of it was found in 2002. Destruction of evidence and loss of evidence recalls the Zodiac case of chapter 4, but it's more understandable in this instance because of the greater age of the case.

TOMB OF THE UNKNOWN MAN

If the testimony at the inquest had given rise to any questions requiring a fresh look at the body, an exhumation would have been required, for the unknown man had been buried on June 14, 1949. The South Australia Grandstand Bookmakers Association stepped forward to pay for the funeral. Initially, a wooden cross bearing the words "UNKNOWN SOMERTON BODY" marked the grave, but this was soon replaced by the tombstone pictured in fig. 6.6.

Despite the marker listing only one person, there are actually three bodies in that particular grave. The cemetery is allowed, once a lease runs out, to place another body on top of the original and change the marker.

Shortly before his funeral, a new pair of pictures were taken and a cast was made of the unknown man's head and shoulders so that a bust could be prepared. As you can see in fig. 6.7, the newer pictures look quite different than the originals because of the time that had passed. They even depict the man wearing a different tie! We can also conclude that the bust doesn't offer as true a likeness as it would had it been made earlier.

FIG. 6.6 Tombstone of the Unknown Man

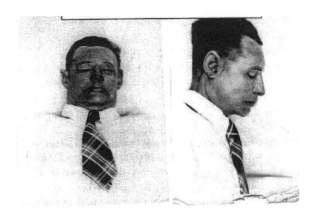

FIG. 6.7 More views of the Unknown Man

When the bust was completed, it was housed at the South Australian Museum. The idea was that someone might be able to make a positive ID from it. More on this later!

THE CIPHER

Before covering the inquest, I described how Dr. Cleland found a piece of paper bearing the words "Tamam Shud" in the unknown man's fob pocket. These words initially baffled the police. They were finally recognized by well-read Frank Kennedy, who pointed out that some versions of *The Rubaiyat*, a collection of poems by Omar Khayyám, end this way, whereas others end with just

"Tamam." With some help from librarians, the police learned that the phrase is Persian (Farsi) and is translated variously as "ended," "finished," "to end," or "to finish."

Detective Leonard Douglas Brown gave a good summary of *The Rubaiyat*:

> The poem itself simply means that we know what this world has in store for us, but we do not know what the other world has in store, and while we are on this earth we should enjoy life to the fullest, and when it is time for us to pass on, pass on without any regrets.[31]

This theme lends some support to the suicide hypothesis, but if this is the case, why didn't the man have the book on him as a sort of suicide note? With the way the "Tamam Shud" paper was tucked away in his fob pocket, it might never have been noticed (it almost wasn't!), and once found, it needn't have been connected to the book from which it came. If a suicidal person wanted to show that he had no regrets, why indicate it in such a roundabout manner?

After the funeral and inquest, the police asked for help finding the specific copy of *The Rubaiyat* from which the "Tamam Shud" paper was torn. They had searched on their own before appealing to the public, but had had no success. The problem was much greater than they had originally guessed, for the book was extremely popular.

Although *The Rubaiyat* was originally written in Persian by Omar Khayyám (1048–1131), more than eight hundred years before Somerton man's death, it was extremely popular at this later time, thanks to a nineteenth century translation by Edward Fitzgerald. Almost every library in Australia had a copy. Bookstores often had multiple editions, and a tremendous number resided in people's homes. Feltus even conjectured that it "may have been one of the most widely read and discussed books of that time"[32] and "probably the single bestselling book of poetry ever to appear in English."[33] Nevertheless, the hunt began. All of the library copies were checked to see if one was missing the words "Tamam Shud." If so, the location of that copy might provide a clue. Alas, this approach didn't work. Not ready to give up, the police decided to release an image of the excised portion of the book to the public in the hope that someone would know which copy originally contained it.

On July 22, a businessman[34] saw the publicity and checked a copy of *The Rubaiyat* in his possession. I use the phrase "in his possession" because the book wasn't really his. Rather, his brother-in-law had left it in his (the businessman's) car's glove compartment when he had been riding with him months earlier—on the day when Somerton Man was found! The businessman saw that the copy was the one being sought and called his brother-in-law, who denied that the book was his. The brother-in-law explained that he had simply found it on the floor in front of the back seat. Assuming that it belonged to the car's owner, he then placed it in the glove compartment for him. The businessman informed the police the next morning.

A microscopic comparison of the "Tamam Shud" paper and the book from which it was presumably torn showed the anticipated match.[35]

Although *The Rubaiyat* was a common book, the relevant copy was from an extremely rare edition. Edward Fitzgerald's translation did not catch on immediately, and this was a first edition put out by Whitcombe and Tombs in New Zealand.[36] Somerton Man's copy went missing again in the 1950s, and none of today's investigators have been able to find one exactly like it. Even Gerry Feltus, who worked for the South Australian Police Force from 1964 to 2004 and who prepared the most in-depth treatment of the case (a book itself), was unable to locate the right printing, despite an intense search. Feltus did manage to find two Whitcombe and Tombs first editions, but neither was a perfect match. One used a different font and didn't even have "Tamam

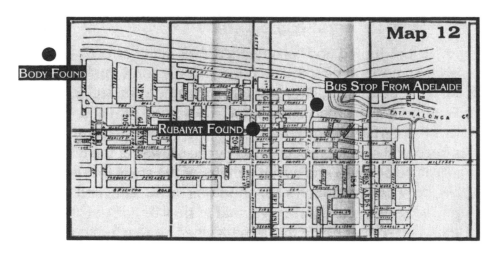

FIG. 6.8 A map showing the close proximity in which
the Unknown Man and *The Rubaiyat* were found

Shud" at the end. Feltus expressed uncertainty as to whether this was the right edition, or if it was instead a Collins Press of England first edition, sent to Australian distributors, although printed in England. The latter was suspected by an investigator before Somerton Man's copy was found. One complicating factor is that a "first edition" can have several print runs, all possibly different in some way. A given printing can even have different states. These occur, for example, when an error (perhaps as minor as a typo) is discovered while making a print run, and the run is paused to correct it. Though different states of a printing can have very subtle differences, different printings of the same edition may appear on different kinds of paper or may have other more easily noticed differences.

Somerton Man's copy of *The Rubaiyat* had a few surprises in store for investigators. One was a handwritten cipher in the back of the book.[37] A little investigation allows us to make some sense of it.

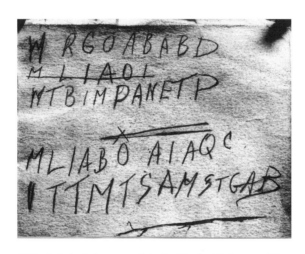

FIG. 6.9 A cipher associated with the Unknown Man

The second line of the cipher is either underlined or crossed out. It is very similar to the fourth line. It could be that whoever created the cipher made a mistake in the second line, crossed it out, and then wrote it correctly on the fourth line. Supporting this argument is the fact that, if we ignore the second line, the letter A is immediately followed by B in half of its appearances. This situation is not likely to occur if the letters are random. Looking at it the other way around, B appears five times in the cipher, and four of these times it comes after the letter A. So, I believe that the second line contained a typo that was corrected in line 4.

Ignoring the crossed-out line, as well as other ambiguities such as the first letter, which may represent W or M (we'll assume W, as these characters do not resemble the unambiguous Ms), and the X above the O (hugs and kisses?), we have the following:

```
WRGOABABD
WTBIMPANETP
MLIABOAIAQC
ITTMTSAMSTGAB
```

Taking a frequency count for the forty-four letters, we find

```
A = 8    G = 2    M = 4    S = 2    Y = 0
B = 5    H = 0    N = 1    T = 6    Z = 0
C = 1    I = 4    O = 2    U = 0
D = 1    J = 0    P = 2    V = 0
E = 1    K = 0    Q = 1    W = 2
F = 0    L = 1    R = 1    X = 0
```

In chapter 1, we saw how useful letter frequencies can be in breaking simple ciphers. For example, recall that the letter E sticks out like a sore thumb, accounting for 12.7% of alphabetic text. However, if we compile frequencies based solely on the initial letter of each word, then the statistics shift and T becomes the most frequent. We find that approximately 18% of the words in a typical English text begin with the letter T. By contrast, only about 2.8% of the words begin with the letter E. So the frequency of a given letter varies with its position in a word.

If we look at the last letter in each word, E rises to the top again. In typical English texts, about 22.3% of words end with an E.

We can simultaneously observe differences in frequency for all of the letters in the alphabet, based on position, with the simple graph in fig. 6.10.

Figure 6.10 shows variability in the popularity of each letter, based on where it occurs within a word. The variability is more extreme for some letters. Examples are E and S, which are far more frequent in the final position, and A and T, which are far more frequent at the initial position.

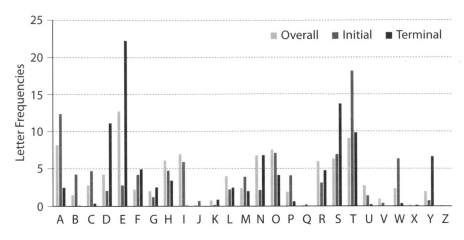

FIG. 6.10 Letter frequencies: overall (light),
initial (medium), and terminal (dark)

We now compare the frequencies of the letters in the cipher from *The Rubaiyat* to each set of columns in the graph in fig. 6.10. We start with the columns representing overall frequency (fig. 6.11).

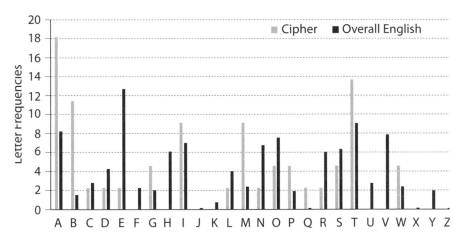

FIG. 6.11 Cipher (light) vs. overall English (dark)

The mismatched columns show that the letters in the cipher do not arise from normal English. Hence, we can rule out a transposition cipher (as seen in the IRA ciphers of chapter 5). If the encipherer had simply re-arranged the letters in the original plaintext message, then the frequencies would be unchanged and the paired columns would all be of roughly the same height.

We might now assume that the system used was some sort of substitution. However, we are better served to make some more comparisons before reaching a conclusion.

In the next graph (fig. 6.12), our cipher letter frequencies are compared with frequencies derived from initial letters in English text.

Here we have a much better match. Compare, for instance, how close to the same height the columns for E are in this graph, with the discrepancy between the same columns in our previous graph.

Finally, we compare the cipher letter frequencies with those derived from terminal letters in English text (fig. 6.13). This graph shows the worst matching of them all![38]

The nice match between initial frequencies of plaintext letters and those appearing in the cipher led many experts to conclude that the message is Somerton Man's personal version of today's LOL, ROTFL, IMHO, or WTF? A much older example is INRI.

These examples are known to a great many people because they have become standard abbreviations. However, ones that aren't standard can be hard to determine, and multiple solutions are almost always possible. Some more examples of varying degrees of popularity follow. How many of them can you recognize or guess? Some have more than one accepted

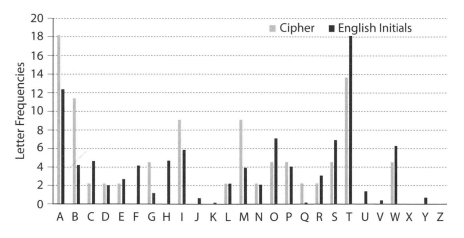

FIG. 6.12 Cipher (light) vs. English initials (dark)

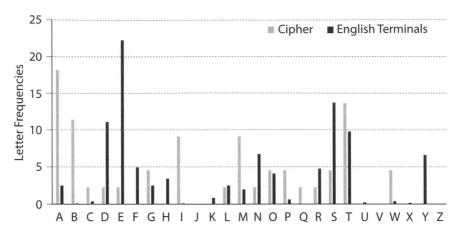

FIG. 6.13 Cipher (light) vs. English terminals (dark)

meaning. The one with periods between the letters is the title of a song by MC Hawking.

AFAIK	FOAF
AFC	FTW
ANFSCD	FWIW
AWGTHTGTATA	F.Y.M.A.S.M.D.
AYBABTU	GMAB
BBW	GMTA
BFD	HMFIC
BICBW	HSIK
BRBGTP	IBTD
BTDT	ICYDK
CIO	IYD
CYA	JIC
DIIK	KUTGW
DILLIGAD	LIRGAS
DIY	LOML
DMML	MCIBTY
EMML	MTFBWY
F2F	NAGI
FCOL	NBTD
FIGMO	NSFW

OFS	TGAL
OP	TMI
PDQ	UIAM
PITA	VFM
QED	WAEF
RTFM	WYP
SOL	XAO
SSEWBA	YBYSA
STBY	YHTBT
TAF	ZA

Answers for all of these, with the exception of the song title, can be found at https://web.archive.org/web/20151108174013/http://www.gaarde.org/acronyms/?lookup=A-Z.

Such abbreviations are sometimes run together. An extreme example was provided as a challenge to readers of *The Phoenician*, a periodical put out for National Security Agency (NSA) retirees.[39]

Can NSA (or The Phoenix Society) Crack This Code??

When teen-age idol Justin Bieber used Twitter to tweet the recording of "At Last" by 13-year-old **Madison Beer** & say "She can sing. Great job," her tweeted response was:

**OMGOGMFHAHDBSBAWHEBSBSHHWEHHDXHSHHAFBBAGEEHYBT
I CANT BELIEVE THIS IS HAPPENING. I AM CRYING.**

The Phoenician editor (who spent some time as a cryppie) had no trouble reading the last 9 words & suspects the first 3 letters of the first part stand "Oh, my gosh!" (or words to the effect) & that the underlying message is in English. But, being one of the few people in the world without a cell phone, he doesn't speak Twitter & has no idea what the next 43 characters stand for. Can some younger person who regularly texts messages tell him what OGMFHAHDBSBAWHEBSBHHWEHHDXHSHHAFB GEEHYBT means? As we said, he doesn't know (& we are sure there are other readers who might appreciate getting the words) .

Please send in your suggestions (even if it's only 3 or 4 letters) to Editor Harry Rosenbluh at **bulbnesor@verizon.net,** or to his Laurel address (back on the **ifc** page), or leave your translation (& name) on his phone answering machine at **301-**

10Q!

PHOENICIAN 16 WINTER '12-13

FIG. 6.14 A challenge for professionals

The conclusion I drew concerning the nature of the cipher in *The Rubaiyat*, that the cipher letters are just the first letters of a string of words, was also arrived at by Navy intelligence in Australia and Jim Gillogly, whose work on IRA ciphers we saw in the previous chapter. Gillogly appears again in chapters 7 and 9. Simon Singh, author of the bestselling *The Codebook*,[40] also put his support behind this sort of system being what was actually used.

Gillogly offered the sample partial decipherments:

`MLIABO` = My Life Is A Bundle Of

`ITTM` = I Think That My

Gerry Feltus put forth this:[41]

`ITTMTSAMSG` = It's Time To Move To South Australia Moseley Street Glenelg

The problem is that many "solutions" are possible. If you took the challenge to identify as many of the sixty abbreviations as possible, you were able get some from having seen them before, but you likely recognized others as having several possible answers. The unknown man's abbreviations have no popular usage to aid in their decryption and may have meant nothing to anyone other than the man himself, or possibly a person or two who were close to him (but unwilling or unable to identify him!). A potential solution that you find may reveal more about you than the unknown man. The cipher letters viewed as initials allow your subconscious thoughts to rise up, acting as a sort of cryptologic Rorschach test!

Though the letter frequencies in the cipher offer strong support for my argument, others have attempted to solve the cipher under the assumption that another sort of system was used.

In his book on the case, Gerry Feltus devoted seventeen pages to various complex, convoluted, and unlikely attacks (made by others) on this cipher.[42] He indicated that he could have included far more. The interested reader may follow his references. The References and Further Reading list at the end of this book also includes YouTube video URLs and other links associated with efforts made to break the cipher by professors and students at the University of Adelaide.

For the moment, I should mention that two of the University of Adelaide students, Lucy Griffith and Peter Varsos (working under the supervision of Professor Derek Abbott and Dr. Matthew Berryman), investigated whether *The Rubaiyat* could have been used as a sort of "one-time pad" to produce the ciphertext. Of course, a one-time pad should be random, not meaningful English. What they looked at should really be described as a running key cipher.

In any case, they did not have any luck with this approach. Even if this was how the cipher was created (which I doubt), the attempt would have been hampered by the fact that they did not have an identical copy of *The Rubaiyat*. Although I do not believe that *The Rubaiyat* has anything to do with the cipher in question, it could have been connected with other ciphers.

If the unknown man was a spy, such a book would be ideal for creating book codes of the sort we'll examine in chapter 9 in connection with the Beale cipher. For now, all you need to understand is that such a book should be rare, but not likely to arouse suspicion, and that both the sender and receiver of a message need a copy because it serves as the key.

Also supporting the idea that the book was intended as a cipher key, whether it was ever actually used for that purpose or not, is the observation that it was the only item connected to the unknown man that wasn't strictly utilitarian. If he had an address book or pictures of friends and family, they were never found. Perhaps his murderer stole those items. The book is the only item that possibly gives us a clue to his personality, yet because of its great popularity, it isn't much of a clue at all. We've seen that the particular edition he had is rare. Was it valuable? He apparently wrote in it, so he wasn't trying to keep it in great condition.

The book may have been selected by a handler for its almost unique combination of being

1. rare (in that particular edition),
2. common (considering the large number of editions), and
3. unlikely to arouse suspicion.

In support of point 3, it must be noted that *The Rubaiyat* was a secular book. If its possessor was suspected of being a Soviet spy, the book would not stick out like a sore thumb, attracting investigators' attention, as would, say the Bible, another common book with many different editions. Investigators would surely be suspicious about why a presumably atheistic communist would possess a Bible.

Items 1 and 2 above seem contradictory, but they are not. The various editions vary greatly. For example, different editions of *The Rubaiyat* change the order of the verses substantially and feature quite different wordings (translations). These changes may have occurred because of the book's poor initial reception. Fitzgerald was likely trying hard to improve the book's appeal with each new edition.

Someone trying to crack a book code based on *The Rubaiyat*, but with the wrong edition, would fail miserably.

A PHONE NUMBER LEADS TO A NURSE AND DEEPENS THE MYSTERY

Having many possible decipherments for the cipher in *The Rubaiyat*, and no way to decide which is correct, we may feel that we are at a dead end. Cryptologically this may be true, but *The Rubaiyat* provided another lead of a totally different nature. In addition to the cipher, it contained a handwritten telephone number.[43]

It was an easy matter for the police to determine to whom the number corresponded. A pseudonym is used in many accounts of this case, but the woman's real name was Jessica Harkness. She is sometimes simply referred to as "the nurse."[44] Although he managed to identify her and speak with her, Feltus gave Jessica the pseudonym Teresa Johnson in his book in order to protect the privacy of her family. As you'll soon see, several of her family members have gone public, so there is no longer any need to continue the masquerade. In any case, the body of the unknown man was found within a quarter mile of Jessica's home, which is also extremely close to where *The Rubaiyat* was found. As a nurse, would Jessica have had knowledge of and access to a poison that could evade the tests conducted by those seeking to determine the man's cause of death? The police certainly had cause to be suspicious, and for whatever reason, Jessica lied to them right from the start. For example, she claimed that she had recently married Prosper Johnson, but this wasn't the case, although they would be wed later. Prosper was married before the nurse, who wed him in early 1950 when his divorce came through. In regard to how her phone number had ended up in the book, she explained that she gave the book to a fellow named Alfred Boxall. Thus, an exciting lead came forth; perhaps Alfred Boxall was the unknown man!

Boxall seemed like a highly patriotic man. He served Australia in World War II, although he didn't have to. On January 12, 1942, he enlisted despite being thirty-five years old with a wife and child. Only men between the ages of nineteen and thirty-three were obligated to serve. His daughter Lesley, a second child, was born on June 14, 1944, and he was lucky enough to have a posting near his family at that time. He was at the army complex on the northern shore of Sydney Harbour, which was also near the Clifton Gardens Hotel. It was at a pub in this hotel that the girlfriend of a fellow officer introduced Boxall to the nurse. The date of this first meeting is not known, but they enjoyed each other's company and discussed, among other things, Boxall's new daughter and *The Rubaiyat*.

At a later chance meeting, at the same spot, the nurse gave Boxall an inscribed copy of *The Rubaiyat* (see fig. 6.15). She signed it Jestyn, but her actual name was Jessica Harkness. Boxall went overseas in June 1945 and apparently hadn't been seen by Jessica since.

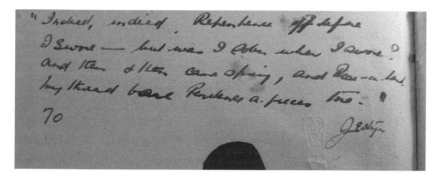

FIG. 6.15 Jessica's inscription in the copy
of *The Rubaiyat* she gave to Alfred Boxall

Police hopes of identifying the unknown man as Alfred Boxall were, alas, dashed when strong evidence to the contrary arose in the form of Alfred Boxall, who was found and confirmed that he received a copy of *The Rubaiyat* from the nurse. Almost as if to pile more frustration upon investigators, he revealed that he still had it! Before going overseas, he regifted the book to his wife. He thought it was in June 1945, but his wife recollected it as Christmas 1944. It turns out that he had two copies of the book in his house, so he may have given his wife the same gift twice (as too many other husbands have done over the years!).

Boxall had a different edition of the book than the unknown man.[45] So if there was a spy ring that had cornered the market on this rare book, he was apparently not part of it!

Because the nurse's phone number dragged her into the case, and she, in turn, dragged Boxall into it, their personal lives became a topic of speculation and suspicion among those interested in the case. Greenwood interprets the nurse's inscription in *The Rubaiyat* as an invitation to begin or continue an affair. As for his passing it on to his wife, Greenwood wrote that it "argues extreme brazen effrontery, complete innocence or something odder."[46] Boxall died in 1995, so it is too late to ask him which it was! However, he never appeared to be hiding anything significant. In a 1978 television interview, he showed the nurse's writing in his copy of the book.

Since Boxall appears to be a dead end, it's natural to shift back to Jessica, whom we've caught lying to police. To add more suspicion to her story, a reconstructed itinerary for the unknown man seems to focus on her. We'll begin our examination of this itinerary with details brought forth at the 1949 inquest. Cleland wrote, in his summary introduction to this inquest,

> It was thought that the deceased must have arrived by train at the Adelaide Railway Station, left his case at the luggage room, purchased a ticket for Henley Beach but missed his train, and then travelled to Somerton by bus. Neither the luggage room attendant, nor the officer who issued the Henley Beach ticket, nor the bus conductor can remember seeing him. No one has come forward to say that he was seen at Somerton between the arrival of the bus and 7 p.m.[47]

Harold Rolfe North, senior porter at the cloak room at Adelaide Railway Station, testified, "At the railway station there is not now a place where a man arriving from the country could bath and shave."[48] It could be that he missed his train because he went to the city baths to get cleaned up before traveling on to his destination. But there is more to consider.

> There was then a strong argument to extend the train service from Henley beach to Glenelg. Unconfirmed reports suggest that maps showing a proposed route from Henley Beach to Glenelg were displayed in the Adelaide Railway Station in 1948.[49]

If these reports are correct, we have yet another possible explanation. The unknown man may have purchased the ticket thinking it would take him to his destination, and then, upon realizing it wouldn't, taken the bus instead.

The unknown man had the phone number of the nurse, who lived in Adelaide, in the back of his copy of *The Rubaiyat*, and Feltus claims that the unknown man called her, but there was no answer. I have not been able to locate any primary documents that indicate a call was actually placed, just newspaper articles, and I'm not sure how this would have been determined, if it was the case. The unknown man would presumably have placed the call from a public phone, and I've seen no record of police obtaining call records from all of the phones he might possibly have used.

As the story goes, the unknown man didn't just call. He also walked to the nurse's house and knocked, getting no response. A neighbor then told him that Jessica and her child were not home. He allegedly looked surprised at hearing this, but again, how do we know? Media accounts were full of errors, contradictions, and speculation presented as fact. If most of the police file hadn't been lost, we might be able to firm up all of this evidence!

If we are only concerned with distances between key locations in the case, a modern map suffices. One is provided in fig. 6.16.

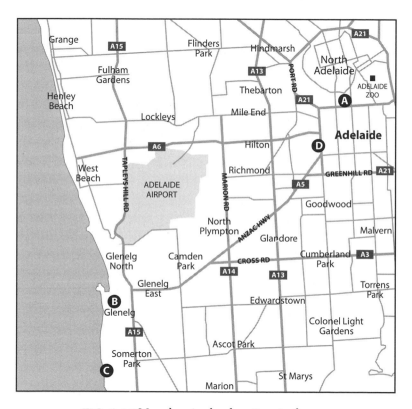

FIG. 6.16 Map showing key locations in the case

A. Adelaide Railway Station
B. Moseley St., Glenelg, where the nurse Jessica Harkness lived
C. Somerton Park Beach
D. The Unknown Man's burial site—West Terrace Cemetery, 161 West Terrace

At this point, you may want to flip back and look again at the map in fig. 6.8 showing where *The Rubaiyat* was found.

The police, having given up Alfred Boxall as an ID for the unknown man, and aware of the information presented above, went back to the nurse and asked her to see if she could identify the deceased. Her reaction was described as "completely taken aback, to the point of giving the appearance that she was about to faint."[50] Yet she claimed that she could not identify the man!

Why would the nurse look like she was going to faint if she didn't know the man? There's no telling how someone might react viewing a corpse for the first time, but nurses quickly become accustomed to such things. And, actually, the situation wasn't really so gruesome—she wasn't even looking at the unknown man, who was in the ground by this time, but just the bust of him that has been prepared!

FIG. 6.17 Bust of the Unknown Man—
Does viewing this make you light-headed?

Gerry Feltus, with a phone number and an old Adelaide telephone directory, managed to identify the nurse. He met with her in 2002, and the two had a long conversation. In his book, Feltus recalled that "She continually diverted the topic from the subject in question."[51] He came away from the interview believing that she knew the identity of the unknown man.

With her story becoming more and more suspicious, the woman held her ground. Perhaps there was an embarrassing story connecting her to the unknown man. It was hoped that she might reveal something after her husband's death, but following this event, she maintained her silence on the matter. Could it be that she feared recriminations more severe than might arise from a disillusioned husband?

FIVE EYES AND MANY SPIES

Today, an intelligence network known as the "Five Eyes," consisting of the United States, United Kingdom, Canada, Australia, and New Zealand, shares a great deal of information. This evolved out of a division of labor that arose between the United States and England during World War II. America focused on breaking Japanese codes and ciphers, and England focused on those of Germany. It was a more efficient way to do things, and over the years, the relationship was strengthened and grew to include the other nations. There are now lesser partners, as well, with whom some intelligence is shared. Australia didn't get off to a great start in this elite club.

In April 1947, U.S. Army Security Agency (ASA) cryptanalyst Meredith Gardner discovered that some of the communications between Soviet intelligence headquarters and the Soviet embassy in Canberra, Australia, were being enciphered by reusing one-time pads. As we'll see in chapter 8, this makes the messages vulnerable. In May, England's codebreaking organization, GCHQ, was asked to collaborate with ASA in attacking these ciphers. The intelligence obtained in this manner was code-named Venona. By November 25, a recovered message, dating to about 1945, revealed that classified British information was being leaked, probably from Australia's Department of External Affairs.

As a consequence of this, in 1948, the United States stopped sharing classified information with Australia. This, in turn, caused Australia to take counterespionage much more seriously.[52]

Frustratingly, Venona evidence could not be used in court without compromising it, so some spies managed to avoid prosecution after discovery. Of course, their careers as spies were over regardless. The Soviets were nervous about what these people might eventually reveal. But were they nervous enough to kill?

Recall that the 1949 inquest suggested digitalis as a possible poison. The Wikipedia article "Taman Shud Case" attempts to build a conspiracy from a pair of sources. The first source firms up the suggested identification of the poison:

> In 1994 John Harber Phillips, Chief Justice of Victoria and Chairman of the Victorian Institute of Forensic Medicine, reviewed the case to determine

the cause of death and concluded that "There seems little doubt it was digitalis."[53]

A sentence later, the piece continues with

Three months prior to the death of the [unknown] man, on 16 August 1948, an overdose of digitalis was reported as the cause of death for United States Assistant Treasury Secretary Harry Dexter White. He had been accused of Soviet espionage under Operation Venona.[54]

The citation given for this last quotation is Willard Edwards, "Hiss spy paper linked to late treasury aid [sic]," (*Chicago Daily Tribune,* November 29, 1949): 1–2. But all this article said about his death was, "At the time, he pleaded that he had a heart ailment. Less than a week later, on August 17, he died at home, reportedly of an overdose of digitalis." Who exactly reported it is never stated, so all we are left with is a dead end. Another researcher found the claim to evaporate upon investigation:

Rumors of mystery surrounding his death were dispelled by Dr. George S. Emerson, in an interview with the *Boston Globe*, which reproduced the death certificate reflecting the doctor's finding that the cause of death was "coronary attack due to disease of coronary arteries of the heart." This account of the causes of White's death is further confirmed in *Harry Dexter White: A Study in Paradox* by David Rees, published by Coward, McCann, Geoghegan in 1973.[55]

The digitalis claim also appears at this site—http://www.globalresearch .ca/us-intelligence-and-the-fate-of-the-private-pre-war-international -banking-system/5378602?print=1—which cites the same source as Wikipedia. However, the reference citation also states

The doctor who signed White's death certificate (and who was not present when White died) later denied that digitalis was a factor. More recently the right-wing author John Koster has charged (with no evidence other than a contemporary movie which White might have seen) that White overdosed on digitalis in order to commit suicide (John P. Koster, *Operation Snow: How a Soviet Mole in FDR's White House Triggered Pearl Harbor* Washington: Regnery Publishing, 2012, 204–05).

Taking everything into consideration, I have to reject a digitalis overdose in the death of White, despite the *Tribune* article by Willard Edwards. After all, if we are willing to accept a claim from the *Chicago Daily Tribune* that is contradicted everywhere else, we could talk about President Dewey![56]

Although some claims evaporate under investigation, the theory of Somerton Man as a spy won't be so easily disproven. For example, if a search through declassified Venona files does not implicate him, it won't mean much. For good as Venona was, it didn't allow the five eyes to identify all of the Soviet spies in Australia. More were revealed in 1954 by Vladimir Petrov, who was working for the KGB while living in Australia. Australian intelligence found out about Petrov the easy way when he defected. Also, the Venona decrypts typically provided only code names for the spies. These then had to be matched with the person's real name. For about half of the code names, the actual identities were never determined.

So what were all these spies doing in Australia? There were two special attractions:

1. In 1944, some uranium deposits were found in South Australia. At this time, the only other known deposits were in Canada, Czechoslovakia, and the Belgian Congo.
2. In November 1946, Australia announced that a site named Woomera, for experimental rocket testing, was being developed in South Australia. The closest big city to Woomera was Adelaide. British-Australian nuclear tests were carried out there in the 1950s. Does the W that begins the unknown man's cipher stand for Woomera?

We saw several conjectures concerning the unused train ticket possessed by the unknown man. If we accept the spy theory, there is yet another.

Perhaps he saw someone following him, or someone about to board the train whom he feared, and he made a last-minute detour to avoid the person. One also imagines him ditching *The Rubaiyat*, which may have been sensitive because of the phone number and cipher, if not for other reasons. A trash can would have been a more obvious place to deposit the book—why place it in a car? Maybe it had to be disposed of very quickly, and no trash can was close enough.

JUST A SMUGGLER?

Greenwood's nonfiction book on Somerton Man included two speculative accounts as to what the unknown man was up to in Australia. One was a reprinting of a (fictional) detective Phryne Fisher short story she wrote in which the unknown man was involved in smuggling arms to Ireland. In the other, which was original to the book, she kept the arms smuggling persona, but shifted his destination to Israel, using the Reynolds persona to be seen shortly. Neither story has much at all to support it, but neither can be absolutely ruled out. Greenwood's evidence includes the unknown man's previously described gurry sores, or scars, his "cargo master" equipment, and a few less impressive tidbits.

Even if he was smuggling arms to Israel, it's unlikely that the unknown man was a Jew; the inquest made note of the fact that he wasn't circumcised. One or more Jews may have honored him after his death, though. Pebbles were piled on his grave, in accordance with Jewish tradition. This tradition doesn't require the recipient to be a Jew and may have simply been their way of thanking a gentile for helping to arm Israel. Readers may recall the same honor being afforded Oskar Schindler, for a different reason, at the end of the Steven Spielberg film *Schindler's List*.

AND A FATHER TOO?

Earlier in this chapter, I alluded to "more peculiar features" of the unknown man. These features offer yet another connection with the nurse. Consider the pictures in fig. 6.18. At a glance, it's just a pair of ears. But looking closer, the ear on the left, belonging to the unknown man, is seen to be quite different. It has a cymba that is larger than the cavum. The ear on the right is an example of the much more common type, in which the cavum is larger. The nurse had an ear like the one on the right.

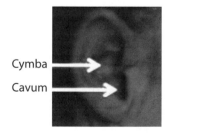 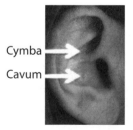

FIG. 6.18 Somerton Man's ear (left) compared
to a far more common type (right)

So how does this feature connect the two? Although the nurse's ear is not a match for the unknown man's, the ear of her son Robin is.

Also, both Robin and the unknown man had a genetic disorder—hyperdontia of both lateral incisors (the condition of having extra teeth). Hyperdontia is present in only two percent of the population, and the unusual ears are a bit rarer, appearing on only one to two percent of whites.

The obvious implication is that the unknown man was Robin's father, and Jessica's claim that she had no idea who he was grows even less believable.

Derek Abbott, a physics professor at the University of Adelaide, estimated the chance of Robin and the unknown man being unrelated, given that they share these characteristics, as somewhere between one in ten million and one in twenty million. Greenwood simply used the longer odds of one in twenty million in her book.

Odds can be stretched one way or the other, depending on how many factors are included. As an illustration, let's take the figures of two percent and one percent for the unusual incisors and the ears, respectively. We could have used two percent for the latter, since the range was one to two percent, or we could have split the difference by using one and a half percent. The smaller figures give us longer odds, but they are still obtained by multiplying the percentages, after converting to probabilities. We have $.02 \times .01 = .0002$, which is just one in five thousand, a long way from the reported odds! Now if we wish to include other factors, we can observe that both men were white. We wouldn't expect them to be related by blood if one was Ethiopian and the other full-blooded Cherokee. What percent of the population is white? This gives us another factor. Of course, if we wanted to be honest, we should take the percentage of whites in South Australia, not the world, but again we have some wiggle room. Another factor that can be incorporated arises from the unknown man having had the nurse's phone number. What percentage of people who have your phone number are not related to you? This number is harder to estimate and clearly depends on how private an individual you are. I have not looked at how Abbott obtained his long odds, for in the end a big number doesn't prove paternity. Nevertheless, I think even the smaller odds of one in five thousand are good evidence in that direction.

Given the ages of the unknown man and Robin, and the implied blood relation, the simplest conclusion is that they are father and son, but it is not the only possibility! Greenwood prefers the theory that the unknown man was the boy's uncle or some other close relation. In her account of the man as a smuggler of arms to Israel, she weaves the nurse in as a sister, niece, or cousin whom he wishes to visit. The unusual features could have skipped Jessica, despite the blood relation.

As with any great mystery, there are many possible interpretations of the evidence. What is your opinion?

SIMILAR DEATHS

If you want to build a much bigger conspiracy theory, here are some ingredients. The first is a newspaper article from August 25, 1945.[57]

PERTH POET-SUICIDE CHOSE OMAR VERSE AS HIS EPITAPH

Some astonishing evidence was given by Samuel Saul Marshall, of Perth, formerly of Singapore, brother of 30-years-old Joseph Haim Saul Marshall, formerly of Perth (also known as George Marshall), who committed suicide by taking poison in the bush close to the water's edge at Taylor Bay, Sydney, on June 3, at the inquest which concluded last week.

Joseph Marshall was found with a copy of the Rubaiyat of Omar Khayyam, the Persian poet, lying on his chest. He had pencilled with a cross the lines:

"Ah, make the most of what we yet may spend,
Before we too, into the dust descend :
Dust under dust, and under dust to lie,
Sans wine, sans song, sans singer and sans end."

His brother said that when he was 7, Joseph met with an accident in Singapore, which affected his health, made him averse to society. At the age of 18, he was sent to France for higher education, after 5 years there, seemed to have a "very strange attitude" towards life.

"He composed and published a book of verse entitled 'Just You and I,' but was extremely discouraged when it was poorly received," his brother said. "We as a family were particularly fond of him, and he of us. When our father lost a fortune in 1929, George unknown to any of us, insured his life for a very large sum against motor-car accidents, planned to drive over a cliff at the end of a year so that we could collect the money. We discovered his intention in time."

Joseph Marshall came to Australia in 1939, lived a life of complete seclusion with books of a philosophical tendency. He later worked for the Prices Commission.

FIG. 6.19 A newspaper article documenting an eerily similar death

---

Bunbury Visit

"In a letter to one of his friends," said his brother, "he described the Prices Commission as being 'turned into a madhouse at his expense.' On January 1 this year he was an inmate of Heathcote after he had been certified as insane by 3 doctors following a visit to Bunbury.

"There he took an overdose of a certain type of pill. It did not take effect, and he then walked to the edge of the sea where he was arrested. He was given intensive insulin treatment, and at the end, the doctor either had to confine him in a mental home or give him the chance of becoming more stabilised.

"I have no doubt that the death he died was a form that would appeal to him as the finest and noblest way of terminating his life. In my opinion the bottle of water found beside him would be a sort of cleansing rite."

Gwenneth Dorothy Graham, hairdresser, of Rockdale, NSW, said, that for about 4 years they had been very friendly. Last time she saw Joseph Marshall was on May 19, when they dined together.

"He had frequently suggested that I should set up in business as a hairdresser," she added. "He said it would cost about £2000. I laughed and said 'No, about £200.' On May 19 he told me he had sent me a cheque for £200. I received it the following Monday."

Coroner Cookson found that Joseph Marshall died from poisoning by a compound of barbituric acid wilfully administered by himself.

FIG. 6.20 The conclusion of the article

This case was also covered in Greenwood's book on Somerton Man. She rejects any connection with the case we are interested in but provides some other information for the interested reader.

Greenwood also mentioned a similar case covered by the *South Australian Register* back in 1881. A heavily clothed man was seen walking on the beach on January 6 of that year and was found dead there on January 10. He was never identified, and although he had a bloody razor and knife on him, the coroner described them as "most inconvenient for inflicting the wound on the throat." He thought it was probably suicide, but left the verdict open.[58] We have similarities to our case, but no copy of *The Rubaiyat* linked to the deceased this time.

So, the possible suicides and/or murders on the beach occurred in 1881, 1945, and 1948. Not the regular spacing we would expect in an episode of *The X-Files*! Also, we need to remember that suicide is, sadly, fairly common.

Given the large number of suicides, we can expect some to have aspects in common with the death of the unknown man.

ENTER THE WEIRD (INTENDED FOR ENTERTAINMENT PURPOSES ONLY!)

One website considered the spy hypothesis (giving it four out of five for credibility) along with two other, much wilder, possibilities, reproduced here, with their credibility ratings.[59]

1. Somerton Man was a time traveller.

This is one of the stranger theories. Some people have suggested over the years that Somerton Man was a time traveller. This explains why his clothes had no labels and why no one could identify him. He was from another time. This could even explain his oddly shaped feet. Perhaps pointy shoes are fashionable for men in his time. It could be that his odd collection of tools were required for maintaining his time machine. Was the code necessary for his return journey?

It's even been suggested that perhaps "Jestyn" was related to Somerton Man (perhaps his great-grandmother), and that's why she was unwilling to reveal more about herself.

Unfortunately, there's no evidence to suggest Somerton Man was a time traveler. And besides, time travel is probably impossible.

Weirld.com[60] **Credibility Rating 1 out of 5**

2. Somerton Man was a genetic anomaly, human–alien hybrid, or experiment.

Somerton Man had some unusual genetic features with his anodontia, large cymba, and well-defined calves. The postmortem pictures also seem to show unusual facial features. It's been suggested that perhaps Somerton Man was genetically altered in some way or even that he wasn't human—perhaps even a human–alien hybrid. This could explain the apparent physiological anomalies.

But, although anodontia can be associated with congenital diseases and syndromes, it is present in around two percent of the general population. The same is true of the comparably large cymba. And his calf muscles could be large for a variety of reasons, including just being extremely fit.

Without genetic testing, it's impossible to rule out the possibility that he was a genetic anomaly, but it seems unlikely. The odd appearance of his face is more likely due to postmortem changes in appearance, bloating, and the effects of embalming.

Weirld.com Credibility Rating 1 out of 5

Greenwood mentioned these theories briefly in her book. To her credit, she also gave them very low credibility ratings.

H. C. REYNOLDS

There are several examples of fresh leads arising in extremely cold cases, when massive attention is brought to the case by an author. In the case of the BTK killer, Robert Beattie deserves a good deal of credit for drawing the murderer out, after decades of inactivity, to the point that police were able to capture and convict him. Although Robert Graysmith's two books on Zodiac, and the film based on the first, did not lead to the killer's identification, they resulted in many new leads arising in the case. As was previously mentioned in chapter 5, thanks to leads provided by their massive audience over the course of more than a thousand episodes, *America's Most Wanted* announced a thousandth capture on the show's official website on May 2, 2008.[61]

Another show claims comparable success:

> *Unsolved Mysteries* has profiled more than one thousand cases in 260 episodes. Over half the cases featuring wanted fugitives have been solved, most as a result of viewer tips. Also, more than one hundred families have been reunited as a result of "Lost Love" segments on the show, and several cases involving missing heirs, murder, fraud and amnesia have also been solved. In seven cases, individuals wrongly convicted of crimes have been released following UM broadcasts.[62]

For the case of the unknown man, media attention had always been present—it was a famous case in Adelaide—but was intensified by the appearance of retired detective Gerry Feltus's book *The Unknown Man*. After this, a woman in Adelaide who wished to remain anonymous came forth with an ID card that had been among her father's belongings. Was the man pictured on the card below destined to die on Somerton Beach in 1948?

This "Seaman's Identification Card" was issued by the United States in 1918 and shows an eighteen-year-old of British nationality. This would make him forty-eight when the unknown man was found, a nice fit with the age estimated for the deceased. A more impressive fit is made when comparing ears and a mole.

FIG. 6.21 A solution to the Unknown Man's identity?

In October 2011, the "unknown woman" who found the ID card took it to Professor Maciej Henneberg, a specialist in biological anthropology and recognized expert in identity verification. Henneberg concluded, "Together with the similarity of the ear characteristics, this mole, in a forensic case, would allow me to make a rare statement positively identifying the Somerton man."[63]

The unknown woman also contacted Australia's *Sunday Mail*. The paper requested searches at the U.S. National Archives, U.K. National Archives, and Australian War Memorial Research Centre, but no further relevant records could be found. If Reynolds is not related to the unknown woman's father, one has to wonder why the man had his ID. Did he take it off of Somerton Man's dead body and keep it as a trophy?

Nick Pelling, whose work on the Voynich manuscript was mentioned in chapter 1, considered the Reynolds identification unlikely, after carrying out his own (somewhat more successful) investigation into Reynolds's life.

MORE LEADS!

But the leads keep coming, and not in the expected manner! Detective Brown thought that someone might reveal information about the unknown man when another member of the household died. But, as mentioned previously, this wasn't the case. When the nurse's husband passed in 1995, she maintained her silence, and the odds against resolving the mystery increased. Surely any bookie would have increased the odds against finding out the truth yet again when the nurse died in 2007, yet years later, another lead arose. It's almost a confession from beyond the grave.

The nurse's daughter, Kate Thomas, claimed in November 2013 that her mother admitted to her that she lied to police and did in fact know the identity of Somerton Man. She made this claim in an interview with the Australian version of *60 Minutes*.[64] Unfortunately, the nurse did not reveal the unknown man's name to her daughter. All she said was "that it is a mystery that was only known to a level higher than the police force."

The narrator described the nurse as a "communist sympathizer," and Kate confirmed that she thought her mother was a spy. She recalled her mother once mentioning, after meeting some Russian immigrants, that she could still understand Russian. However, she wouldn't explain where she learned it when her daughter naturally asked. Kate didn't express affection for her mother in the interview, but instead remarked, "She had a dark side, a very strong dark side." She even went so far as to say, "There's always that fear that I've thought maybe she was responsible for his death."

Kate thinks that Somerton Man was also a spy and that her mother had an affair with him. Although Kate was fathered by Prosper Johnson, Jessica was initially raising Kate's older brother Robin as a single mother. The boy, born in July 1947, was sixteen months old at the time of the unknown man's death.

The *60 Minutes* program brought even more potential relatives of the unknown man into the picture. Roma Egan married Robin and believes that his father was, in fact, Somerton Man. She bases her opinion on the physical resemblance mentioned previously—Robin and Somerton Man had the same ear and dental abnormalities. If Roma is correct, her daughter Rachel is the unknown man's granddaughter. Rachel and Roma both want to know the truth and favor an exhumation and DNA testing.

This is complicated by the fact that the unknown man was embalmed with formalin.[65] He was actually the first person in South Australia to have this new embalming technique applied, and it would make the testing more difficult. The best chances of obtaining viable DNA would be from his teeth, bones, or hair, rather than the degraded soft tissue.

DNA testing can determine if Rachel is a descendant of the unknown man and can tell us what part of the world he came from, through comparison with databases containing samples from around the world. However, it cannot reveal the unknown man's name because there isn't even any suspect DNA with which to make a comparison. Complicating matters further, Kate Thomas doesn't want the body exhumed.

There's yet another problem. According to University of Adelaide physicist Dr. Derek Abbott, the clock is ticking. He commented in the *60 Minutes* piece that "The problem is if we leave the Somerton Man any longer, if we let him go by another ten years, we could lose some of the non-chromosomal information in the DNA, which we really need to link to close relatives."

Well before these potential relatives came forth, Abbott requested that the body be exhumed. However, Attorney General John Rau rejected his first request in October 2011, writing that there needed to be "public interest reasons that go well beyond public curiosity or broad scientific interest."[66] Believing that the potential relatives strengthen his case, Abbott appealed to Rau again in November 2013. He wrote,

> The imperative to identify this unknown man is on par with the current practice of identifying unidentified WWI and WWII graves for bringing closure to their families, and there is a considerable general public interest in the case to do so.[67]

> In terms of specific public interest, there is a potential descendant of the unknown man living in Australia and I am now able to verify a compelling likelihood of this based on both historical proximity and anatomical evidence.[68]

Alas, Abbott's luck was no better the second time. Rau declined his request again. Feltus, who spent a tremendous amount of time on the case, reflected on this decision in a posting on his website:

> I received a letter dated 03/06/2014 from The Honourable John Rau Deputy Premier Attorney-General advising that after a number of considerations he had declined to grant consent to Professor Abbott to exhume the remains of the unidentified man. I must say that I, along with a large number of people agree with his decision. This is the second time Derek Abbott has had his application to exhume the body of the "Unknown Man" declined. I do hope that he finally accepts the decision and devotes his expertise to his chosen career.[69]

Abbott is not motivated by any sort of conspiracy theory. In fact, he even rejects the idea that the unknown man was a spy. He said, "Everyone loves the spy angle but I can't find any evidence whatsoever. You can access old ASIO records through the national archives—and we have done a run on the nurse's names and Alf Boxall—and nothing comes up. Everything in that era is now out in the open. It just doesn't stack up as a spy thing."[70]

I respectfully disagree. An example illustrates why. James Bamford filed a Freedom of Information Act (FOIA) request with the National Security Agency (NSA) in the United States for any files they might have on him. The result of the search was that there was nothing. Bamford later learned that he himself had been given a code name. When he filed a request for files containing the code name, he got what he was looking for! Perhaps the intelligence community in Australia has something, and it just hasn't been asked for in the right way yet. Or maybe the records were destroyed or lost, as was the case with so much of the material from the police investigation.

Whether the DNA testing is ever carried out or not, we still have hope for an eventual identification of the unknown man. If his identity really was known at a higher level, as Kate reports her mother claiming, then Australian declassification efforts may reveal it publicly in years to come. And if he was a Soviet spy ... The Cold War is over, and much material has come forth from the archives of formerly communist nations. Perhaps a solution will arise from that side of the story. My prediction is that his identity will eventually become known publicly, but that his cipher will remain ambiguous.

Case 2: Cryptology and Cyanide

AT THE AIRPORT

A bit before 10:00 a.m., on the morning of January 20, 1953, U.S. Army Private Madrid King was walking to the Philadelphia International Airport, when he discovered the body of a man at the bottom of a steep twelve-foot ditch.

About twenty minutes before his body was found, the man was seen walking in the area by Rev. Robert M. Anderson, who described him as a "stocky man" "between 25 and 30 years old"[71] and "wild-eyed."[72] Anderson described the encounter as follows:

> He was on a ramp between the terminal building and the air express office. I said, "Good morning." He did not reply, but continued pacing up and

down. I noticed then that he was staring straight ahead and that the pupils of his eyes were dilated.[73]

In another newspaper article, Anderson said "he seemed nervous."[74]

Did this agitated man know his life was about to end?

This is a case for which we have few answers, and many more questions. In fact, even the basic facts outlined above are uncertain. According to another account, it was a taxicab driver who found the body. Smaller details also vary. For example, the ditch was also described as ten feet deep.[75] Many more contradictory claims are presented in the pages that follow in what should be simple matters of fact.

The man in the ditch had no identification on him. Coroner Joseph Ominsky and Sgt. James McTague, who headed the police intelligence unit, estimated the man's age as twenty-five to thirty years. His height was reported variously as five feet ten inches, or six feet, or somewhere in between. He weighed 190 to 200 pounds, and his hair and eyes were both brown. There were no signs of violence. The man's sports jacket had a label reading "Witte Brothers, 5th Avenue."[76] This led investigators to New York.

DEATH BY CYANIDE

An autopsy was unable to determine the cause of death, but the chemists had no doubts.

Dr. Edward Burk, of the police crime laboratory, discovered signs of potassium cyanide poisoning. According to the coroner, Joseph Ominsky, there was enough in his body to "kill 10 men."[77] Death was instantaneous. Ominsky also said, "This may not be a suicide. The cause of death may not have been self-inflicted. The body was too immaculately arranged; even with his thick-lensed glasses in proper position on his face. This could not be possible if he had fallen. It gives rise to the possibility that the body might have been placed there."[78]

Without elaborating further, Ominsky noted that the cyanide "was not of the usual type—it was mixed with some other agent."[79]

A white handkerchief was found on the bank of the ditch, ten feet from the bottom. Did this fall out as the man was being carried to the bottom of the ditch?

Besides the man's size, there was another memorable detail, the thick-lensed glasses mentioned by Ominsky. Descriptions varied:

"thick plastic-frame glasses"[80]

"very thick horn-rimmed glasses"[81]

"tortoise shell glasses"[82]

But the thickness must have been extreme. Detective McTague said, "His glasses were odd. They were so thick they were almost magnifying glasses."[83]

If you are up to no good and are concerned about potential witnesses offering a description of you to authorities, one solution is to be sure to have some feature that is so dramatic it will be remembered to the exclusion of all else. Odd glasses are one possibility.

How the cyanide was administered remained a mystery.

It was reported, "No vial or container of any kind was found at the scene or in the youth's pockets, according to Ominsky."[84] But the next day, an account described a "five-inch long test tube" that was found about five feet from the body.[85] How could this have been missed? Oddly, no traces of poison were found in the tube, and no cork or stopper was located.

Yet another account claims "Police said a broken glass vial was found on the man."[86]

Ominsky said that the poison could have been taken in capsule or powder form. He also remarked, "So far, we have not been able to find evidence to indicate a capsule was used, and Dr. John S. Shinn will perform a second and more thorough autopsy on the body to try and discover capsule traces."[87] The deadly dose was apparently not taken in capsule form.

The FBI sent the man's fingerprints to Washington and New York, based on the sports coat label, but no matches were found.

A HIDDEN CIPHER

Police obtained a clue from another part of the man's body. There was a six-inch-long, two-inch-wide strip of tape on his abdomen. When it was removed, a piece of paper that was rolled up under the tape fluttered to the floor. Upon unrolling, the seven-inch by three-inch piece of paper revealed an enciphered message.[88]

ENTER THE FBI

A pair of names that appeared in the cipher, in apparent plaintext, caused police to contact the FBI. The names were Dulles and Conant. John Foster Dulles was designated by President Eisenhower to become the new Secretary

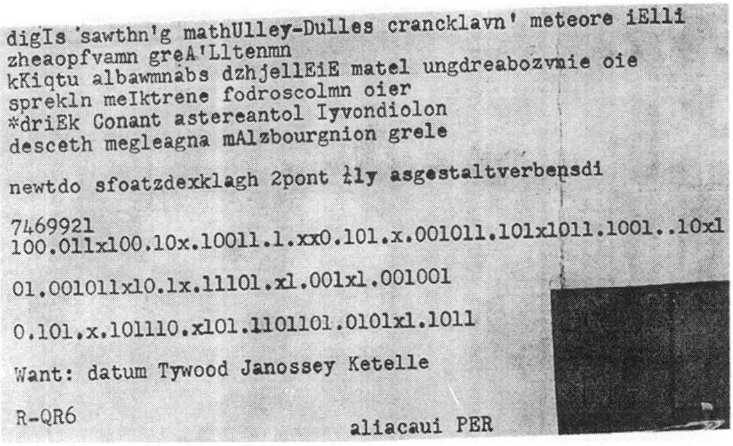

```
digIs 'sawthn'g mathUlley-Dulles crancklavn' meteore iElli
zheaopfvamn greA'Lltenmn
kKiqtu albawmnabs dzhjellEiE matel ungdreabozvmie oie
sprekln meIktrene fodroscolmn oier
*driEk Conant astereantol Iyvondiolon
desceth megleagna mAlzbourgnion grele

newtdo sfoatzdexklagh 2pont tly asgestaltverbensdi

7469921
100.011x100.10x.10011.1.xx0.101.x.001011.101x1011.1001..10x1

01.001011x10.1x.11101.x1.001x1.001001

0.101.x.101110.x101.1101101.0101x1.1011

Want: datum Tywood Janossey Ketelle

R-QR6
                                        aliacaui PER
```

FIG. 6.22 The cipher found taped to the mystery man's abdomen

of State at the time, and Dr. James B. Conant was President of Harvard University and slated to become American High Commissioner in West Germany.

These identifications, repeated without variation in every newspaper article on the case, are no more certain than other "facts" in the case. NSA historian David Hatch pointed out other possible identifications for "Dulles," namely John Foster's brother Allen, incoming Director of Central

FIG. 6.23 A U.S. postage stamp from 1960 honoring Dulles

FIG. 6.24 James B. Conant as he appeared on February 18, 1953

Intelligence (DCI), and Eleanor Dulles, the State Department desk officer for West Germany.[89]

MORE CONNECTIONS WITH GERMANY

Another connection to Germany was found in the man's fake alligator skin wallet.[90] Instead of a picture of a girlfriend, there was one of an airplane with a Nazi swastika on its tail assembly. The photo carried the notation "France Field, Panama" on the back. The only other picture in the wallet was of Rodin's famous sculpture *The Thinker*.

If the mystery man committed suicide with the cyanide, it fit in with how top Nazis ended their lives less than a decade earlier. Erwin Rommel used it in 1944, and Heinrich Himmler, Adolf Hitler, and Eva Braun all used it in 1945, although Hitler also shot himself for good measure. Hermann Göring used it in 1946. Horrifyingly, Goebbels had it given to his six children. He then took it himself (which we don't mind).

Recalling the lack of response the man gave to Rev. Anderson's greeting of "Good morning," we have a possible explanation, if the man had a strong German accent that he did not want anyone to hear.

OTHER POSSESSIONS

Other items found on the man's body included the following:[91]

1. a plastic cylinder containing a signal fuse;
2. the casing of a spent .38 caliber bullet (found in a pocket of his topcoat);

3. a fountain pen gun;
4. forty-seven cents; and
5. four keys on a chain, one possibly for a Buick automobile.[92]

Item 3 deserves a closer look! Sgt. James McTague provided a description:

> There was also a fountain pen or pencil gun. We're checking to find out just what it is. It looks like a cross between a gas gun and something else. It's blunt on both ends, with a firing pin attached.[93]

It had to be a pretty rare item if a police sergeant had to check to find out what it was!

A much less suspicious item found with the man was the February issue of *Galaxy Science Fiction*. It was reported that "Officers said that one article in the magazine, entitled "Mystic Numbers," and written by a Dr. Bell, was concerned with cryptography."[94] E. T. Bell wrote science fiction under the pseudonym John Taine, but he didn't have an article in the given issue, nor was anything dealing with codes or ciphers contained therein. It was also reported that, according to detectives, Conant's and Dulles's names appeared in this issue in different stories, but this is incorrect as well.[95]

AN IDENTIFICATION

Based on the estimated age of the man, he would have been somewhere between seventeen and twenty-two at the end of World War II. The physically imposing mystery man could have been a war-hardened veteran who went on to serve the fatherland as a spy in America. He was finally positively identified. It turns out that he was a teenage Jew from Brooklyn. Why he had a picture of a Nazi aircraft in his wallet has never been explained.

The identification came about when Bessie Rubin of Brooklyn read about the body being found in a New York paper on January 20. Her son had disappeared on Monday, January 19. She last saw him at 9:30 a.m. on that date.

Bessie called the Philadelphia police and said that the description of the man found at the airport was similar to that of her son. She mentioned that her son had thick-lensed eyeglasses, as did the dead man, and said he was five feet eleven inches and weighed 175 pounds. This was a bit lighter than the reported weight of the mystery man, and at eighteen, Bessie's son was far younger than the police believed the deceased to be. Bessie said that her son

FIG. 6.25 A science fiction magazine
whose contents don't match reports

looked older than his age. The clothing worn by the deceased also somewhat matched her son's attire.

It was reported that pictures of the corpse were taken at the city morgue and "transmitted by the local FBI office via wire photo machine to New York headquarters," so that agents there could show them to Bessie and possibly get a definite identification.[96]

However, the next step that is documented is Bessie, her husband Samuel, and her brother Max Gerstman arriving in Philadelphia. Samuel identified the body of his son in the Philadelphia morgue at 3:00 p.m. on January 21. Bessie collapsed before viewing the body.

With the man identified as eighteen-year-old Paul Emanuel Rubin,[97] one mystery was solved, but another remained—what was the teen doing?

According to his parents, Rubin had had a habit since he was a boy of sending enciphered messages to friends. So he may have been the creator of the cipher, rather than a recipient. Also, Rubin's mother recalled seeing her son cut some strips of adhesive tape before leaving home that morning. However, neither parent could make any sense of his last cipher.

The police spoke with an unidentified friend of Paul Rubin, in the hope that he could explain the cipher.

> The friend told police that he, Rubin, and other young men had been working with codes and that he probably could decipher the message found on Rubin's body if he could find the proper code books to work with. He said it might take him a week. "They're very complicated," he added. "Anyone who reads science fiction will know what I mean." ... The friend had no explanation for Rubin's death and said he doubted that the note would give any answer.[98]

It is not known how much effort the unidentified friend exerted attempting to break the cipher, but the FBI office in Philadelphia, at least, took it seriously and forwarded it to its cryptanalysts in Washington.

The parents believed that their son was heading off to New York University when he left the house Monday morning. He was a sophomore there studying chemistry. He almost certainly either had access to cyanide or knew someone who did. Thus, suicide would seem to be a possibility. But Bessie said that her son was in good mental and physical health when he left home and did not appear to be worried about anything. Also, she said that Rubin had fifteen dollars with him at that time. Why did he only have forty-seven cents left when he was found?[99] Where did $14.53 go? And what about the other odd items found on his body?

Samuel and Bessie provided explanations for some of the items.

> The parents said the long cylindrical fuse, with wire attached, which was found in Paul's pockets, may have been a device he used in one of the magic tricks with which he experimented frequently.[100]

> They identified one of the four keys on a key-ring found in his pocket as belonging to an old automobile the family no longer owned. An empty shell cartridge also found in the same pocket, they said, was probably one he had taken from a friend, interested in target-shooting, to reload for him with his own mixture of gunpowder.[101]

In addition to magic tricks, Bessie said that her son was fond of "heavy literature" and "airplane lore."[102] The latter could offer some explanation of the picture of the plane with the swastika in his wallet. But why that one of all possible planes, and what was the meaning of the notation on the back?

Other details Bessie revealed about her son were that he was a brilliant student, an avid chess player, and interested in foreign affairs. He was familiar with many foreign students in New York City and often spoke of his friendships with those at New York University (NYU). His parents said that he never brought any of these foreign students to their home. He had, however, "frequently brought home jars filled with strange liquids," according to Bessie, who said she had no knowledge of their contents.[103]

Conspicuously absent from the parents' explanations was any comment about the fountain pen gun. With that and many other important questions left unanswered, Paul Rubin's body was sent back to Brooklyn for an Orthodox Jewish burial.

THE NYU CONNECTION AND OTHER LEADS

On January 23, *The Philadelphia Inquirer* reported that Sergeant Michael Schwartz of the Philadelphia homicide squad said New York police had been asked to talk to NYU students and faculty to see if Rubin made it to campus or to class Monday. Another paper followed up on this, reporting, "Police said they had learned that some of his student friends saw him cut a strip of adhesive tape with a scissors shortly before he vanished."[104] This is a bit strange. Rubin's mother saw him with the tape before he left the house. Did it come off and have to be redone?

Other valuable information came from the NYU interviews. On January 27, Ominsky raised the possibility that Rubin may have committed suicide over a hazing incident. He said that he and Detective Frank A. Levinthal had many "interesting conversations" with NYU faculty members and friends and relatives of the deceased. Ominsky thought it possible that Rubin was depressed over failing to be accepted by a fraternity he hoped to join.[105]

Launching my own investigation decades later, I contacted NYU archives, asking for any information they might have on the case. Claire Ashley Wolford, a graduate assistant, replied, "I've taken a look through some newspapers and found nothing about this incident in school papers." She went on to mention an article in the *New York Times*[106] that indicated Rubin was at the Washington Square College of Arts and Sciences, which she explained was a separate campus from the University Heights campus at the time.[107] I had imagined that the case would have made for much bigger news, even if it was a separate campus. It was even covered in *Life*![108]

Ominsky pursued other avenues of investigation. A quote from him follows:

Keeping in mind my opinion that this death is no suicide, I call upon the FBI to explore the connections of the deceased in New York and Philadelphia to determine if he was a messenger for some unusual activity. That, I assume, is being followed through and until this question is answered I cannot set a time for an inquest.[109]

A tip, if accurate, provided part of Rubin's itinerary:

A truck driver telephoned Detectives Edward Schriver and Frank Kelly he had picked up a youth answering to Rubin's description on Monday at Baltimore Pike and Church Lane, East Lansdowne. The hitchhiker told the driver he was going to the inauguration in Washington. He was dropped off at Chester, the driver said.[110]

This only raises more questions. If Rubin intended to go to the inauguration, why wouldn't he tell any of his friends or family? If any of them came forth with such information, it was not presented in any newspaper articles I could find.

The inquest was finally held on February 23, 1953, but none of the big questions were answered. Ominsky, conducting the inquest, said that no motive had been established and that Rubin died under "circumstances presently unknown." No evidence showed how Rubin got to Philadelphia or how the cyanide was administered.[111]

There are other questions that could be easily answered, but if they were, I've been unable to find the results. For example, Paul Rubin had a typewriter, and investigators said that the cipher would be checked against it to see if he was the composer or the recipient. With his habit of making ciphers, it seems likely that he was the composer, but a definitive answer would be nice.

One newspaper article reporting on the inquest actually served to make the death even more mysterious by contradicting previously unchallenged details of the case. It reported, "The youth, who lived at 122 Vernon St. Brooklyn, was found by Harry Kessler, of 973 N. 6th st., a taxi driver, in a ditch near the airport Jan. 20."[112]

All previous accounts agreed that Rubin was found by an army private walking to the airport. If the private had been traveling to the airport in a cab, then this alternate story would make a bit more sense, but it was always reported that he was on foot!

The same newspaper article reports that Edwin S. Schriver, a homicide squad detective, said at the inquest that Rubin was failing his classes and was in danger of being expelled. Schriver was convinced it was a suicide. But Samuel Rubin, the teen's father, said that his son started college at age sixteen, "hardly did any homework," yet was a good student and was in good spirits when he left home the day before.[113]

ALAN TURING

During World War II, England's top cryptology expert was Alan Turing. He played an important role in breaking Nazi Enigma ciphers, and the intelligence that was gathered as a consequence made a huge difference. It shaved at least a year off the war and may even have been a game changer. In the 1950s, all of this information was still highly classified. Although Turing was greatly regarded for his mathematical work, the general public was completely unaware that he was a war hero. In 1952, he was arrested for being a homosexual and was given a choice of hormone treatments to "cure" him or a prison sentence. He chose the treatments. The official story is that he became depressed, put cyanide on an apple, ate it, and died. Turing's mother never accepted her son's death as a suicide, and there are conspiracy theories surrounding it.

Aware of the conspiracy theories, though not actually believing them, when I first learned about Rubin's strange cyanide death, I had to do a quick check of dates. Rubin died on January 20, 1953, and Turing died on June 7, 1954. So the deaths were separated by about seventeen and a half months, not close enough in time, in my opinion, to be significant.

A paper reporting on Rubin's death noted, "About 40 to 50 deaths a year from cyanide are reported in New York City. Most of them are suicides."[114] Worldwide the numbers would be even more distressing. With such a large number of cyanide deaths, we are sure to find elements in common between some, but this does not imply a connection. The paper also stated, "cyanide has been used by a number of famous scientists to commit suicide, principally because it kills so quickly."[115]

Attempts to substantiate a conspiracy in the death of Turing continue. For example, Peter Tatchell wrote to Prime Minister David Cameron requesting a fresh investigation. Tatchell obviously wanted to make a strong argument in support of his request, but honesty apparently compelled him to comment, "Although there is no evidence that Turing was murdered by state agents, the fact that this possibility has never been investigated is a major failing."[116]

MORE TO COME

Henry Langen, past president of the American Cryptogram Association (ACA), singled out the Rubin cipher as the greatest challenge he faced. At least I presume it was the Rubin cipher. Notice how what he relates differs from our case in extremely important ways, yet has enough in common that it is unlikely to be a separate case.

> Cryptology has been a fascinating hobby for me for the past 22 years, and I've spent up to six months on a single message before I could decipher it. There's one I've never cracked—and neither has the FBI and all the other experts who have worked on it. It turned up several years ago, when the body of a young man was fished out of the Delaware River near Philadelphia. There was a bullet hole in his back, between his shoulder blades, so it couldn't have been a suicide. And taped across his bare stomach was a coded message![117]

> After local cryptologists failed to decipher it, it was given to the FBI and various skilled cipher-hobbyists around the country. I gave up on it at the end of two years, but as far as I know others are still working away. If and when it's solved, it may explain a murder.[118]

We might be intimidated to make our own efforts against the cipher, considering how much effort was expended by others with no payoff, but I don't believe anyone has worked on it since the 1950s, back when graph paper and a number 2 pencil were the cryptanalyst's main tools. It turned into a forgotten case. I think there's a good chance that a reader making use of today's computing power can crack it.

With all the contradictions in the newspaper accounts, another avenue of investigation is highly desired. Interestingly enough, one of the papers provided such a lead.[119]

We find something of great potential value to researchers typed at the top of this photo: "Cipher Case File # 540226-15." This is an FBI case file number. Unlike Somerton Man, the Rubin case quickly dropped from the public interest. It was possible that the police and FBI files on the case were long lost. If they still existed, they were likely collecting dust for decades. With this number and a few 1950s newspaper articles, I filed

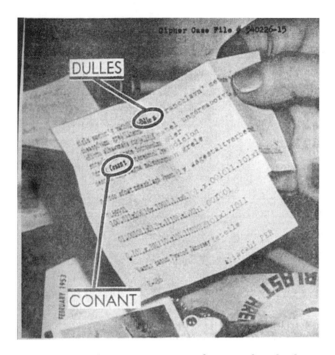

FIG. 6.26 A newspaper image that provides a lead

Freedom of Information Act (FOIA) requests with the Philadelphia police, the FBI, and NSA. The police claimed not to retain files so old, and the NSA search didn't yield anything. However, the FBI replied, directing me to material at the National Archives and Records Administration (NARA). The public may access a tremendous amount of material at NARA, but the Rubin files had not yet been reviewed for release. Naturally, I requested a review, and a short time later received a response that included the following paragraph:

> You requested access to Federal Bureau of Investigation (FBI) Headquarters Case File 65-HQ-61458, which was identified by the FBI as potentially responsive to your request for files pertaining to Paul E. Rubin. We located the file in box 1123 of record entry 569603 (UD-07D 2); identified as Record Group 65: Records of the Federal Bureau of Investigation; Classification 65 (Espionage), Headquarters Case Files, 1936–1978. Paul Emanuel Rubin is the subject of the file, which consists of two parts (Section 1 and Section Sub A) and totals an estimated 160 pages. The records were compiled as

part of an internal security investigation conducted between January 1953 and March 1953.

The description of the file, by itself, tells us a great deal! Would the FBI devote three months to investigating the case, generating 160 pages in the process, if it was a simple suicide? It seems unlikely. The paragraph above even tells us the FBI classified the case under espionage.

The cover letter was followed by sixteen pages from the file in question. Sadly, they were all copies of newspaper articles. The file needed to be reviewed before it could be released, but the newspaper articles were obviously not going to present a problem, since they already existed on the outside.

As much as we would like to know if the FBI made any progress on Rubin's cipher, we can only wait and hope the rest of the file is approved for release. However, we can, in the meantime, attack it ourselves.

The numerical portion of the cipher is unlike anything I've seen before. I look forward to hearing readers' thoughts! For now, I show how we can form a conjecture about the alphabetic portion from a few simple statistics.

If we count the letters only and ignore case, the top half of the cipher (before getting to the mostly numerical part) gives frequencies that closely match normal English, as the graph in fig. 6.27 shows.

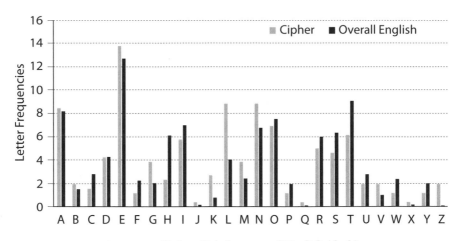

FIG. 6.27 Cipher (light) vs. overall English (dark)

The match is not perfect. We have, for example too few of H and T, and too many of K, L, and Z in the cipher. The missing T and H are explained if the message was written in telegram style. Because people sending telegrams typically paid by the word, they would often leave out THE where it would appear in normal English. The meaning would remain clear, and a bit of cash would be saved. Leaving this word out repeatedly would depress the frequencies of the letters contained in it. In the cipher, E is nevertheless a bit more frequent than normal. Still, this hypothesis seems reasonable. With most of the cipher letters having a close match to their frequencies in normal English, and two of the exceptions having a simple explanation, the obvious conclusion would be that we are dealing with a transposition cipher. But there's another statistic to consider!

Notice that blank spaces were included in the ciphertext. Ignoring the punctuation and numbers again, the beginning of the ciphertext consists of 261 letters spread over 31 "words." So we get an average word length of 8.4 letters. This is way too high. As was mentioned in chapter 1, the average word length in English is only 5.1 letters.[120] Transposing the letters and spaces does not change this average length.

A hypothesis that accounts for this statistic is that nulls were inserted into the ciphertext. Recall that nulls are letters that have nothing to do with the message and are ignored by the intended recipient. Although any letter might be a null, the composer of the cipher may have used K, L, and Z a bit too often, making them more frequent in the cipher than they are in normal English. Perhaps by coincidence, if you look at a keyboard (or a 1950s typewriter), you will see that the second line of letters ends with K and L. The letters then continue on the third line with Z. Extending this group of three letters in each direction on the keyboard, we have J K L Z X. All five of these letters are overrepresented in the cipher, as compared with normal English frequencies. Maybe it isn't a coincidence!

If I had a complete solution, the Rubin cipher would be solved, and hence, not included in this book! The statistics above lead me to suspect a method, though. My best guess is that the encipherer took the original message and did not change any letters, but rather inserted many nulls, roughly in accordance with their frequencies. That is, he used common letters as nulls more often than rare letters but, for some reason, overused J, K, L, Z, and X. This method of encipherment caused the average word length to greatly increase, while giving the illusion of a transposition cipher.

Deleting possible nulls, we can obtain many different messages. The first line could be, for example, I SAW DULLES RAN TO ELI. I don't believe this is the correct solution, but I would be interested in seeing other possibilities achieved in this manner.

One word in particular jumped out at me. "mAlzbourgnion" is close to Mauborgne, the codiscoverer of the one-time pad. Perhaps an extra layer of security was added by misspelling some words before inserting nulls.

Also, it is possible that "Conant" might not be a name, but rather "Cant" or even "Cat," "Cot," or "Can."

Some readers may notice a bit of German mixed into the cipher. We find "sprekln," which resembles "sprechen" (to speak) and "asgestaltver-bensdi," which contains the German words "gestalt" (form) and "ver-ben" (verbs). If the cipher included liberal use of nulls, the encipherer could have gotten playful and inserted nulls that combined with message letters to make actual words in English and German.

It is my conjecture that Somerton Man used too few letters and his message will never be read, whereas Rubin used too many (perhaps a hundred nulls). I think Rubin's message will be recovered. Are you the one who will meet this challenge?

Perhaps the message hidden in this cipher will shed light on why Paul Rubin never made his classes on January 20, 1953.

Update! As this book was being copyedited, I received the FBI's file on the case in response to my FOIA request. There were only minor redactions, protecting the identity of a CIA employee who became involved in the investigation. Some new twists were revealed, such as Rubin's use of an alias, but despite the FBI's best efforts, they were unable to crack the cipher. Rather than summarize the file, I have had it placed, in its entirety, on the website Princeton University Press maintains for this book: http://press.princeton.edu/titles/10949.html. The interested reader is encouraged to download it and investigate further.

Case 3: An Illiterate Encipherer?

IN THE GHETTO

Our third victim, Ricky McCormick, was born on June 14, 1958, grew up poor, and had essentially no hope of being able to improve his situation. Frankie Sparks, Ricky's mother, described him as "retarded." A cousin, who was very close to Ricky, suspected that Ricky was bipolar or had schizophrenia. From childhood on, he had to deal with asthma and chest pains. Barely able to read or write, he somehow made it to Martin Luther King High School in St. Louis, but he never graduated. As a dropout, only menial jobs were available to him. He held a variety of these, but also received disability checks because of chronic heart problems. His habit, begun around age ten, of smoking at least a pack of cigarettes a day didn't help his health, nor did his drinking about twenty caffeinated beverages per day.

He seems to have initially tried to avoid criminal activity. While in his teens, and even as an adult, he would often take a bus or hitchhike to get away from neighborhood thugs who started fights and dealt drugs near his home.

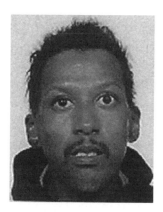

FIG. 6.28 Ricky McCormick

Although he remained nonviolent, with his low IQ and limited options, he made some horrible choices. St. Louis police arrested him in November 1992. McCormick was then thirty-four years old and had fathered two children with an underage girl. He had been sleeping with her since she was eleven. McCormick pleaded guilty and received a three-year prison sentence.[121] After thirteen months, he was released.[122]

Major Tom O'Connor of the Major Case Squad said that McCormick actually had at least four children, although he didn't reveal anything about the mother(s) of the other two that he knew of. McCormick never married.

McCormick had a St. Louis address, but also ones in nearby Belleville and Fairview Heights, as well as several in Illinois. A person who bounces around many addresses is sometimes actually homeless and simply staying where he can. However, St. Charles County Sherriff's Sergeant Kevin Wilson said that McCormick was unemployed and on disability welfare, but not homeless.

LAST DAYS

By the late 1990s, McCormick found more trouble. He was making trips to Florida, and his girlfriend, Sandra Jones, said he was picking up marijuana. He brought the marijuana into an apartment that he and Jones had in a housing project, where it was divided up into Ziploc bags. He told Jones that he was holding them for Baha Hamdallah. All indications are that McCormick didn't abuse drugs or even alcohol himself, but Baha, who went by "Bob," was more dangerous to be entangled with than either.

Contradicting Wilson's claim that McCormick was unemployed, Christopher Tritto, in the most detailed article on the case, reported that McCormick worked for Baha at the Amoco gas station at 1401 Chouteau Avenue in downtown St. Louis. Perhaps McCormick took on an honest job and then learned that Baha had a better paid position to fill as well.

In any case, Baha had a violent history. In 1997, a police officer saw him shoot at another man from his car. Baha was arrested but not prosecuted. In March 1998, he allegedly shot his brother Bahjat following an argument. Bahjat claimed that he had never seen the shooter before and gave a description that contradicted other witnesses, including one from a man who knew Baha. (Two months after McCormick's death, Baha was allegedly shot by another Hamdallah brother, Juma. This time, it would seem, it was Baha's turn to not press charges.) Also in March 1998, Baha was arrested for beating a man with a hammer. Adding a bit of variety to the violence, the victim was not a relative, and no shots were exchanged. However, the victim was gunned down two weeks before Baha was set to appear in court about the beating. Thus, Baha avoided conviction. Police informants said the murder was "at the behest of Baha Hamdallah." Yet another Hamdallah brother, Jameil, is a registered sex offender.[123]

Although some states have now legalized marijuana, in the 1990s it was banned in all fifty states. McCormick was not only facing harsh penalties from the authorities if caught transporting the drug, but he was also risking the wrath of his boss, Baha Hamdallah, if he screwed up, and other drug dealers, if they tired of the competition.

His last trip to Orlando, Florida, began early on June 15, 1999, when he bought a one-way ticket at the downtown St. Louis Greyhound bus terminal.

McCormick called the Amoco gas station at least once while in Orlando. Jones didn't say what she thought that call might have been about, but she did indicate that she thought McCormick might have done something wrong on this last trip. She said that he seemed scared when he got back to St. Louis.

His actions upon returning to St. Louis matched those of a man afraid to be found. Although his meanderings in the St. Louis area have not been reconstructed in full detail, what was learned gives an indication of his unease.

On the afternoon of June 22, McCormick entered the emergency room of Barnes-Jewish Hospital claiming that he had chest pains and breathing difficulties. He was admitted and kept for two days before being released on June 24. His next stop was his Aunt Gloria's apartment. He took the bus there but only stayed for about an hour.

McCormick went to another emergency room (Forest Park Hospital) at about 5:00 p.m. on June 25, again claiming that he had trouble breathing. If he was hoping to spend the night, he was disappointed, as he was released in less than an hour. His Aunt Gloria heard that he spent the night in the waiting room anyway. Normally, people can't wait to get out of the hospital. Why would he stay if he wasn't afraid to go home?

The next morning, around 11:30 a.m., Jones spoke with McCormick on the phone. He said that he was heading to the Amoco gas station to get food. It is unclear whether he got there that day or not, but he was there on June 27, according to an employee who spoke with police.[124] This is the last place that police can place McCormick before his death. Medical examiners were positive that he died sometime that day. If McCormick was on the run from Baha, as his preference for spending the night in a hospital waiting room instead of home seems to indicate, why did he finally return to the gas station? Perhaps he received false assurances that any mistakes he might have made in Orlando had been forgiven.

If she heard any such stories, Sandra Jones kept them to herself. When she learned of her boyfriend's death, she suspected Baha, but she did not provide any details of how exactly she thought he got at McCormick. Perhaps he had someone else carry out his dirty work this time.

A police informant said that Gregory Lamar Knox was responsible for the murder. Knox was a drug dealer whose territory included the housing project where McCormick lived. He was also a suspect in several other homicides, so the informant's claim seemed reasonable. The police had a possible connection between Baha and Knox, but they needed more. In spite of all their efforts, they were unable to gather enough evidence to arrest Baha or Knox (or anyone else) in connection with the McCormick case.

But more evidence might come from McCormick himself.

McCORMICK'S BODY

On June 30, 1999, a woman driving on a field road near U.S. Route 67 saw a body near a cornfield and reported it to police. Despite decay, fingerprints identified the deceased as forty-one-year-old Ricky McCormick.[125] He was face down in the field, which was in St. Charles County, twenty miles from his home and job in downtown St. Louis. Had he reverted to his habit of fleeing town to avoid trouble? With his health problems, he might then have died of natural causes. But if he hitched a ride, no witnesses have come forth to say that they provided it; if he took a bus and then walked a bit, there is also no evidence of it. Still, authorities did not rule out the possibility that his poor health led to his death.

The body of an alleged prostitute was found in an abandoned house on the same section of U.S. Route 67 in 1995. In that instance, it was definitely not a natural death, as her body was riddled with bullets. In 2001, the naked bodies of two more women were found three hundred yards from where McCormick was discovered. Was there any connection among these cases? Could the same psychopath have murdered McCormick?

Exposure to the elements took a heavy toll on McCormick's body. It was a suspicious death, but after an examination, pathologists declared the cause of death to be "undetermined."

By this point, we are used to contradictions, and yet another was provided concerning McCormick's injuries. On June 30, when he was found, authorities said that he seemed to have a head injury. But the next day, O'Connor claimed that the body had suffered too much decomposition to determine what injuries he may have had.

McCORMICK'S CIPHER

Although his body was too far gone to yield much information, two pages of ciphertext found on McCormick's body were legible. Investigators kept the existence of these pages secret but made no progress in deciphering them. Lieutenant Craig McGuire of the St. Charles County sheriff's office said that his office and the Major Case Squad solicited help from the FBI within a year of McCormick's death.

By this time, the FBI codebreakers were no longer in Washington, D.C., but rather in Quantico, Virginia. The FBI agreed with the St. Charles County sheriff's detectives, who thought that McCormick had been murdered, but despite their great expertise, they had no better luck cracking the ciphers.

The FBI is fond of top ten lists. They even have one for unsolved ciphers, and the McCormick cipher made that list. At this time, the only one of the ten that was known to the public was the Zodiac cipher seen in chapter 5. The other nine were all kept secret.

Years passed with no solution, and finally the FBI shared the ciphers with members of the American Cryptogram Association at their national convention. Still, no solution was found.

The FBI's next step was to ask everyone. On March 29, 2011, they released the cipher and provided a few basic details of the case:[126]

On June 30, 1999, sheriff's officers in St. Louis, Missouri discovered the body of 41-year-old Ricky McCormick. He had been murdered and dumped in a field. The only clues regarding the homicide were two encrypted notes found in the victim's pants pockets.

Despite extensive work by our Cryptanalysis and Racketeering Records Unit (CRRU), as well as help from the American Cryptogram Association, the meanings of those two coded notes remain a mystery to this day, and Ricky McCormick's murderer has yet to face justice.

"We are really good at what we do," said CRRU chief Dan Olson, "but we could use some help with this one."

In fact, Ricky McCormick's encrypted notes are one of CRRU's top unsolved cases. "Breaking the code," said Olson, "could reveal the victim's whereabouts before his death and could lead to the solution of a homicide. Not every cipher we get arrives at our door under those circumstances."

The more than 30 lines of coded material use a maddening variety of letters, numbers, dashes, and parentheses. McCormick was a high school dropout, but he was able to read and write and was said to be "street smart." According to members of his family, McCormick had used such encrypted notes since he was a boy, but apparently no one in his family knows how to decipher the codes, and it's unknown whether *any-one* besides McCormick could translate his secret language. Investigators believe the notes in McCormick's pockets were written up to three days before his death.

Over the years, a number of CRRU's examiners—who are experts at breaking codes—have puzzled over the McCormick notes and applied a variety of analytical techniques to tease out an answer. "Standard routes of cryptanalysis seem to have hit brick walls," Olson noted. Our cryptanalysts have several plausible theories about the notes, but so far, there has been no solution.

To move the case forward, examiners need another sample of McCormick's coded system—or a similar one—that might offer context to the mystery notes or allow valuable comparisons to be made. Or, short of new evidence, Olson said, "Maybe someone with a fresh set of eyes might come up with a brilliant new idea."

That's where the public comes in. The FBI has always relied on tips and other assistance from the public to solve crimes, and although breaking a code may represent a special circumstance, your help could aid the investigation. Take a look at McCormick's two notes. If you have an idea how to break the code, have seen similar codes, or have any information about the Ricky McCormick case, send them to us online at http://forms.fbi.gov/code or write to CRRU at the following address:

FBI Laboratory
Cryptanalysis and Racketeering Records Unit
2501 Investigation Parkway
Quantico, VA 22135
Attn: Ricky McCormick Case

There is no reward being offered, just a challenge—and the satisfaction of knowing that your brain power might help bring a killer to justice.

"Even if we found out that he was writing a grocery list or a love letter," Olson said, "we would still want to see how the code is solved. This is a cipher system we know nothing about."

As with the Rubin case, contradictory statements in simple matters of fact appeared in the press. The request for help issued by the FBI referred to the cipher being "found in the victim's pants pockets," but according to a newspaper account, "In his [McCormick's] shirt pocket, investigators found two coded notes."[127]

So were the notes in his jeans or in his shirt?

The FBI request also noted that McCormick "was able to read and write" and "According to members of his family, McCormick had used such encrypted notes since he was a boy." This was backed up by McGuire, who said,

"We asked the family, and they said he did it quite often. Nobody really knows what it means. It's kind of like private diary writing."[128]

Yet, a newspaper reported "family members say they never knew of Ricky to write in code. They say they only told investigators he sometimes jotted down nonsense he called writing, and they seriously question McCormick's capacity to craft the notes found in his pockets."[129] McCormick's mother, who only found out about the cipher when the FBI went public with it years after it was discovered, said, "The only thing he could write was his name. He didn't write in no code." And McCormick's cousin, Charles McCormick, said the man "couldn't spell anything, just scribble."[130]

Bearing all of this in mind, we now examine the ciphers.

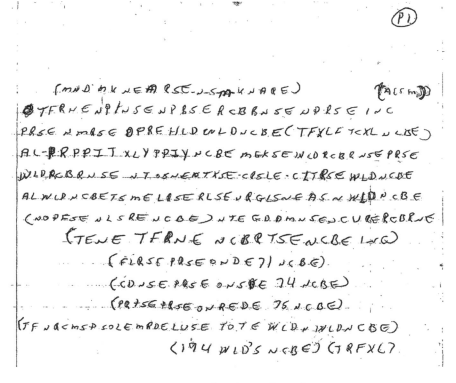

FIG. 6.29 A page of McCormick's cipher

The three cases covered in this chapter are in not only chronological order, but also "Goldilocks and the Three Bears order." The first cipher had too few letters retained, the second had too many, and the third? The third was just right.

NOTES

FIG. 6.30 The other page of McCormick's cipher

The notes seem to display elements of secret languages and simplified phonetic spellings. For example, "MLSE" could be code for "miles."[131]

A nice explanation for another part of the note is that TRFXL = Traffic Light. Also, at the bottom of the first page reproduced above, we have WLD'S. The letter S frequently follows an apostrophe in normal English. If the note represented a cipher, be it substitution or transposition, the apostrophe would just as likely be followed by any other letter of the alphabet.

So, if McCormick's ciphers really are just highly abbreviated English, should we be able to read them? Well, it depends on what percent of letters is missing. In the case of Somerton Man, where only initial letters appear to have been left, we cannot be certain about obtaining the correct solution—there are too many possibilities. However, if more than half of the letters are present in McCormick's cipher, we should be able to read it. This is because English is about 50% redundant. Consider the following sentence with some letters (indicated by underscores) removed.

```
T_E Q_I_K B_O_N F_X J_M_S _V_R T_E L_Z_ D_G.
```

The original had thirty-five letters, and I removed fourteen of them, but you were likely able to read it without much difficulty. In other words, twenty-one out of the original thirty-five letters, or 60%, was enough to keep it readable.

Several examples follow as challenges to the reader, with varying percentages of the letters removed. How far down the list can you read? The first one is long, to help you get warmed up!

25% Removed:

```
HER_'S TO _HE C_AZY _NES. _HE M_SFI_S. TH_ REB_LS.
T_E TR_UBL_MAK_RS. T_E RO_ND P_GS I_ THE _QUA_E
HO_ES. T_E ON_S WH_ SEE _HIN_S DI_FER_NTL_. THE_'RE
N_T FO_D OF _ULE_. AND _HEY _AVE _O RE_PEC_ FOR
_HE S_ATU_ QUO. _OU C_N QU_TE T_EM, D_SAG_EE W_TH
T_EM, G_ORI_Y OR _ILI_Y TH_M. AB_UT T_E ON_Y TH_NG
Y_U CA_'T DO _S IG_ORE _HEM. _ECA_SE T_EY C_ANG_
THI_GS. T_EY P_SH T_E HU_AN R_CE F_RWA_D. AN_ WHI_E
SO_E MA_ SEE _HEM _S TH_ CRA_Y ON_S, WE _EE G_NIU_.
BEC_USE _HE P_OPL_ WHO _RE C_AZY _NOU_H TO _HIN_
THE_ CAN _HAN_E TH_ WOR_D, AR_ THE _NES _HO D_. -
APP_E IN_.
```

25% Removed:

```
IF Y_U WA_T YO_R CH_LDR_N TO _E IN_ELL_GEN_, REA_
THE_ FAI_Y TA_ES. I_ YOU _ANT _HEM _O BE _ORE
_NTE_LIG_NT,  R_AD  T_EM  M_RE  F_IRY  _ALE_
- ALB_RT E_NST_IN
```

33% Removed:

MY _OU TR_ IS _HE _OR D, A_D M_ RE_IG ON _S T_ DO
_OO_ - TH_MA_ PA_NE

33% Removed:

I E_JO_ GE_TI G P_OP E A GR_ AN_ GE_TI G U_DE NE_TH
_HE_R S_IN, _SP CI_LL_ PE_PL_ WH_ DO_'T T_IN_.
- JE_LO _IA_RA

50% Removed:

I _O_'T _O _R G_. I _M _R G_. - _A V_D R _A I

50% Removed:

H_L_ O_ L_F_ I_ F_C_IG U_, T_E _T_E_ H_L_ I_ D_A_I G
_I H _T. - _E R_ R_L_I_S

Remember, it doesn't matter how long it takes to fill in any of the above. If it can be done at all, even if it takes months, the missing letters are redundant!

In some instances, removing a very small percentage of letters can make the message ambiguous. Consider ATTACK FROM THE _O_TH AT NOON. Only two out of twenty-four letters were removed. We have more than ninety-one percent of the message, but we do not know if the attack is to come from the NORTH or the SOUTH.

Another sort of example in which a small percentage of deleted letters leaves us with an ambiguous message is the following:

MY FAVORITE MUSICIAN IS _ _ _ _ _   _ _ _ _ _ _.

Here, I deleted the last two words, which were of lengths five and six. With twenty letters remaining, we have more than sixty-four percent of the message, and you will certainly be able to find meaningful solutions to finish the sentence, but you will not know for sure if the answer you obtained is what I intended.

For McCormick's cipher, we don't have conveniently placed underscores to let us know the locations of the deleted letters, but in another way, the solution is easier to find than for some of my earlier challenges because of context.

Reconstructing my examples, you had no context within which to work. For the McCormick ciphers, you can expect names of streets and such from a specific area. On the other hand, he likely misspelled many words, in addition to leaving letters out. I encourage you to have a go at it anyway! It will be interesting to see if different readers can obtain similar solutions. Good luck!

At the beginning of this chapter, I summarized the plot of Anthony Boucher's short story "QL696.C9." To refresh your memory, a murdered librarian typed out the call number given in the story's title as her life drained away, and police narrowed the suspect list to junior librarian Stella Swift, children's librarian Cora Jarvis, library patron James Stickney, and high school teacher Norbert Utter. You were asked to determine who the murderer was. One way to obtain the answer is by considering the subject that would be covered by a book with the given call number. "QL" refers to zoology, the 600s cover vertebrates, with 696 designated specifically for birds, and .C9 addresses Cypseli (swifts). Hence, Stella Swift was the killer. But the more elegant solution requires neither having the Library of Congress classification system memorized nor taking the time to look it up. Presuming that the subject indicated by the call number ties in to the killer's name, it would have to be Swift, for the other names (Jarvis, Stickney, and Utter) fail to correspond to anything that would be categorized under the system.

7

From Beyond
the Grave?

Although I'm skeptical of the ability of the victims in the previous chapter, or any other deceased individuals, to ever be able to communicate with investigators, others have attempted to find ways to prove that such communication is possible. Interestingly, their methods have sometimes made use of cryptology and have created more challenges in the form of unsolved ciphers.

The Great Escape

Perhaps the first example was that of Harry Houdini. Early in his career, he established a secret code with his wife Bess so that they could convey messages to each other without anyone in the audience being aware of it. This is the code he would have used if he was able to communicate with her after his death:

```
Pray = 1 = A
Answer = 2 = B
Say = 3 = C
Now = 4 = D
Tell = 5 = E
Please = 6 = F
Speak = 7 = G
Quickly = 8 = H
Look = 9 = I
Be quick = 10 or 0 = J
```

In this code, each of the first ten letters and numbers are represented by words. So, for example, the word BAD could be represented, letter by letter, as "Answer, Pray, Now." This code might look insufficient to express words with letters beyond J, but the numbers make it possible. To get an S, which is the nineteenth letter, Houdini simply used the words indicated by 1 and 9. That is, "Pray-Look" represented S. After Harry Houdini's untimely death on October 31, 1926, Bess waited for a message in this code.

Arthur Ford often gets credit for contacting Houdini, but how legitimate is this claim? The first points Ford scored weren't from using this code. Rather he came up with the word "FORGIVE," allegedly from Houdini's mother, who died before he did. Bess knew that this was an important word for Houdini, one that he longed to hear from his mother, and some took this as confirmation that the message was genuine, but dates are important here.

Ford relayed the "FORGIVE" message on February 8, 1928. Almost a year before this, on March 13, 1927, the *Brooklyn Eagle* had quoted Bess as saying that any authentic message coming from her husband would have to include the word "forgive."

Ford later conveyed a message allegedly from Houdini himself. It was

Rosabelle, answer, tell, pray-answer, look, tell, answer-answer, tell.

"Rosabelle" was a song Houdini's wife used to sing. The rest of the words followed the code she shared with her husband and spelled out BELIEVE.

Bess accepted this as a genuine message from her deceased husband, but once again dates are important. Bess had not kept the code secret. She revealed it to Harold Kellock, who included it in an authorized biography that saw print in 1928, well before Ford conveyed the BELIEVE message on January 8, 1929.

Bess later changed her mind about the communication being real, but stories of Ford's success persist.

Thouless's Three Ciphers

The story of cryptology and the afterlife continued in the 1940s with Cambridge don Robert H. Thouless. He created an enciphered message that he thought could not be solved by anyone not possessing the key. He then

FIG. 7.1 Robert H. Thouless (1894–1984)

planned to try to communicate the key to some living person after his own death. The event of someone obtaining the correct key would then, according to Thouless, confirm communication from beyond the grave.

We'll call Thouless's encrypted message, which he published in *Proceedings of the Society for Psychical Research*, Cipher A. It's reproduced here:[1]

```
CBFTM HGRIO TSTAU FSBDN WGNIS BRVEF BQTAB QRPEF BKSDG
MNRPS RFBSU TTDMF EMA BIM
```

All Thouless revealed about it was that

> it uses one of the well-known methods of encipherment with a key-word which I hope to be able to remember in the after life. I have not communicated and shall not communicate this key-word to any other person while I am still in this world, and I destroyed all papers used in enciphering as soon as I had finished.[2]

Thouless wasn't unaware of cryptanalysis. After expressing concerns about someone obtaining the key while he was still alive by reading his mind (psi cognition), he wrote

A further doubt may be felt as to whether the key might not be found by the normal process of rational inference, since it is commonly supposed that a sufficiently skilled cipher expert can decipher any passage without knowing the key if he is given long enough time. The idea is however, erroneous unless certain conditions are fulfilled. The cipher must be a simple one (such as a mono-alphabetic monographic substitution) or else the passage enciphered must be of considerable length. A short passage enciphered by a simple type of substitution does not indeed need an expert; the reading of such passages is often made a puzzle to amuse children. A long passage (or a number of passages) enciphered by a more complex method is more difficult to decipher without knowledge of the key, but it is said that an expert can do so with messages enciphered by most (but not all) systems if he is given sufficient time. The passage I have given is neither enciphered by a simple process of substitution or transposition nor, I think, long enough for it to be possible to use the methods adapted for breaking the less simple systems. I do not think that it could be deciphered be a cipher expert without knowledge of the key while it could easily be deciphered by anyone with an elementary knowledge of ciphers if he were told the system and given the key.[3]

Although he doubted the success of straightforward cryptanalysis, Thouless did consider the possibility that someone might find a process through which his cipher could be turned into a meaningful message that he had never intended. (Recall the false decipherments we saw in previous chapters for the Voynich manuscript, the Dorabella cipher, and Zodiac's 340 cipher.) To guard against this, Thouless revealed that his message was "an extract from one of Shakespeare's plays."[4] So, although an erroneous decipherment might be found, Thouless figured the chances were minuscule that it would happen to be a passage from Shakespeare.

Trying to cover all the bases, Thouless prepared for the possibility that he might not remember his key after death, despite his best efforts. To address this, he had a clue placed in a sealed envelope and deposited with the Society for Psychical Research. The envelope would only be opened if attempts following his death failed without it.

As it was, there was no need to open it, for a solution came well before Thouless's death. An unidentified "cipher expert" took Cipher A as a challenge and came up with the solution after working on it, in his spare time, for two weeks.

This was not Thouless's first failure. Another even earlier cipher he created was broken by Denys Parsons Esq., but this one hadn't been published, so there was no public embarrassment over it.[5] In his paper, Thouless thanked "D. Parsons" and another unidentified individual for demonstrating how a running key cipher can be broken. From this, we can conclude that his very first failed attempt was of this type. Such ciphers are discussed in chapter 8. In any case, the early attempt was broken in forty-eight hours.

Although Thouless didn't specify what method was used for Cipher A, it was a Playfair cipher. Such ciphers make use of scrambled alphabets written in 5 × 5 grids. The scrambling is often accomplished with a keyword or key phrase. For example, if the keyword is MACHETE, we have the following grid:

M	A	C	H	E
T	B	D	F	G
I/J	K	L	N	O
P	Q	R	S	U
V	W	X	Y	Z

It's normal to combine I and J in the grid, or to omit J, or to omit the rarest letter, Z, in order to get down to a twenty-five-letter alphabet that can fit into a 5 × 5 grid.

To see how this grid can be used to encipher, consider the message

IT'S TOO HOT FOR CLOTHES.

As a preliminary step, any doubled letters should be broken up by placing an X between them. In the message above, we have to change only TOO to TOXO. The reason for this will be made clear later.

The message is next split into pairs of letters:

IT ST OX OH OT FO RC LO TH ES

Then some simple rules are followed to encipher each pair of letters. The rules depend on the relative positions of the letters in our grid. There are three possibilities:

1. Both letters appear in the same row.
2. Both letters appear in the same column.
3. The letters are from different rows and different columns.

Our first pair, IT, falls into case 2. In this case, we simply take the letters appearing below the message letters. That is, IT enciphers as PI.

Our second pair, ST, falls into case 3. In this case, we find the smallest rectangle within our grid that contains S and T and use the other two corners as our ciphertext pair. Both the message and cipher pairs are **boldfaced** below, with the cipher pair also **underlined**, to make them stand out, and make the small rectangle clear.

M	A	C	H	E
T	B	D	**F̲**	G
I/J	K	L	N	O
P̲	Q	R	**S**	U
V	W	X	Y	Z

To eliminate the ambiguity of whether FP or PF should be the ciphertext pair, the rule specifies that we begin with the letter that appears in the same row as the first letter in our plaintext pair. Thus, we have PF in this case. If our plaintext pair was TS, our ciphertext pair would be FP.

We continue on, using these rules and get the ciphertext pairs LZ NE IG GN and XD.

Case 1 doesn't arise until we arrive at the plaintext pair LO. These letters appear in the same row, so we take the letters appearing to the right of the message letters. So, L becomes N, but there is nothing to the right of O, so we have to wrap around (as in a Pac-Man video game) to the other side of our grid to get the letter I. That is, the pair LO enciphers as NI.

Our complete ciphertext is PI PF LZ NE IG GN XD NI FM HU.

The reason that we broke up the double O in TOO before enciphering is to avoid the necessity of a fourth case. A doubled letter would be an instance of both case 1 and case 2, so it would be unclear which rule should be followed. Also, breaking up such patterns is typically a good idea anyway!

Because the Playfair cipher replaces the letters two at a time, it's known as a digraphic substitution cipher. It's a leap forward from monoalphabetic substitution ciphers (MASCs), but it's still weak. Repeated letter pairs may sometimes encipher differently, depending on their positions in the message (see the example below), but they'll often encipher identically. Thus, just as single-letter frequencies help us defeat MASCs, digraph frequencies help to defeat the Playfair cipher.

Example: Consider the Danny Trejo quote "EVERYTHING GOOD THAT HAS HAPPENED TO ME HAS HAPPENED AS A DIRECT RESULT OF HELPING SOMEONE ELSE."

To encipher this with the Playfair system, we first break up every doubled letter with an X:

EVERYTHING GOXOD THAT HAS HAPXPENED TO ME HAS HAPXPENED AS A DIRECT RESULT OF HELPING SOMEONE ELSE

Then we break the letters into pairs.

EV ER YT HI NG GO XO DT HA TH AS HA PX PE NE DT OM EH
AS HA PX PE NE DA SA DI RE CT RE SU LT OF HE LP IN GS
OM EO NE EL SE

The word HAS appears twice in the quote and is enciphered in two different ways because of how the letters got split into pairs. But HAPPENED also appears twice and is enciphered the same way each time. We don't need to complete the enciphering process to see that this is a weakness. If an attacker can correctly guess a word that appears repeatedly in the message, he or she then stands a good chance of correctly lining it up with repeated ciphertext letters and using this to begin unraveling the rest of the cipher.

Whatever attacks Thouless may have been aware of, he made it clear that he thought his message was too short for any of them to be applicable. He was wrong. Thouless identified neither the cipher expert who broke it, nor the techniques he used, but some observations can be made on weaknesses that could have led to the solution.

The first thing to observe is that any keyword can be used to scramble the grid of letters used for enciphering, but many words do not contain the letters V, W, X, Y, and Z. When one of these words is used, the last line of the grid preserves these letters in order. That is, the grid has the form

```
?   ?   ?   ?   ?
?   ?   ?   ?   ?
?   ?   ?   ?   ?
?   ?   ?   ?   ?
V   W   X   Y   Z
```

The positions marked by ? are still unknown, but twenty percent of the grid is correctly identified. A cryptanalyst sometimes assumes that this partially filled in grid is correct, gambling on the keyword not containing V, W, X, Y, or Z. In the case of Thouless's cipher, the cryptanalyst wins the bet.

Actually looking at Cipher A suggests some more possibilities. The cipher was

```
CBFTM HGRIO TSTAU FSBDN WGNIS BRVEF BQTAB QRPEF BKSDG
MNRPS RFBSU TTDMF EMA BIM
```

Splitting this into pairs of letters, we get

```
CB FT MH GR IO TS TA UF SB DN WG NI SB RV EF BQ TA BQ
RP EF BK SD GM NR PS RF BS UT TD MF EM AB IM
```

We see that some digraphs (BQ, EF, SB, and TA) appear twice. These are likely to represent high-frequency digraphs in normal English. A lot of trial and error may be involved in matching these up correctly, but cryptanalysts are patient. And there's at least one more weak spot that can be exploited.

The ciphertext pairs include what's known as a reversal. We see both BS and SB. In normal English, some pairs of letters are common in one order, but very rare, or even nonexistent, in the other order. Consider QU and UQ, for example. A pair of letters that stands out from the rest by being of high frequency in either order is ER/RE. So, the attacker is wise to assume that BS represents ER and SB represents RE. It could just as well be the other way around (BS represents RE and SB represents ER). As it turns out, the first possibility is correct. Looking at reversals is a useful technique in attacking a variety of cipher systems. In chapter 5, I pointed out how it could be of value against the unsolved ciphers made by Henry Debosnys.

Having identified three weak spots in Thouless's cipher, we're still far from reading it. There are many paths the cipher expert may have taken from here, and he almost certainly made some guesses that turned out to be wrong (the decipherment did take two weeks). The ideas outlined above just give an idea of how the attack is likely to have begun and show that the cipher was not nearly as secure as Thouless thought.

The cipher expert eventually determined that the keyword was SUR-PRISE. This was the word that Thouless had planned to communicate from beyond the grave, allowing his message to be revealed with much less effort. Forming the enciphering/deciphering square with this keyword, we have

```
S  U  R  P  I    (The letter J was left out by Thouless.)
E  A  B  C  D
F  G  H  K  L
M  N  O  Q  T
V  W  X  Y  Z
```

Applying this to Cipher A gives

```
Cipher   CB FT MH GR IO TS TA UF SB DN WG NI SB RV EF BQ TA
Message  BA LM OF HU RT MI ND SG RE AT NA TU RE SX SE CO ND


Cipher   BQ RP EF BK SD GM NR PS RF BS UT TD MF EM AB IM
Message  CO UR SE CH IE FN OU RI SH ER IN LI FE SF EA ST
```

Inserting word spacing and appropriate punctuation, gives

```
BALM  OF  HURT  MINDS  GREAT  NATURE'S  X  SECOND  COURSE
CHIEF  NOURISHER  IN  LIFE'S  FEAST.
```

The X appeared only in order to separate the repeated letters SS. The message was, as promised, from Shakespeare. It's given below, in context, as it appears in *Macbeth*.

> Sleep that knits up the ravelled sleave of care
> The death of each day's life, sore labour's bath
> Balm of hurt minds, great nature's second course,
> Chief nourisher in life's feast.

Thouless knew that the Playfair cipher couldn't be expected to hold secure for long messages, but he thought his message, at only sixty-five letters, was short enough to be safe. Whoever broke it didn't set any records, though. For Playfair ciphers, the records belong to Alf Mongé.

After spending the first nineteen years of his life in his native country of Norway, Mongé came to America and served in the army. During World War II, he spent a great deal of time working for the British in London. In appreciation of this, the British government awarded Mongé the Order of the British Empire. Before the war, back in 1936, Mongé broke a Playfair cipher of only thirty letters that had been issued as a challenge three years earlier. Mongé was able to solve Playfair ciphers just a bit shorter than Thouless's, ones with fifty to sixty letters, in just thirty minutes.

The question of how short a cipher needs to be, in a given system, to resist decipherment by having more than one key that yields a meaningful plaintext was addressed in chapter 5. Recall that we termed this length the unicity distance and represented it by U in the formula $U = \log_2(K)/D$, where K is the number of possible keys and D is the redundancy per letter of the message. To apply this formula to a Playfair cipher, we have to decide in how many ways the grid can be scrambled. If we allow scrambling only by real words, then K depends on the size of the dictionary the words come from. At an extreme, the *Oxford English Dictionary*, second edition (1989), which came in twenty volumes with a total weight of 137.72 pounds (62.6 kilograms), had 615,100 word forms defined and/or illustrated.[6] Applying the formula for this dictionary gives $U = \log_2(615,100)/0.7 \approx 27$ letters. The true value would be a bit lower because there are instances of distinct words that result in the same Playfair grid. One example is the pair *CARE* and *CAREER*. Using all possible scramblings of the grid, we get $U = \log_2(26!)/0.7 \approx 126$ letters. So you see what a difference a big keyspace can make!

For modern military ciphers, an important design criterion is to have a unicity point that exceeds the length of the messages that will be sent using the system.

Plan B

With his cipher demolished, Thouless wasn't out of the game. He had actually prepared for this eventuality. In an appendix to his paper titled "A Supplementary Test," Thouless wrote

> It is theoretically possible, although I think unlikely, that a sufficiently skillful cipher expert may decipher the passage I have given as my first test without previous knowledge of the key if he works at the problem hard enough. The cipher system used is not theoretically "unbreakable" although the shortness of the passage will be, I hope, an insurmountable obstacle to the breaking of it. It seems, therefore, worthwhile to add a second message in an absolutely unbreakable cipher system, *i.e.* in one which cannot be deciphered without knowledge of the key although the system of ciphering is fully known. A system fulfilling this condition, which also fulfils the condition that the key can be easily communicated cannot be a simple one. I will give the passage and then explain the system used and the kind of key that should be communicated after my death.[7]

Thouless's Cipher B was created in a manner completely different from his failed attempt, Cipher A, and is reproduced below.[8]

```
INXPH CJKGM JIRPR FBCVY WYWES NOECN SCVHE GYRJQ TEBJM
TGXAT TWPNH CNYBC FNXPF LFXRV QWQL
```

In contrast to Cipher A, Thouless revealed his method for creating this cipher. It was based on a sixteenth-century system known as the Vigenère cipher, although Blaise de Vigenère (1523–1596) was not the actual inventor. Vigenère ciphers are discussed in chapter 9, but for the current chapter, only Thouless's modification is relevant.

In Thouless's version, the key was simply some published work. Thouless took some string of words from this work and eliminated words that already appeared earlier in the string. His original string would have to have been

long enough to yield at least seventy-four distinct words after eliminating du-
plicates, for each word would then be used to encipher a single letter of his
seventy-four character message.

So, how can a word be used to encipher a single letter? The first step was
to convert each word to a number. To do this, each letter of the word was
replaced by its numerical value using the scheme A = 1, B = 2, C = 3, . . .,
Z = 26. Then, all of these values were added together. For example, BACON
would yield 2 + 1 + 3 + 15 + 14 = 35. Any time Thouless got a result larger
than or equal to 26, he would reduce it by 26. Thus, 35 becomes 35 − 26 = 9.
For a long word, 26 might have to be subtracted more than once. Mathematicians refer to this process as reducing modulo 26 or "mod 26" for short. It's
equivalent to dividing the given number by 26, ignoring how many times it
goes in evenly, and then simply jotting down the remainder. So, following this
process for each word, Thouless would obtain a sequence of numbers between 0 and 25, inclusive. Suppose that his message was GREETINGS FROM
THE DEAD and the numbers obtained from the words of the published work
he used to generate his key were 14, 8, 21, 5, 2, 11, ... Then he would line
them up like so

```
 G    R    E    E    T    I    N   . . .
14,   8,  21,   5,   2,  11,  17  . . .
```

and shift each letter forward in the alphabet by the amount indicated by the
number written below it. This process gives

```
 U    Z    Z    J    V    T    E   . . .
```

The last calculation above, N + 17 = E, requires a bit of explanation. If we
count forward from N, we reach Z upon getting to 12. We can't go beyond Z,
but we must count to 17. Therefore, we start back at A and move forward
through the alphabet again. When we get to 17, we're at the letter E, which
is our ciphertext letter. Another way to think about this process is to imagine
the alphabet written as a big circle, like the numbers on a clock face. Adding a
number to a letter just causes the clock hand to move that many positions to
some other letter.

To avoid ambiguity, Thouless instructed that "Words joined by a hyphen are
treated as two words" when generating the key.[9] This, perhaps unintentionally,

provided a small clue. Why would Thouless think to make such a specification if his key generator did not contain a hyphenated word?

One might also assume that the quotation behind Cipher B would again be from Shakespeare. But before flipping through your copy of the First Folio looking for passages containing hyphenated words, consider the following passage from Thouless's paper:

> It is not necessary to give any indication of the nature of the original passage. The possibilities of arbitrary decipherment are unlimited, but the distinction between a correct and incorrect decipherment is sufficiently indicated by the fact that the key-letter series in a correct decipherment must be a series of letters obtained in the way described above from some identifiable passage in a printed work. That more than one such passage would give a meaningful interpretation of the enciphered passage is almost infinitely improbable. My aim will be to indicate after my death the passage from which the key series of letters has been derived.[10]

So, Thouless was unconcerned about the "almost infinitely improbable" event of an erroneous decipherment for Cipher B. That means we can't expect the plaintext to be from Shakespeare again. However, it's unlikely that he would pick an extremely obscure work. If he were able to communicate the title to someone after his death, he would then want that person to be able to find the work without too much trouble so that the message could be deciphered and the communication from the afterlife confirmed.

Thouless encouraged others to create ciphers, as he had, and for the same purpose, but Denys Parsons thought the method of Cipher B involved too much calculation for many people. Therefore, he suggested that others use the simpler scheme of selecting a passage to serve as the key, but not using all of it (as a running key cipher would—see chapter 8), but rather taking every third letter. These letters would then be converted to their numerical equivalents and used to encipher in the same manner as in Cipher B's last step.

Thouless recognized that taking every third letter from some meaningful text would not yield a random string, but rather one with frequencies matching normal English. However, he mistakenly stated that frequencies of two- or three-letter combinations would be "near enough to randomness to be of no help in breaking the cipher."[11] Thouless's unidentified cipher expert thought this simplified system was "quite insecure."[12]

Thouless did not propose a Plan C, or Cipher C, at this time, in case both Cipher A and Cipher B were broken. Part of his reason for not doing so went beyond his confidence in the security of the first two ciphers. He wrote, "I think it would reduce the chance of the success of the experiment if I tried to burden my memory with a third key."[13]

Two keys seems to have been the magic number for Thouless. In the aftermath of Cipher A being broken, Thouless came up with a replacement. We'll call it Cipher C. He presented it in a paper in 1948, immediately after Cipher A was broken.

Cipher C was an improved version of the method used for Cipher A. Known as a double Playfair cipher,[14] the method begins with encipherment using Playfair with a grid scrambled according to some keyword, but then continues with more steps. Following this initial encipherment, some particular letter is then placed both in front of and at the end of the ciphertext. The slightly lengthened ciphertext is then enciphered again using Playfair, but with a different keyword used to scramble the grid. The reason for adding the letter to the start of the ciphertext before the second round of Playfair is to break the previously enciphered pairs in half. The letter added at the end is simply to keep the total number of characters even. There was no need for the two additional letters to be the same, but they were. And this created a weak spot! In any case, when Thouless applied these steps to his message, his final cipher was[15]

```
BTYRR OOFLH KCDXK FWPCZ KTADR GFHKA HTYXO ALZUP PYPVF
AYMMF SDLR UVUB
```

This is what he published as his last cipher, which I'm calling Cipher C. As the decades passed, both Cipher B and Cipher C remained unbroken. Thouless left the world of the living in 1984, at age 90, and the real challenge/experiment began.

There must have been much disappointment among Thouless's loved ones and his colleagues in psychical research, for all attempts at communicating the keys to his ciphers were deemed failures. More years passed until, in 1996, James J. Gillogly and Larry Harnisch obtained the key to Cipher C. They announced that the two Playfair keywords were BLACK and BEAUTY. The letter added to both ends of the first Playfair encryption was T. The message was revealed to be

```
THIS IS A CIPHER WHICH WILL NOT BE READ UNLESS I GIVE
THE KEYWORD.
```

The paper by Gillogly and Harnisch explaining how they learned the key-words ended with the following paragraph.

The Survival Research Foundation offered a $1000 reward to the first person to send the key to either cipher B or C to their offices, but unfortunately the Toshiba laptop was not ectoplasmically receptive—or even built—when the offer expired in 1987. Our successful computational séance must be its own reward.[16]

In case you didn't recall the name Gillogly from chapter 5's discussion of his solutions to IRA ciphers, I remind you now that he's a cryptanalyst and not a medium. Thus, as of 1996, Thouless has one chance left, Cipher B, to prove that something of him survives and can communicate.

Despite their masterly solution of Cipher C, Gillogly and Harnisch couldn't crack Cipher B. They wrote that they "believe that solving this cipher will require finding the right text."[17] They made a mighty attempt, testing hundreds of books, but the one Thouless actually used was apparently not among them. Perhaps someone can take a psychological approach, gaining a deep enough understanding of his personality to figure out what published work he used to generate his key.

Back in 1948, Thouless couldn't anticipate the incredible power that computers would bring to bear on cracking ciphers. What new tools might the future offer?

Inspired by Thouless

Although Thouless's experiments failed, he did succeed in inspiring others to create ciphers of their own for the same purpose. Thomas Eugene Wood, a retired British solicitor, was among them.[18] He chose wisely in picking the method of Cipher B for his own encryption.[19] His prize was a bit smaller, though. He offered £20. Wood published his cipher in 1950 as this:[20]

```
FVAMI NTKFX XWATB OIZVV X
```

Wood noted,

> The key passage to the decipherment of my message is in an accessible book. After my death, I shall try to communicate the key passage. I shall also try to communicate in what foreign language the key passage is. The message, if and when deciphered, will not be in any one language. I shall also try to communicate what languages are used in the message. The use of foreign languages is to meet the point mentioned in the paragraph commencing "Additional Note" on page 262 of the [Thouless] article as to listing the commonest words in the English language and so to increase the difficulty of decipherment.[21]

Like Thouless, Wood destroyed all papers used in the enciphering process and did not reveal his key to anyone.[22] There is a letter, though, in which he gives a sample solution.[23] Although this is not the correct solution, it shows the range of possibilities he envisioned. His sample key was "To be or not to be, that is the question," but not as Shakespeare wrote it in *Hamlet*. Instead, Wood used a Spanish translation. With this key, the message was revealed as KUBBA DAJES UISIC IHOLZ. It isn't obvious where to insert word spacing here to get a meaningful message, but we must remember that Wood indicated the message would not be in English. The correct spacing is KUBBADA JE SUIS ICI HOLZ. The first word is Urdu for "Attention," the next three are French meaning "I am here," and the last is German for "Wood," which served as a signature. So, the recovered message is "Attention. I am here.—Wood."

A cryptanalyst trying the right key could easily miss the solution, if he didn't recognize any of the languages used. Presumably, someone confident that he or she had really communicated with Wood after his demise would take the time to look up the words in various foreign language dictionaries or to consult with linguists. If you're investigating only one proposed solution, you can spend much more time on it!

Wood, who was born in 1887, died in 1972, and his cipher remains unsolved, casting great doubt on his being able to communicate from beyond the grave. Thouless must have been saddened by this failure, but he lived to see another researcher take up his cause.

Like Wood, Frank Tribbe was a lawyer, but his career was carried out in the United States. His investigations weren't limited to considerations of the afterlife. He wrote the books *Creative Meditation* (1975), *Portrait of Jesus? The*

FIG. 7.2 Frank C. Tribbe (1914–2006)

Illustrated Story of the Shroud of Turin (1983), *Guidebook for the Study of the Paranormal* (*with Robert H. Ashby, 1988*), *Denny and the Mysterious Shroud* (for teens and "casual readers," 1998), *An Arthur Ford Anthology* (editor, 1999), *I, Joseph of Arimathea* (2000), *The Holy Grail Mystery Solved* (2003), and *Spirit Images: Pictorial Proof of the Fourth Dimension* (2008). Tribbe also wrote scores of papers, delivered lectures on similar topics, and edited the journal *Spiritual Frontiers*.

In 1980, he had a paper published in the *Journal of the Academy of Religion and Psychical Research* with the following abstract:

> The technique for developing a simple, easy-to-prepare but difficult-to-decipher code to prove survival of the human soul is outlined and procedures for its use are given.[24]

In this case, the "difficult-to-decipher" code is simply a monoalphabetic substitution cipher (MASC). The idea is that the individual, while still alive, uses a key phrase to encipher a message and then sends the ciphertext in a sealed envelope to the Survival Research Foundation in Phoenix, Arizona. The envelope is to remain sealed until the author is deceased and someone contacts the foundation claiming to have learned the key via communication from beyond the grave.

I find it interesting that these individuals who devoted so much time to considerations of an afterlife actually were able to spend much more than the average amount of time on Earth. The next researcher, Arthur S. Berger, is certainly no exception. Born in 1920, he's still alive as of this writing.

FIG. 7.3 Arthur S. Berger (1920–)

Like many researchers we have already met, Berger was a lawyer. He was also a veteran of World War II and published his most recent book in 2012, showing impressive activity at an advanced age. Most of Berger's books focus on questions concerning the afterlife.

In 1983, Berger came up with another approach to the problem, which though not cryptologic in nature, did at least involve numbers. His approach depended on first finding a spirit that appeared to be in genuine contact through some medium. The deceased would then be asked to come up with three numbers, one of which was the sum of the other two, and keep them completely secret for the moment. Then, other mediums would be visited. Through each medium, the deceased would try to communicate one of the numbers. Berger wrote, "If the third number received by a medium was the sum of the two smaller numbers received by each of two other mediums, the experiment would be a success."[25]

One of the problems with this experiment is that some mediums share information. To see how this can work, consider a group of three mediums, Alice, Bob, and Carl. Now suppose that Victor goes to Alice and, suspicious of her abilities, is vague about the lost loved one he wants to get in touch with. Alice guesses that it is his father and says, "I'm sensing that there's an older man who wants to speak with you." This vagueness on the part of the medium is intentional and covers the possibilities of the deceased being a father, grandfather, uncle, or even an older brother. Victor might snort in derision that all of his older male relatives are still alive, that he was trying to reach his sister, or he might sit through the session without revealing much more, but he might leave obviously dissatisfied. In any case, Alice is likely to know that her guess missed the mark. Victor is unlikely to return, but Alice shares everything she has learned about Victor with Bob and Carl. When Victor visits Bob, Bob might open with "I sense there's a woman who wants to get in touch with you" or he might be more specific if Victor revealed to Alice that he wanted to talk to a deceased sister. If Bob ends up having a good session, Victor is likely to return. If Bob makes some statements too far off the mark, he'll lose Victor's future business, but he'll have learned a bit more about Victor that can then be passed on to Carl. If Victor goes on to visit Carl, the session may go quite well. The accumulated knowledge from Victor's reactions to the wrong guesses of Alice and Bob should be enough for Carl to fool Victor and form a profitable relationship.

Victor may go home and tell his father that Carl, whom he had never met before, "knew things he couldn't possibly have known."

If you're wondering why Alice and Bob would help Carl, when he could be viewed as competition, the reason is that clients don't typically visit mediums in alphabetical order, as in the example above. Some visit Alice first, some visit Bob first, and some visit Carl first. So, if all three share information with each other, each regularly benefits from the failures of the others.

Organized groups of mediums have long shared information on clients, especially wealthy ones. This information sharing is much easier in the digital age. Imagine a password-protected Wiki serving as a database for mediums, each of whom pays a monthly fee for access to continually updated information.

In performances where a medium does readings for audience members, be aware that those who received the readings, in some cases, purchased their tickets in advance with credit cards and received tickets for specific seats. It's not hard for a medium with trusted researchers in his or her employ to know a great deal about some of these people in advance of the show.

Despite the critical flaw in Berger's experiment (of assuming that the mediums were not in contact with each other), when he first carried out the experiment, it failed. More positively, it could be said to have shown that not all mediums are in cahoots!

The "By the Numbers" Test

Another method proposed by Berger (in his 1987 book *Aristocracy of the Dead*) did make use of cryptology. Before dying, the individual who will attempt to communicate from beyond the grave randomly selects a word from a particular dictionary. In Berger's example, the word was "builder." The predeceased then numbers all of the letters in that particular entry of the dictionary, from the word itself, including its pronunciation, and on through all definitions. Only nonalphabetic characters are skipped over. See the example reproduced in fig. 7.4.

1	2	3	4	5	6	7	8	9	10	11	12	13
B	u	i	l	d	e	r	(b	i	l	d	e	r)

14	15	16	17	18	19	20	21	22	23	24	25	26
1. o	n	e	w	h	o	b	u	i	l	d	s	o

27	28	29	30	31	32	33	34	35	36	37	38	39
r	o	v	e	r	s	e	e	s	b	u	i	l

40	41	42	43	44	45	46	47	48	49	50	51	52
d	i	n	g	o	p	e	r	a	t	i	o	n

53	54	55	56	57	58	59	60	61	62	63	64	65
s;	o	n	e	w	h	o	s	e	o	c	c	u

66	67	68	69	70	71	72	73	74	75	76	77	78
p	a	t	i	o	n	i	s	t	o	b	u	i

79	80	81	82	83	84	85	86	87	88	89	90	91
1	d,	a	s	a	c	a	r	p	e	n	t	e

92	93	94	95	96	97	98	99	100	101	102	103	104
r,	a	s	h	i	p	w	r	i	g	h	t,	o

105	106
r	a

FIG. 7.4 An example from Berger

Then the predeceased enciphers a message by replacing each letter with one of the numbers it was paired with above. When letters appear more than once in his message, he will often be able to encipher them differently. For example, the dictionary entry for builder allows E to be enciphered as 6, 12, 16, 30, 33, 34, 46, 56, 61, 88, or 91. Berger's sample message and encipherment are shown here:[26]

```
Test message: W    E    H    A    V    E    S    O    U    L    S
Encipherment: 98   6    18   106  29   12   32   26   21   39   53
```

Enciphered messages would be stored along with the title and edition of the dictionary that was used. Someone guessing the word (or claiming that it was communicated to him or her by the deceased) could apply it and see if a meaningful message resulted.

Even though homophones allow for varied encipherments, this system is weak. Berger didn't see the weakness. He wrote that the homophonic feature would "make breaking the encipherment without knowledge of the key exceedingly difficult."[27] He also wrote

> Since the Merriam-Webster's used in the example is an unabridged dictionary which contains approximately 600,000 entries, the chance of finding the key "builder" by random guesses is 1 in 600,000, odds great enough to eliminate chance.[28]

In 1991, Berger proposed using this sort of test as a way to confirm reincarnation. As an example, he offered 98 6 18 106 29 12 32 26 21 39 53 again. He didn't give the corresponding plaintext this time, but rather wrote, "This jumble of numbers is, of course, meaningless, and the message behind them cannot be read by anyone without the correct key."[29]

Berger found a way to put a positive spin on previous experimenters' failures, of which he was well aware. He wrote, "Also, experience has shown that all the people who have been involved in posthumous experiments in the past, such as F.W.H. Myers, R.H. Thouless and J.G. Pratt, and who left test messages were thoroughly honest, because no one since their deaths has come up with their secret keys."[30] I agree with this assessment and respect these researchers for their integrity.

We've seen some variety in the ciphers and mathematics that have been brought to bear on the problem, but another, equally unproductive, direction has been pursued as well. It's of interest because it also connects with security issues, this time physical.

Combination Lock Tests

In 1968, Dr. Ian Stevenson of the Department of Psychiatry at the University of Virginia Medical School proposed the "combination lock test." The idea was that combination locks (Sargent and Greenleaf Model 8088, to be specific), with combinations set by various individuals, would be sent to mediums. The mediums would then try to open them and, depending on whether the locks were opened before or after the death of the individual who set a particular combination, thereby confirm telepathy/clairvoyance, or communication with the dead.

FIG. 7.5 Ian Stevenson (1918–2007)

In describing this experiment, Stevenson wrote, "Since three numbers each selected out of fifty possibilities (six digits) are included in the combination, the chances of finding the correct combination through random trials are 1 in 125,000, odds sufficiently high to justify the use of the padlock for the purpose intended."[31]

However, the experiment would prove nothing, as there are many methods of opening such locks that require neither mind reading nor communication with the dead.[32] Stevenson was aware of some of these techniques and addressed the matter briefly.

> Participants are warned against the use of certain cheap combination locks whose combinations can be reset by the owner. Such locks sometimes have defects and may be opened by jarring or otherwise than by using the exact combination. The manufacturers of the lock I have myself adopted (Sargent and Greenleaf, Inc.) assert that this lock cannot be opened by jarring, shaking, or vibration; it also cannot be opened by listening to changes in the movement as the dial is turned.[33]

In a similar manner, many vendors of crypto products assure potential buyers that their products are secure. Let the buyer beware!

Stevenson saw his combination lock test as a more user-friendly version of Thouless's test. Although Wood pursued Thouless's path, many others were apparently intimidated by the idea of learning a bit about cryptology.

Indeed, it seems that Stevenson's experiment attracted more participants. Stevenson returned to the topic with a paper in 1976 that provided readers of his first paper with an update. At that time, he had eleven locks in his office, including two of his own and one with a combination set by Thouless. Interestingly, Thouless used numbers derived from the key to one of his ciphers (presumably Cipher C) as the combination for his lock. All of the locks were given or mailed to Stevenson by people wishing to participate in the experiment. Two of the owners died before this second paper was published, yet their locks remained locked. Stevenson also reported on another pair of locks in England that were part of such an experiment, although they weren't registered with him. These also remained locked after their owners' deaths.

Another update came from Stevenson and two coauthors in 1989. By this time, Thouless had died, as had parapsychologist J. G. Pratt, who also left a lock behind. Both of these remained locked like the others, and it wasn't from lack of trying. By the time this second update saw print, there had been forty-two attempts made on Pratt's lock.

In this same paper, an update was given on Thouless's Cipher B (142 failed attempts) and Cipher C (87 failed attempts—Gillogly and Harnisch's solution came later).

Stevenson's own lock remained shut after his death in 2007.

The efforts made by the professionals (mostly lawyers) were far outdone in terms of unlikeliness by another lawyer from Rochester, New York. Please bear in mind, when reading the following paragraphs, that there is often a world of difference between articles that appear in academic journals (from which the material above was drawn) and claims put forth in self-published books.

Radar Love

The most bizarre approach to communication with the dead having a cryptologic connection is definitely that of Joseph Martin Feely. One of many to have claimed a solution to the Voynich manuscript, Feely didn't stop there. He went on to use transposition as a means of communicating with people such as Leonardo da Vinci, Thomas Edison, and Franklin Delano Roosevelt. Whereas other researchers carried out experiments or tests, as far as Feely was concerned, science had already moved on to the next step. He wrote

> That postmortuary intercommunication does occur is an accepted probability in scientific circles. There the only open questions now are as to how it is accomplished, and also as to how the purported sender can be identified satisfactorily.[34]

As Feely explained it in his 1954 book *Electrograms from Elysium*, he had found out not only how it was accomplished, but also a way to positively identify the senders.

> The theory throughout the foregoing discussion is that the departed mind in sending messages from the afterlife through his name code did, in order to identify himself, pick out of the immense medley of possible anagrammatic combinations of the letters in his name those intelligible sequences or sentences that disclose from his career in the flesh some hidden, recondite fact, which very few other persons would be likely to

know; and that he did then by his electrical brain, or rather mind waves, radiate instantaneously the items of that identifying secret directly to the electrical terminals of nerves in the inner ear of the still enfleshed recipient, where they are received telepathically; and there they at once join with some of the letters in the eye of the person thus clairvoyantly beholding electrically the letters of the sender's name, laid out in view, as they fall one by one into line, and altogether finally compose the sentence meant to be conveyed. Then the recipient having thus perceived the indicated means of identification, proceeds to look outside for extrinsic evidence of the hidden fact suggested by the sender for the purpose of becoming identified. Then the recipient, upon having discovered in the concrete the hidden fact, concludes the more probable explanation of it all to be that the electrogram has been sent by the true owner of the name that served as his code. In some cases reliable identification may be supplied only by context, pertinency or cross-correspondences. In general it may be well said that now while extrasensory perception can make itself more outstanding through the discoveries of modern electricity, parapsychology as copartner can make itself more articulate through the personal name anagram.[35]

He believed that the method allowed for two-way communication.

It is self-evident ... that the human brain broadcasts electrical waves which can, in some cases, travel wirelessly through world-wide distances; and that such waves likewise can be received by the electrical capability of the brain; and that the information thus communicated is intelligible and intimately pertinent to the participants; and that much of it may have been designed by the senders to identify extrinsically themselves as the senders of the messages[36]

Messages Feely received from Thomas Alva Edison included "TO LIVE ON AS HAD MAD," "STAMINA SHOVE LOAD," and "HO! LOVE ADS STAMINA."[37]

One of Feely's messages from Arthur Conan Doyle was presented as an anagram challenge in chapter 3. Some of the other Doyle anagram messages were "ON, SIR, CLEAR THY AROUND," "HAND, SIR, ANY TRUE COLOR," "AIRS HOLD COUNTRY NEAR," "AIRS CHAT: ONLY ROUNDER," "AIRS

RECOUNT LADY HONOR," "AIR RECOUNT LADYS' HONOR," "LADY AIR CURETH NO SON," and "AIRS CON HEART ROUNDLY."[38]

As you can see, these messages require a great deal of explanation if any sense is to be tortured out of them. And I didn't pick out the worst of the lot; these actually represent some of Feely's better efforts! Most of his electrograms have horrendous misspellings (because of the paucity of letters in certain names), and foreign words are frequently introduced. In addition to this, Feely frequently fails to use all of the letters in a name or has to fill in missing letters he doesn't have.

Oddly, for SIR ARTHUR CONAN DOYLE, Feely missed the anagram HOR-RENDOUS CARNALITY, which would have put a somewhat nontraditional, but not necessarily undesirable, slant on the afterlife.

In all, Feely generated more than 1,500 such electrograms from a group of thirty-eight people.

Believing that he received messages from former U.S. President Franklin D. Roosevelt about a fall the man had and kept secret, Feely wrote to his widow, Eleanor, seeking confirmation. Amazingly, she took the time to write back, "I have no idea of what my husband was supposed to be referring to, nor where your quotation comes from."[39]

Undeterred, Feely wrote to her again with more information. New electrograms revealed that FDR had been in an elevator that suddenly fell seven stories, resulting in an injury to his left ankle. Once again, he heard back. Eleanor wrote, "I am sure the incident you describe never happened."[40]

A few pages later in Feely's recounting of this episode, he announced that he finally received confirmation of the injury from the former president's nephew (who was also deceased and sent his confirmation as an electrogram).

A Modern Approach

In 1994, a researcher attempted to move survival research into the world of modern cryptology. Michael Levin, a member of Harvard Medical School's Genetics Department, wrote

> The method depends on the existence of trap-door or one-way algorithms that can be used to encode a message but cannot be reversed. The method works as follows: The original person uses such an algorithm on a

computer to encode his phrase into a string (sequence) of symbols, *never leaving a record of the original phrase*. Then he publishes the encrypted form, along with the algorithm used to encrypt it. Once he is deceased, the only way to obtain a "hit" is to have the message communicated by paranormal means. When someone thinks he has received the message, he simply runs it through the published algorithm, and the result is compared to the published encoded form. If it matches, the message is a 100% hit.[41]

Such one-way algorithms are not actually known to exist. We have algorithms that we think might be one-way, but there's no proof yet. Just as Thouless couldn't have imagined someone breaking his Cipher C decades later with a computer, so we cannot predict the future of today's "one-way" algorithms. A later paragraph in Levin's paper indicated some confusion.

There are several candidates for appropriate algorithms. One of the most common is RSA. It uses the fact that it is easy to find large prime numbers but that it is computationally very difficult to factor large integers into prime factors. There is also DES and the Blum-Blum-Shub cryptosystem. Good descriptions of these algorithms can be found in Denning (1982), Gaines (1944), Merckle [sic] (1982), Pfleeger (1989), Price and Davies (1984), and Sinkov (1968).[42]

RSA is covered in chapter 11 of this book. The others are not, but none of them date back beyond the 1970s, so it's unclear why Levin cites books from 1944 and 1968 as containing "Good descriptions of these algorithms." Levin also wrote

The final potential pitfall of this technique is that someday, when mathematical knowledge and computer power grow beyond a certain capability, the algorithm may be reversed, so that the message can be decoded; also, an all-out search among all possible strings may become feasible. Some of these algorithms have been mathematically proven to be irreversible. In addition, most common cryptographic algorithms are being continually subjected to cracking tests; any compromise of an algorithm quickly becomes public knowledge among the computer security and cryptography community. If a "hit" is encountered, it is easy to check with the relevant professional group and to ignore it if they have utilized a method that has been reversed. All of these issues notwithstanding, encryption algorithms provide a useful tool for survival experiments.[43]

Contrary to what Levin claims, none of the encryption algorithms have been proven to be irreversible. As I wrote above, one-way algorithms are not actually known to exist.

Levin referenced papers by Thouless and Stevenson, so although he got some details wrong, he was likely inspired by them to update their experiments for the 1990s. However, his effort doesn't seem to have succeeded. Unlike Stevenson, he provided no updates, and Google Scholar doesn't list any other authors as citing his work.[44] If he actually prepared a cipher to be read after his death, it is not readily available.

Still, the basic idea has not died!

Another Attempt

Of the attempts covered thus far, all except Feely admitted to failure. And we can safely dismiss Feely's claimed success as a combination of wishful thinking and self-delusion. However, this hasn't stopped others from repeating the experiments in altered forms. A recent attempt was carried out by Susy Smith, founder of the nonprofit Survival Research Foundation. She created a pair of cipher tests before her death on February 11, 2001.

While alive, she believed that she was in contact with her deceased mother, as well as the prominent American philosopher and psychologist William James. The latter also had a strong interest in parapsychology and was a founder of the American Society for Psychical Research.

So Smith probably had high hopes for being able to communicate with the living herself after her demise. She shared her life story, as well as her hopes for the afterlife and her plan to communicate, in a book titled *The Afterlife Codes*.[45] She even offered a $10,000 reward for whoever was able to receive her key and break her code.

Smith's website www.afterlifecodes.com is no longer up, but the Internet Archive Wayback Machine[46] has preserved it from various points in time. The most recent that shows something useful is from May 17, 2001.[47] A link on this page gives the following explanation of Smith's approach to the problem.[48]

Explaining the Code and Questions

Shortly after the Survival Research Foundation was established I asked board members Frank Tribbe and Clarissa Mulders to create a foolproof code that people could leave with the hope of being able to send a message that would break it after their death. Frank was a lawyer with the United States Government and Clarissa was a teacher with a Phi Beta Kappa key. They used their brilliant minds devising an encoding system that has been judged by code authorities to be accurate, successful, and scientifically well designed.[49]

The Tribbe-Mulders Code involves a juggling of several alphabets that is extremely complicated for an individual to prepare by hand, I've done it and I know; but for today's computers it's just a breeze. The procedure goes like this: First you decide on a brief secret message you would like to leave. (Tribbe and Mulders referred to it as a "key phrase"). It should be of enough significance to you that you will remember it after your death.

It should also be fairly well known to the public so that someone here could pick it out of the atmosphere (or wherever) when you send it from there. An example of a key phrase you might use is "the power of positive thinking". It could be a line from a poem or a nursery rhyme or the title of a book like *Life is Forever*. This is not your code, remember, it is your secret message that is the key that will unlock your code. When you have chosen this and are sure you won't forget it and have definitely not told it to another living soul, you are ready to tackle the Questionnaire and when you have finished that successfully you may Register.

In the registration process the computer in its secret heart will combine your key phrase with the standard alphabet, and that combination is what it retains. It will then randomly choose a line from a book whose pages have been hidden within it for that purpose, and it will encode this line using the new alphabet. The result will be a jumble of letters that looks like gobble-dygook, and this is your code. Nobody knows what it is or what it means. It can only be unscrambled when your original secret message is presented to the computer again.

Since the possibility of fraud has reared its ugly head, you will be presented with 46 questions which you will answer and the computer will store hidden away alongside your code. In the future, sometime after your passing, a friend or relative or well-meaning strangers or mediums and psychics might purport to receive your secret message. When they bring it to us, each will have to take the tests and register. Then the computer will compare the words he brings with your hidden phrase and if ... the big IF ... it should happen to be the identical secret message that breaks your code, there is going to be considerable excitement. I am not putting out false hopes, but neither am I discounting the possibility that this strange joyous event could possibly happen to someone sometime.

Now the person who is registering his code should also do things to protect himself. So that investigation may be made if someone really thinks he has received contact with you from the hereafter, you must leave in your will or with your valuable papers the address: The Susy Smith Project, Human Energy Systems Laboratory, Department of Psychology, University of Arizona, Tucson. And you must not offer a reward. I've already told you that isn't a good idea.[50] But here's something you can do. You can have a control, which is considered of value scientifically. Tell it around that you have a secret phrase hidden in your head for future reference many years hence after you pass on. If any friend or acquaintance guesses it or attains it telepathically, or even if it comes from a medium or psychic, while you are still with us, that's all right. You can just e-mail the SS Project for another code. (And be sure to tell us the details of how it happened. Such information can be valuable to us when standardizing statistics later on.) But if no one can break it now when you are here and then some day in the future someone comes up with the secret phrase that checks out against the one you left, the evidence will be much more convincing that it actually came from you and it will be a very good indication that you have survived death. Controls can give just a bit more protection.

Remember that when you register your code you are telling the world you love those you will be leaving behind. But don't hurry off.

Now, click here and go to the **Questionnaire/Registration**.

Despite heavy promotion in the book *The Afterlife Codes* and the offer of a $10,000 reward, Smith's effort, like the others before her, failed. The website from which the quoted material above was taken has long since come down, and the fate of the messages enciphered through it is uncertain. It is known that there were about 1,000 people registered.

A Skeptic's Challenge

Klaus Schmeh wrote about some of the ciphers discussed here in his 2012 book *Nicht zu Knacken*[51] (available only in German) and, despite being a skeptic, put forth a challenge of his own, based on one of today's top ciphers, the Advanced Encryption Standard (AES).

Undoubtedly, there will be more attempts in the future to use ciphers to prove life after death, but it's time now to move on to other challenges.

8

A Challenge Cipher

Is there a manner of enciphering that is so secure it can never be broken, no matter how clever the attackers, regardless of how much computing power is brought against it, with no limits on how much time is allowed? Yes, there is. During the years 1917 to 1918, Joseph O. Mauborgne and Gilbert Vernam discovered such a system.[1] It's called the one-time pad, sometimes abbreviated as OTP. Here's how it works.

Suppose you can generate a random sequence of numbers between 0 and 25. As a sample, we'll use 6 3 6 25 16 6 10 6 18 1 3 7 24 25 18 21 8 10 12 1 13 0 2 17 0 13 22 16 4 24 4 10 4 13.

Now we'll use this to encipher the message YOU MET ME AT A VERY STRANGE TIME IN MY LIFE.[2]

We simply line the letters up with our random sequence of numbers and shift by the appropriate amount in each case, looping back to the start of the alphabet if the shift takes us past Z.

```
Y O U  M  ET ME  A T A V  E  R  Y  ST R  AN GE T  I M
6 3 6  25 16 6 10 6 18 1  3  7  24 25 18 21 8 10 12 1 13 0  2  17 0
E R A  L  UZ WK  S U D C  C  Q  Q  NB B  MO TE V  Z M
```

```
 E  I  N  M  Y  L  I  F  E.
13 22 16  4 24  4 10  4 13
 R  E  D  Q  W  P  S  J  R
```

So our ciphertext is ERALU ZWKSU DCCQQ NBBMO TEVZM REDQW PSJR.

Anyone who can guess the key can then shift each letter back by the appropriate amount to get the original message. Given this fact, how can we claim that the system is unbreakable? Well, why would anyone guess that particular key? If instead someone guesses the key 23 3 4 25 0 23 15 9 7 6 15

25 6 8 5 2 3 13 18 22 12 0 18 6 24 25 11 3 18 22 4 10 14 13, the message will come out as

HOW MUCH BLOOD WILL YOU SHED TO STAY ALIVE?[3]

And if someone guesses the key 6 3 6 9 20 12 3 3 24 3 10 16 24 3 2 20 5 19 19 7 7 6 19 18 8 13 12 25 9 18 4 6 5 24, the message will come out as

YOU CAN'T HURT ME, NOT WITH MY CHEESE HELMET![4]

To see how one cipher can be turned into such wildly different messages, let's look closely at the first ciphertext letter, E. If we guess that the key starts with 0, then this E will decipher to E. If we guess that the key starts with 1, it will decipher to D. Similarly, guessing that the keys starts with 2, 3, 4, 5, 6, 7, 8, 9, and 10, gives us the first message letter as C, B, A, Z, Y, X, W, V, and U, respectively. I didn't list all of the possibilities, but the twenty-six possible numbers the key might begin with lead to each of the twenty-six letters of our alphabet as the first letter of the message. That is, depending on the key, the first letter of the message could be anything. The situation is the same for the second letter of the ciphertext, and the third, etc. That is, any letter of the ciphertext can decipher to any letter of the alphabet for some particular key. In other words, we can get out any message imaginable, for some key. The only restriction is that the message has to be the same length as the original ciphertext.

Because there's no way for the would-be solver to distinguish the intended message from the very large set of all possible messages of the given length, we say that the cipher is unbreakable. Although the proof is more mathematical in nature, this is essentially the approach Claude Shannon later used to show that not only is this system unbreakable, but that any unbreakable system must be equivalent to it. That is, another unbreakable system, while looking different, would, at a fundamental level, have to be the same as using a one-time pad.

Spies often carried very small pads filled with numbers or letters. These were used to encipher messages in the manner described above. The version that uses letters is simply a more compact way to represent the numbers between 0 and 25. We have A = 0, B = 1, C = 2, ... Z = 25. We refer to such a system, whether it uses numbers or letters, as a one-time pad because if the same sequence of numbers is ever reused, both messages can be recovered (and have been in some instances). Having to use a different key every time is a nuisance, especially for very long messages (think of enciphered sound or video files), which is why this system isn't used for everything.

Also, it's important that the key be random. If the key has a pattern to it, the method can be broken. This has happened.

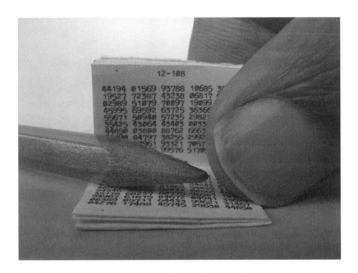

FIG. 8.1 A one-time pad

In recent years, a tremendous number of new cipher systems have been put forth. It's time-consuming to analyze them and see if they are indeed strong or if they have subtle weaknesses and should be avoided. In general, experts are unwilling to take the necessary time unless the person proposing it already has a good reputation in the field. This sounds like a catch-22. How can you earn a strong reputation if experts won't look at your work? The answer is that they will look at attacks. When a new attack is proposed on some existing system, it's usually a lot easier to evaluate than a new cipher system. And when someone is coming up with good attacks, there's a much greater willingness among experts to look at systems he or she proposes that are claimed to resist such attacks. In a nutshell, a crook is sometimes the best at figuring out how to make things secure.

Back in 1914, Mauborgne made his reputation by showing that the Playfair cipher could be broken. We examined this system in the previous chapter, but Mauborgne's work on it was the first published attack. He also solved the same challenge cipher that Edward Elgar did, the one posed by Schooling, upon seeing it re-presented in a 1939 issue of *The Signal Corps Bulletin*.[5] The very next issue contained Mauborgne's solution.

FIG. 8.2 Joseph O. Mauborgne (1881–1971)

As previously mentioned, Mauborgne put forth the one-time pad in 1917–1918. But between his attack on Playfair and his presentation of an unbreakable alternative, he tested another system by presenting a challenge to the troops. It remains unsolved. Although several other unsolved ciphers may well be hoaxes with no underlying meaning, a cipher created by Mauborgne seems unlikely to fall into that category. With his credentials, when Joseph O. Mauborgne discusses ciphers, he is to be taken seriously.

The challenge cipher was put forth by Mauborgne in 1915. It was rediscovered by Louis Kruh in 1991 (or perhaps a bit earlier) in the Rare Book Division of the New York Public Library, among some old Army Signal School material. It appears below, in the context of the documents.[6]

The following cipher message and explanatory letter from First Lieut. J. O. Mauborgne, 8th Infantry, is submitted to the Signal Class. Officers interested in Cryptography are invited to translate this cipher message, and turn in the translations to the Director's Office.

Guartel de Espana, Manila, P.I.,
October 23rd, 1915.

From: 1st Lieutenant J. O. Mauborgne, 8th Infantry.
To: The Director, Army Signal School.
Subject: New type of military cipher.

1. Having evolved a cipher which I believe to be suitable to military use, and superior in many ways to the regular cipher disk or the "Playfair" cipher in the matter of ease of employment by the enlisted man, and extreme simplicity, coupled with indecipherability, when properly used, I request that this cipher be turned over to the instructors and student officers of the Signal School for attack from the point of decipherability, and for discussion with a view to its adoption as a standard cipher for use in the service.

2. To the end just stated, I suggest that the Director permit the enclosed cipher message to be put on the bulletin board where it can be copied and worked upon by the officers at the school, in their leisure time; a time limit of three months being allowed for solution, inasmuch as it is well known that the leisure time of the student is a very small quantity.

3. At the expiration of the time mentioned, the sealed envelope herewith, marked "Cipher No. 1, Translation" might be opened if the Director is willing, and the translation of the message placed upon the bulletin board to show why the standard method of attacking a substitution cipher fails in this case, and to see if the officers at the school can develop any method by which translation might be obtained from the cipher given. It is believed that this will stimulate invention, and perhaps, result in something of service to the army.

4. After another month has elapsed, the other sealed envelope enclosed herewith and marked, "Cipher No. 1, Method of enciphering and key used" may be opened, and, with the consent of the Director, submitted to the School for discussion, with the object stated in paragraph 1.

5. I am told that the Signal School has developed some men who have become exceptionally keen in deciphering captured cryptograms, and I trust that the Director will consider my request favorably and permit these officers to attack this message in the manner outlined.

6. If any attack upon this cipher is successful, I shall be glad to hear of it.

J. O. Mauborgne, 1st Lieut. 8th Infantry.

CIPHER NO. 1

This is a simple, single-letter substitution cipher adapted to military use, which is believed to be practically indecipherable when used in the manner employed in the production of the cipher given below, and which can be used by the enlisted man with greater precision and facility, after five minutes instruction, than either the "Playfair" or the Cipher Disk in use in our service at the present time. If you are able to decipher the following message, please inform me of the number of hours necessary, and the method used. I believe this cipher will be found to dodge all the rules of deciphering now in our possession.

```
PMVEB DWXZA XKKHQ RNFMJ VATAD YRJON FGRKD TSVWF TCRWC
RLKRW ZCNBC FCONW FNOEZ QLEJB HUVLY OPFIN ZMHWC RZULG
BGXLA GLZCZ GWXAH RITNW ZCQYR KFWVL CYGZE NQRNI JFEPS
RWCZV TIZAQ LVEYI QVZMO RWQHL CBWZL HBPEF PROVE ZFWGZ
RWLJG RANKZ ECVAW TRLBW URVSP KXWFR DOHAR RSRJJ NFJRT
AXIJU RCRCP EVPGR ORAXA EFIQV QNIRV CNMTE LKHDC RXISG
RGNLE RAFXO VBOBU CUXGT UEVBR ZSZSO RZIHE FVWCN OBPED
ZGRAN IFIZD MFZEZ OVCJS DPRJH HVCRG IPCIF WHUKB NHKTV
IVONS TNADX UNQDY PERRB PNSOR ZCLRE MLZKR YZNMN PJMQB
RMJZL IKEFV CDRRN RHENC TKAXZ ESKDR GZCXD SQFGD CXSTE
ZCZNI GFHGN ESUNR LYKDA AVAVX QYVEQ FMWET ZODJY RMLZJ
QOBQ-
```

J. O. Mauborgne, 1st Lieut. 8th Infantry.
Manila, P. I., October 21st, 1915.
Army Signal School, Fort Leavenworth, Kansas, December 13, 1915.

▼ ▼ ▼

Kruh published this cipher in both *The Cryptogram* and *Cryptologia*, yet no solutions were turned in by the readers of either. In the *Cryptologia* piece, Kruh wrote "The archive where I found this document did not contain the solution. But it was probably solved or otherwise deemed unsuitable for use because there is no knowledge of a new cipher being adopted by the Army around that time."[7]

So not only did Mauborgne discover the only unbreakable cipher (the one-time pad), but he also concocted a different sort of cipher that, though breakable in theory, has remained unsolved. If he were alive today, I can imagine him saying, "I don't always create ciphers. But when I do, they are unbreakable."

Mauborgne's career went on to last many more years, until he finally retired in 1941 at the rank of major general. Some highlights include his role in the development of an aircraft radiotelephone, allowing communications between air and ground for the first time, and serving as the army's twelfth chief signal officer, in command of the Signal Corps from October 1937 until he reached the mandatory retirement age in 1941. As a retiree, he didn't serve in World War II, but when he was active he supported the development of radar, which played a pivotal role in the war. For all of this, Mauborgne was inducted into the Military Intelligence Corps Hall of Fame in 1988.

Toward a Solution

I looked at various statistics associated with the ciphertext that Mauborgne put forth as his challenge, but this isn't what led me to recognize the system he used. Rather, it was his comments that it was "a simple, single-letter substitution cipher," superior to other systems "in the matter of ease of employment by the enlisted man, and extreme simplicity," and that it could be "used by the enlisted man with greater precision and facility [than systems currently in use] after five minutes instruction."

I concluded that Mauborgne had used a cipher wheel, like the one pictured in fig. 8.3. It's a simple device to use but much stronger than many more complicated systems.

FIG. 8.3 A cipher wheel used by the U.S. military

Cipher wheels have been invented, used, forgotten, and reinvented several times. They are now known to go back to the fifteenth century, with rediscoverers including American President Thomas Jefferson in the eighteenth century and the French cryptanalyst Étienne Bazeries at the end of the nineteenth century.

The number of disks has varied, but what they all have in common is a set of movable disks with the entire alphabet written on each. They are significantly stronger if each alphabet appears in a different order. To send a message, the user simply turns the individual disks so that the message appears in some row along the length of the device. He or she then picks some other row and copies down those letters to send as the ciphertext.

The recipient of the ciphertext has an identical cipher wheel device, and he or she turns the wheels of it to make the ciphertext appear in some row. The recipient then looks at the other twenty-five rows, searching for a line that's readable. When the recipient finds it, he or she has found the original message.

Kruh commented on no new system being implemented in 1915 and then concluded that Mauborgne's system must have been solved. But since when is the military known to quickly implement good suggestions? How many officers ask those below them if there's anything they'd like to change? The cipher wheel method that I'm proposing was behind the 1915 cipher was actually adopted; it just took a while!

The U.S. Army adopted the cipher wheel after World War I, calling it "Cipher Device M-94," with instructions for use published in February 1922. The U.S. Navy adopted it in October 1926 under the name CSP-488.

Another Possibility

The following passage, written by Herbert O. Yardley, opens the door to another possibility for Mauborgne's unsolved cipher—a door that I will attempt to close!

> My opinion of Colonel Mauborgne can best be explained by reciting briefly some of my experiences with him. For a period of two or three years before the war Colonels Mauborgne and [Parker] Hitt had advertised through the Signal Corps an invulnerable method of using the U.S. Army cipher disk, namely the running key. The running key is a key that is not composed of a group of letters or a word but of a paragraph or a page of some book that is as long as the message to be enciphered. This affords a key that never repeats and was believed by Mauborgne and Hitt to be indecipherable.[8]

Could the 1915 cipher have been created using a running key? We'll consider this possibility, but let's look at an example first in case Yardley's vague description wasn't clear enough.

The first step is for the sender and receiver to agree on a book, or some other long piece of writing. The important part is that it be at least as long as the message that's to be sent. Let's say the book chosen is *Night in the Lonesome October* by Richard Laymon (it's good!). It begins

> I was twenty years old and heartbroken the night it started.
> My name is Ed Logan.
> Yes, guys can be heartbroken, too. It isn't an affliction reserved for women only.

Okay a slow start, but it really picks up later.

Now, to send the message "When I was a little kid my mother told me not to stare into the sun. So, when I was six I did,"[9] we begin by simply lining up the letters in the message with the letters in our book, like so

```
WHENIWASALITTLEKIDMYMOTHERTOLDMENOTTOSTAREINTOTHESUNSO
IWASTWENTYYEARSOLDANDHEARTBROKENTHENIGHTITSTARTEDMYNAM
```

```
WHENIWASSIXIDID
EISEDLOGANYESGU
```

We then continue by shifting each letter in the top line forward by the numerical value of the letter in the second line. This works the same way as an alphabetical key for the one-time pad. For convenient reference, the numerical values of each letter are reproduced here:

A	B	C	D	E	F	G	H	I	J	K	L	M	N	O	P	Q	R	S	T
0	1	2	3	4	5	6	7	8	9	10	11	12	13	14	15	16	17	18	19

U	V	W	X	Y	Z
20	21	22	23	24	25

In our message, W has the letter I below it, so W gets shifted forward by eight positions. This takes us past the end of the alphabet, but that's okay; we just start again at the beginning when this happens. So, W becomes E. The letter H is shifted forward by twenty-two to become D. Then we get to E, which is shifted forward by zero. That is, E stays put! Continuing in this manner, our cipher message becomes

EDEFBSEFTJGXTCWYTGMLPVXHVKUFZNQRGVXGWYATZXAGTFMLHE SASA
APWRLHOYSVVMVOX

Now let's compare the original message to our ciphertext by lining them up.

```
WHEN I WAS A LITTLE KID MY MOTHER TOLD ME NOT TO STARE
EDEF B SEF T JGXTCW YTG ML PVXHVK UFZN QR GVX GW YATZX

INTO THE SUN SO WHEN I WAS SIX I DID
AGTF MLH ESA SA APWR L HOY SVV M VOX
```

Notice that the first three words, WHEN I WAS, became EDEF B SEF, but the second time these words appeared they were enciphered as APWR L HOY. This is because the words lined up with different letters from the Richard Laymon book when they reappeared.

This variation can fool people into thinking it's a better cipher than it really is. We'll see some weaknesses soon, though.

There are several reasons that I don't think Mauborgne's challenge from 1915 was a running key cipher.

1. Although the system is not incredibly complex, it isn't as simple as the system Mauborgne used in 1915 is made out to be. I've explained how the running key cipher works to many groups of students, but I've never been able to do it in five minutes. If you are seeing it for the first time and read through the explanation in under five minutes with full understanding, keep in mind that it was made easier here by its similarity to the one-time pad, which was covered first.
2. Mauborgne thought that the solution could be displayed alongside the cipher, without anyone being able to figure out how he got from the message to the ciphertext. In the case of a running key cipher, the connection would be obvious.
3. The frequencies of the letters in the ciphertext don't match what would be expected to arise from a running key cipher. For example, in English, the most common word is THE. Because both the message and the key are meaningful English texts, we can expect THE in the message to line up with THE in the key from time to time, so that the resulting encipherment would be MOI. But these letters never

appear in Mauborgne's ciphertext. By itself, this isn't proof that a running key wasn't used. Perhaps an alignment of THE and THE was just avoided by chance.

What did arise often in Mauborgne's cipher was the letter R. It appeared forty-seven times, making it by far the most frequent character. How often would we expect this character to appear in a running key cipher? Well, it could arise from an A in the message combining with an R in the key. Abbreviating this as A + R, we can express the other possibilities as B + Q, C + P, D + O, E + N, F + M, G + L, H + K, I + J, S + Z, T + Y, U + X, and V + W.

Now, let's just focus on the first pairing. How likely is it that an A in the message will line up with an R in the key? It's simply the probability of A times the probability of R. Looking back at the table of probabilities given in chapter 1, this is seen to be $(0.08167)(0.05987)$ ≈ 0.004889. But we could also get an R from the pair combining in the other order. That is an R in the message and an A in the key. In our notation, that would be R + A. So, we should double our previously calculated value. That is, the probability of an R arising from an A and an R (in either order) is $(2)(0.08167)(0.05987) \approx 0.009779$. Repeating this calculation for the other dozen pairs of letters that can give rise to R, if a running key cipher was used, and adding them all together, gives a total probability of about 0.04247. In other words, we can expect, on average, about 4.247% of the letters in a cipher produced by a running key to be R. Looking at Mauborgne's cipher, we see 499 characters, and 4.247% of that number is about 21. That is, if a running key cipher was used, we should see R about 21 times. But this letter actually appeared 47 times! This big difference suggests it was not a running key cipher. One could make this comparison for each of the twenty-six letters in the ciphertext. Some would be close, probably by coincidence, but the ones that are way off are enough to make us look in another direction.

4. Running key ciphers aren't too difficult to solve. In fact, point 3 above is the reason why. Each cipher letter is likely to arise from one of the more probable combinations that can produce it. By looking at a group of cipher letters and jiggling the best bets for each, one can simultaneously form words of the message and the key, and thus unravel the cipher. I put forth two attacks in papers I coauthored

with undergraduates, and another student (although not mine) built on one of these papers and presented a truly devastating attack. If it were just a running key cipher, it should have been broken by now. But if it isn't a running key cipher, as I'm arguing, how in the world could we solve it? The techniques described above won't work against a cipher wheel, and solving it with a cipher wheel may be the only way to prove that a cipher wheel was used in the first place!

Breaking the Cipher Wheel

First, it should be known that the security of cipher wheels is not based on keeping the order of the alphabets on each disk secret. It's hoped that this can be done, but realistically, it's assumed that the enemy can obtain such details.

Periodically, users of such devices should remove all of the wheels from the shaft and place them back on in a different order. The order of the disks serves as the key. If an enemy has gained details of the alphabets, either through capture of a device or through a traitor or spy, he or she still has to contend with a large number of possible orderings for the disks. For a version with twenty-five disks, there are 25! = 15,511,210,043,330,985,984,000,000 possibilities.

This is far too many to check one by one, even with powerful computers, but a more clever modern attack can handle it!

An attack on a cipher system is more valuable if it works on a wide range of systems. One such attack is known as "hill climbing." The paragraphs that follow illustrate how it can be made to work against cipher wheels, but the same approach can defeat MASCs, Playfair ciphers, and more.

To start our attack, we place the wheels on the shaft in some random order. It's highly unlikely that it's the correct order, but that's okay! Pick a particular line of text going across the device. Assign a score to this result based on how well it matches normal English. There are a number of ways to do this. One decent method is based on the frequencies of all consecutive three-letter groups (trigraphs). For example, if the line of text is INEVERTAKEMYSKATE SOFF, we take the trigraphs INE NEV EVE VER ERT RTA TAK AKE KEM EMY MYS YSK SKA KAT ATE TES ESO SOF OFF and sum their frequencies (or probabilities) in normal English.

If the wheels were in the correct order and we picked the right line of text, the score should be very high. At this first step, though, it's probably low. Dissatisfied with the score, we pick two disks at random and switch them. We then compute our new score. If it's better, we keep the change. If it's worse, we switch the two disks back to their previous positions. Either way, we then make another random switch of two disks. Once again, we keep the change only if it led to a higher score.

Now imagine doing this tens or hundreds of thousands of times. As we keep only changes that make the solution look more like English (statistically), the various improvements should lead us to the highest possible score. Imagine the score represented on a graph. As we travel up this graph, keeping only changes that improve the score, we eventually find ourselves on the top of a "hill," where any change whatsoever serves only to lower our score. When the disks are in this position, we should have the correct solution.

This would be incredibly tedious to do by hand. That's why I used the word "imagine" after just a few disk switches. It became a thought experiment! Now, all you have to do is imagine a *computer* carrying out the switches and scoring, and the process goes from a thought experiment back to reality. It's no problem for a computer to make hundreds of thousands of switches; it can do it very quickly. This is a practical attack.

There are a few possible pitfalls that need to be addressed, though. First, it's possible to get to a point where any change lowers the score, yet the solution is not correct. Continuing the analogy with climbing an actual hill, this event is like finding oneself atop a small peak from which one can see a higher peak but cannot get to it without going downhill first. There are some simple changes that can be made to our program that will help avoid this situation.

First, have early switches involve more than two disks. Early on, it's not likely that we have more than a small number of disks correctly positioned. So why just make small changes? It would be better to move something like six disks at a time early in the process. After thousands or tens of thousands of iterations, we could move a small number of disks, and then after another large number of iterations, an even smaller number of disks. Finally, we'd get down to just two disks being switched at each iteration. This can be compared to taking bigger steps as we try to climb the hill. Once we're close to the top, smaller steps are more appropriate. Otherwise, we could miss the top and step clear over it to the other side. The big steps early on could help us clear the small hills and get on the one that goes the highest.

Another improvement is not always keeping just the changes that led to a better score. That is, we should sometimes undo a change that improved the score and sometimes keep a change that made the score lower! This sounds like a bad idea, but it isn't. If we find ourselves scaling the wrong peak, we need to descend before we can start climbing up the larger hill. The way to allow occasional downhill movement is to assign a percent chance of going in the higher scoring direction. For example, initially we might keep only changes that lead to a higher score sixty percent of the time. After a large number of iterations, we could raise this to sixty-five percent or seventy percent. After another large number of iterations, we could stay with better scoring changes eighty percent of the time. Eventually, we can raise it to close to one hundred percent.

In this latter approach, the percentage of the time that we keep "improving" changes is known as the temperature. So, we raise the temperature as we carry out our huge number of switches. The term "temperature" was motivated by comparisons to a method of tempering swords by heating them repeatedly to different temperatures and then cooling them down again. This is known as *annealing*. Because we're not actually tempering swords, but rather applying a somewhat similar approach to solving a cipher, the process in our context is often called *simulated* annealing. If you want to throw around some jargon, you can talk about breaking ciphers through *hill climbing with simulated annealing*. I think we need an abbreviation for this, just like I abbreviated monoaphabetic substitution cipher as MASC. Sadly, HCWSA doesn't have much of a ring to it.

The last detail that needs to be addressed is the simplest. In the initial explanation, I suggested looking at a particular row of letters on the cipher wheel. What if the wrong row is chosen? It doesn't do any good to look two rows down from the ciphertext for the plaintext, if it's actually fifteen rows down. On the other hand, switching disks can improve the score for one row but lower it for another. What's to be done about this? Well, as we're doing this on a computer, it's no problem to simply follow the entire process twenty-five times, focusing on a different row each time. At the end, the computer will have results for the tops of twenty-five "hills" and can then see which is the highest and present it as the final result.

Which Cipher Wheel?

We ought to be able to apply our HCWSA attack to the cipher from 1915. American cipher wheels are no longer classified, but rather just expensive collectibles. They can sometimes be purchased in online auctions, and they can also be seen in museums such as the National Cryptologic Museum, located just a bit down the road from NSA.

The problem is that they don't all have the alphabets in the same order. In the pages that follow, I will present all of the distinct versions that I've been able to find that are close in time to 1915. If one of them matches that used by Mauborgne, an attack based on it ought to work.

We first examine a record provided by William F. Friedman of a set of alphabets attributed to Parker Hitt, a friend of Mauborgne's with a stellar career of his own. Hitt authored *Manual for the Solution of Military Ciphers* (1916), the first book on cryptology written by an American, and became chief signal officer for the 1st Army.

Friedman wrote

This appears to be the original model of Colonel Parker Hitt's strip cipher device with the so-called "Star Cipher" alphabets. Colonel Hitt once told me that he independently conceived this system in 1915. This antedates Mauborgne's device which the latter originally called the QDEYAUB cipher because the sequences QD, EY, AU, and UB were the only repeated ones in his alphabets.

Friedman's memory wasn't quite perfect—Mauborgne really called it the DEYAUB cipher (no Q), but the basic reason for the name was recalled correctly. It was only DE, YA, and UB that repeated in Mauborgne's alphabets.

The Hitt alphabets that Friedman's paragraph was paired with are shown in fig. 8.4.

As you can see, the alphabets are numbered at the bottom. In some versions, there was a row of stars separating the ends of the alphabets from the numbering. This clear divider would prevent a 0 or 1 beneath it from being confused with the letters O and I. It was probably this row of stars that led to Hitt's alphabets becoming known as the "star cipher" alphabets.

Of course, for enciphering, the alphabets could be used in any order. This order provided the key. But why are the alphabets each written out twice? Hitt intended his device to be used differently than I described it here, although the net effect is the same. Instead of having each alphabet distributed around the edge of a disk, he had each running down a strip of paper. An image of his device from 1916 is shown in fig. 8.5.

With this device, the user pulls a strip up (or pushes it down) until the desired letter shows through the window in the horizontal guide. Once the message is filled in going across, the user selects some other horizontal line of characters as the ciphertext.

ALPHABETS FOR THE STAR CIPHER

```
A X Q J X H F C G D B W M Y P L O E U I X S N T R  0
E U V K Z N T D B F C G P L W M Y I A O Z V R H S  1
I A O Q J R H S C G D B F M Y P L W E U J X Q N T  2
O E U I K S N T R B F C G D L W M Y P A K Z V K H  3
U I A O E T R H S N G D B F C Y P L W M Q J X Q J  4
H O E U I L S N T R A F C G D V W M Y P V K Z V K  5
N T I A O M Y R H S E U D B F X Q P L W B Q J X Q  6
R H S E U P L W N T I A O C G Z V K M Y C G K Z V  7
S N T R A W M Y P H O E U I B J X Q J L D B F J X  8
T R H S N Y P L W M U I A O E K Z V K Z F C G D Z  9
B S N T R A W M Y P Q O E U I Q J X Q J G D B F C 10
C G R H S E U P L W V K I A O H K Z V K L F C G D 11
D B F N T I A O M Y X Q J E U N T J X Q M Y D B F 12
F C G D H O E U I L Z V K Z A R H S Z V P L W C G 13
G D B F C U I A O E J X Q J X S N T R X W M Y P B 14
L F C G D K O E U I K Z V K Z T R H S N Y P L W M 15
M Y D B F Q J I A O H J X Q J B S N T R A W M Y P 16
P L W C G V K Z E U N T Z V K C G R H S E U P L W 17
W M Y P B X Q J X A R H S X Q D B F N T I A O M Y 18
Y P L W M Z V K Z V S N T R V F C G D H O E U I L 19
J W M Y P J X Q J X T R H S N G D B F C U I A O E 20
K Z P L W B Z V K Z L S N T R A F C G D H O E U I 21
Q J X M Y C G X Q J M Y R H S E U D B F N T I A O 22
V K Z V L D B F V K P L W N T I A O C G R H S E U 23
X Q J X Q F C G D Q W M Y P H O E U I B S N T R A 24
Z V K Z V G D B F C Y P L W M U I A O E T R H S N 25
A X Q J X H F C G D B W M Y P L O E U I X S N T R  0
E U V K Z N T D B F C G P L W M Y I A O Z V R H S  1
I A O Q J R H S C G D B F M Y P L W E U J X Q N T  2
O E U I K S N T R B F C G D L W M Y P A K Z V K H  3
U I A O E T R H S N G D B F C Y P L W M Q J X Q J  4
H O E U I L S N T R A F C G D V W M Y P V K Z V K  5
N T I A O M Y R H S E U D B F X Q P L W B Q J X Q  6
R H S E U P L W N T I A O C G Z V K M Y C G K Z V  7
S N T R A W M Y P H O E U I B J X Q J L D B F J X  8
T R H S N Y P L W M U I A O E K Z V K Z F C G D Z  9
B S N T R A W M Y P Q O E U I Q J X Q J G D B F C 10
C G R H S E U P L W V K I A O H K Z V K L F C G D 11
D B F N T I A O M Y X Q J E U N T J X Q M Y D B F 12
F C G D H O E U I L Z V K Z A R H S Z V P L W C G 13
G D B F C U I A O E J X Q J X S N T R X W M Y P B 14
L F C G D K O E U I K Z V K Z T R H S N Y P L W M 15
M Y D B F Q J I A O H J X Q J B S N T R A W M Y P 16
P L W C G V K Z E U N T Z V K C G R H S E U P L W 17
W M Y P B X Q J X A R H S X Q D B F N T I A O M Y 18
Y P L W M Z V K Z V S N T R V F C G D H O E U I L 19
J W M Y P J X Q J X T R H S N G D B F C U I A O E 20
K Z P L W B Z V K Z L S N T R A F C G D H O E U I 21
Q J X M Y C G X Q J M Y R H S E U D B F N T I A O 22
V K Z V L D B F V K P L W N T I A O C G R H S E U 23
X Q J X Q F C G D Q W M Y P H O E U I B S N T R A 24
Z V K Z V G D B F C Y P L W M U I A O E T R H S N 25
★ ★ ★ ★ ★ ★ ★ ★ ★ ★ ★ ★ ★ ★ ★ ★ ★ ★ ★ ★ ★ ★ ★ ★ ★ ─
0 0 0 0 0 0 0 0 0 0 1 1 1 1 1 1 1 1 1 1 2 2 2 2 2 2
1 2 3 4 5 6 7 8 9 0 1 2 3 4 5 6 7 8 9 0 1 2 3 4 5
```

FIG. 8.4 Hitt's twenty-five alphabets

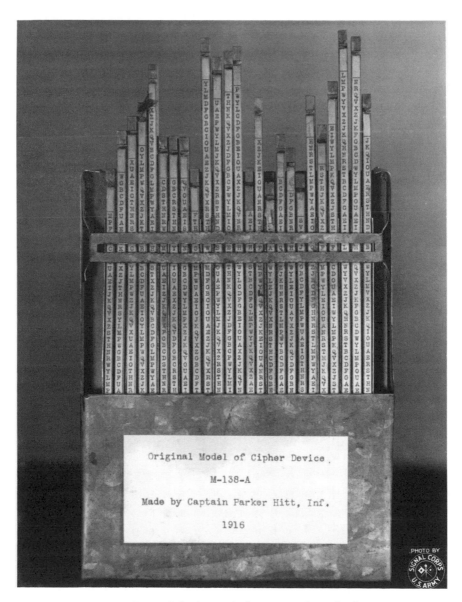

FIG. 8.5 A device equivalent to a cipher wheel

The image from 1916 shows the device being used to decipher a message. The ciphertext appears in the window of the horizontal guide, and the message is revealed on the last horizontal line. The reason for showing the entire alphabet twice on every strip is to guarantee that there are other horizontal lines showing letters in each position. If the alphabet wasn't repeated, imagine what would happen if the user needed the last letter on the first strip and the first letter on the second strip to begin the message. There wouldn't be any other horizontal row of letters across these two strips, and enciphering would be impossible. Although a simple device, a version of the strip cipher, officially called M-138-A, saw a great deal of use, and lasted into World War II.

The Hitt alphabets are transcribed here as single alphabets, as they would be used in the equivalent cipher wheel, instead of Hitt's intended strip cipher.

```
A A A A A A A A A A A A A A A A A A A A A A A A A
E E E N E E E V E E E E X E E E E M E E E N
I I I R R I I I X X I I I Z Z I I I P P I I I R R
O O S S S O O Z Z Z O O J J J O O W W W O O S S S
U T T T U J J J J U K K K K U Y Y Y U T T T T
H H H H H K K K K K Q Q Q Q Q L L L L L H H H H
N N N C Q Q Q Q Q V V V V V M M M M Z N N N N J
R R R D D V V V V C X X X X N P P P J J R R R K K
S S F F F X X X D D Z Z Z R R W W K K K S S Q Q Q
T G G G Z Z F F F J J S S S Y Q Q Q Q T V V V V
B B B B B J G G G G K T T T T V V V V V X X X X X
C C C M B B B B B H H H H H X X X X X Z Z Z Z Z
D D D P P C C C C N N N N N M Z Z Z Z N J J J J C
F F W W W D D D R R R R R P P J J J R R K K K D D
G Y Y Y Y F F S S S S S W W W K K S S S Q Q F F F
L L L L L G T T T T T Y Y Y Y Q T T T T V G G G G
M M M M Q H H H H H L L L L L H H H H B B B B B
P P P V V N N N N M M M M M C N N N C C C C M
W W X X X R R R P P P P P D D R R R D D D D P P
Y Z Z Z Z S S W W W W W F F F S S F F F F F W W W
J J J J J T Y Y Y Y Y G G G G T G G G G G Y Y Y
K K K K K L L L L L B B B B B B B B B B B L L L L
Q Q Q Q E M M M M E C C C C E C C C C E M M M M E
V V V I I P P P I I D D D I I D D D I I P P P I I
X X O O O W W O O O F F O O O F F O O O W W O O O
Z U U U U Y U U U U G U U U U G U U U U Y U U U U
```

As was indicated previously, Mauborgne recognized the repetition in Hitt's alphabets as a weakness. He attempted to construct alphabets that were as different as possible. When he was done, the only pairs of letters that appeared consecutively on more than one disk were DE, YA, and UB. Thus was born the DEYAUB cipher.

Mauborgne's twenty-five alphabets are reproduced below. The question is, did he have these alphabets in some form (strip or cylinder) early enough to use in the 1915 challenge? It's known that he had two cipher wheel devices with these alphabets made in the Army Signal School shop in early 1917, but did he have a less elegant prototype before that?

```
A A A A A A A A A A A A A A A A A A A A A A A A A
B C D E F G H I J K L M N O P Q R S T U V W X Y Z
C D K D N P X H D E T N C D B J M D O T N V K J D
E E O C Q O J P S L M F J W V N Y M J R K S W P N
I H M B U C E J K B S L I P H U O C Y Z H F R X B
G F J I K I Z O Q D X H L K I B F N L X R D E M U
D I U F D X B B O F V Q D J Y T T E F Q G L V V H
J J B G O L N W I J Q G H V K G H Q X L O I D K Y
F K G J P U I K V G P C B I S I E B N Y X E T B F
V T E H I R K C T H N U M U G M U O G I E B U Q W
U L P L T N P V Z O O J K Q U W S Z W O Y H F W J
Y M H K J D V F E N H T G H E Z Z P H V B K O U L
M O S M B Y R Z F M U B X Z N R J L V B F N Y G V
H U C R R Z O L H T W Y U C T V X G C P S R H L G
T V Z U H H G Q G P D P Z T C L D V M E J J M O T
Q Y I O C W S E Y R I Z T X X X P J I S M Q L S C
K G N Q Y B Y R U Q Z K S B O C C R R N U Z S T Q
Z Z X V S J D Y N S Y X W L W S W K B H D G I E M
O N F P L S U N L V C I Q E F H G Y S J Q M Q C P
L P Y T W Q L S P Z G S Y G Q D Q T E W C X N H S
R Q Q N E F C U M U K R V N D E I F K M L P J N O
X X R W M K F M B X R D O Y R O B U U D Z U C Z E
S R T Y Z V M G X Y F V R R L K K I P G W C P F X
P W V X V M Q T W W B E P S J F L W D F T O G R T
W S W Z X E T D C I E W F M Z P N X Z C I T B I K
N B L S G T W X R C J O E F M Y V H Q K P Y Z D I
1 2 3 4 5 6 7 8 9 · · · · · · · · · · · · · · · ·25
```

Mauborgne's disks were also numbered with letters, but with only twenty-five disks, one letter had to be left out. It was A that was omitted, so the first disk was B, the second was C, and so on, until the last, which was Z. This scheme makes a lot of sense, as the letter identification of each disk is the letter that appears below A on it.

Did you notice that the seventeenth alphabet begins ARMY OF THE US ...? When the U.S. Navy adopted the device, they gave it a new name, but they were stuck with the alphabets on the disks, the seventeenth of which was a constant reminder that it was an army man who had (re)invented it.

It's also worth noting that Mauborgne's cipher wheel had twenty-five disks, and his 1915 ciphertext, with 499 letters, is just one character shy of being a multiple of twenty-five. So, if he used a cipher wheel, he had to form his message on the wheel, and then read across another row of it twenty times to get his entire message enciphered.

Someone managing to select the right alphabets to attack Mauborgne's 1915 cipher may have another hurdle to deal with. This is detailed in the section below.

A Later Challenge

About a decade before publishing Mauborgne's 1915 challenge, Lou Kruh presented a challenge that Mauborgne issued later, in 1918. This challenge is known to have been created by Mauborgne's cipher wheel alphabets provided above. It consisted of twenty-five ciphertexts, all arising from a cipher wheel with the disks keyed in the same order.

The twenty-five ciphertexts were these:

```
 1 VFDJL QMMJB HSYVJ KCJTJ WDKNI
 2 CGNJM ZVKQC JPRJR CGOXG UCZVC
 3 CSTDT SSDJN JDKKT IXVEX VHDVK
 4 OZBGF VTUEC UGTZD KYWJR VZSDG
 5 QIRMB FTKBY CGAQV DQCVQ AHZGY
 6 VQWRM IHDHB RQBWU LKJCS KEYUU
 7 SSEIQ DWHNH QHGIK HAADN GNFBY
 8 VXDVX NIGJO PCOTN GKWAX YTNWL
 9 QJRLH AWTWU CYXVM BGJCR SBHWF
10 DULPK UXMVL XFUPS ULRZK PDALY
```

```
11 DCAIY LUPMB NACQE OPTLH KKRGT
12 MGODT VGUYX NHKBE WPOUR VTQOE
13 TBVEB QDXGP LCPUY AVVBK ZEOZY
14 FIJDW WBKTY GBSMB PZWYP RRZCW
15 DYVPJ CLNXE SCMFO YPIZF PEBHM
16 MYYTJ RFMEP PHDXP ODFZO WLGLA
17 EYKKD XHTEV TRXWK CJPSG MASCY
18 LGQLV HTUIP YAUGJ PGDLH UZTKV
19 BRKTJ RGGTB HMLXX FRHOA AZVWU
20 CDUDV DBZUA ELRPO SPUJD XRZWA
21 EUFBT TWNIY HHTNW QNFVE NYGBY
22 TUTVY NGLPG TYOLI HXZQT XSGOJ
23 PBTJC CJONJ UNIXB UAQBI WNIHL
24 VHNKR XVZMD KFHUY XRNDD KXXVM
25 NNHBF VQH0B LXCYM AKFLS SSJXG
```

This challenge was originally directed at Herbert O. Yardley and William F. Friedman. Neither expert could make anything of it. Mauborgne then provided a hint—the words ARE YOU appeared back to back in one of the messages. But even with this crib, Yardley and Friedman couldn't make any progress. This is what led to the army adopting the cipher wheel in 1921.

Friedman finally found the solutions to the ciphers in 1941, but not through cryptanalysis! Rather, he found them by going through the files in Mauborgne's office after his retirement.

When Kruh re-presented the challenge, he offered a prize (a new edition of Yardley's lively book *The American Black Chamber*) to the first person to provide all twenty-five messages, along with an explanation as to how they were found.

Kruh didn't have to wait long. Solutions to all twenty-five ciphers were found by TRIO (American Cryptogram Association [ACA] pen name for Fenwick Wesencraft of London[10]) in time to see print in the November–December issue of that same year (1982). The solution was also found by REEY (Mark E. R. Eyre, also of London) and THE BRUIN (Moshe Rubin of Jerusalem), but these gentlemen were not first. The article by TRIO, in which he presented the solutions, began with the comment "Mauborgne wasted his time dreaming up ridiculous test phrases."

The phrases were revealed to be these:

```
 1 14 Chlorine and oxygen have not b
 2  8 Where did you meet each other
 3  1 Drink this potion quickly for
 4 20 Well, make me the same shape but
 5 13 Cyanogen is a colorless gas in
 6 25 Phenols are benzene derivati
 7  2 Xylonite and artificial ivor
 8 12 I went to a new theatre, the Pala
 9 24 Picric acid is explosive and i
10 18 Llangollen is a town in Wales a
11  6 Yvette, are you going shopping
12 23 Orthophosphoric is the compo
13 16 Caoutchouc is closely allied
14 17 Olefiant gas, ethene or ethyle
15  3 See the terrible tank tackle a
16  7 It is a thin limpid liquid that
17 22 If it is insoluble in water it i
18 11 Silver has been known from rem
19 21 Hot concentrated sulphuric a
20  4 Small coefficient of expansi
21  9 Palladium possesses a power o
22 19 Absorbing and condensing thi
23 15 Compounds of platinum form tw
24 10 Gold occurs widely dristribut
25  5 Oxidation caused by it probab
```

The author also noted "and very dull reading they make."

The numbers between the cipher numbers and the solutions indicate the distance on the cipher wheel from the row making the message to the row selected for use as the ciphertext.

For the record, the order of the disks was 4, 14, 5, 6, 2, 13, 3, 24, 21, 11, 19, 8, 7, 16, 15, 17, 10, 9, 1, 12, 25, 23, 22, 20, 18.

This was nice work, but it didn't mean that the solvers were better than Yardley and Friedman. The government experts didn't know the alphabets that were used by Mauborgne, whereas Kruh presented them in his challenge. He did, however, make one error, writing a K where an L belonged. This tiny detail will become relevant soon!

Readers interested in seeing details of how the decipherments were obtained are referred to the original explanation published in *The Cryptogram*, but be warned, the solver described his method as "simple but extremely tedious."[11] He did not make use of a computer!

Trust, But Verify

In the April 1993 *Cryptologia* piece in which he presented Mauborgne's 1915 unsolved cipher, Kruh wrote, "The cipher was published in *The Cryptogram*, the official publication of the American Cryptogram Association, early last year." But it wasn't in a 1992 issue. It actually appeared in the March–April 1991 issue, p. 7, under Kruh's ACA pen name MEROKE.

So he made a mistake in a citation. He probably anticipated having his *Cryptologia* paper see print in 1992, and when it was delayed to the following year, he forgot to update the "early last year" reference to the previous piece. Big deal. Surely he would take care in the more important task of transcribing the cipher, right?

Rather than simply accept Kruh's presentation of Mauborgne's cipher, as a suspicious cryptologist, I wanted to see the original document. Perhaps Kruh left out something that might give us further insight into the solution, or more seriously, perhaps he made an error in copying the cipher letters for his article. Sadly, there's precedent for the latter.

Kruh and two coauthors wrote about yet another challenge cipher, actually a set of three of them, in a paper published in 1990. These ciphers were created using a system John F. Byrne designed in 1918 and called Chaocipher.[12] Byrne tried repeatedly for almost forty years to interest the U.S. government in his system, without success. He died in 1960, but his son, John Byrne, knew how the secret system worked and wished to continue promoting his father's work with the new challenge. Although the third cipher was presented correctly, the other two contained errors.

The most serious errors occurred in the second cipher. It was presented as

```
ENWSC EAQGI VIDEM WUMSN ZMNTV UFDLB JKKMR HHSNB KTJBH
VPTWH FMQQJ PGRWF FVJMD HFUZO XEOZT MKZSA MJYRL SQSXU
ZYEKR JBFRE SGGFX FEGXL PWTWL ZAVIM TBDTQ BLVRZ VEMMT
LXITZ
```

The deciphered message should have been

```
Our  own  memories  are  mysterious  enough  even  to
ourselves  and  we  should  recognize  that  they  do  not
exist  in  space  only  in  time  and  that  they  are  in  all
respects  immaterial.
```

But this couldn't be obtained. The third grouping of ciphertext letters, `VIDEM`, was wrong. It should have been written `UIOEM`. Misrepresenting `U` as `V` and `O` as `D` in the ciphertext meant that even someone applying the correct deciphering algorithm (which was kept secret) with the correct key (also secret!) would get only this:

```
OUROW NMEMO OIPXA AEJZG GRYTF NYMRS DJOIM USHNV LLFIN
WOJWK JAGTA MUQVR CBGGS HJAMW CCRGY JKAEY SPFAP PMHPZ
VXWFQ DGHXA HKHRJ EWLST ALBPT JSZAQ HQXVI WBLIC QTHQJ
MZVWU¹³
```

The paper with this serious error was not written by Kruh alone. He had two coauthors, one of whom was John Byrne, the son of the creator of the system! Knowing that this cipher was misrepresented by Kruh, and that he also made an error in reproducing the cipher alphabets for Mauborgne's later challenge, it doesn't take a leap of faith to imagine that Mauborgne's 1915 cipher could have been misrepresented as well.

So I contacted a librarian at the New York Public Library's Rare Book Division, asking for scans of the relevant papers, just to make sure I was presenting an accurate version of Mauborgne's 1915 cipher. Here's the response I got:

> *Dear Craig,*
> Thanks for your email regarding the cipher question. I have searched in our catalogs and have scanned our shelves—I cannot locate an item that matches this in our collection. The description is well intended, but as you point out there is no call number (or exact title) and that makes tracking this down a bit difficult. If you do come across any other references, please do let me know, and I will continue the search for you. There is a 1912 imprint that we have in the regular collection that is authored by Mauborgne, but this pre-dates your reference. Please let me know if I can be of any more assistance and best of luck.

Best,

Kyle R. Triplett, Librarian
Brooke Russell Astor Reading Room for
Rare Books and Manuscripts
The New York Public Library
Archives, Manuscripts, and Rare Books
Stephen A. Schwarzman Building,
Room 328
476 Fifth Avenue, New York, NY 10016
manuscripts@nypl.org

So, until these pages turn up, or a copy of them is found elsewhere, we have yet another doubt to contend with!

More Challenge Ciphers

Mauborgne's cipher challenge was intended to prove a point—that his system was superior to those in use at the time. However, there are many more cipher challenges that were put forth for a wide range of reasons. Some were simply challenges to entertain people, much like crossword puzzles and Sudoku. Some appear to have served as talent searches. Others are likely hoaxes. This chapter details examples from each category that remain challenges today, for solutions have yet to be found.

The Forgotten Cipher of Alexander d'Agapeyeff

Our first example appeared in 1939, as the world headed to war. The years that followed were a strange time for cryptology. The war generated an increased interest in codes and ciphers, but there were also, at least in the United States, attempts made by the government to suppress the interest. As new books on cryptology appeared to fill the public's demand, Secretary of War Henry L. Stimson decided that such knowledge was too dangerous to be disseminated. He sent off the following letter requesting a secret sort of ban.[1]

▼ ▼ ▼

CONFIDENTIAL
WAR DEPARTMENT
WASHINGTON

WD 461 (8-10-42)MS   AUG 13 1942
American Library Association
750 Michigan Avenue
Chicago, Illinois.

Gentlemen:

It has been brought to the attention of the War Department that member libraries of the American Library Association have received numerous requests for books dealing with explosives, secret inks, and ciphers.

It is requested that these books be removed from circulation and that these libraries be directed to furnish their local office of the Federal Bureau of Investigation with the names of persons making requests for same.

Your cooperation in this matter will be appreciated.

Sincerely yours,

Henry L Stimson (signed)
Secretary of War

This document contains information affecting the national defense of the United States within the meaning of the Espionage Act, U.S.C. 50; 31 and 32. Its transmission or the revelation of its contents in any manner to an unauthorized person is prohibited by law.

CONFIDENTIAL

▼ ▼ ▼

Censorship often results in a greater demand for the banned items. However, in this case, the general public wasn't even aware of the censorship because of the paragraph that followed Stimson's signature.

The backgrounds of the authors whom Stimson wanted to suppress were varied. Alexander d'Agapeyeff (1902–1955) was one of the most qualified. His dust jacket biography is reproduced here:

Mr. Alexander d'Agapeyeff is well qualified to write about Codes and Ciphers, for after joining the British army as a private at the age of sixteen, and getting a commission in the field at seventeen, he became intelligence officer with the North Russian Expeditionary Force, and was with the

British Military Mission in the Baltic till 1921, where he became au courant with the spying and counter-spying activities between Germany and Russia. After a short post-war time at Cambridge he went to Africa, made a rough survey of the route to Lake Chad, and became 'blood brother' of a wild tribe; and as surveyor and administrator of a tract of bush country in the back of beyond, had an opportunity of studying native customs, mythology and religion.[2]

FIG. 9.1 First edition of *Codes & Ciphers*

Although it wasn't mentioned in this brief biography, d'Agapeyeff was born in St. Petersburg, Russia. It would be interesting to know more about his time in Africa, but the only glimpse we are given is a statement in the book that he "spent several years in the African bush" and learned how the natives conveyed messages by beating drums. His native name was Nkoy.[3] Further biographical details are surprisingly difficult to find on this man who got around so much. As it turns out, his activities were much farther ranging than his wife would have liked. She divorced him in 1929, claiming that he had "frequently committed adultery." The court found her charges convincing.[4]

In 1940, the British formed a new organization called the Special Operations Executive (SOE). Its purpose was to infiltrate occupied Europe under cover and wreak havoc, or as Winston Churchill put it, "set Europe ablaze." D'Agapeyeff may have been well qualified for the SOE in some respects, but he was turned down. His betrayal of his wife may have had something to do with this.[5] Who could ever trust a man capable of betraying the person closest to him? In any case, he certainly found a way to do his share during World War II. He served as an RAF Wing Commander.

D'Agapeyeff's challenge cipher appeared on page 158 of his book *Codes and Ciphers*. He introduced it with the words, "Here is a cryptogram upon which the reader is invited to test his skill." The challenge is reproduced here:

```
75628 28591 62916 48164 91748 58464 74748 28483 81638
18174 74826 26475 83828 49175 74658 37575 75936 36565
81638 17585 75756 46282 92857 46382 75748 38165 81848
56485 64858 56382 72628 36281 81728 16463 75828 16483
63828 58163 63630 47481 91918 46385 84656 48565 62946
26285 91859 17491 72756 46575 71658 36264 74818 28462
82649 18193 65626 48484 91838 57491 81657 27483 83858
28364 62726 26562 83759 27263 82827 27283 82858 47582
81837 28462 82837 58164 75748 58162 92000
```

It's reasonable to assume that the method d'Agapeyeff used wasn't too far removed from those he discussed in the book. Ruling out systems not covered could save a lot of time. In particular, his book hardly mentions machine ciphers, which were in wide use at the time of publication. Such systems occupy just a tiny bit more than one page, and d'Agapeyeff showed that he was behind the times by ending this minimal coverage with "All these machines, however, even those with a typewriting keyboard, do not provide the degree of secrecy required for government use, and are therefore not used to any large extent."[6]

Even though d'Agapeyeff's book didn't cover the most advanced systems and was only 160 pages long, it still managed to describe a wide range of systems. These included (but were not limited to!) Caesar shift, skytale, Polybius, shorthand, mason marks, Oghams, Athbash, Tritheim's code (replacing each letter with one of fourteen words to create another fake text), trellis (I called it a grille in chapter 1), Vigenère, Solomon's alphabet (a MASC), Francis Bacon's biliteral cipher, jargon, transposition with multiple keys, Louis XIV's *grand chiffre*, Napoleon's *petit chiffre*, double columnar transposition, signaling codes (flags, torches, and semaphore), Morse code, "bush telegraph" or drum code in Africa (conveying messages by beating a drum), tramp's code, marked cards, shop codes (indicating what the owner paid for each item), Braille, hand alphabets, boy scout signs, commercial codes, St. Cyr slide, MASC, columnar transposition, cipher disk, Playfair, combined substitution-transposition, route transposition, code books, enciphered code, book code using a dictionary, cipher wheel, and a musical code (disguised as sheet music).

Several of the systems included the use of nulls.

We can quickly dismiss almost all of the systems listed above as possibilities for the challenge cipher because they don't convert messages to numbers. Polybius would be a good possibility, but that would yield a MASC, which would have been broken by now.

Something similar to Polybius appeared on page 127 and is shown in fig. 9.2.

	1	2	3	4	5	6
1	AB	AH	AN	AG	AM	AT
2	EC	EK	EG	EF	EM	ES
3	ID	IE	IR	ID	IL	IW
4	OF	OM	OS	OC	OK	OG
5	UG	UN	UB	UT	UF	UP

FIG. 9.2 A Polybius-like cipher

This grid allows us to substitute numbers for syllables instead of just individual letters. For example, the message A HUGE CAT ATE SAM is broken into AH UG EC AT AT ES AM and enciphered as 12 51 21 16 16 26.

It took me a while to come up with this lame example. The number of choices for the syllables is way too small and prevents the vast majority of messages from being enciphered. Also notice that the entry ID appears twice. This was probably a typo and not intended to introduce a homophone. Could d'Agapeyeff have used a larger version of this grid for his challenge cipher?

Another cipher he showed that works like Polybius, in how it converts text to letters, is shown in fig. 9.3.[7]

But it wouldn't be fair to use something like this as the challenge. How in the world would the reader be expected to determine what words appeared in the grid, given such a short message?

GERMAN ARMY CIPHER

2nd figure.

1st figure.		3	6	0	7	4	8	1	9	5	2
	2	a	ä	ai	au	aü	b	c	ch	ck	d
	6	e	ei	eu	f	ff	g	h	i	ie	j
	3	k	l	ll	m, mm	n,'nn	o	ö	p	pp	r
	7	s	sch	sp	spr	ss	st	str	t	tt	u
	4	ü	v	w	x	y	z	0	1	2	3
	0	4	5	6	7	8	9	.	,	;	?
	8	section	army	artillery	Battalion	Battery	Brigade	bridge	Division	railway	squadron
	1	field	flight	flying corps	engineers	rifles	trench mortars	group	infantry	scouts	guns
	9	cavalry	company	command	corps	men	morse	muni-tions	officer	horse	pioneer
	5	regiment	Red Cross	snipers	sappers	staff	tanks	train	troop	watch	balloons

Example:

Message: Three Companies to attack bridge-head at once.

Plain text: 3 Companies t o a tt a ck bridge h e a d a t o n c e .

Cipher: 42, 96, 79, 38, 23, 75, 23, 25, 81, 61, 63, 23, 22, 23, 79, 38, 34, 21, 63, 01.

Cryptogram: 42967 93823 75232 58161 63232 22379 38342 16301

FIG. 9.3 More Polybius-inspired cryptology

On pages 102 and 104, d'Agapeyeff showed a cipher that used the following as its key:

A	B	C	D	E	F	G	H	I	J	K	L	M	N	O	P	Q	R
21	27	12	29	35	19	26	36	11	28	37	18	25	31	13	38	17	24

S	T	U	V	W	X	Y	Z
32	14	23	33	16	34	22	15

This is a MASC and can be disregarded as too easy to be a possibility.

The only possibility left to consider from his book appeared on pages 106 and 107. It began by assigning numerical values to the letters like so:

A	B	C	D	E	F	G	H	I	J	K	L	M	N	O	P	Q	R	S	T
1	2	3	4	5	6	7	8	9	10	11	12	13	14	15	16	17	18	19	20

U	V	W	X	Y	Z
21	22	23	24	25	26

With this assignment, the message ENEMY MOVES EAST became 5 14 5 13 25 13 15 22 5 19 5 1 19 20.

But this was only the first step of the enciphering process. Another alphabet with A = 101, B = 102, C = 103, and so on, was used to encipher a keyword. In the example given, the keyword was HONOUR, which became 108 115 114 115 121 118.

These new numbers were then lined up with the previously enciphered message (repeating as necesary) and a subtraction was carried out. The result was

```
108 115 114 115 121 118 108 115 114 115 121 118 108 115
−  5  14   5  13  25  13  15  22   5  19   5   1  19  20
───────────────────────────────────────────────────────
103 101 109 102  96 105  93  93 109  96 116 117  89  95
```

Collecting the digits into groups of size five, the ciphertext became

```
10310 11091 02961 05939 31099 61161 17899 5.
```

This is similar to the Nihilist cipher put forth as a challenge by Schooling and broken by Elgar. The same sort of approach Elgar used would also work here.

Given that the book's coverage was well behind the state of the art, the solution to d'Agapeyeff's challenge should not have proved too elusive. Yet, nobody managed to find it. Even d'Agapeyeff was unable to help, for by the time frustrated enquiries reached him, he had forgotten the method he used to create the ciphertext in the first place! Perhaps in an effort to stave off more enquiries, d'Agapeyeff omitted the challenge from future editions of *Codes and Ciphers*. It appeared only in the first edition.

But his deletion didn't cause the challenge cipher to be forgotten. David Shulman, a member of the American Cryptogram Association (ACA), wrote about it in the April/May 1952 issue of the group's publication *The Cryptogram*. He offered a "suitable award" for the first person to send in the solution. This tactic had worked for Shulman once before. At a meeting of the New York Cipher Society, he presented an unsolved cipher from a novel by W. E. Woodward which, like d'Agapeyeff's challenge, the author himself had forgotten how to recover. A member of the ACA (pen name Yum Yum) solved it and received an autographed copy of the book as a prize.

Shulman even provided some preliminary analysis to help the reader. He ignored the last three zeros as nulls and divided the rest of the digits into groups of two. He then presented the following frequency table:

04 = 1	72 = 9	84 = 11
61 = 0	73 = 0	85 = 17
62 = 17	74 = 14	91 = 12
63 = 12	75 = 17	92 = 3
64 = 16	81 = 20	93 = 2
65 = 11	82 = 17	94 = 1
71 = 1	83 = 15	95 = 0

Notice that the first digit of each pair is always 6, 7, 8, 9, or 0 and the second digit is always 1, 2, 3, 4, or 5. It suggests a Polybius square set up like so:

	1	2	3	4	5
6	A	B	C	D	E
7	F	G	H	I	K
8	L	M	N	O	P
9	Q	R	S	T	U
0	V	W	X	Y	Z

The letters could be placed inside in any order though. It would be almost unbelievable that all of d'Agapeyeff's readers could fail to decipher something so simple. Indeed, Shulman pointed out that the frequency distribution doesn't resemble that of a MASC. We have too sharp a divide between the frequent and the rare, for example.

After Shulman's article was published, years passed before more was revealed. Finally, in 1959, Shulman provided an update. He explained that an army captain wrote to him offering further insight. The captain described arranging the pairs of digits (with the last three 0s discarded) into a 14 by 14 grid, as shown below. The pairs were placed in the grid vertically, column by column, as in the columnar transposition ciphers seen in chapter 5.

```
75 67 64 63 85 64 63 74 62 65 65 83 72 72
62 74 75 65 74 85 75 81 62 83 62 83 63 84
82 74 83 65 63 85 82 91 85 62 64 85 82 62
85 82 82 81 82 63 81 91 91 64 84 82 82 82
91 84 84 63 75 82 64 84 85 74 84 83 72 83
62 83 91 81 74 72 83 63 91 81 91 64 72 75
91 81 75 75 83 62 63 85 74 82 83 62 83 81
64 63 74 85 81 83 82 84 91 84 85 72 82 64
81 81 65 75 65 62 85 65 72 62 74 62 85 75
64 81 83 75 81 81 81 64 75 82 91 65 84 74
91 74 75 64 84 81 63 85 64 64 81 62 75 85
74 74 75 62 85 72 63 65 65 91 65 83 82 81
85 82 75 82 64 81 63 62 75 81 72 75 81 62
84 62 93 92 85 64 04 94 71 93 74 92 83 92
```

He noted that all of the low-frequency numbers (92, 93, 94, and 04) appeared only in the last row. He presumed these to be nulls used to complete some of the columns. Removing these leaves us with just thirteen distinct two-digit pairs.

But where do we go from there?

Shulman's approach of offering a prize for the solution of a cipher had worked in the past, and it did result in some progress being made this time, but ultimately it failed to yield a solution. Other researchers continued to make attempts in the decades that followed.

In 1978, Wayne G. Barker had his analysis of the cipher published. He was likely unaware of the work done by the army captain, for he simply got to the same point in a slightly different way.

Most recently, Gordon Rugg (whom we already encountered in chapter 1 and see again in chapter 10) and Gavin Taylor have been examining the cipher, as well as looking closely at a solution proposed by Armand Van Zandt. They've been publishing their progress online in a series of articles.[8]

Van Zandt thinks the decipherment is simply seven copies of the alphabet, in order.

```
ABCDEFGHIJKLMNOPQRSTUVWXYZ
ABCDEFGHIJKLMNOPQRSTUVWXYZ
ABCDEFGHIJKLMNOPQRSTUVWXYZ
```

```
ABCDEFGHIJKLMNOPQRSTUVWXYZ
ABCDEFGHIJKLMNOPQRSTUVWXYZ
ABCDEFGHIJKLMNOPQRSTUVWXYZ
ABCDEFGHIJKLMNOPQRSTUVWXYZ
```

The problem is that Van Zandt gave too cursory an account of how he obtained his solution. There may be some flexibility in the substitutions he makes in one step, and this usually indicates an incorrect solution. But Rugg and Taylor (and many others, including me) suspect that d'Agapeyeff made mistakes in the enciphering process. So, they're trying to find simple errors that d'Agapeyeff could have made that would, in turn, have led to the inconsistent substitutions Van Zandt had to make to get his solution.

Rugg and Taylor would like to discuss the proposed solution with Van Zandt, but they've been unable to find contact information for him. In the draft version of installment six of their series they wrote, "Armand—if you're reading this, we're impressed by the work that we've seen, and we'd very much like to hear from you; Gordon's easily contactable if you Google his Keele email address."[9]

The pair of researchers promised that more installments in this series will appear. I look forward to seeing them!

The Feynman Ciphers

The public has an appetite for eccentric physicists, and Richard Feynman filled the gap for some of the years between Albert Einstein and Stephen Hawking. In addition to professional credentials, like winning a Nobel Prize for his work on quantum electrodynamics, and having served on the team that developed the atomic bomb at Los Alamos, he had a pair of best-selling books consisting of entertaining tales from his life. Part of one became the basis for the feature film *Infinity* starring Mathew Broderick as Feynman.

While at Los Alamos, Feynman enjoyed testing security. One way he did this was by picking up safecracking skills. He read several books on the topic and was later able to truthfully say

I thought to myself, "Now *I* could write a safecracker book that would beat every one, because at the beginning I would tell how I opened safes whose

FIG. 9.4 Feynman (1918–1988) honored on
a U.S. postage stamp

contents were bigger and more valuable than what any safecracker anywhere had opened—except for a life, of course—but compared to the furs
or the gold bullion, I have them all beat: I opened the safes which contained
all the secrets to the atomic bomb: the schedules for the production of the
plutonium, the purification procedures, how much material is needed, how
the bomb works, how the neutrons are generated, what the design is, the
dimensions—the entire information that was known at Los Alamos: *the
whole schmeer!*"[10]

He also aggravated the postal censors at Los Alamos by using ciphers in
letters exchanged with his wife. If a later report is to be believed, Feynman
may have had the tables turned on him. On December 20, 1987, a little less
than two months before Feynman died, Chris Cole posted the following to sci.
crypt:

When I was a graduate student at Caltech, Professor Feynman showed me
three samples of code that he had been challenged with by a fellow scientist at Los Alamos and which he had not been able to crack. I also was
unable to crack them. I now post them for the net to give it a try.[11]

The first was labeled "Easier" and is this:

```
MEOTAIHSIBRTEWDGLGKNLANEAINOEEPEYST
NPEUOOEHRONLTIROSDHEOTNPHGAAETOHSZO
TTENTKEPADLYPHEODOWCFORRRNLCUEEEEOP
GMRLHNNDFTOENEALKEHHEATTHNMESCNSHIR
AETDAHLHEMTETRFSWEDOEOENEGFHETAEDGH
RLNNGOAAEOCMTURRSLTDIDOREHNHEHNAYVT
IERHEENECTRNVIOUOEHOTRNWSAYIFSNSHOE
MRTRREUAUUHOHOOHCDCHTEEISEVRLSKLIHI
IAPCHRHSIHPSNWTOIISISHHNWEMTIEYAFEL
NRENLEERYIPHBEROTEVPHNTYATIERTIHEEA
WTWVHTASETHHSDNGEIEAYNHHHNNHTW
```

The letter frequencies are about what they should be in normal English. This indicates that it's a transposition cipher. Jack Morrison, of NASA's Jet Propulsion Laboratory, recognized this fact and solved the cipher. In his sci.crypt post from December 21, 1987, he wrote

> Here is what I make of the first one. It's a pretty standard transposition: split the text into 5-column pieces, then read from lower right upward. I assume the odd language is something from literature I'm not familiar with; maybe someone can correct the punctuation and word breaks . . .[12]

His solution followed. The literature that he wasn't familiar with was an excerpt from Chaucer's *Canterbury Tales*. To see how difficult inserting word spacing and punctuation can be, consider the raw result of undoing the transposition.

```
WHANTHATAPRILLEWITHHISSHOURESSOOTET
HEDROGHTEOFMARCHHATHPERCEDTOTHEROOT
EANDBATHEDEVERYVEYNEINSWICHLICOUROF
WHICHVERTUENGENDREDISTHEFLOURWHANZE
PHIRUSEEKWITHHISSWEETEBREFTHINSPIRE
DHATHINEVERYHOLTANDHEETHTHETENDRECR
OPPESANDTHEYONGESONNEHATHINTHERAMHI
SHALVECOURSYRONNEANDSMALEFOWELESMAK
ENMELODYETHATSLEPENALTHENYGHTWITHOP
ENYESOPRIKETHHEMNATUREINHIRCORAGEST
HANNELONGENFOLKTOGOONONPILGRIM
```

Because of the many changes that English has gone through since Chaucer's time, this passage is still hard to read in spots, even after inserting word breaks.

```
Whan that Aprille, with his shoures soote
The droghte of March hath perced to the roote,
And bathed every veyne in swich licour
Of which vertu engendred is the flour;
Whan Zephirus eek with his sweete breeth
Inspired hath in every holt and heeth
The tendre croppes, and the yonge sonne
Hath in the Ram his half coursy ronne,
And smale foweles maken melodye,
That slepen al the nyght with open ye
(So priketh hem nature in hir corages),
Thanne longen folk to goon on pilgrim[ages,]¹³
```

Although Morrison rapidly solved the "Easier" cipher, he had nothing to say about the second cipher Cole posted, which was labeled "Harder." It's given here:

```
XUKEXWSLZJUAXUNKIGWFSOZRAWURORKXAOS
LHROBXBTKCMUWDVPTFBLMKEFVWMUXTVTWUI
DDJVZKBRMCWOIWYDXMLUFPVSHAGSVWUFWOR
CWUIDUJCNVTTBERTUNOJUZHVTWKORSVRZSV
VFSQXOCMUWPYTRLGBMCYPOJCLRIYTVFCCMU
WUFPOXCNMCIWMSKPXEDLYIQKDJWIWCJUMVR
CJUMVRKXWURKPSEEIWZVXULEIOETOOFWKBI
UXPXUGOWLFPWUSCH
```

This one stumped not only Feynman, Cole, and Morrison, it also stumped everyone else and remains unsolved today. It's clearly not a transposition cipher because U and W are the two most frequent letters.

Looking at groups of three or more letters reveals that several appear more than once. We have

Letters	Distance Between
WUR	193
RKX	186
VTW	63
UFP	87
WUF	76
MVR	6
CMUW	102
CMUW	26
WUID	39
CJUMV	6
JUMVR	6

Such repeats can often be exploited to help break ciphers. They make cracking Vigenère ciphers (which are covered in this chapter) especially easy. But it's unlikely that the cipher above is a Vigenère because the repeats don't seem to help in this case.

The third cipher was labeled "New Message." But this designation gives no indication of how hard it was supposed to be. Apparently, it's pretty hard, as it remains unsolved. It appears here:

```
WURVFXGJYTHEIZXSQXOBGSVRUDOOJXATBKT
ARVIXPYTMYABMVUFXPXKUJVPLSDVTGNGOSI
GLWURPKFCVGELLRNNGLPYTFVTPXAJOSCWRQ
DORWNWSICLFKEMOTGJYCRRAOJVNTODVMNSQ
IVICRBICRUDCSKXYPDMDROJUZICRVFWXIFP
XIVVIEPYTDOIAVRBOOXWRAKPSZXTZKVROSW
CRCFVEESOLWKTOBXAUXVB
```

For this cipher, V is the most common character, so, once again, we can conclude that it wasn't created using just transposition. The repeated pieces and the distances between them are

Letters	Distance Between
WUR	72
RVF	165
GJY	115

Letters	Distance Between
RUD	125
PYT	49
PYT	92
ICR	4
ICR	19

As with the second cipher, these repeats don't seem to suggest anything useful.

For the Feynman ciphers, the world of codebreakers has a record of one victory and two defeats. Turning to a more recent set of ciphers, we'll see a more respectable record.

The CIA's Cipher Mystery (1990)

With the aim of having a "pleasing work environment," the Central Intelligence Agency (CIA) set aside funds in the late 1980s to commission original art for its new headquarters building. A panel was established to solicit entries from American artists. In one of their meetings, they reviewed more than seven hundred slides submitted by 275 artists.[14]

The process resulted in James Sanborn being awarded a contract to develop a proposal. The proposal had to gain final approval from some big names in the agency before construction could begin, including the Director of Central Intelligence. In the end, the artist earned $250,000, and the agency received a unique sculpture bearing a cipher, part of which is pictured in fig. 9.5. It's called *Kryptos*.

The sculpture was dedicated on November 5, 1990. Director of Central Intelligence William H. Webster made remarks at the ceremony. Included in these were the following:

> What we have always wanted—in our buildings and in the art that makes them live—is to create an ideal place. For us, that would be one that is right for both reflection and challenge.[15]

and a comment directed at Sanborn:

> I know that all intelligence professionals will enjoy the opportunity to tackle the code you've presented us in this courtyard.[16]

FIG. 9.5 *Kryptos*

Webster was the one person who could read the sculpture's encrypted text without doing any cryptanalysis. At the ceremony, Sanborn gave him two sealed envelopes. One had the keywords that allowed decipherment, and the other had the plaintext. Sanborn said, "I hope you can keep a secret," as he handed them over.[17] Webster was the only one with whom Sanborn ever shared the secret of *Kryptos*.

The sculpture consists, in part, of four panels. Two panels on the left provided encrypted text. The two on the right showed the alphabet and another alphabet scrambled by the word KRYPTOS, along with shifts representing it in all possible starting positions. Because the sculpture is not accessible to those without business at CIA headquarters, an image was posted on the agency's website, along with the text appearing on the curved panels. These images are reproduced in figs. 9.6 and 9.7.

PANEL 1

```
E M U F P H Z L R F A X Y U S D J K Z L D K R N S H G N F I V J
Y Q T Q U X Q B Q V Y U V L L T R E V J Y Q T M K Y R D M F D
V F P J U D E E H Z W E T Z Y V G W H K K Q E T G F Q J N C E
G G W H K K ? D Q M C P F Q Z D Q M M I A G P F X H Q R L G
T I M V M Z J A N Q L V K Q E D A G D V F R P J U N G E U N A
Q Z G Z L E C G Y U X U E E N J T B J L B Q C R T B J D F H R R
Y I Z E T K Z E M V D U F K S J H K F W H K U W Q L S Z F T I
H H D D D U V H ? D W K B F U F P W N T D F I Y C U Q Z E R E
E V L D K F E Z M O Q Q J L T T U G S Y Q P F E U N L A V I D X
F L G G T E Z ? F K Z B S F D Q V G O G I P U F X H H D R K F
F H Q N T G P U A E C N U V P D J M Q C L Q U M U N E D F Q
E L Z Z V R R G K F F V O E E X B D M V P N F Q X E Z L G R E
D N Q F M P N Z G L F L P M R J Q Y A L M G N U V P D X V K P
D Q U M E B E D M H D A F M J G Z N U P L G E W J L L A E T G
```

PANEL 2

```
E N D Y A H R O H N L S R H E O C P T E O I B I D Y S H N A I A
C H T N R E Y U L D S L L S L L N O H S N O S M R W X M N E
T P R N G A T I H N R A R P E S L N N E L E B L P I I A C A E
W M T W N D I T E E N R A H C T E N E U D R E T N H A E O E
T F O L S E D T I W E N H A E I O Y T E Y Q H E E N C T A Y C R
E I F T B R S P A M H N E W E N A T A M A T E G Y E E R L B
T E E F O A S F I O T U E T U A E O T O A R M A E E R T N R T I
B S E D D N I A A H T T M S T E W P I E R O A G R I E W F E B
A E C T D D H I L C E I H S I T E G O E A O S D D R Y D L O R I T
R K L M L E H A G T D H A R D P N E O H M G F M F E U H E
E C D M R I P F E I M E H N L S S T T R T V D O H W ? O B K R
U O X O G H U L B S O L I F B B W F L R V Q Q P R N G K S S O
T W T Q S J Q S S E K Z Z W A T J K L U D I A W I N F B N Y P
V T T M Z F P K W G D K Z X T J C D I G K U H U A U E K C A R
```

FIG. 9.6 The left-hand side panels of *Kryptos*, according to the CIA website

I reproduced these images in my first book, thinking I was using the best possible source, but I later learned that there are some mistakes in the text that the CIA put forth. One mistake is especially relevant, and we return to it later.

Because of its mystery, *Kryptos* has garnered a tremendous amount of attention. I believe that it must be one of the best known sculptures of its era. Unfortunately, it attracted some negative attention at the agency before it was even completed. Part of a petition signed by sixteen CIA employees is reprinted here.[18] The signatures were redacted before this document was released in response to a Freedom of Information Act (FOIA) request:

PANEL 3

```
A B C D E F G H I J K L M N O P Q R S T U V W X Y Z A B C D
A K R Y P T O S A B C D E F G H I J L M N Q U V W X Z K R Y P
B R Y P T O S A B C D E F G H I J L M N Q U V W X Z K R Y P T
C Y P T O S A B C D E F G H I J L M N Q U V W X Z K R Y P T O
D P T O S A B C D E F G H I J L M N Q U V W X Z K R Y P T O S
E T O S A B C D E F G H I J L M N Q U V W X Z K R Y P T O S A
F O S A B C D E F G H I J L M N Q U V W X Z K R Y P T O S A B
G S A B C D E F G H I J L M N Q U V W X Z K R Y P T O S A B C
H A B C D E F G H I J L M N Q U V W X Z K R Y P T O S A B C D
I B C D E F G H I J L M N Q U V W X Z K R Y P T O S A B C D E
J C D E F G H I J L M N Q U V W X Z K R Y P T O S A B C D E F
K D E F G H I J L M N Q U V W X Z K R Y P T O S A B C D E F G
L E F G H I J L M N Q U V W X Z K R Y P T O S A B C D E F G H
M F G H I J L M N Q U V W X Z K R Y P T O S A B C D E F G H I
```

PANEL 4

```
N G H I J L M N Q U V W X Z K R Y P T O S A B C D E F G H I J
O H I J L M N Q U V W X Z K R Y P T O S A B C D E F G H I J L
P I J L M N Q U V W X Z K R Y P T O S A B C D E F G H I J L M
Q J L M N Q U V W X Z K R Y P T O S A B C D E F G H I J L M N
R L M N Q U V W X Z K R Y P T O S A B C D E F G H I J L M N Q
S M N Q U V W X Z K R Y P T O S A B C D E F G H I J L M N Q U
T N Q U V W X Z K R Y P T O S A B C D E F G H I J L M N Q U V
U Q U V W X Z K R Y P T O S A B C D E F G H I J L M N Q U V W
V U V W X Z K R Y P T O S A B C D E F G H I J L M N Q U V W X
W V W X Z K R Y P T O S A B C D E F G H I J L M N Q U V W X Z
X W X Z K R Y P T O S A B C D E F G H I J L M N Q U V W X Z K
Y X Z K R Y P T O S A B C D E F G H I J L M N Q U V W X Z K R
Z Z K R Y P T O S A B C D E F G H I J L M N Q U V W X Z K R Y
A B C D E F G H I J K L M N O P Q R S T U V W X Y Z A B C D
```

FIG. 9.7 The right-hand side panels of *Kryptos*, according to the CIA website

We, the undersigned, register our strong disapproval of the alleged art-work going up around the agency headquarters. It is unconscionable to us, in an era of massive budget deficits, the homeless, and likely reductions in the defense budget, that funds should be so grossly misappropriated. The expenditure of funds on any type of art would be inappropriate under these circumstances. The particular examples of art being forced upon us in this case come in for special scorn and ridicule because of their meaninglessness. Such meaningless art is out of place at an institution whose mission is to defend the meaning of the nation in the form of the constitution. The displeasing qualities of this art degrade rather than enhance our workplace and we have more important things to discuss than expensive conversation pieces.

We suggest that, if there is any real value to them at all, the sculptures be sold and our tax dollars be put to better use. The various geometric patterns strewn about the corridors would also be well disposed of in this manner. Some examples of employee art from the various exhibits that have appeared would better serve to decorate our halls and use our limited funds morally and wisely.

What constitutes art can be debated, but the petition definitely erred on one count. *Kryptos* was not "meaningless."

Jim Gillogly was the first to publicly reveal part of the meaning, despite his not even working for the CIA. He announced his solution on June 16, 1999, less than five months shy of a decade from the dedication. The long wait for a solution meant that he got a lot of media coverage, including an article in the *New York Times*.

The problem was more complicated than some realized. It was not a matter of finding a single key to read the entire ciphertext. Rather, Gillogly found that the left panels consisted of four distinct ciphers.

The first *Kryptos* ciphertext, now sometimes referred to as K1, was short. It's provided here:

```
EMUFPHZLRFAXYUSDJKZLDKRNSHGNFIVJ
YQTQUXQBQVYUVLLTREVJYQTMKYRDMFD
```

To understand how it worked, I need to introduce another cipher system, the Vigenère cipher. It was mentioned earlier in this book, including in the discussion of the Feynman ciphers, but it's now time to look at the details.

MASCs are weak because they consistently replace letters with the same substitutions. By contrast, the Vigenère cipher offers a way to systematically vary the substitutions. An example makes this clear. Suppose our message is

```
If you're going to threaten me with a knife, you may
as well cut me a little.
```

To encipher with Vigenère, we need to select a keyword. For this example, I'll use JIGSAW. The letters in JIGSAW provide us with six different substitution alphabets, like so

```
ABCDEFGHIJKLMNOPQRSTUVWXYZ (plaintext)
JKLMNOPQRSTUVWXYZABCDEFGHI (cipher alphabet 1)
IJKLMNOPQRSTUVWXYZABCDEFGH (cipher alphabet 2)
GHIJKLMNOPQRSTUVWXYZABCDEF (cipher alphabet 3)
STUVWXYZABCDEFGHIJKLMNOPQR (cipher alphabet 4)
ABCDEFGHIJKLMNOPQRSTUVWXYZ (cipher alphabet 5)
WXYZABCDEFGHIJKLMNOPQRSTUV (cipher alphabet 6)
```

I simply wrote JIGSAW down the left-hand side and then horizontally wrote out the rest of the alphabet, beginning with each of the letters in JIGSAW. When the end of the alphabet is reached, the letters start at the beginning again, until all twenty-six letters are listed.

To encipher the message, we go letter by letter, enciphering the first using cipher alphabet 1, the second with cipher alphabet 2, and so on. After using cipher alphabet 6 to encipher the sixth letter of the message, we're all out of alphabets. So we return to the top and use cipher alphabet 1 again, followed by 2, etc.

To help with the bookkeeping and make sure that the appropriate alphabet is used for each letter, before enciphering, the keyword is usually written out repeatedly beneath the letters in the message, like so:

```
IFYOUREGOINGTOTHREATENMEWITHAKNIFEYOUMAYASWELLCUT
JIGSAWJIGSAWJIGSAWJIGSAWJIGSAWJIGSAWJIGSAWJIGSAWJ

MEALITTLE
IGSAWJIGS
```

Now, carefully enciphering gives us this:

```
IFYOUREGOINGTOTHREATENMEWITHAKNIFEYOUMAYASWELLCUT
JIGSAWJIGSAWJIGSAWJIGSAWJIGSAWJIGSAWJIGSAWJIGSAWJ
RNEGUNNOUANCCWZZRAJBKFMAFQZZAGWQLWYKDUGQAOFMRDCQC
```

```
MEALITTLE
IGSAWJIGS
UKSLECBRW
```

Rewriting the ciphertext above without the key, and inserting word spacing allows us to notice some nice features of this cipher.

```
IF YOU'RE GOING TO THREATEN ME WITH A KNIFE, YOU MAY AS
RN EGU NN OUANC CW ZZRAJBKF MA FQZZ A GWQLW  YKD UGQ AO

WELL CUT ME A LITTLE.
FMRD CQC UK S LECBRW
```

The first time we had YOU (as part of YOU'RE), it was enciphered as EGU, but when it reappeared, it was enciphered as YKD. This is because the key lined up with YOU differently the second time, so each letter of it was enciphered with a different cipher alphabet.

If it happens to line up the same way, then the word will be enciphered exactly as it was the first time. Then, the distance between these identical pieces of ciphertext will be a multiple of the length of the keyword. This is a weakness, because once an attacker knows the length of the keyword, he or she can separate out letters that were all enciphered with the same cipher alphabet and take advantage of letter frequencies to determine it.

For example, if an attacker found that the keyword had length six, as above, then he or she would know that every sixth letter (positions 1, 7, 13, 19, 25, etc.) all used cipher alphabet 1. Looking at these selected letters as a group, he or she could see which is the most frequent and tentatively identify it as representing E. Now, even though the cipher alphabets begin with different letters, the other letters follow in order. So, if, for example, M is identified as E, then N is F, O is G, etc. The whole cipher alphabet is determined by one letter! In a similar manner, the letters in positions 2, 8, 14, 20, 26, etc. of the cipher would all have made use of cipher alphabet 2. Continuing as before, the entire cipher collapses.

A way to block against this attack (at least a little!) is by scrambling the order of the letters in the cipher alphabets. We'll still have the length of the key given away as before, but identifying one letter won't reveal the rest of them, because the attacker won't know how the scrambling was done.

This is how the first cipher on the *Kryptos* sculpture was created. The alphabets were scrambled using the keyword KRYPTOS, as seen on the panels on the right, and the keyword for selecting which alphabets to use was PALIMPSEST.

So, the cipher alphabets were these:

```
ABCDEFGHIJKLMNOPQRSTUVWXYZ (plaintext)
PTOSABCDEFGHIJLMNQUVWXZKRY (cipher alphabet 1)
ABCDEFGHIJLMNQUVWXZKRYPTOS (cipher alphabet 2)
LMNQUVWXZKRYPTOSABCDEFGHIJ (cipher alphabet 3)
IJLMNQUVWXZKRYPTOSABCDEFGH (cipher alphabet 4)
MNQUVWXZKRYPTOSABCDEFGHIJL (cipher alphabet 5)
PTOSABCDEFGHIJLMNQUVWXZKRY (cipher alphabet 6)
SABCDEFGHIJLMNQUVWXZKRYPTO (cipher alphabet 7)
EFGHIJLMNQUVWXZKRYPTOSABCD (cipher alphabet 8)
SABCDEFGHIJLMNQUVWXZKRYPTO (cipher alphabet 9)
TOSABCDEFGHIJLMNQUVWXZKRYP (cipher alphabet 10)
```

Gillogly was actually used to solving ciphers of this sort. As a member of the American Cryptogram Association, he regularly saw them in their periodical, *The Cryptogram*, where they're referred to as a Quagmire III ciphers. The panels on the left-hand side would lead someone aware of such ciphers to try using them as the cipher alphabets, but Gillogly still had to figure out the other part of the key—the word PALIMPSEST. The decipherment he finally obtained was

```
BETWEEN SUBTLE SHADING AND THE ABSENCE OF LIGHT LIES
THE NUANCE OF IQLUSION.
```

The Q in IQLUSION was an intentional "error" made by Sanborn. We'll come back to this.

The second cipher, K2, consists of the following characters:

```
VFPJUDEEHZWETZYVGWHKKQETGFQJNCE
GGWHKK?DQMCPFQZDQMMIAGPFXHQRLG
TIMVMZJANQLVKQEDAGDVFRPJUNGEUNA
QZGZLECGYUXUEENJTBJLBQCRTBJDFHRR
```

```
YIZETKZEMVDUFKSJHKFWHKUWQLSZFTI
HHDDDUVH?DWKBFUFPWNTDFIYCUQZERE
EVLDKFEZMOQQJLTTUGSYQPFEUNLAVIDX
FLGGTEZ?FKZBSFDQVGOGIPUFXHHDRKF
FHQNTGPUAECNUVPDJMQCLQUMUNEDFQ
ELZZVRRGKFFVOEEXBDMVPNFQXEZLGRE
DNQFMPNZGLFLPMRJQYALMGNUVPDXVKP
DQUMEBEDMHDAFMJGZNUPLGEWJLLAETG
```

Gillogly discovered that K2 was also a Quagmire III, using the keys KRYP-TOS (for scrambling the alphabets) and ABSCISSA (for selecting the cipher alphabets to be used and specifying their order). The decipherment Gillogly found was

```
IT WAS TOTALLY INVISIBLE. HOW'S THAT POSSIBLE? THEY
USED THE EARTH'S MAGNETIC FIELD. X THE INFORMATION
WAS  GATHERED  AND  TRANSMITTED  UNDERGRUUND  TO  AN
UNKNOWN  LOCATION.  X  DOES  LANGLEY  KNOW  ABOUT  THIS?
THEY SHOULD: IT'S BURIED OUT THERE SOMEWHERE. X WHO
KNOWS THE EXACT LOCATION? ONLY WW. THIS WAS HIS LAST
MESSAGE. X THIRTY EIGHT DEGREES FIFTY SEVEN MINUTES
SIX POINT FIVE SECONDS NORTH, SEVENTY SEVEN DEGREES
EIGHT MINUTES FORTY FOUR SECONDS WEST. ID BY ROWS.
```

This portion contained another intentional spelling error, undergru-und, but it also contained an error that Sanborn hadn't planned. The real mistake came at the end and was owned up to by Sanborn in 2006. The last few lines of the ciphertext were EWJLLAETG, but they should have been ESWJLLAETG. The decipherment given by the flawed portion of the cipher was "ID BY ROWS." The addition of the missing S changes the result completely. It becomes "X LAYER TWO." Normally, an error, in the enciphering process or in writing out the final cipher, results in gibberish being produced at that point in the deciphered message. It was just an incredible coincidence that Sanborn's accidental omission of the letter S led to a meaningful message completely different from what he intended.

When Gillogly moved on to the third cipher, K3, shown below, he noticed a change. The question mark gave him an obvious potential stopping point, but the letters leading up to that point had frequencies that matched normal English. This could not be a Quagmire III. The artist must have switched systems and used transposition to encipher this portion.

```
ENDYAHROHNLSRHEOCPTEOIBIDYSHNAIA
CHTNREYULDSLLSLLNOHSNOSMRWXMNE
TPRNGATIHNRARPESLNNELEBLPIIACAE
WMTWNDITEENRAHCTENEUDRETNHAEOE
TFOLSEDTIWENHAEIOYTEYQHEENCTAYCR
EIFTBRSPAMHHEWENATAMATEGYEERLB
TEEFOASFIOTUETUAEOTOARMAEERTNRTI
BSEDDNIAAHTTMSTEWPIEROAGRIEWFEB
AECTDDHILCEIHSITEGOEAOSDDRYDLORIT
RKLMLEHAGTDHARDPNEOHMGFMFEUHE
ECDMRIPFEIMEHNLSSTTRTVDOHW?
```

Following Gillogly's solution, at least two more ways of obtaining the plaintext for K3 were found. In the paragraphs below, the method that I think is the easiest to explain is shown.[19] It's not known which method was actually used by Sanborn.

Ignoring the question mark, arrange the letters into fourteen rows of twenty-four letters each:

```
ENDYAHROHNLSRHEOCPTEOIBI
DYSHNAIACHTNREYULDSLLSLL
NOHSNOSMRWXMNETPRNGATIHN
RARPESLNNELEBLPIIACAEWMT
WNDITEENRAHCTENEUDRETNHA
EOETFOLSEDTIWENHAEIOYTEY
QHEENCTAYCREIFTBRSPAMHHE
WENATAMATEGYEERLBTEEFOAS
FIOTUETUAEOTOARMAEERTNRT
IBSEDDNIAAHTTMSTEWPIEROA
GRIEWFEBAECTDDHILCEIHSIT
EGOEAOSDDRYDLORITRKLMLEH
AGTDHARDPNEOHMGFMFEUHEEC
DMRIPFEIMEHNLSSTTRTVDOHW
```

Then rotate the whole rectangle of letters ninety degrees to the right to get twenty-four rows of fourteen letters, like this:

```
DAEGIFWQEWRNDE
MGGRBIEHONAOYN
RTOISONEEDRHSD
IDEEETAETIPSHY
PHAWDUTNFTENNA
FAOFDEACOESOAH
ERSENTMTLELSIR
IDDBIUAASNNMAO
MPDAAATYERNRCH
ENREAEECDAEWHN
HEYCHOGRTHLXTL
NODTTTYEICEMNS
LHLDTOEIWTBNRR
SMODMAEFEELEEH
SGRHSRRTNNPTYE
TFIITMLBHEIPUO
TMTLEABRAUIRLC
RFRCWETSEDANDP
TEKEPEEPIRCGST
VULIIREAOEAALE
DHMHETFMYTETLO
OELSRNOHTNWISI
HEEIORAHEHMHLB
WCHTATSEYATNLI
```

Now arrange the characters into forty-two rows of eight letters each:

```
DAEGIFWQ
EWRNDEMG
GRBIEHON
AOYNRTOI
SONEEDRH
SDIDEEET
AETIPSHY
```

```
PHAWDUTN
FTENNAFA
OFDEACOE
SOAHERSE
NTMTLELS
IRIDDBIU
AASNNMAO
MPDAAATY
ERNRCHEN
REAEECDA
EWHNHEYC
HOGRTHLX
TLNODTTT
YEICEMNS
LHLDTOEI
WTBNRRSM
ODMΛEFEE
LEEHSGRH
SRRTNNPT
YETFIITM
LBHEIPUO
TMTLEABR
AUIRLCRF
RCWETSED
ANDPTEKE
PEEPIRCG
STVULIIR
EAOEAALE
DHMHETFM
YTETLOOE
LSRNOHTN
WISIHEEI
ORAHEHMH
LBWCHTAT
SEYATNLI
```

Rotate once more by ninety degrees to get this:

```
SLOWLYDESPARATLYSLOWLYTHEREMAINSOFPASSAGED
EBRISTHATENCUMBEREDTHELOWERPARTOFTHEDOORWA
YWASREMOVEDWITHTREMBLINGHANDSIMADEATINYBRE
ACHINTHEUPPERLEFTHANDCORNERANDTHENWIDENING
THEHOLEALITTLEIINSERTEDTHECANDLEANDPEEREDI
NTHEHOTAIRESCAPINGFROMTHECHAMBERCAUSEDTHEF
LAMETOFLICKERBUTPRESENTLYDETAILSOFTHEROOMW
ITHINEMERGEDFROMTHEMISTXCANYOUSEEANYTHINGQ
```

With word spacing and punctuation inserted, and changing the final Q to ?, the decipherment is this:

```
SLOWLY,  DESPARATLY  SLOWLY,  THE  REMAINS  OF  PASSAGE
DEBRIS  THAT  ENCUMBERED  THE  LOWER  PART  OF  THE  DOORWAY
WAS REMOVED. WITH TREMBLING HANDS I MADE A TINY BREACH
IN  THE  UPPER  LEFT-HAND  CORNER.  AND  THEN,  WIDENING
THE  HOLE  A  LITTLE,  I  INSERTED  THE  CANDLE  AND  PEERED
IN.  THE  HOT  AIR  ESCAPING  FROM  THE  CHAMBER  CAUSED  THE
FLAME  TO  FLICKER,  BUT  PRESENTLY  DETAILS  OF  THE  ROOM
WITHIN EMERGED FROM THE MIST. X CAN YOU SEE ANYTHING?
```

An intentional error here is the misspelling `desparatly`.

At this point, Gillogly was left with just K4. It's shown below and remains unsolved:

```
                     OBKR
UOXOGHULBSOLIFBBWFLRVQQPRNGKSSO
TWTQSJQSSEKZZWATJKLUDIAWINFBNYP
VTTMZFPKWGDKZXTJCDIGKUHUAUEKCAR
```

So, when all was said and done, Gillogly had a record of three and one against *Kryptos*. That's pretty damn good, considering that all other would-be solvers were zero and four. Well, almost all ...

After Gillogly put forth his solutions, it was revealed that David Stein, a CIA physicist and senior analyst, had found them first, in February 1999. His results were published in the classified version of the CIA journal *Studies in Intelligence*, now available online. Stein didn't use a computer for his analysis of the sculpture, but rather relied on pencil and paper.

John Markoff, reporting for the *New York Times,* wrote:

> Stein sounded a bit miffed when he learned that Gillogly had used a computer in his pursuit of the hidden codes.
>
> "*Kryptos* was meant to be solved with pencil and paper," he said.[20]

Gillogly disagreed! He said:

> There were no written rules in this contest. As far as I'm concerned a crack is where you find it. The choice of tool isn't the important part, but rather the decisions about how to use the tools.[21]

The former director of the CIA, William Webster, went on the record as well:

> For his part, Webster, the former Director, said yesterday that he had long since forgotten the answer. "I have zero memory of this," he said. "It was philosophical and obscure."
>
> But he sided with Gillogly on using a computer. "Who set the rules here?" he asked. "This is precisely what the agencies do when they try to break codes."[22]

Not only was Gillogly defended by Stein's former boss, but he also got to see Stein's recognition of having been the first solver destroyed, just as had happened to Gillogly.

The CIA's chief competitor, the National Security Agency (NSA), eventually revealed that its employees had the solution before Stein.

The fourth cipher remains, as far as is publicly known, unsolved.

> Sanborn, the artist, who has designed a number of sculptures that are puzzles, has said he believes that the ultimate secret hidden in the text of *Kryptos* will never be deciphered. It was designed by Edward M. Scheidt, a former chairman of the C.I.A.'s Cryptographic Center.[23]

Interest in *Kryptos* has grown tremendously over the past decade. In large part, this interest has been fueled by two of Dan Brown's novels. The dust jacket of the hardcover edition of *The Da Vinci Code* bore numbers giving the almost exact latitude and longitude of the sculpture. Brown's follow-up novel, *The Lost Symbol*, however, was much less subtle and provided a strange interpretation of a deciphered portion that was not intended by the artist and, in fact, angered him. The relevant portion of plaintext was Who knows the exact location? Only WW. In the novel, the WW portion was turned upside down to get MM, which was said to stand for Mary Magdalene. Sanborn had really intended it to stand for William Webster, the Director of Central Intelligence at the time the sculpture was both commissioned and dedicated. As was mentioned above, he is the only one to whom Sanborn revealed the secrets of the sculpture. By now, though, a few others have learned the secret. WW passed the solution on to his successors.

Another boon to the popularity of *Kryptos* was provided by Elonka Dunin, who organized a tremendous amount of *Kryptos* material, both primary and secondary sources, on her website and is also a frequent and enthusiastic lecturer on the topic of the sculpture, its ciphers, and the artist behind them.

Despite all of this publicity and the tremendous interest it generated, the fourth cipher remains unsolved. It seems likely that it made use of a method completely different from those that produced the first three ciphers. Some progress has been made, although mostly by Sanborn, as you'll see!

On November 20, 2010, Sanborn revealed that the "NYPVTT" portion of the ciphertext decrypts to "BERLIN."

For reasons that I will make clear, I suspected that matrix encryption may have been used to produce K4. Sanborn's clue facilitated some testing in this direction. But before showing the results, I need to show how matrix encryption works. As is so often the case, an example seems to be the clearest way to do this.

For our purposes, it will suffice to define a matrix as a rectangular array of numbers. This is oversimplifying things, but it won't make any difference. Square matrices can be useful for encryption. I'll use $M = \begin{pmatrix} 8 & 11 \\ 15 & 3 \end{pmatrix}$. The letter M stands for matrix here. Because **M** has two rows and two columns, we say it's a "2 by 2 matrix." Sometimes this is written as 2×2. **M** will be used to encipher

```
A REPO MAN SPENDS HIS LIFE GETTING INTO TENSE
SITUATIONS.
```

As an initial step, the letters are replaced with numbers using the scheme $A = 0, B = 1, C = 2, ... Z = 25$. This gives us

A	R	E	P	O	M	A	N	S	P	E	N	D	S	H	I
0,	17,	4,	15,	14,	12,	0,	13,	18,	15,	4,	13,	3,	18,	7,	8,

S	L	I	F	E	G	E	T	T	I	N	G	I	N	T	O	T
18,	11,	8,	5,	4,	6,	4,	19,	19,	8,	13,	6,	8,	13,	19,	14,	19,

E	N	S	E	S	I	T	U	A	T	I	O	N	S	X
4,	13,	18,	4,	18,	8,	19,	20,	0,	19,	8,	14,	13,	18,	23

An X was added at the end for reasons that become clear later. Now we begin to encipher the first pair of numbers, 0 and 17, like so

$$\begin{pmatrix} 8 & 11 \\ 15 & 3 \end{pmatrix}\begin{pmatrix} 0 \\ 17 \end{pmatrix} = \begin{pmatrix} 8 \times 0 + 11 \times 17 \\ 15 \times 0 + 3 \times 17 \end{pmatrix} = \begin{pmatrix} 187 \\ 51 \end{pmatrix}.$$

What's happening here? We took the first row in our matrix, 8 11, and multiplied each number by the corresponding value in the column of message numbers that was placed on its right. That is, the first number in the first row of our matrix, 8, was multiplied by the first number in the column that followed it, 0. Then, the second number in the first row of our matrix, 11, was multiplied by the second number in the column that followed it, 17. These two products were finally added together. That's how we got $8 \times 0 + 11 \times 17$. Next, we did the same thing with the second row of the matrix to get $15 \times 0 + 3 \times 17$.

Actually doing the multiplication and addition gives $\begin{pmatrix} 187 \\ 51 \end{pmatrix}$.

We want to convert our final answer back to letters, but their numerical values only go up to 25. So, whenever we get an answer greater than or equal to 26, we subtract 26 (as many times as necessary) until our answer is in the

proper range. For the example above, we get $\begin{pmatrix} 187 \\ 51 \end{pmatrix} = \begin{pmatrix} 5 \\ 25 \end{pmatrix}$. This is then converted to the letters F and Z by using our original substitutions.

The process of subtracting 26 repeatedly until we end up in the range 0 to 25 is known as "modulo 26 arithmetic," or "mod 26" for short. Another way to accomplish the same result is to divide 26 into the given number, ignoring how many times it goes in evenly, and then write down the remainder.

To encipher the next pair of letters in our message, we take the next pair of numbers and perform the exact same calculation with them. We have

$$\begin{pmatrix} 8 & 11 \\ 15 & 3 \end{pmatrix}\begin{pmatrix} 4 \\ 15 \end{pmatrix} = \begin{pmatrix} 8 \times 4 + 11 \times 15 \\ 15 \times 4 + 3 \times 15 \end{pmatrix} = \begin{pmatrix} 197 \\ 105 \end{pmatrix} = \begin{pmatrix} 15 \\ 1 \end{pmatrix}$$

This gives us the pair of letters P B. Continuing in this manner, the complete ciphertext is

FZPBKMNNXDTVOVOZFRPFUAHNGXOFXDUPOLQPWKNVEOGXVPHB.

Comparing this to the original message, we see the variety of letters that end up replacing repeated letters.

A REPO MAN SPENDS HIS LIFE GETTING INTO TENSE
F ZPBK MNN XDTVOV OZF RPFU AHNGXOF XDUP OLQPW

SITUATIONSX.
KNVEOGXVPHB.

This cipher has a feature in common with the Playfair cipher. Both replace the characters two at a time, albeit in completely different ways. But matrix encryption can do more. We can use matrices that are 3 by 3, or 4 by 4, or even larger to encipher letters three, four, or more at a time. A really quick example is given now for how the multiplication and addition is done for a 3 by 3 matrix. The beginning of our original message A R E = 0, 17, 4 is used as the plaintext. We have

$$\begin{pmatrix} 13 & 2 & 3 \\ 4 & 0 & 5 \\ 7 & 21 & 2 \end{pmatrix}\begin{pmatrix} 0 \\ 17 \\ 4 \end{pmatrix} = \begin{pmatrix} 13 \times 0 + 2 \times 17 + 3 \times 4 \\ 4 \times 0 + 0 \times 17 + 5 \times 4 \\ 7 \times 0 + 21 \times 17 + 2 \times 4 \end{pmatrix} = \begin{pmatrix} 40 \\ 20 \\ 365 \end{pmatrix} = \begin{pmatrix} 20 \\ 20 \\ 1 \end{pmatrix} \ (\text{mod } 26)$$

Converting back to letters gives us U U B.

I think 3 by 3 matrix encryption is far more likely to have been used for K4 than the 2 by 2 variety.

The idea that matrix encryption was used for K4 was not original to me. It's been discussed every now and then online for years, but it's not clear who was the first to suggest it. I have three reasons for suspecting its use. They're given below, and the best is saved for last!

Reason 1: Intentional errors in previous messages.

Sanborn said that the previous parts of *Kryptos* (K1-K3) have some role in decrypting K4. This made me suspect that the errors intentionally introduced into the previous parts might somehow be turned into numerical values to be used as entries for a decipher-ing matrix. Each error has an x and y coordinate in the rectangular array of ciphertext. These are $(25, 2)$ for the K in the second line of the sculpture; $(24, 6)$ for the R in the sixth line, and $(11, 15)$ and $(15, 15)$ for the L and E, respectively, in the fifteenth line. Four intentional errors thus provide eight numbers (ignoring the numerical values of the letters themselves, which would increase our count to 12). This is twice the number needed for 2 by 2 matrix encryption, but not enough for a 3 by 3 matrix. The larger number of 12 also fails to match the number of entries in any square matrix. Maybe these numbers provide only part of the matrix. I don't have a complete answer for this, but the intentional errors are known to be important.

Reason 2: Sanborn's clue.

If 3 by 3 matrix encryption was used, then to decipher, the letters would be converted to numbers, in groups of three, and multiplied by the deciphering matrix. Looking at how the cipher breaks into groups of three, we see that the portion Sanborn revealed to mean BERLIN consists of two whole groups, NYP and VTT. It's not split over any incomplete groups.

```
OBK RUO XOG HUL BSO LIF BBW FLR VQQ PRN GKS SOT
WTQ SJQ SSE KZZ WAT JKL UDI AWI NFB NYP VTT MZF
PKW GDK ZXT JCD IGK UHU AUE KCA R
```

This may be a coincidence. It might also be an attempt by Sanborn to provide a nicer clue than a portion of ciphertext that does not represent any encrypted blocks in their entirety.

Reason 3: "HILL".

The best indicator that matrix encryption was used is that the word "HILL" appears halfway down the right-hand side of *Kryptos*. You won't see it if you look at the representation given at the CIA's webpage for the texts. This is the relevant error I mentioned earlier. Fig. 9.8 shows it in context.

<table>
<tr><th colspan="2">THE CODE</th><th>THE KEY</th></tr>
<tr><td>K1</td><td>EMUFPHZLRFAXYUSDJKZLDKRNSHGNFIVJ</td><td>ABCDEFGHIJKLMNOPQRSTUVWXYZABCD</td></tr>
<tr><td></td><td>YQTQUXQBQVYUVLLTREVJYQTMKYRDMFD</td><td>AKRYPTOSABCDEFGHIJLMNQUVWXZKRYP</td></tr>
<tr><td>K2</td><td>VFPJUDEEHZWETZYVGWHKKQETGFQJNCE</td><td>BRYPTOSABCDEFGHIJLMNQUVWXZKRYPT</td></tr>
<tr><td></td><td>GGWHKK?DQMCPFQZDQMMIAGPFXHQRLG</td><td>CYPTOSABCDEFGHIJLMNQUVWXZKRYPTO</td></tr>
<tr><td></td><td>TIMVMZJANQLVKQEDAGDVFRPJUNGEUNA</td><td>DPTOSABCDEFGHIJLMNQUVWXZKRYPTOS</td></tr>
<tr><td></td><td>QZGZLECGYUXUEENJTBJLBQCRTBJDFHRR</td><td>ETOSABCDEFGHIJLMNQUVWXZKRYPTOSA</td></tr>
<tr><td></td><td>YIZETKZEMVDUFKSJHKFWHKUWQLSZFTI</td><td>FOSABCDEFGHIJLMNQUVWXZKRYPTOSAB</td></tr>
<tr><td></td><td>HHDDDUVH?DWKBFUFPWNTDFIYCUQZERE</td><td>GSABCDEFGHIJLMNQUVWXZKRYPTOSABC</td></tr>
<tr><td></td><td>EVLDKFEZMOQQJLTTUGSYQPFEUNLAVIDX</td><td>HABCDEFGHIJLMNQUVWXZKRYPTOSABCD</td></tr>
<tr><td></td><td>FLGGTEZ?FKZBSFDQVGOGIPUFXHHDRKF</td><td>IBCDEFGHIJLMNQUVWXZKRYPTOSABCDE</td></tr>
<tr><td></td><td>FHQNTGPUAECNUVPDJMQCLQUMUNEDFQ</td><td>JCDEFGHIJLMNQUVWXZKRYPTOSABCDEF</td></tr>
<tr><td></td><td>ELZZVRRGKFFVOEEXBDMVPNFQXEZLGRE</td><td>KDEFGHIJLMNQUVWXZKRYPTOSABCDEFG</td></tr>
<tr><td></td><td>DNQFMPNZGLFLPMRJQYALMGNUVPDXVKP</td><td>LEFGHIJLMNQUVWXZKRYPTOSABCDEFGH</td></tr>
<tr><td></td><td>DQUMEBEDMHDAFMJGZNUPLGEWJLLAETG</td><td>MFGHIJLMNQUVWXZKRYPTOSABCDEFGHI</td></tr>
<tr><td>K3</td><td>ENDYAHROHNLSRHEOCPTEOIBIDYSHNAIA</td><td>NGHIJLMNQUVWXZKRYPTOSABCDEFGHIJL</td></tr>
<tr><td></td><td>CHTNREYULDSLLSLLNOHSNOSMRWXMNE</td><td>OHIJLMNQUVWXZKRYPTOSABCDEFGHIJL</td></tr>
<tr><td></td><td>TPRNGATIHNRARPESLNNELEBLPIIACAE</td><td>PIJLMNQUVWXZKRYPTOSABCDEFGHIJLM</td></tr>
<tr><td></td><td>WMTWNDITEENRAHCTENEUDRETNHAEOE</td><td>QJLMNQUVWXZKRYPTOSABCDEFGHIJLMN</td></tr>
<tr><td></td><td>TFOLSEDTIWENHAEIOYTEYQHEENCTAYCR</td><td>RLMNQUVWXZKRYPTOSABCDEFGHIJLMNQ</td></tr>
<tr><td></td><td>EIFTBRSPAMHHEWENATAMATEGYEERLB</td><td>SMNQUVWXZKRYPTOSABCDEFGHIJLMNQU</td></tr>
<tr><td></td><td>TEEFOASFIOTUETUAEOTOARMAEERTNRTI</td><td>TNQUVWXZKRYPTOSABCDEFGHIJLMNQUV</td></tr>
<tr><td></td><td>BSEDDNIAAHTTMSTEWPIEROAGRIEWFEB</td><td>UQUVWXZKRYPTOSABCDEFGHIJLMNQUVW</td></tr>
<tr><td></td><td>AECTDDHILCEIHSITEGOEAOSDDRYDLORIT</td><td>VUVWXZKRYPTOSABCDEFGHIJLMNQUVWX</td></tr>
<tr><td></td><td>RKLMLEHAGTDHARDPNEOHMGFMFEUHE</td><td>WVWXZKRYPTOSABCDEFGHIJLMNQUVWXZ</td></tr>
<tr><td></td><td>ECDMRIPFEIMEHNLSSTTRTVDOHW?OBKR</td><td>XWXZKRYPTOSABCDEFGHIJLMNQUVWXZK</td></tr>
<tr><td>K4</td><td>UOXOGHULBSOLIFBBWFLRVQQPRNGKSSO</td><td>YXZKRYPTOSABCDEFGHIJLMNQUVWXZKR</td></tr>
<tr><td></td><td>TWTQSJQSSEKZZWATJKLUDIAWINFBNYP</td><td>ZZKRYPTOSABCDEFGHIJLMNQUVWXZKRY</td></tr>
<tr><td></td><td>VTTMZFPKWGDKZXTJCDIGKUHUAUEKCAR</td><td>ABCDEFGHIJKLMNOPQRSTUVWXYZABCD</td></tr>
</table>

FIG. 9.8 A representation of *Kryptos* showing the "HILL" message

This could be another unintentional error in the sculpture, similar to the one that appeared in an early part of the ciphertext, but this seems unlikely. Such an error would be more likely to produce gibberish or a word with no obvious connection to cryptology.

So, what is the connection to cryptology? Matrix encryption was discovered by Lester Hill in 1929. For this reason, it's often called the Hill cipher. Thus, Sanborn's clue isn't subtle at all.

Now that I've provided my best evidence for the use of matrix encryption, it's time to look at the results of an investigation I made into this possibility with a former colleague, Greg Link, and a former student, Dante Molle.

Given a numerical assignment for the letters in the alphabet, A = 0, B = 1, C = 2, ... Z = 25, as shown above, it's not hard to write a computer program to test all invertible 2 by 2 matrices to see if any yield a meaningful message. The qualifier "invertible" refers to the matrices for which there exist other matrices that can undo the changes they put the pairs of numbers through. There are only 157,248 such matrices.

None of them worked. Thanks to Sanborn's clue, we didn't have to examine all 157,248 results. We simply had the computer search for the word BERLIN among them. It never appeared.

But there are other obvious numerical assignments to test. Some cryptologists prefer to number the alphabet as A = 1, B = 2, C = 3, ... X = 24, Y = 25, Z = 0. Remember that in mod 26, we have 26 = 0. Sadly, this assignment also failed to get us BERLIN for any 2 by 2 matrix.

Next we tried values inspired by the right panels of *Kryptos*. The scrambled alphabet

K R Y P T O S A B C D E F G H I J L M N Q U V W X Z

suggests K = 0, R = 1, Y = 2, ... Z = 25, or possibly K = 1, R = 2, Y = 3, ... W = 24, X = 25, Z = 0.

Neither of these worked for the 2 by 2 case. We then tried all shifts of the alphabet scrambled with the word KRYPTOS. In other words, we tried all of the alphabets shown on the right panels of the sculpture. We tried numerical assignments beginning with both 0 and 1. We also tried every shift of the completely ordered alphabet, A B C D E F G H I J K L M N O P Q R S T U V W X Y Z, starting with both 0 and 1. None of these worked. The word BERLIN never appeared.

Thinking the word HILL on the right panels might be significant in more than one way, we went back to the 1929 paper in which Lester Hill first proposed matrix encryption and looked at the numerical assignment he used. It was a random scrambling.[24]

K	P	C	O	H	A	R	N	G	Z	E	Y	S	M	W	F	L	V	I	Q
0	1	2	3	4	5	6	7	8	9	10	11	12	13	14	15	16	17	18	19

D	U	X	B	T	J
20	21	22	23	24	25

We tried this, but the results were no better.

For the 3 by 3 case, the number of invertible matrices jumps to 1,634,038,189,056. Still, a program written by Link and Molle that took about ten hours to run for any given alphabet got the job done. We tried all of the alphabets described above.

Although none of the 2 by 2 matrices we tested yielded BERLIN in the given position (or anywhere else for that matter), a great many 3 by 3 matrices did. We typically found between 17,000 and 18,000 such matrices for each alphabet tested.[25] Although the word BERLIN appeared, it was surrounded by gibberish in all of the instances we examined. But with so many possibilities, we looked at only a tiny percentage of them. We needed a better way to proceed!

Fortunately, on November 20, 2014, while we were thinking about this, Sanborn released another clue. It represented an expansion of his previous clue. He said that NYPVTTMZFPK deciphered to BERLIN CLOCK. That is, he expanded on the portion he had previously deciphered. This was very useful! We hoped it would decrease our tremendous number of possible solutions down to just one. The reality was that it reduced it to zero. The words "BERLIN CLOCK" weren't generated by any of the numerical assignments we tried for 3 by 3 matrices.

Why did Sanborn keep providing clues? *Kryptos* has generated a great deal of publicity for him, but there are indications that he's ready for it to be solved. For example, there have been two break-ins at his studio where the criminals sought not jewelry or cash, but the solution to K4. Despite these break-ins, Sanborn has succeeded in keeping the secret of K4 hidden. He certainly kept it hidden from my coresearchers and me.

Even though the attack described above failed, I'm not ready to rule out matrix encryption. Only a minuscule percentage of numerical assignments for the alphabet was tested. Also, it's possible that matrix encryption was used in a more sophisticated manner than was investigated. If we let $\mathbf{M}$ represent the matrix and P and C represent three letters of plaintext and ciphertext each, then the method considered had the form $\mathrm{C} = \mathbf{M}\mathrm{P}$. That is, a matrix $\mathbf{M}$ was multiplied by values of the plaintext letters P to yield the ciphertext C. Maybe the correct equation was of the form $\mathrm{C} = \mathbf{M}\mathrm{P} + \mathrm{B}$. Or there could have been some sort of chaining used, where the encryption of each block depends not only on the matrix used, but also on the values of the letters in a previous block of plaintext or ciphertext. Many modern ciphers are implemented in chaining modes like this.

Maybe a 4 by 4 matrix was used. There are more than 12 sextillion of these suitable for encryption, and larger matrices offer far greater possibilities! Would Sanborn have made his challenge that difficult? Another possibility that could have complicated matters is the numerical assignment for the letters changing over the course of the message. Perhaps each letter is assigned a numerical value in accordance with the shifting alphabets scrambled by the keyword KRYPTOS.

Or maybe a pair of 2 by 2 matrices were used. We ruled out their being used in alternation, but they could have been used in some other way. The four intentional errors, when their x and y coordinates are considered, yield eight numbers. That's the right amount for a pair of 2 by 2 matrices.

But even when a solution is found to this last cipher, the mystery won't be over.

Mr. Sanborn said this week that the sculpture contains a riddle within a riddle—one that will be solvable only after the four encrypted passages are known.[26]

Sanborn spoke of this in an earlier interview.

They will be able to read what I wrote, but what I wrote is a mystery it-self. There are still things they have to discover once it's deciphered. There are things in there they will never discover the true meaning of. People will always say, 'What did he mean by that?' What I wrote out were clues to a larger mystery.[27]

While the solved portions of *Kryptos* indicate that the sculpture is firmly rooted in the world of paper and pencil encryption, our next challenge in-cludes a mix of old methods and very modern techniques, embracing more than two thousand years of cryptologic history.

Cicada 3301

It has been called "the most elaborate and mysterious puzzle of the inter-net age," and is listed as one of the "Top 5 eeriest, unsolved mysteries of the Internet" by the *Washington Post*, and much speculation exists as to its purpose. Many have speculated that it is a recruitment tool for the NSA, CIA, MI6, or a cyber mercenary group. Others have claimed it is an Alter-nate Reality Game (ARG), but the fact that no company or individual has taken credit or tried to monetize it, combined with the fact that none who have solved the puzzles have ever come forward, has led most to feel that it is not. Others have claimed it is run by a bank working on cryptocurrency.[28]

The mystery began on January 4, 2012, with a post to www.4chan.org. This website allows users to anonymously post small amounts of text and include a picture, if they like, or respond to someone else's post in the same manner. There are various boards for particular interests; one of the more popular is called /b/, where the content is "random."

> The idea is post something that can never be unseen. Half of the posts on /b/ are there specifically to make people not want to come back to /b/.
>
> —Gregg Housh, Internet activist associated
> with the hacker group Anonymous.[29]

The image from /b/ that concerns us was not of this nature. It's reproduced in fig. 9.9.

FIG. 9.9 The beginning of Cicada 3301 in January 2012

Opening an image in a text editor wouldn't normally yield any interesting reading matter, but in this case, at the very end, there was a message appended:

TIBERIVS CLAVDIVS CAESAR says "lxxt>33m2mqkyv2gsq3q=w]O2ntk"

The reference to the fourth Caesar, followed by what looks like a cipher, makes it natural to wonder if a Caesar shift cipher was used. Shifting each of the symbols back by four takes the first few letters, lxxt, to http, so it looks like the cipher is going to reveal a website.

It's obvious how to shift the letters, but what's to be done with the numbers and the symbols >, =, and] ? The answer is provided by something computer scientists are familiar with, namely ASCII. This is an abbreviation for "American Standard Code for Information Interchange." It assigns numerical values to a wide range of characters. Table 9.1, on the next page, is an ASCII table.

So we have numerical representations for upper- and lowercase letters, numbers, punctuation, and other characters. Number 32 designates a blank space, and there are numbers for other keyboard actions that do not result in something appearing on the screen. These include Tab (number 9) and Escape (number 27). There's also "extended ASCII," which provides numerical values in the range of 128 through 256 to represent another 128 (less common) characters.

So, using the ASCII chart, shifting > back by four gives :, and shifting] by the same amount, we get Y. The numbers are also easy to shift. We finally get the website fully deciphered as

http://i.imgur.com/m9sYK.jpg

Going to this website yielded another image, shown in fig. 9.10.

FIG. 9.10 A dead end?

Table 9.1 Regular ASCII Chart

Dec	Name	Char	Dec	Char	Dec	Char	Dec	Char
0	Null	NUL	32	Space	64	@	96	`
1	Start of heading	SOH	33	!	65	A	97	a
2	Start of text	STX	34	"	66	B	98	b
3	End of text	ETX	35	#	67	C	99	c
4	End of xmit	EOT	36	$	68	D	100	d
5	Enquiry	ENQ	37	%	69	E	101	e
6	Acknowledge	ACK	38	&	70	F	102	f
7	Bell	BEL	39	'	71	G	103	g
8	Backspace	BS	40	(	72	H	104	h
9	Horizontal tab	HT	41	)	73	I	105	i
10	Line feed	LF	42	*	74	J	106	j
11	Vertical tab	VT	43	+	75	K	107	k
12	Form feed	FF	44	,	76	L	108	l
13	Carriage feed	CR	45	-	77	M	109	m
14	Shift out	SO	46	.	78	N	110	n
15	Shift in	SI	47	/	79	O	111	o
16	Data line escape	DLE	48	0	80	P	112	p
17	Device control 1	DC1	49	1	81	Q	113	q
18	Device control 2	DC2	50	2	82	R	114	r
19	Device control 3	DC3	51	3	83	S	115	s
20	Device control 4	DC4	52	4	84	T	116	t
21	Neg. acknowledge	NAK	53	5	85	U	117	u
22	Synchronous idle	SYN	54	6	86	V	118	v
23	End of xmit block	ETB	55	7	87	W	119	w
24	Cancel	CAN	56	8	88	X	120	x
25	End of medium	EM	57	9	89	Y	121	y
26	Substitute	SUB	58	:	90	Z	122	z
27	Escape	ESC	59	;	91	[	123	{
28	File separator	FS	60	<	92	\	124		
29	Group separator	GS	61	=	93	]	125	}
30	Record separator	RS	62	>	94	^	126	~
31	Unit separator	US	63	?	95	_	127	DEL

This was not a dead end, but another clue.

Information can be hidden in any digital image in a much subtler way than appending text to the end of the file. The color of each pixel in an image is represented by a string of zeros and ones. Changing the least significant of these (like changing the weight of something from 2,743 ounces to 2,744 ounces) results in an image that can't be distinguished from the original by the human eye. One can change as many pixels as desired in such a way that the new least significant bits, when taken together, contain a message.

This is just one example of how a message may be hidden or "embedded" in an image. There are many more ways to do it. In general, this area is known as digital steganography. Although the details are all classified, it's rumored that 9/11 terrorists communicated by embedding messages in pornographic images and sharing them over the Internet.

One tool for embedding and extracting messages in images is called Out-Guess (see http://www.outguess.org/ for details). The image reproduced above contained the words "out" and "guess." So, for those aware not only of digital steganography in general, but of the OutGuess software in particular, the next step wasn't hard to see. Anyone without this knowledge would be unlikely to progress to the next clue!

Using OutGuess yielded a message that was simply another website, http://pastebin.com/aXYZzzcv. It contained the following

Here is a book code. To find the book, and more information, go to http://www.reddit.com/r/a2e7j6ic78h0j/

1:20	14:4	27:5
2:3	15:8	28:1
3:5	16:4	29:2
4:20	17:5	30:18
5:5	18:14	31:32
6:53	19:7	32:10
7:1	20:31	33:3
8:8	21:12	34:25
9:2	22:36	35:10
10:4	23:2	36:7
11:8	24:3	37:20
12:4	25:5	38:10
13:13	26:65	39:32

40:4	53:18	66:18
41:40	54:4	67:45
42:11	55:6	68:10
43:9	56:4	69:2
44:13	57:24	70:17
45:6	58:64	71:9
46:3	59:5	72:20
47:5	60:37	73:2
48:43	61:60	74:34
49:17	62:12	75:13
50:13	63:6	76:21
51:4	64:8	
52:2	65:5	

Good luck.

3301

The Reddit page for which the link was provided contained several posts that looked to be enciphered text. These were more difficult to crack than the Caesar shift, though. The top of the page provided clues. One of these is reproduced in fig. 9.11.

FIG. 9.11 A clue for deciphering part of the 2012 Cicada 3301 challenge

These are Mayan numbers. The dots each represent one, and every bar represents five. The football-looking object second from the right designates zero. Thus, the image yields

10, 2, 14, 7, 19, 6, 18, 12, 7, 8, 17, 0, 19

The page also contained the string a2e7j6ic78h0j7eiejd0120. Lining this up with our Mayan numbers, we notice a nice fit:

10,2,14,7,19,6,18,12, 7, 8,17,0,19
 a, 2, e,7, j,6, i, c,7, 8, h, 0, j,7,e,i,e,j,d,0,1,2,0

The single-digit numbers match perfectly, and larger numbers are converted to letters. That is a = 10, c = 12, e = 14, etc. Our string of letters and numbers is longer, but the relationship is clear. We can fill in the missing numbers across the top row as

```
10, 2, 14, 7, 19, 6, 18, 12, 7, 8, 17, 0, 19, 7, 14,
 a, 2,  e, 7,  j, 6,  i,  c, 7, 8,  h, 0,  j, 7,  e,

18, 14, 19, 13, 0, 1, 2, 0
 i,  e,  j,  d, 0, 1, 2, 0
```

This sequence of numbers served as the key for deciphering the posts on the Reddit page. For example, the first comment to be posted was

```
Ukbn Txltbz nal hh Uoxelmgox wdvg Akw; hvu ogl rsm ar
sbv ix jwz[30]
```

Lining the key up under these letters and then shifting each letter backward by the appropriate amount gives this:

```
 U   k   b   n    T   x   l   t   b   z ...
10,  2, 14,  7,  19,  6, 18, 12,  7,  8 ...
 K   i   n   g    A   r   t   h   u   r ...
```

So we have the beginning of a message about King Arthur.

Deciphering all of the posts on the Reddit page in this manner provided the "book" from which the book code encountered earlier can be decoded. The "book" begins

1. King Arthur was at Caerlleon upon Usk; and one day he sat in his
2. chamber; and with him were Owain the son of Urien, and Kynon the son
3. of Clydno, and Kai the son of Kyner; and Gwenhwyvar and her
4. handmaidens at needlework by the window. And if it should be said
5. that there was a porter at Arthur's palace, there was none. Glewlwyd
6. Gavaelvawr was there, acting as porter, to welcome guests and
7. strangers, and to receive them with honour, and to inform them of the

8. manners and customs of the Court; and to direct those who came to the

9. Hall or to the presence-chamber, and those who came to take up their lodging.

10. In the centre of the chamber King Arthur sat upon a seat of green

11. rushes, over which was spread a covering of flame-coloured satin, and

12. a cushion of red satin was under his elbow.

13. Then Arthur spoke, "If I thought you would not disparage me," said

14. he, "I would sleep while I wait for my repast; and you can entertain

15. one another with relating tales, and can obtain a flagon of mead and

16. some meat from Kai." And the King went to sleep. And Kynon the son

17. of Clydno asked Kai for that which Arthur had promised them. "I,

18. too, will have the good tale which he promised to me," said Kai.

19. "Nay," answered Kynon, "fairer will it be for thee to fulfill

20. Arthur's behest, in the first place, and then we will tell thee the

21. best tale that we know." So Kai went to the kitchen and to the mead-

22. cellar, and returned bearing a flagon of mead and a golden goblet,

23. and a handful of skewers, upon which were broiled collops of meat.

24. Then they ate the collops and began to drink the mead. "Now," said

25. Kai, "it is time for you to give me my story." "Kynon," said Owain,

26. "do thou pay to Kai the tale that is his due." "Truly," said Kynon,

27. "thou are older, and art a better teller of tales, and hast seen more

28. marvellous things than I; do thou therefore pay Kai his tale."

29. "Begin thyself," quoth Owain, "with the best that thou knowest." "I

30. will do so," answered Kynon.

31. "I was the only son of my mother and father, and I was exceedingly

32. aspiring, and my daring was very great. I thought there was no

33. enterprise in the world too mighty for me, and after I had achieved

34. all the adventures that were in my own country, I equipped myself,

35. and set forth to journey through deserts and distant regions. And at

36. length it chanced that I came to the fairest valley in the world,

37. wherein were trees of equal growth; and a river ran through the

38. valley, and a path was by the side of the river. And I followed the

39. path until mid-day, and continued my journey along the remainder of

40. the valley until the evening; and at the extremity of a plain I came

41. to a large and lustrous Castle, at the foot of which was a torrent.

42. And I approached the Castle, and there I beheld two youths with

43. yellow curling hair, each with a frontlet of gold upon his head, and

44. clad in a garment of yellow satin, and they had gold clasps upon

45. their insteps. In the hand of each of them was an ivory bow, strung
46. with the sinews of the stag; and their arrows had shafts of the bone
47. of the whale, and were winged with peacock's feathers; the shafts
48. also had golden heads. And they had daggers with blades of gold, and
49. with hilts of the bone of the whale. And they were shooting their daggers.
50. "And a little way from them I saw a man in the prime of life, with
51. his beard newly shorn, clad in a robe and a mantle of yellow satin;
52. and round the top of his mantle was a band of gold lace. On his feet
53. were shoes of variegated leather, fastened by two bosses of gold.
54. When I saw him, I went towards him and saluted him, and such was his
55. courtesy that he no sooner received my greeting than he returned it.
56. And he went with me towards the Castle. Now there were no dwellers
57. in the Castle except those who were in one hall. And there I saw
58. four-and-twenty damsels, embroidering satin at a window. And this I
59. tell thee, Kai, that the least fair of them was fairer than the
60. fairest maid thou hast ever beheld in the Island of Britain, and the
61. least lovely of them was more lovely than Gwenhwyvar, the wife of
62. Arthur, when she has appeared loveliest at the Offering, on the day
63. of the Nativity, or at the feast of Easter. They rose up at my
64. coming, and six of them took my horse, and divested me of my armour;
65. and six others took my arms, and washed them in a vessel until they
66. were perfectly bright. And the third six spread cloths upon the
67. tables and prepared meat. And the fourth six took off my soiled
68. garments, and placed others upon me; namely, an under-vest and a
69. doublet of fine linen, and a robe, and a surcoat, and a mantle of

The numbers in the book code each indicate a line and a character. For example, the code begins with 1:20, so we look at line 1 and take the 20th character (counting blank spaces), which is c. The code continues with 2:3, so we look at line 2 and take the 3rd character, which gives us an a. The full decipherment is this:

Call us at us tele phone numBer two one four three nine oh nine six oh eight

But before seeing the result of calling (214) 390-9608, there are a few more secrets in the Reddit page left to ferret out. It contained a pair of images, reproduced in figs. 9.12 and 9.13.

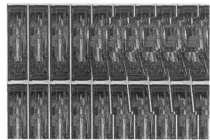

FIG. 9.12 Those making it to the Reddit page found a welcome mat ...

FIG. 9.13 ... and a wallpaper autostereogram

The Welcome image hid another message steganographically. Like the one in the duck image, it could be recovered with OutGuess. It read

```
-----BEGIN PGP SIGNED MESSAGE-----
Hash: SHA1
- From here on out, we will cryptographically sign all messages
with this key.
It is available on the mit keyservers. Key ID 7A35090F, as
posted in a2e7j6ic78h0j.
Patience is a virtue.
Good luck.
3301
-----BEGIN PGP SIGNATURE-----
Version: GnuPG v1.4.11 (GNU/Linux)
```

```
iQIcBAEBAgAGBQJPBRz7AAoJEBgfAeV6NQkP1UIQALFcO8DyZkecTK5pAIcGez7k
ewjGBoCfjfO2NlRROuQm5CteXiH3Te5G+5ebsdRmGWVcah8QzN4UjxpKcTQRPB9e
/ehVI5BiBJq8GlOnaSRZpzsYobwKH6Jy6haAr3kPFK1lOXXyHSiNnQbydGw9BFRI
fSr//DY86BUILE8sGJR6FA8Vzjiifcv6mmXkk3ICrT8z0qY7m/wFOYjgiSohvYpg
x5biG6TBwxfmXQOaITdO5rO8+4mtLnP//qN7E9zjTYj4Z4gBhdf6hPSuOqjh1s+6
/C6IehRChpx8gwpdhIlNf1coz/ZiggPiqdj75Tyqg88lEr66fVVB2d7PGObSyYSp
HJl8llrt8Gnk1UaZUS6/eCjnBniV/BLfZPVD2VFKH2Vvvty8sL+S8hCxsuLCjydh
skpshcjMVV9xPIEYzwSEaqBq0ZMdNFEPxJzC0XIS1WSfxROm85r3NYvbrx9lwVbP
mUpLKFn8ZcMbf7UX18frgOtujmqqUvDQ2dQhmCUywPdtsKHFLc1xIqdrnRWUS3CD
eejUzGYDB5lSflujTjLPgGvtlCBW5ap00cfIHUZPOzmJWoEzgFgdNc9iIkcUUlke
e2WbYwCCuwSlLsdQRMA//PJN+a1h2ZMSzzMbZsr/YXQDUWvEaYI8MckmXEkZmDoA
RLOxkbHEFVGBmoMPVzeC
=fRcg
```

```
-----END PGP SIGNATURE-----
```

The gibberish that follows this message was not another cipher to solve, per se. To understand it, some modern cryptology needs to be detailed.

By the end of the twentieth century, much more was expected of cryptology than securely enciphering messages. The bar had been raised to also include authenticity and integrity. That is, it's now expected that cryptology can guarantee not only that the message came from the person claiming to have sent it (authenticity), but also that it hasn't been changed by anyone while it was en route (integrity). The first of these considerations is addressed in a bit more detail in chapter 11, but I give a little background now.

Some modern systems tackle these greater demands by using two distinct keys, one that's revealed to the public and one that's kept private. If you want to send a message to someone using one of these systems, all you have to do is look up his or her public key and use it to encipher the message. The recipient will then use his or her private key to decipher it. Because it's extremely difficult to figure out the private key from the public key, the enciphered message you send should be almost impossible to break for anyone intercepting it.

This allows people to communicate securely, even if they haven't met ahead of time to agree on a key to use. But used another way, it does even more. When the message above was sent by whoever is behind Cicada 3301, a "signature" followed it. To oversimplify a bit, this was generated by the sender enciphering the original message above it with his secret *private* key. The sender referred the reader to the location on the MIT keyservers where his public key could be found. Applying this to the apparent gibberish of the signature reveals another copy of the message. Anyone can decipher it, but only someone knowing the private key could have created it. Thus, it serves as a signature for the message.

Okay, I indicated that I was oversimplifying. I'll now make it right. If a message is long, it's time-consuming to encipher the whole thing with a private key to create a signature. It also doubles the length of what needs to be sent. More typically, the person wanting to sign forms a sort of fingerprint of the message to represent it. It's this much shorter document that he or she then signs by enciphering it with his or her private key. It's easy to take a person's fingerprint, but how do we get a message reduced to a smaller form? Mathematical tools known as hash functions do this. Detailing their design would take us too far afield, but they basically convert a text to a much shorter length in a way that makes it extremely difficult to find another message that converts to the same result. If the recipient changes part of the message and tries to claim the altered document as having been signed, the signature reveals the lie. Changing the message almost certainly results in a different value being produced for it by the hash function, and enciphering this different value with a private key gives a different final signature. The attempted fraud will be detected. It's important in this process that the particular hash function that was used be known. This is not something that's meant to be kept secret.

The signature method employed here began by using SHA-1 (Secure Hash Algorithm 1), followed by encipherment with a private key belonging to Cicada 3301. All of the real Cicada messages that appeared over the years were also signed, and this served as an excellent way to distinguish them from imposters. By consistently using the same private key to sign messages, they are confirmed to all have come from the same person (or group, if the key was shared among several members of Cicada).

You might wonder why the signature is so long, if the message was hashed first. The reason is that the Cicada 3301 group used a very strong key to encipher the hashed message, and powerful keys in this system give longer final results.

Moving to the other image on the Reddit page, solvers discovered that it hid two things. First, without any cryptology being applied, one can see that it's a wallpaper autostereogram. Like the better known images in the Magic Eye books, which are of a different type—random-dot autostereograms—relaxing the focus of one's eyes while staring at the image causes an object to appear to be floating in front of the background pattern. In the current example, it's a cup.

This same image also yielded a message via OutGuess. It read

```
-----BEGIN PGP SIGNED MESSAGE-----
Hash: SHA1
The key has always been right in front of your eyes.
This isn't the quest for the Holy Grail. Stop making
it more difficult than it is.
Good luck.
3301
-----BEGIN PGP SIGNATURE-----
Version: GnuPG v1.4.11 (GNU/Linux)
```

```
iQIcBAEBAgAGBQJPCBl3AAoJEBgfAeV6NQkPo6EQAKghp7ZKYxmsYM96iNQu5GZV
fbjUHsEL164ZLctGkgZx2H1HyYFEc6FGvcfzqs43vV/IzN4mK0SMy2qFPfjuG2JJ
tv3x2QfHMM3M2+dwX30bUD12UorMZNrLo8HjTpanYD9hL8WglbSIBJhnLE5CPlUS
BZRSx0yh1U+wbnlTQBxQI0xLkPIz+xCMBwSKl5BaCb006z43/HJt7NwynqWXJmVV
KScmkpFC3ISEBcYKhHHWv1IPQnFqMdW4dExXdRqWuwCshXpGXwDoOXfKVp5NW7Ix
9kCyfC7XC4iWXymGgd+/h4ccFFVm+WWOczOq/zeME+0vJhJqvj+fN2MZtvckpZbc
CMfLjn1z4w4d7mkbEpVjgVIU8/+KClNFPSf4asqjBKdrcCEMAl80vZorElG6OVIH
aLV4XwqiSu0LEF1ESCqbxkEmqp7U7CHl2VW6qv0h0Gxy+/UT0W1NoLJTzLBFiOzy
QIqqpgVg0dAFs74SlIf3oUTxt6IUpQX5+uo8kszMHTJQRP7K22/A3cc/VS/2Ydg4
o6OfN54Wcq+8IMZxEx+vxtmRJCUROVpHTTQ5unmyG9zQATxn8byD9Us070FAg6/v
jGjo1VVUxn6HX9HKxdx4wYGMP5grmD8k4jQdF1Z7GtbcqzDsxP65XCaOYmray1Jy
FG5OlgFyOflmjBXHsNad
=SqLP
```

```
-----END PGP SIGNATURE-----
```

Many solvers of the Cicada 3301 puzzles missed interesting details and clues like these as they raced ahead. In some cases, it became necessary to backtrack before it was possible to move forward again. Now, as for the phone number revealed by the book code, those who called (214) 390-9608 heard a computer-generated voice say

> Very good. You have done well. There are three prime numbers associated with the original final.jpg image. 3301 is one of them. You will have to find the other two. Multiply all three of these numbers together and add a .com to find the next step. Good luck. Goodbye.[31]

The prime numbers are those that have exactly two distinct positive numbers that go into them evenly. For example, 17 is divisible only by 1 and 17, so it's a prime. An example of a number that is not prime is given by 18 because it's divisible by 1, 2, 3, 6, 9, and 18. The first few primes are 2, 3, 5, 7, 11, 13, 17, 19, and 23. We do not call 1 a prime because it is divisible only by 1. Remember, to be prime, a number must have exactly *two* distinct positive divisors.

Primes arose in many ways in the Cicada 3301 challenges. The very name "Cicada 3301" embraces several primes. We have not only 3301 as a prime, but also its reversal 1033. The terms "reversible prime" and "emirp" are synonymous and are used to describe numbers that have this characteristic. More primes arise from the insect the group gets its name from. Depending on which species we look at, we find that a cicada spends either 13 or 17 years in the ground before coming out to reproduce and die. Both of these are prime numbers.

The image for which solvers were expected to associate two more primes measured 509 pixels wide and 503 pixels tall. Both of these numbers are prime. The product of 3301, 509, and 503 is 845,145,127.

Solvers also found this number when they looked up the public key for Cicada 3301 on the MIT keyserver. It can still be seen there and is reproduced here:[32]

Search results for '0x7a35090f'

Type	bits/keyID	Date	User ID
pub	4096R/7A35090F	2012-01-05	Cicada 3301 (845145127)

The website http://845145127.com/ had a picture of a cicada and a countdown. OutGuess once more served to extract a message from the image. It read

```
-----BEGIN PGP SIGNED MESSAGE-----
Hash: SHA1
You have done well to come this far.
Patience is a virtue.
Check back at 17:00 on Monday, 9 January 2012 UTC.
3301
-----BEGIN PGP SIGNATURE-----
Version: GnuPG v1.4.11 (GNU/Linux)
```

```
iQIcBAEBAgAGBQJPCKDUAAoJEBgfAeV6NQkPf9kP/19tbTFEy+ol/vaSJ97A549+
E713DyFAuxJMh2AY2y5ksiqDRJdACBdvVNJqlaKHKTfihiYW75VHb+RuAbMhM2nN
C78eh+xd6c4UCwpQ9vSU4i1Jzn6+T74pMKkhyssaHhQWfPs8K7eKQxOJzSjpDFCS
FG7oHx6doPEk/xgLaJRCt/IJjNCZ912kYinmOm7c0QdRqJ+VbV7Px41tP1dITQIH
/+JnETExUzWbE9fMf/eJ1/zACF+gYii7d9ZdU8RHGi14jA2pRjc7SQArwqJOIyKQ
IFrW7zuicCYYT/GDmVSyILM03VXkNyAMBhG90edm17sxliyS0pA06MeOCjhDGUIw
QzBwsSZQJUsMJcXEUOpHPWrduP/zN5qHp/uUNNGj3vxLrnB+wcjhF8ZOiDF6zk7+
ZVkdjk8dAYQr62EsEpfxMT2dv5bJ0YBaQGZHyjTEYnkiukZiDfExQZM2/uqhYOj3
```

yK0J+kJNt7QvZQM2enMV7jbaLTfU3VZGqJ6TSPqsfeiuGyxtlGLgJvd6kmiZkBB8
Jj0Rgx/h9Tc4m9xnVQanaPqbGQN4vZF3kOp/jAN5YjsRfCDb7iGvuEcFh4oRgpaB
3D2/+Qo9i3+CdAq1LMeM4WgCcYj2K5mtL0QhpNoeJ/s0KzwnXA+mxBKoZ0S8dUX/
ZXCkbOLoMWCUfqBn8QkQ
=znly

```
-----END PGP SIGNATURE-----
```

At the designated time, the site displayed the following:

```
Find our symbol at the location nearest you:
Coordinates:
52.216802, 21.018334
48.85057059876962, 2.406892329454422
48.85030144151387, 2.407538741827011
47.664196, -122.313301
47.637520, -122.346277
47.622993, -122.312576
37.577070, 126.813122
37.5196666666667, 126.995
36.0665472222222, -94.1726416666667
33.966808, -117.650488
29.909098706850486 -89.99312818050384
25.684702, -80.441289
21.584069, -158.104211
--33.90281, 151.18421
```

The hidden content of the Cicada image also changed to provide twelve of the above fourteen coordinates:

```
-----BEGIN PGP SIGNED MESSAGE-----
Hash: SHA1
52.216802, 21.018334
48.85057059876962, 2.406892329454422
48.85030144151387,2.407538741827011
47.664196, -122.313301
47.637520, -122.346277
47.622993, -122.312576
37.5196666666667, 126.995
33.966808, -117.650488
29.909098706850486 -89.99312818050384
25.684702, -80.441289
21.584069, -158.104211
- -33.90281, 151.18421
3301
```

```
-----BEGIN PGP SIGNATURE-----
Version: GnuPG v1.4.11 (GNU/Linux)
```

iQIcBAEBAgAGBQJPChn7AAoJEBgfAeV6NQkPZxMP/05D9TkSpwRaBXPqYthuyqxx
uo+ZDyr/yVIlAdurTBiWb3aGxKJjtWg/vlcHcatK0TGL2qaHwB/FFZQAaqOyU7Zf
DXdpWr8PWoWhpWNYUK8IrOaYu1SmWlJnkTdUSzGrX01bwjwMmJJoPNS7CJuO6MaA
2GIwpv2G7lYqnH3xeX3kzGlPMsVb/wucKRjobsbdbreh1SNuQuRnhfe4s+oHTTqs
XjtGL/VhBI0DUAdfLqW7z4C+Gvbx6okC8x5Sj2N2UTJOiyMYXz5+QyHoA6fo9g5V
6zodNpx/RvxuZP2Ssc9TqERgTo5FjRBpON1vjDalHgg0H2Fus2LK3gh+NZfj1i5b
Oqa4Cqd9epI2pe+glXn86j9crS+2BEAr1cguqAFepvI9sdFEornDja4VXwDtUdM8
9hMVkU5NiTUYfvxZbL6W7rHIF7wxjGUwpe1ViuixG+cKNfv0enrt60PrtDByBOWI
9LLIUE0cB5HDT1xrczZ/55CtuM3Zf07/10nLFdmgR0oa8KUA9gWcPs6S1EpBa185
VcyOTqbpIPiT8neiJEkXarbJeFk15m1P73Fr8XZxdj7EHK0aOwGYcc8e4PmW/dSh
gcrSNXiePCbcRVRD2n9L47C0LkNyRpoBkmjvtpcRyp5ISe+0xcx/QI+gc1lkSijC
89qV+ymCHae1RiSDxVbd
=ZJ37

```
-----END PGP SIGNATURE-----
```

The fourteen locations spanned the globe:

Oleandrów 6, 01-001 Warsaw, Poland

89-91 Rue de la Plaine, 75020 Paris, France

36 Rue des Maraîchers, 75020 Paris, France

4739 University Way NE, Seattle, WA 98105, USA

514 Crockett St, Seattle, WA 98109, USA

428 15th Ave E, Seattle, WA 98112, USA

South Korea, Seoul, Gangseo-gu, Banghwa-dong, 830-8

South Korea, Seoul, Yongsan-gu, Seobinggo-dong, 287-1

853-899 W Dickson St, University of Arkansas, Fayetteville, AR 72701, USA

15717-15735 Euclid Ave, Chino, CA 91708, USA

State Highway 407, New Orleans, LA 70131, USA

8718-8798 SW 152nd Ave, Miami, FL 33193, USA

66-420 Kamehameha Hwy, Haleiwa, HI 96712, USA

143 George St, Erskineville NSW 2043, Australia

Each of these locations was found to have a clue wrapped around a telephone or light pole, as shown in fig. 9.14.

This is the point in the challenge where the participants realized that a group with a lot of resources was behind it.

Despite there being fourteen locations with QR codes, there were only two distinct messages among them. One of the messages was this:

FIG. 9.14 A Cicada 3301 clue in Warsaw, Poland

```
-----BEGIN PGP SIGNED MESSAGE-----
Hash: SHA1
In twenty-nine volumes, knowledge was once contained.
How many lines of the code remained when the Mabinogion paused?
Go that far in from the beginning and find my first name.
1:29
6:46
the product of the first two primes
2:37
14:41
17:3
27:40
the first prime
2:33
1:1
7:45
17:29
21:31
12:17
the product of the first two primes
22:42
```

```
15:18
24:33
27:46
12:29
25:66
7:47
```

You've shared too much to this point. We want the best,
not the followers. Thus, the first few there will receive
the prize.
Good luck.
3301

```
-----BEGIN PGP SIGNATURE-----
Version: GnuPG v1.4.11 (GNU/Linux)
```

iQIcBAEBAgAGBQJPB11uAAoJEBgfAeV6NQkP9oAP+gLu+FsRDf3aRcJtBk-
COU2MX
r/dagOTvCKWtuV+fedy0enWUZ+CbUjXOr98m9eq2z4iEGqKd3/MBXa+DM9f6YGUE
jPum4wHtQDSJlZMazuYqJOVZGw5XmF25+9mRM6fe3H9RCiNDZpuXl3MzwdivYhcG
B5hW14PcdHHteQf3eAUz+p+s06RDs+q1sNGa/rMQIx9QRe71EJwLMMkMfs81kfJC
tCt21+8ud0Xup4tjUBwul7QCcH9bqKG7cnR1XWsDgdFP6a4x9Jl2/IUvp1cfeT7B
YLS9W31CM8thMemJr+ztQPZrpDlaLIitAT2L0B3f/k4co89v5X2I/toY8Z3Cdvoi
hk0AdWzMy/XLDgkPnpEef/aFmnls53mqqe9xKAUQPMrI73hiJ+5UZWuJdzCpvt+F
BjfQk15EJoUUW16K2+mBA1cSd+HJlnkslUTsjkq0E36XKChP+Cvbu/p6DLUMM2Xl
+n3iospCkkHR9QDcHzE4Rxg9A435yHqqJ/sL2MXG/CY8X4ec6U0/+UCIF9spuv8Y
7w66D05pI2u9M/081L7Br0i0Mpdf9fDblO/6GksskccaPkMQ3MRtsL+p9o6Dnbir
6Z2wH2Kw1Bf0GfX4VcpHBikoWJ5blCc6tfvT+qXjVOZjWAL7DvReavSEmW1/fubN
C3RWcjeI4QET2oKmV2NK
=LWeJ
```
-----END PGP SIGNATURE-----
```

If someone found a complete interpretation of this clue, he or she never shared it publicly. A possible partial solution was posted by Swedish security expert Joel Eriksson. It begins

"In twenty-nine volumes, knowledge was once contained" may refer to the 11th edition of *Encyclopedia Britannica*, which consisted of exactly 29 volumes and that is now in the public domain and available for download since it was released back in 1910–1911.[33]

The other message was similar:

```
-----BEGIN PGP SIGNED MESSAGE-----
Hash: SHA1
A poem of fading death, named for a king
Meant to be read only once and vanish
Alas, it could not remain unseen.
```

1:5
152:24
the product of the first two primes
14:13
7:36
12:10
7:16
24:3
271:22
10:7
13:28
12:7
86:17
93:14
the product of the first two primes
16:7
96:4
19:13
47:2
71:22
75:9
77:4
You've shared too much to this point. We want the best,
not the followers. Thus, the first few there will receive
the prize.
Good luck.
3301
-----BEGIN PGP SIGNATURE-----
Version: GnuPG v1.4.11 (GNU/Linux)

iQIcBAEBAgAGBQJPB/nmAAoJEBgfAeV6NQkPEnEQAKl5qtb3ZE5vs+cO8KuzAi4a
tQEE71fvb65KQcX+PP5nHKGoLd0sQrZJw1c4VpMEgg9V27LSFQQ+3jSSyan7aIIg
SDqhmuAcliKwf5ELvHM3TQdyNb/OnL3R6UvavhfqdQwBXCDC9F0lwrPBu52MJqkA
ns93Q3zxec7kTrwKE6Gs3TDzjlu39YklwqzYcUSEusVzD07OVzhIEimsOVY+mW/C
X87vgXSlkQ69uN1XAZYp2ps8zl4LxoaBl5aVtIOA+T8ap439tTBToov19nOerusB
6VHS192m5NotfQLnuVT4EITfloTWYD6X7RfqspGt1ftb1q6Ub8Wt6qCIo6eqb9xm
q2uVzbRWu05b0izAXkHuqkHWV3vwuSfK7cZQryYA7pUnakhlpCHo3sjIkh1FPfDc
xRjWfnou7TevkmDqkfSxwHwP5IKo3r5KB87c7i0/tOPuQTqWRwCwcWOWMNOS7ivY
KQkoEYNmqD2Yz3Esymjt46M3rAuazxk/gGYUmgHImgcu1zzK7Aq/IozXI7EFdNdu
3EoRJ/UL9Y0l0/PJOG5urdeeTyE0b8bwgfC2Nk/c8ebaTkFbOnzXdAvKHB03KEeU
PtM6d6DngL/LnUPFhmSW7K0REMKv62h9KyP/sw5QHTNh7Pz+C63OO3BsFw+ZBdXL
hGqP6XptyZBsKvz2TLoX
=aXFt

-----END PGP SIGNATURE-----

The numbers in both of these messages had the same format as the numbers in the earlier book code. Indeed, these were book codes as well. The initial text portion of each message provided clues as to what books were used.

The message with

```
A poem of fading death, named for a king
Meant to be read only once and vanish
Alas, it could not remain unseen.
```

led to a poem by William Gibson, the author who coined the term "cyberspace" and created the science fiction subgenre of cyberpunk.

The poem was titled *Agrippa* (*A Book of the Dead*), and it was not intended to last. It was put out on a 3.5-inch disk that would encipher the poem after it was used for the first time. The disk was embedded in a book that was treated with photosensitive chemicals so that the contents would begin to fade when exposed to light.

In spite of all of this, in 2012 the poem was available online.[34] Using it as the book for the code provided in the corresponding message yielded the new message

```
sq6wmgv2zcsrix6t.onion
```

Popular websites end in .com (for commerical), .gov (for government), .edu (for educational), among others, but .onion is less well known and operates in a completely different manner. Such websites cannot be accessed by regular web browsers. For access, one must first install special software, which can be found at https://www.torproject.org/. The software makes several things possible.

1. Accessing .onion websites.
2. Browsing the Internet anonymously (concealing both location and identity).
3. Hosting a website anonymously (as in 2 above).

The name Tor, in the web address given here, stands for "The Onion Router" because the process it uses to realize the listed items has many layers, like an onion.

When a user connects to the Internet through onion routing, he or she is not immediately connected to the website he or she wants to visit by a proxy server. Instead, the request is sent through a chain of randomly selected proxy servers, until it finally arrives at the destination. With the exception of this last exit node connection, all of the connections are encrypted. The nested encryptions correspond to the layers of an onion and ensure that any given computer knows the address of only the computer that connected with it, and the computer it, in turn, connects with. This is how anonymity is preserved.

There are thousands of proxy servers in the Tor network, and the path a given person's data follows through this network varies from session to session. In fact, it even changes every ten minutes for individual sessions. The servers are run by volunteers who believe in privacy and anonymity as human rights.

Those taking advantage of the ability to host websites anonymously are said to offer "hidden services." Such sites have been around since 2004. Collectively, they're part of the Dark Web, which in turn is part of the Deep Web (websites that aren't indexed by search engines). The most famous Tor hidden service was Silk Road, through which a tremendous volume of illegal drugs was sold. Others use Tor websites as a way to speak out against the repressive governments in power where they live. These sites can serve as powerful voices for freedom and help dissidents secretly organize.

The Tor website Cicada 3301 led solvers to didn't offer drugs or inspiration for the politically oppressed. Instead, it provided more instructions, although, oddly, with some misspelled words this time:

```
-----BEGIN PGP SIGNED MESSAGE-----
Hash: SHA1
Congratulations!
Please create a new email address with a public, free, web-based
service. Once you've never used before, and enter it below. We
recommend you do This while still using Tor, for anonymity.
We will email you a number in the next few days (in the order
in which you've arrived to this page). Once you've received it,
come back to this site and append a slash and then the num-
ber you received to this url. (For example, if you received
"3894894230934209", then you would go to "http://sq6wmgv2zcsri
x6t.onion/3894894230934209")
3301
-----BEGIN PGP SIGNATURE-----
Version: GnuPG v1.4.11 (GNU/Linux)
```

iQIcBAEBAgAGBQJPCikYAAoJEBgfAeV6NQkPtCkQAIEa2HC2OSduG8zWtH/0LSJ9
dnQLfjTI/MAPsnp0KXpzcQJ4p2UbNdRb5qtqp3HZQ5qwKK3b/MT+2eB5X+h5q5v/
bVOsPKXN7W+krTi8a2v5KA1fRj8mKcFTXkx9hdq6z6qeePVpeOGXeitU0jgO4uXu
1N9hor4AQFQoXvuELnlqS4YX+/nNgK+Y1Vu1ekvyn9LusidMpSxS/ciiZLD5d3T1
TJeUOQzdw9T1UmmdeigfBRoqZDnm5UESnS8qS1vd7zg9U1pw2GfCV8QQl0zWp01S
KCrArG0Qg4K4shZRQ2BopGM95oFVw2pOnzYr9pp8kpipBWQ6URRF7IEcuG1k5fBt
sfiz7Eb7WWr4Cwk/1Ri625FMyo0Z2KDc1Wo8nkFXZ0/Xle6tyyhhJUZGRbZuHLau
LR5+oQEKKYTRrJFNhqswcXYsCQBBYM2XS+v3nbmTDVKyGNmMpp3/yC/RR91hQ7fr
FqMd5qHKSD0pc5tU5G29J+Ez3CDVN7E60Xy3I1I5ugGE8GP5DmqHn4k/x272lroX
BtiIRTh4f+ghcC6DLdPBc1PRAfLtRbpVR3GVTSCJ/EwBZrH4QJTEGcuXfPQryHbF
0BO795yyXxM61Q9QnU4dkAAQepKLHgcvD1rEJYJzMWo+BOCu7MChoOyo1Zq4eeuI
cPx04KJd4sGtMVWolA3/
=zGQW

-----END PGP SIGNATURE-----

With the communication between the Cicada 3301 organizers and leading contestants becoming more private at this point, the puzzles that followed are less certain. There are all sorts of rumors. Eventually, the message in fig. 9.15 was posted to Reddit.

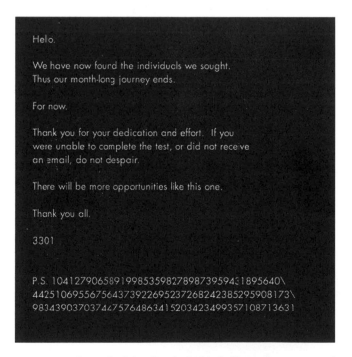

FIG. 9.15 The end of the Cicada 2012 challenge is announced

A full accounting of Cicada 3301 would fill a book by itself. To leave room for other material, I omit details of the challenges from 2013 and 2014, but be aware that the puzzles got harder!

The year 2015 didn't see a Cicada 3301 challenge, but the organization did put out a message denying any involvement with a claimed hack of Planned Parenthood by a group calling itself 3301. In January 2016, Cicada 3301 released another message, but again there was no new challenge. It might return in 2017, but it's also possible that it will never be seen again. However, there's another Internet mystery that's appeared twice so far and is, as of this writing, claiming that it will come again.

PCCTS and Jahbulonian

The story of this cipher challenge began sometime in 2010 when a strange website appeared. The address was www.pccts.com. This domain name is presently up for sale and bears no trace of its past mystery, but an archived version from October 20, 2010, is preserved at https://web .archive.org/web/20101020055419/http://pccts.com/. It shows us that those visiting the site on that day saw the image and text that are in fig. 9.16.

PAUPERES COMMILITONES CHRISTI TEMPLIQUE SOLOMONICI

EST. MCXIX

FIG. 9.16 From the official website of the Knights Templar?

The Latin text translates to "Poor Fellow-Soldiers of Christ and of the Temple of Solomon" and is another name for the Knights Templar. Below this is the year in which the group was formed, 1119. Clicking on the shields, including the small one held by the knight in the center image, leads to other pages with enciphered messages, hidden texts, and hidden links. As an example, the letters "itanimullI" appear on one. When read backwards, this gives "Illuminati."

One of the linked pages (now preserved at https://web.archive.org/web /20100809120304/http://www.illuminatiorder.info/illuminati/babilu-d. html) showed four sphinxes looking at an area in the middle of the screen that was solid black, like the rest of the background. However, the area could be highlighted, revealing the German text

Am Osten der Welt
steht der Mitternachtsberg.
Ewiglich wirkt sein Licht.
Des Menschen Auge kann ihn nicht sehen,
und doch ist er da.

Ueber dem Mitternachtsberg strahlt
Die schwarze Sonne.
Des Menschen Auge kann sie nicht sehen,
Und doch ist sie da.

Im Inneren leuchtet ihr Licht.
Einsam sind die Tapfren und die Gerechten—
Doch mit ihnen ist die Gottheit

—Inschrift aus Babilu

An English translation of this is

At the east of the world stands
The Dark Midnight Mountain.
Eternally his light is performing.
The human being's eye cannot see him,
And still the giant is there.

Above the Midnight Mountain shines
The Black Sun.
The human being's eye cannot see her,
And still she is there.

In the interior her light lights up.
The brave and the just ones are secluded—
Still the deity is with them.

—Inscription of Babilu

Instead of detailing the links and puzzles further, I'll allow you to explore the archived version on your own. Whether the site had anything to do with an actual secret society or not doesn't really matter; it's still fun to explore.

However, things were to become very serious for those running the site. On July 25, 2011, they posted the following:

DISCLAIMER (25-7-2011): WE ARE IN NO WAY CONNECTED TO THE IDIOT IN NORWAY! THIS WEBSITE WAS BUILD IN 2010 TO SHOW THE WORLD HOW EASY IT IS TO START A CONSPIRACY! CARIPS.COM ALSO IS PART OF THE HOAX!! THIS PROJECT IS PART OF QFF - QUO FATA FERUNT - HTTP://WWW.QUOFATAFERUNT.COM—AND QFF IS A CONSPIRACY / ALTERNATIVE MEDIA WEBSITE /FORUM

THE PAIN AND SUFFERING THAT THIS IDIOT HAS CAUSED IS ENORMOUS.
OUR HEARTS GO OUT TO ALL THE VICTIMS,
THEIR FAMILY MEMBERS, LOVED ONES AND FRIENDS.

The "idiot in Norway" was Anders Behring Breivik (pronounced AHn-ders BRAY-vick). He was described as a "right-wing Christian extremist, with a hatred of Muslims."[35] On July 22, 2011, he set off a fertilizer bomb outside the building in which Prime Minister Jens Stoltenberg's office was located. The Prime Minister was not injured, but eight others died. Breivik then took a ferry to Utoya Island, where, dressed in a police uniform, he began shooting the attendees of a Labour Party youth camp. By the time he was captured by an elite Norwegian police unit, he had killed sixty-nine people on the island.[36]

Breivik said that he was a member of PCCTS.

On the same day that Breivik committed the murders, a book was posted online consisting of more than 1,500 hate-filled pages. Breivik's lawyer said that his client had spent years working on it.[37] The text includes many lines like the following:

The European Military Order and Criminal Tribunal (the PCCTS—Knights Templar) have found all European category A, B and C traitors guilty of a series of crimes against Europeans (charges 1-8). All European category A and B traitors have been sentenced to death. Punishment is pending awaiting effectuation by Justiciar Knights.

As might be expected, some strange statements are mixed in with the hate.

Although the PCCTS, Knights Templar is a pan-European indigenous rights movement we give all Europeans, regardless of skin colour, the opportunity to become a Justiciar Knight as long as the individual is either a Christian, Christian agnostic or a Christian atheist.

Let us be perfectly clear: if you are unwilling to martyr yourself for the cause, then the PCCTS, Knights Templar is not for you.

Under these circumstances, the PCCTS, Knights Templar will for the future consider working with the enemies of the EU/US hegemony such as Iran (South Korea is unlikely), al-Qaeda, al-Shabaab or the rest of the devout fractions of the Islamic Ummah, with the intention to deploy small nuclear, radiological, biological or chemical weapons in Western European capitals and other high-priority locations. Justiciar Knights and other European Christian martyrs can avoid the scrutiny normally reserved for individuals of Arab descent and we can ensure successful deployment and detonation in the location of our choice.

Nevertheless, we cannot, under any circumstances, accept deployment of nuclear weapons for surface detonation above 0.2 kilotons, as it would involve too many civilian casualties. However, smaller devices up to approximately 0.2 kilotons are ideal for annihilating concentrations of category A and B traitors (traitor HQs). This scenario will involve the destruction of up to 1-2 city blocks/HQs with high concentrations of traitors, and therefore suits our purpose.

As a Justiciar Knight of Knights Templar Europe, we are anti-Nazi and approximately 40% of our policies do not coincide with NSDAP's policies.

The book also describes setting up a website, although the details don't match the one we've been looking at.

By the end of August 5, 2011, the PCCTS website under discussion here was suspended.

Version 2.0?

In late 2013, a similar site appeared at a brand new address, http://jahbulonian.byethost7.com/.

The group announced itself as being "back." A screenshot of the website appears in fig. 9.17. Like the PCCTS site, it gives the year 1119 as the date of establishment. Following this are the letters KT PCCTS, so we have the Knights Templar once again.

FIG. 9.17 Screenshot announcing the return of the PCCTS

No-Lub-Haj looks like a phrase in some foreign language, but it's just Jahbulon (a name present in the URL) written backward, with hyphens added. Hebrew is normally read right to left, so the text at the very bottom of the page might have been meant to provide a clue for interpreting No-Lub-Haj. The Hebrew characters אלף־בית, put in the order English uses are Aleph, Lamed, Pe, Beth, Yod, Tav, and together they mean "alphabet." The numbers above the Hebrew word are addressed shortly.

The website changed a few times. The next version is depicted in fig. 9.18. The small print near the center states "What you are reading is correct. PCCTS is back. The last time, most of you have heard of us was somewhere [sic] in 2011. There were many videos of people that were trying to debunk our story. Nobody managed to reveal our secret, so we would love to see you all give it another try."

The implication is that the site was put up by the same group that presented the 2010–2011 hoax. If so, it's incredible that they hadn't learned their lesson. Even worse, the new site carried anti-Islamic statements. If the site was put up by a new group or individual, the referral to 2011 indicates

FIG. 9.18 Screenshot of an updated version of the website

at least a familiarity with the old site. Also, the common images make for an unlikely coincidence if the creator had not seen the old site. Why in the world would anyone aware of what happened in Norway start this up again? We'll look at some of the puzzles and ciphers that arose in this iteration and then look a bit deeper at the mystery of who was behind the 2013–2014 iteration.

At the bottom of the page, the small print challenges the viewer to "break the code," which brings us to the number 415211519221222213, which appeared in both of the images above. It's actually a simple cipher. It can be broken up as 4 15 21 15 19 22 1 22 22 13. When each of these is replaced by the letter of our alphabet in the indicated position, we get

D O U O S V A V V M

But if you can figure out what this means, you will have solved a bigger mystery!

The mystery was created by Thomas Anson (1695–1773), who had become one of the wealthiest men in England. He renovated the family estate at Shugborough Hall in Staffordshire, and among the many monuments and works of art he had constructed or placed on the property was one known as the Shepherd's Monument, shown in fig. 9.19.

FIG. 9.19 The Shepherd's Monument

This monument bears what's believed to be an enciphered dedication (fig. 9.20), but for whom?

FIG. 9.20 The Shugborough inscription

Dave Ramsden put forth a decipherment that, though it cannot be proven correct, seems like a decent possibility. Anyone intrigued by this cipher should read his book.[38]

Other researchers have come to conclusions different than Ramsden's. What did the person behind the new PCCTS website think it meant?

Eventually, the website began running a clip from the TV show *Seinfeld* in which George Costanza is thrilled with the results of doing everything opposite his natural inclinations. Creepy music was added to the soundtrack and, at the end, the video changed to a black background with the palindromic words "Satenetas Rotenetor" appearing in white.

FIG. 9.21 A clip from Seinfeld as it appeared on the Jahbulonian website

The video was overlaid with four alternating messages. They were

```
1.17 is he who comes from East and West...
  The magic of the number 7 and the number 10
2.D17
  Stones can roll...
  What did they hide in Rolde?
3.Round number 2
  Only 11 people knew what we know
  G between lines and hooks
4.Crisis?
  What Crisis??!!
  VAT TOG SUNK
```

You may have fun trying to make sense of these clues. I'll point out only that in the third message "G between lines and hooks" may well be a reference to the masonic symbol depicted in fig. 9.22.

In general, this group's challenge feels more like something put together by a Masonic version of the Riddler than the work of experts in computer security. For example, moving the cursor around the screen and clicking on certain very specific spots, which were not indicated in any way to draw attention to them as links, takes the user to various "hidden links." Some people may have had fun trying to find all of these hidden links, but an easy shortcut

FIG. 9.22 The Masonic square and compass

made finding them trivial. All the visitor has to do is right-click anywhere on the page and select "View source." The code for the website then appears, and all of the addresses for all of the links (and other information, as well) can be seen at once. The natural question is, were the designers unaware of this shortcut? The people behind Cicada 3301 made the contestants work a bit harder than this!

The linked website http://jahbulonian.byethost7.com/666partsofhell.html had a screen that was entirely black, but there was a black image on it that could be copied. It had the name "lookhereevenifitlooksdark" and is reproduced in fig. 9.23.

FIG. 9.23 A black box

Pasting it into Microsoft Office Picture Manager and clicking "Auto Correct" produced fig. 9.24.

FIG. 9.24 The hidden image revealed

This was much easier than having to use OutGuess, as in the Cicada 3301 challenge.

The runes decipher to YOU ARE CLOSER.

The Seinfeld clip was eventually replaced by another video that claimed "We have the spear." This apparently referred to the legendary Spear of Destiny purportedly used by the Roman soldier Casca Longinus to pierce the side of Jesus Christ.

Clicking "View source" for the new page, I noticed the following among the content:

```
<meta name="keywords" content="knights templar, pccts,
templars, templar, templer, templars, templar knights,
crusades, p.c.c.t.s., 1119, carips, saint martin, anti
islam, anti=islamic, pro christ, christus, christ,
war, last crusade, 322 ">
```

This is the part that really concerns me. Why is "anti islam" listed here? More should have been learned from the Norway incident. This made me more interested to find out who was behind the site.

Also, the claims of being a return from 2011 made me wonder what content the present site had at that time. Using the Wayback Machine, at https://archive.org/web/, I found the answer to that question.

On July 19, 2011, August 20, 2011, and September 19, 2011 (the only captures from that year), http://jahbulon.com contained a discussion of stock market investing, but the following was also present.[39]

So just what does Jahbulon mean?

This is said to be the secret Masonic name of god, and it has come in for a great deal of criticism recently by non-Masons as it is said to have been derived from three god-names. The word is supposed to be divisible into the syllables Jah hy Baal leb and On Na, which are said to be representative of the three deities of Yahweh, Baal and Osiris. It is this presumed presence of the 'pagan' gods Baal and Osiris in a Judaeo-Christian Masonic context that tends to get some commentators in a bit of a lather. However, the link between the syllable On and Osiris is tentative in the extreme; Osiris' Egyptian name was Asar, and to transliterate this pronunciation into 'On' is somewhat cavalier. Even the Greeks, whose translations are often suspect, managed to derive Osiris, which is a reasonable attempt at pronouncing Asar.

Instead, the syllable 'On' most likely referred to the biblical On, or Heliopolis, a name that was derived from the Egyptian An or On. In its turn, the syllable Baal could refer as much to a 'lord' or 'king' as the god Baal. This new interpretation would then derive a phrase something like 'Yahweh is Lord of Heliopolis', which is probably not much more acceptable to fundamentalist Judaeo-Christian sensitivities, and so Grand Lodge predictably stays silent on this topic.

The captures from 2012 do not show this and simply deal with stocks. However, a linked page from December 10, 2012, does contain it and also explains the purpose of the main site.[40]

Mission: The mission is simply the maximization of my returns through the use of aggressive fundamental and technical analysis. I accomplish this by identifying the trends and fundamentals of the current market on a daily basis and reacting in a manner that will increase my returns while ensuring capital preservation. This site serves as my trading diary.

Goal: To make $5 million dollars in 15 years, or rather by January 2025. This site was not developed to tell YOU what to do with your account. The goal is for me to reach $5 million. Do not follow me! I am a not giving out investment advice. This is my trading diary for me to learn from.

It also contained the following.

> **Privacy Policy:** I am obsessive about my privacy, and I assume you are also. I will
> **never** share or sell readers names or email addresses, PERIOD.

There are many hackers who would take this individual's obsession with his privacy as a challenge.

On a version of the site archived on March 2, 2013, a contact address is given, HolyGrailResearchLab@yahoo.com. This address shows up a few times elsewhere on the web. Digging a little deeper might reveal the identity of the person behind Jahbulonian.

As of this writing, http://jahbulonian.byethost7.com/ has little content. It simply bears a link "No Lub Haj!" which when clicked on leads to http://www.jahbulon.com/. This second website displays an image of a skateboard tipped over on its side. Beneath this is the text "we will be online soon." Viewing the source for the page reveals no hidden links but does show that the image is 375 pixels tall and 666 pixels wide.

I think it's safe to say that the site is a hoax, as the similar site was admitted to be in 2011. The only challenge is determining who is behind it, and that can be pursued at any time.

I think the next challenge cipher is also likely in the hoax category, but you should decide for yourself!

The Beale Ciphers—A Challenge with No Solution

Let's consider the units we use to measure various goods for a moment. We buy gallons of gasoline (okay, some of us buy liters), cases of beer (okay, some of us buy kegs), bags of mulch, pounds of shrimp. But if it's gold we're buying, the unit of measurement is much smaller. Gold is measured in troy ounces, each of which is one twelfth of a pound. We certainly don't measure gold in tons. That is, if we aren't talking about Fort Knox or the Beale ciphers! The Beale ciphers allegedly conceal the location of more than a ton of buried gold (2,921 pounds to be precise) and more than two tons of silver (5,100 pounds). There were a lot of jewels too, as you'll see!

The story begins with the publication of a twenty-three-page pamphlet in 1885. The pamphlet was published by James Ward, and the identity of the author was not revealed, but several other players in this drama are named. One of these is Robert Morriss (1778–1863), a hotel keeper in Lynchburg, Virginia. According to the pamphlet, it was in 1862 that Morriss first revealed

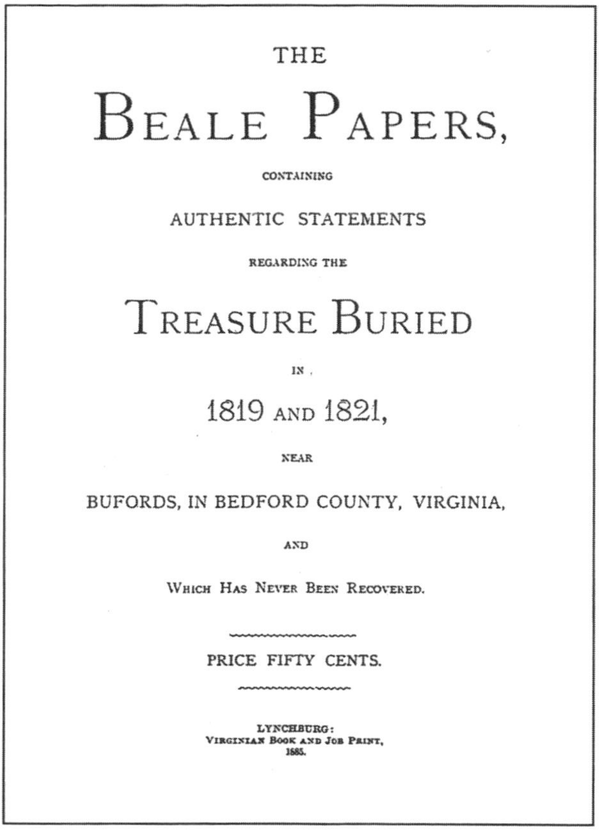

THE

BEALE PAPERS,

CONTAINING

AUTHENTIC STATEMENTS

REGARDING THE

TREASURE BURIED

IN .

1819 AND 1821,

NEAR

BUFORDS, IN BEDFORD COUNTY, VIRGINIA,

AND

WHICH HAS NEVER BEEN RECOVERED.

PRICE FIFTY CENTS.

LYNCHBURG:
VIRGINIAN BOOK AND JOB PRINT,
1885.

FIG. 9.25 Birth of a myth?

the story of the tremendous treasure to the anonymous author. The story is provided in Morriss's own words. The author claims that he had Morriss put it in writing, and the pamphlet reproduced his words inside quotation marks. This is an important detail, as you'll soon see.

Morriss's story is of a Thomas J. Beale, who first came to his hotel in January 1820. This guest left near the end of March, but returned in January 1822. He left, as before, in March, but this time he gave Morriss a box and a story before departing. Morriss recounted that according to Beale, the box "contained papers of value and importance; and which he desired to leave in my charge until called for hereafter." Morriss later received a letter from Beale, sent from St. Louis, and dated May 9, 1822. It was the last he was to hear from the man, but he didn't know that at the time. He placed the letter with the box. The letter was reproduced, in full, in the pamphlet published decades later by Ward. Only the most important part is reproduced here.

> With regard to the box left in your charge, I have a few words to say, and, if you will permit me, give you some instructions concerning it. It contains papers vitally affecting the fortunes of myself and many others engaged in business with me, and in the event of my death, its loss might be irreparable. You will, therefore, see the necessity of guarding it with vigilance and care to prevent so great a catastrophe. It also contains some letters addressed to yourself, and which will be necessary to enlighten you concerning the business in which we are engaged. Should none of us ever return you will please preserve carefully the box for the period of ten years from the date of this letter, and if I, or no one with authority from me during that time demands its restoration, you will open it, which can be done by removing the lock. You will find, in addition to the papers addressed to you, other papers which will be unintelligible without the aid of a key to assist you. Such a key I have left in the hands of a friend in this place, sealed, addressed to yourself, and endorsed not to be delivered until June, 1832. By means of this you will understand fully all you will be required to do.

If Morriss had followed Beale's instructions, he would have opened the box in 1832, but for some reason he didn't actually do so until 1845, at which time he had to have the lock broken to get at the contents. But he didn't share what he found with anyone else until 1862, when he finally revealed the secret to the anonymous author of the pamphlet.

The author, in turn, waited until 1885 to tell the world. His pamphlet detailed how the contents began with a letter from Beale to Morriss, written months before leaving the hotel for the final time. The letter is reproduced in full in the pamphlet, but here it is summarized with only the most important portions quoted.

In the letter, Beale told how he had organized a group of thirty people to visit the great Western plains to hunt buffalo, grizzly bears, and other game. The group left "old Virginia" for St. Louis, Missouri, in April 1817. In that city, they procured everything they needed and headed out again on May 19 for Santa Fe, New Mexico, intending to arrive in the fall. Their guide recommended that because of potential dangers, the group should form a military-type organization with a leader, whom the others must obey. Beale was chosen as the leader. Santa Fe was reached on about December 1.

In early March, some of the group left on what was intended to be a short excursion of a few days. After a month or more, two of the party returned and explained their long absence to the worried men they'd left behind. They told how they traveled north for some days, finding sufficient game, but when they were ready to return encountered an immense herd of buffalo, so they followed them instead, for two weeks or longer, securing many. On one of these days, while camping in a small ravine about two hundred fifty to three hundred miles north of Santa Fe, one of the men found gold. It was to tell of the find that the two men returned to the main group.

Beale returned with them to where the gold was still being extracted. The men agreed to work as equal partners, dividing whatever they could gather evenly. The work continued for eighteen months or more, and a great deal of gold and silver was recovered. Finally in the summer of 1819, they were ready to transport what they had found up to that point to somewhere more secure. They feared that having so much wealth in such a dangerous area put their lives at risk. Ultimately, they decided that the treasure should be sent to Virginia under Beale's charge and buried in a cave near Buford's Tavern in Bedford County.

Beale and the ten men who traveled all the way to Virginia with him found that local farmers visited that cave too often in order to store potatoes and other vegetables. So the men found another place in the county and secured the treasure there. Beale had one other task to carry out on behalf of his group while in Bedford County. He was to find some reliable person who could be depended on in the event that the group met with a hard fate while trying to recover yet more gold. This person would be tasked with dividing the gold and silver recovered thus far among the heirs of the group.

Beale's letter revealed that it was at this time that Morriss was chosen for the task, although he wouldn't be given any hint of it until later. Beale first returned to the source of the gold and made another haul back to Virginia, placing it with the original batch. It was after this second deposit that Beale gave Morriss the box, although he gave him no hint of the contents.

The next line of the letter becomes important in our analysis, so it is quoted. Beale wrote,

I intend writing you, however, from St. Louis, and impress upon you its importance still more forcibly.

He then continued,

The papers enclosed herewith will be unintelligible without the key, which will reach you in time, and will be found merely to state the contents of our depository, with its exact location, and a list of the names of our party, with their places of residence, etc.

Recall that Morriss previously received a letter from Beale that stated, "Such a key I have left in the hands of a friend in this place, sealed, addressed to yourself, and endorsed not to be delivered until June, 1832. By means of this you will understand fully all you will be required to do."

The reference to "this place" meant St. Louis, for this is where the quoted letter was mailed from. So apparently, Beale intended to have a friend mail Morriss the key at the end of the ten years. This seems like a safe enough strategy. Asking someone to mail a letter at a specific time shouldn't arouse too much suspicion. Beale could even have given the purported mailer some bogus explanation as to what the letter was about. And what motivation would the man have had to open it? He couldn't have expected anything fabulous to have been contained within. However, if this was Beale's plan, it didn't work out. Morriss never received the key.

Beale explained in his letter to Morriss that because ten years had passed (actually it was many more years because Morriss did not follow the instructions) without Beale returning, his party must all be dead. It then asked Morriss to retrieve the gold and silver from where it was hidden and divide it into thirty-one equal shares. Thirty of these were to be given to the parties named in one of the ciphers, and the last share was for Morriss, in appreciation of his help. The final paragraph repeated the way the treasure was to be divided. It is quoted here, for reasons that will become clear:

You will find in one of the papers, written in cipher, the names of all my associates, who are each entitled to an equal part of our treasure, and opposite to the names of each one will be found the names and residences of the relatives and others, to whom they devise their respective portions. From this you will be enabled to carry out the wishes of all by distributing the portion of each to the parties designated. This will not be difficult, as their residences are given, and they can easily be found.

This letter was accompanied by pages filled with numbers, the ciphers. But without the key, he could make nothing of them. However, he did eventually confide in the anonymous author of the pamphlet that revealed this story to the world. This author figured out the trick to making sense of the list of numbers headed by "No. 2." That is, he solved the second cipher. The numbers he saw were these:

115, 73, 24, 807, 37, 52, 49, 17, 31, 62, 647, 22, 7, 15, 140, 47, 29, 107, 79, 84, 56,
239, 10, 26, 811, 5, 196, 308, 85, 52, 160, 136, 59, 211, 36, 9, 46, 316, 554, 122, 106,
95, 53, 58, 2, 42, 7, 35, 122, 53, 31, 82, 77, 250, 196, 56, 96, 118, 71, 140, 287, 28,
353, 37, 1005, 65, 147, 807, 24, 3, 8, 12, 47, 43, 59, 807, 45, 316, 101, 41, 78, 154,
1005, 122, 138, 191, 16, 77, 49, 102, 57, 72, 34, 73, 85, 35, 371, 59, 196, 81, 92, 191,
106, 273, 60, 394, 620, 270, 220, 106, 388, 287, 63, 3, 6, 191, 122, 43, 234, 400, 106,
290, 314, 47, 48, 81, 96, 26, 115, 92, 158, 191, 110, 77, 85, 197, 46, 10, 113, 140, 353,
48, 120, 106, 2, 607, 61, 420, 811, 29, 125, 14, 20, 37, 105, 28, 248, 16, 159, 7, 35,
19, 301, 125, 110, 486, 287, 98, 117, 511, 62, 51, 220, 37, 113, 140, 807, 138, 540,
8, 44, 287, 388, 117, 18, 79, 344, 34, 20, 59, 511, 548, 107, 603, 220, 7, 66, 154, 41,
20, 50, 6, 575, 122, 154, 248, 110, 61, 52, 33, 30, 5, 38, 8, 14, 84, 57, 540, 217, 115,
71, 29, 84, 63, 43, 131, 29, 138, 47, 73, 239, 540, 52, 53, 79, 118, 51, 44, 63, 196, 12,
239, 112, 3, 49, 79, 353, 105, 56, 371, 557, 211, 505, 125, 360, 133, 143, 101, 15, 284,
540, 252, 14, 205, 140, 344, 26, 811, 138, 115, 48, 73, 34, 205, 316, 607, 63, 220, 7,
52, 150, 44, 52, 16, 40, 37, 158, 807, 37, 121, 12, 95, 10, 15, 35, 12, 131, 62, 115, 102,
807, 49, 53, 135, 138, 30, 31, 62, 67, 41, 85, 63, 10, 106, 807, 138, 8, 113, 20, 32, 33,
37, 353, 287, 140, 47, 85, 50, 37, 49, 47, 64, 6, 7, 71, 33, 4, 43, 47, 63, 1, 27, 600, 208,
230, 15, 191, 246, 85, 94, 511, 2, 270, 20, 39, 7, 33, 44, 22, 40, 7, 10, 3, 811, 106, 44,
486, 230, 353, 211, 200, 31, 10, 38, 140, 297, 61, 603, 320, 302, 666, 287, 2, 44, 33,
32, 511, 548, 10, 6, 250, 557, 246, 53, 37, 52, 83, 47, 320, 38, 33, 807, 7, 44, 30, 31,
250, 10, 15, 35, 106, 160, 113, 31, 102, 406, 230, 540, 320, 29, 66, 33, 101, 807, 138,
301, 316, 353, 320, 220, 37, 52, 28, 540, 320, 33, 8, 48, 107, 50, 811, 7, 2, 113, 73, 16,
125, 11, 110, 67, 102, 807, 33, 59, 81, 158, 38, 43, 581, 138, 19, 85, 400, 38, 43, 77,
14, 27, 8, 47, 138, 63, 140, 44, 35, 22, 177, 106, 250, 314, 217, 2, 10, 7, 1005, 4, 20,
25, 44, 48, 7, 26, 46, 110, 230, 807, 191, 34, 112, 147, 44, 110, 121, 125, 96, 41, 51,
50, 140, 56, 47, 152, 540, 63, 807, 28, 42, 250, 138, 582, 98, 643, 32, 107, 140, 112,
26, 85, 138, 540, 53, 20, 125, 371, 38, 36, 10, 52, 118, 136, 102, 420, 150, 112, 71, 14,
20, 7, 24, 18, 12, 807, 37, 67, 110, 62, 33, 21, 95, 220, 511, 102, 811, 30, 83, 84, 305,
620, 15, 2, 10, 8, 220, 106, 353, 105, 106, 60, 275, 72, 8, 50, 205, 185, 112, 125, 540,
65, 106, 807, 138, 96, 110, 16, 73, 33, 807, 150, 409, 400, 50, 154, 285, 96, 106, 316,
270, 205, 101, 811, 400, 8, 44, 37, 52, 40, 241, 34, 205, 38, 16, 46, 47, 85, 24, 44, 15,
64, 73, 138, 807, 85, 78, 110, 33, 420, 505, 53, 37, 38, 22, 31, 10, 110, 106, 101, 140,
15, 38, 3, 5, 44, 7, 98, 287, 135, 150, 96, 33, 84, 125, 807, 191, 96, 511, 118, 40, 370,
643, 466, 106, 41, 107, 603, 220, 275, 30, 150, 105, 49, 53, 287, 250, 208, 134, 7, 53,
12, 47, 85, 63, 138, 110, 21, 112, 140, 485, 486, 505, 14, 73, 84, 575, 1005, 150, 200,
16, 42, 5, 4, 25, 42, 8, 16, 811, 125, 160, 32, 205, 603, 807, 81, 96, 405, 41, 600, 136,
14, 20, 28, 26, 353, 302, 246, 8, 131, 160, 140, 84, 440, 42, 16, 811, 40, 67, 101, 102,
194, 138, 205, 51, 63, 241, 540, 122, 8, 10, 63, 140, 47, 48, 140, 288

The anonymous author made sense of them by using the Declaration of Independence as the key. The pamphlet claims that the ciphers were created by "Thomas J. Beale," but never specifies what the J stood for. Some later authors writing about this interesting tale refer to "Thomas Jefferson Beale." This is an unjustified assumption. The Beale of the story has never been convincingly identified, and it is not known what his middle initial stood for. The, perhaps false, identification may have been accidental, as the Declaration of Independence was written by the cryptographer (and U.S. President) Thomas Jefferson. A hasty writer may have unconsciously filled in Beale's middle name while thinking about the Declaration, leaving later writers to reproduce his or her error.

So, how did the Declaration of Independence serve to unravel the numerical cipher? Very easily! The anonymous author simply numbered the words in the Declaration and replaced each of the numbers in the cipher with the first letter of the word bearing that number.

institued (115), hold (73), another (24), ...

The recovered message turned out to be

I have deposited in the county of Bedford, about four miles from Buford's, in an excavation or vault, six feet below the surface of the ground, the following articles, belonging jointly to the parties whose names are given in number "3," herewith:

The first deposit consisted of one thousand and fourteen pounds of gold, and three thousand eight hundred and twelve pounds of silver, deposited November, 1819. The second was made December, 1821, and consisted of nineteen hundred and seven pounds of gold, and twelve hundred and eighty-eight pounds of silver; also jewels, obtained in St. Louis in exchange for silver to save transportation, and valued at $13,000.

The above is securely packed in iron pots, with iron covers. The vault is roughly lined with stone, and the vessels rest on solid stone, and are covered with others. Paper number "1" describes the exact locality of the vault so that no difficulty will be had in finding it.

An exciting message to recover! And Beale promised that the location would be revealed by deciphering the first set of numbers, marked "1." However, the anonymous author found that the Declaration of Independence didn't yield anything meaningful—it was not the key to all of the ciphers! He indicated spending a tremendous amount of time trying to break cipher 1, but admitted to complete failure despite these efforts.

In consequence of the time lost in the above investigation, I have been reduced from comparative affluence to absolute penury, entailing suffering upon those it was my duty to protect, and this, too, in spite of their remonstrances.
—the anonymous author

Rather than sharing the story with one trusted friend, as Morriss had done, he gave the puzzle of the remaining ciphers to the world by publishing them. They are reproduced below.

THE LOCALITY OF THE VAULT.

71, 194, 38, 1701, 89, 76, 11, 83, 1629, 48, 94, 63, 132, 16, 111, 95, 84, 341, 975, 14, 40, 64, 27, 81, 139, 213, 63, 90, 1120, 8, 15, 3, 126, 2018, 40, 74, 758, 485, 604, 230, 436, 664, 582, 150, 251, 284, 308, 231, 124, 211, 486, 225, 401, 370, 11, 101, 305, 139, 189, 17, 33, 88, 208, 193, 145, 1, 94, 73, 416, 918, 263, 28, 500, 538, 356, 117, 136, 219, 27, 176, 130, 10, 460, 25, 485, 18, 436, 65, 84, 200, 283, 118, 320, 138, 36, 416, 280, 15, 71, 224, 961, 44, 16, 401, 39, 88, 61, 304, 12, 21, 24, 283, 134, 92, 63, 246, 486, 682, 7, 219, 184, 360, 780, 18, 64, 463, 474, 131, 160, 79, 73, 440, 95, 18, 64, 581, 34, 69, 128, 367, 460, 17, 81, 12, 103, 820, 62, 116, 97, 103, 862, 70, 60, 1317, 471, 540, 208, 121, 890, 346, 36, 150, 59, 568, 614, 13, 120, 63, 219, 812, 2160, 1780, 99, 35, 18, 21, 136, 872, 15, 28, 170, 88, 4, 30, 44, 112, 18, 147, 436, 195, 320, 37, 122, 113, 6, 140, 8, 120, 305, 42, 58, 461, 44, 106, 301, 13, 408, 680, 93, 86, 116, 530, 82, 568, 9, 102, 38, 416, 89, 71, 216, 728, 965, 818, 2, 38, 121, 195, 14, 326, 148, 234, 18, 55, 131, 234, 361, 824, 5, 81, 623, 48, 961, 19, 26, 33, 10, 1101, 365, 92, 88, 181, 275, 346, 201, 206, 86, 36, 219, 324, 829, 840, 64, 326, 19, 48, 122, 85, 216, 284, 919, 861, 326, 985, 233, 64, 68, 232, 431, 960, 50, 29, 81, 216, 321, 603, 14, 612, 81, 360, 36, 51, 62, 194, 78, 60, 200, 314, 676, 112, 4, 28, 18, 61, 136, 247, 819, 921, 1060, 464, 895, 10, 6, 66, 119, 38, 41, 49, 602, 423, 962, 302, 294, 875, 78, 14, 23, 111, 109, 62, 31, 501, 823, 216, 280, 34, 24, 150, 1000, 162, 286, 19, 21, 17, 340, 19, 242, 31, 86, 234, 140, 607, 115, 33, 191, 67, 104, 86, 52, 88, 16, 80, 121, 67, 95, 122, 216, 548, 96, 11, 201, 77, 364, 218, 65, 667, 890, 236, 154, 211, 10, 98, 34, 119, 56, 216, 119, 71, 218, 1164, 1496, 1817, 51, 39, 210, 36, 3, 19, 540, 232, 22, 141, 617, 84, 290, 80, 46, 207, 411, 150, 29, 38, 46, 172, 85, 194, 39, 261, 543, 897, 624, 18, 212, 416, 127, 931, 19, 4, 63, 96, 12, 101, 418, 16, 140, 230, 460, 538, 19, 27, 88, 612, 1431, 90, 716, 275, 74, 83, 11, 426, 89, 72, 84, 1300, 1706, 814, 221, 132, 40, 102, 34, 868, 975, 1101, 84, 16, 79, 23, 16, 81, 122, 324, 403, 912, 227, 936, 447, 55, 86, 34, 43, 212, 107, 96, 314, 264, 1065, 323, 428, 601, 203, 124, 95, 216, 814, 2906, 654, 820, 2, 301, 112, 176, 213, 71, 87, 96, 202, 35, 10, 2, 41, 17, 84, 221, 736, 820, 214, 11, 60, 760

The next cipher was supposed to contain the names of the thirty men who had a stake in the treasure, and the names and addresses of their heirs.

NAMES AND RESIDENCES.

317, 8, 92, 73, 112, 89, 67, 318, 28, 96, 107, 41, 631, 78, 146, 397, 118, 98, 114, 246, 348, 116, 74, 88, 12, 65, 32, 14, 81, 19, 76, 121, 216, 85, 33, 66, 15, 108, 68, 77, 43, 24,

122, 96, 117, 36, 211, 301, 15, 44, 11, 46, 89, 18, 136, 68, 317, 28, 90, 82, 304, 71, 43, 221, 198, 176, 310, 319, 81, 99, 264, 380, 56, 37, 319, 2, 44, 53, 28, 44, 75, 98, 102, 37, 85, 107, 117, 64, 88, 136, 48, 151, 99, 175, 89, 315, 326, 78, 96, 214, 218, 311, 43, 89, 51, 90, 75, 128, 96, 33, 28, 103, 84, 65, 26, 41, 246, 84, 270, 98, 116, 32, 59, 74, 66, 69, 240, 15, 8, 121, 20, 77, 89, 31, 11, 106, 81, 191, 224, 328, 18, 75, 52, 82, 117, 201, 39, 23, 217, 27, 21, 84, 35, 54, 109, 128, 49, 77, 88, 1, 81, 217, 64, 55, 83, 116, 251, 269, 311, 96, 54, 32, 120, 18, 132, 102, 219, 211, 84, 150, 219, 275, 312, 64, 10, 106, 87, 75, 47, 21, 29, 37, 81, 44, 18, 126, 115, 132, 160, 181, 203, 76, 81, 299, 314, 337, 351, 96, 11, 28, 97, 318, 238, 106, 24, 93, 3, 19, 17, 26, 60, 73, 88, 14, 126, 138, 234, 286, 297, 321, 365, 264, 19, 22, 84, 56, 107, 98, 123, 111, 214, 136, 7, 33, 45, 40, 13, 28, 46, 42, 107, 196, 227, 344, 198, 203, 247, 116, 19, 8, 212, 230, 31, 6, 328, 65, 48, 52, 59, 41, 122, 33, 117, 11, 18, 25, 71, 36, 45, 83, 76, 89, 92, 31, 65, 70, 83, 96, 27, 33, 44, 50, 61, 24, 112, 136, 149, 176, 180, 194, 143, 171, 205, 296, 87, 12, 44, 51, 89, 98, 34, 41, 208, 173, 66, 9, 35, 16, 95, 8, 113, 175, 90, 56, 203, 19, 177, 183, 206, 157, 200, 218, 260, 291, 305, 618, 951, 320, 18, 124, 78, 65, 19, 32, 124, 48, 53, 57, 84, 96, 207, 244, 66, 82, 119, 71, 11, 86, 77, 213, 54, 82, 316, 245, 303, 86, 97, 106, 212, 18, 37, 15, 81, 89, 16, 7, 81, 39, 96, 14, 43, 216, 118, 29, 55, 109, 136, 172, 213, 64, 8, 227, 304, 611, 221, 364, 819, 375, 128, 296, 1, 18, 53, 76, 10, 15, 23, 19, 71, 84, 120, 134, 66, 73, 89, 96, 230, 48, 77, 26, 101, 127, 936, 218, 439, 178, 171, 61, 226, 313, 215, 102, 18, 167, 262, 114, 218, 66, 59, 48, 27, 19, 13, 82, 48, 162, 119, 34, 127, 139, 34, 128, 129, 74, 63, 120, 11, 54, 61, 73, 92, 180, 66, 75, 101, 124, 265, 89, 96, 126, 274, 896, 917, 434, 461, 235, 890, 312, 413, 328, 381, 96, 105, 217, 66, 118, 22, 77, 64, 42, 12, 7, 55, 24, 83, 67, 97, 109, 121, 135, 181, 203, 219, 228, 256, 21, 34, 77, 319, 374, 382, 675, 684, 717, 864, 203, 4, 18, 92, 16, 63, 82, 22, 46, 55, 69, 74, 112, 134, 186, 175, 119, 213, 416, 312, 343, 264, 119, 186, 218, 343, 417, 845, 951, 124, 209, 49, 617, 856, 924, 936, 72, 19, 28, 11, 35, 42, 40, 66, 85, 94, 112, 65, 82, 115, 119, 236, 244, 186, 172, 112, 85, 6, 56, 38, 44, 85, 72, 32, 47, 63, 96, 124, 217, 314, 319, 221, 644, 817, 821, 934, 922, 416, 975, 10, 22, 18, 46, 137, 181, 101, 39, 86, 103, 116, 138, 164, 212, 218, 296, 815, 380, 412, 460, 495, 675, 820, 952

The pamphlet ends with the anonymous author stating that he has put everything he knows about the matter in the pamphlet and that the publisher knew nothing of it until seeing his manuscript. So, even though no one knew his identity, he made an effort to stave off questions.

With such a tremendous amount of gold and silver potentially available to the person who can break the ciphers shown above, they have attracted an equally tremendous amount of attention! However, I'm convinced that there is no treasure. There are many reasons to hold this opinion, but I'll just present the ones that arise from the ciphers themselves and a mathematical analysis of the writing in the pamphlet.

Reason 1: The Declaration of Independence exists in many versions.
In the many reprintings this document saw over the years, printers

sometimes introduced small errors accidentally. At other times, they attempted to "improve" the document by making intentional changes. Some changes would have only a small effect on a decipherment attempt. An example of this type is "inalienable" vs. "unalienable." However, other changes could have a devastating effect. In some cases, a word was deleted, or an extra word inserted. Such a change would then cause every word after it to be misnumbered, if the wrong version was used. The result of applying such a document as the key, if the error didn't occur near the end of the Declaration, would be gibberish. What are the odds that the author of the pamphlet would use the same version of the Declaration as Beale had?

NSA mathematician Todd Mateer found an online version of the original Declaration of Independence and carefully numbered the words. Using this as the key for deciphering Cipher 2, he got the following.

```
ahaie depos otedt nttte ointt oaitd strsa boapt hrrmi lestr
oabaa ottst tafep coiat ionor iaalt snpti ntbea owtht ssram
wnhst hhfbh ntdth ntoof hcang mttaw fntbt tonat fgphi otatt
ttheo attoe swott tttdt sabea tiiti ntndb ththr tfttb ewnth
thotn tttde posit eotta stedo otinh itdrt tanda oigte tfoos
estth btlds tittt httia ghtta odbed indtw efief ornds oosaa
ieroi posit tdnhi iiaht eeftt nttee nthes econd watmw dnttu
tifht eentw entth ntwfs aonst ottot htine ttfnh aetre dcnos
eienp oands hsaoa owtot welii oafob edato eigtt teiah ttfti
aierm fsoti wtfso btain idtns tehaa tinep chang etotw intgo
ntftt tatto nttdi alail tattt gttit rhtss andtt aaigs theab
oieit secrb tatpc eohdi nibhf ohthw ittdg ttchi ittth eiitf
tnith aftlt fined wstts tonta ndthi itste atres tonth ltdst
onett oitee tintt twuar htttb hpioe rtamd eront ditcg tlttt
heopi atloc alitt tsttt imtdt tttha tftti ostea lttwi lafit
adstt indtn gtn
```

This is quite far removed from readable text. A would-be solver might look at this and then move on to try some other key.

Reason 2: Even the correct version of the Declaration of Independence won't reveal Cipher 2, as the author of the pamphlet claimed.

Suppose, by chance, that the author of the pamphlet happened to use the same version of the Declaration of Independence as Beale. If he carefully numbered all of the words and then used this numbering to try to solve the cipher, he would still get gibberish! This is because whoever enciphered the document made several mistakes in numbering the words in his copy of the Declaration. His mistakes were the following:[41]

1. Treating "self-evident" as two words (as it would be if the hyphen wasn't present).

2. Around position 480, the encipherer somehow skipped ten words in the numbering!

3. Also, the number 480 was applied to two different words.

4. A word was miscounted around position 630.

5. A word was miscounted around position 670.

Because of these mistakes in numbering by the encipherer, a careful decipherer would generate a much different key, one that couldn't turn the cipher into anything sensible. But the author of the pamphlet got a meaningful message out. So, if we take him for an honest man, he would have to not only have used the same version of the Declaration as Beale, but also to have made the same errors in numbering the words. How likely is that?

Reason 3: We get a highly improbable result when applying the Declaration of Independence to Cipher 1.

The numbers in Cipher 1 go up to 2,906, whereas numbering the words of the Declaration of Independence, we must stop with 1,322. So, there are many numbers in Cipher 1 that have no alphabetical equivalent when trying to apply the Declaration as the key. A normal person would reject it as a possible key for this reason. If for some reason someone tried to apply it anyway, he would get a message that begins SCS?E TFA?G CDOTT, where ? represents a number too high to be paired with a letter from a word in the Declaration. Because this is gibberish, a normal person would stop and try a different key.

It's a good thing that James Gillogly isn't normal. He kept going, or rather, his computer did. He programmed it to complete the task, and he got a very unlikely result. Later in the decipherment, the message became ABFDE FGHII JKLMM NOHPP. That is, Gillogly got most of the alphabet out, in order! Even some of the errors had neat explanations. The F that appears where the alphabet has a C may have arisen because the encipherer wanted a C (position 194), but accidentally wrote the number 195, to represent a neighboring word that starts with F.

The Declaration doesn't work as a key for Cipher 1, so if the cipher is real, this incorrect key should yield gibberish. The chances of an incorrect key yielding the alphabetical sequence we saw above is extremely minute. The conclusion we are drawn to is that Cipher 1 is a meaningless hoax. Whoever created it was trying to jot down a list of random numbers that would look like a cipher, but not actually be one. Either he got bored and wrote out an almost alphabetical sequence to occupy his mind, or he realized that it's hard to make up numbers that are really random, and thought that the alphabetical sequence would give him numbers that fit the bill better. His reasons don't matter. What matters is that he provided those skeptical about the legitimacy of Cipher 1 with strong evidence to support their view.

Reason 4: Cipher 3 is too short to be what it claims.

In 1927, Kendell Foster Crossen, the author behind a regular feature titled "Solving Cipher Secrets" in the pulp detective magazine *Flynn's Weekly*, found a way to shoot down Cipher 3 without even trying to solve it. His efficient approach involved looking at the number of letters in Cipher 3 and the amount of information it was supposed to contain.[42] Could 618 letters really convey the names and addresses of the heirs of the thirty men who collected the gold and silver? He presumed sixty heirs, but even if we assume that each man left everything to a single person, we still have thirty names and thirty addresses, so we'd have 10.3 letters, on average, for each name and address combination. Can your name and address be expressed so tersely?

Reason 5: Stylometrics/Stylometry

Three styles of writing are represented in *The Beale Papers*. We have not only the author's words, but also long quotations of letters from Beale and Morriss. We can analyze their writing styles statistically and, if they are in fact distinct, real people, each should give us different values. Several investigators compared these styles, including NSA codebreaker Solomon Kullback, whose service included great successes during World War II and earned him a spot in NSA's Hall of Honor. Let's look at some of the data.

The results presented here were published by Joe Nickell, who is best known as a skeptical investigator of claims of the paranormal. He thought that the unknown author was Ward and labeled the writings not attributed to Beale or Morriss as such. I simply designate these writings as "Author." Nickell made several comparisons between Author and Beale, but also included three other nineteenth century Virginians (Chief Justice John Marshall, John Randolph of Roanoke, and John Randolph Tucker) to show how samples known to represent different people look. The numbers in tables 9.2 to 9.6 speak for themselves.

Table 9.2 Average Number of Words per Sentence Suggests Beale = Author[a]

	Average Number of Words per Sentence	Standard Deviation
Beale	29.95	12.76
Author	29.74	13.75
Marshall	19.13	11.40
Randolph	20.11	15.09
Tucker	34.49	22.03

[a]Nickell, Joe, "Discovered, The Secret of Beale's Treasure," The Virginia Magazine of History and Biography 90, no. 3 (July 1982), 310–24.

Next, Nickell presented the percentage of words for each author that were "the," "of," and "and."

Table 9.3 Percentage of Occurrences of Common Words[a]

	the	of	and
Beale	5.10%	3.51%	3.66%
Author	5.20%	3.26%	3.47%
Marshall	3.08%	1.92%	2.69%
Randolph	4.55%	2.93%	2.87%
Tucker	7.35%	5.28%	3.85%

[a]Nickell, Joe, "Discovered, The Secret of Beale's Treasure," The Virginia Magazine of History and Biography 90, no. 3 (July 1982), 310–24.

Table 9.3 was for overall percentage, but it can be restricted to the first word in each sentence instead. For this special case, the result is in table 9.4.

Table 9.4 Percentage of Occurrences of Common Words at the Start of Sentences[a]

	the	of	and
Beale	6.80%	0.00%	0.00%
Author	6.12%	0.00%	0.00%
Marshall	8.93%	2.86%	0.00%
Randolph	10.34%	0.00%	1.82%
Tucker	2.90%	1.73%	1.54%

[a]Nickell, Joe, "Discovered, The Secret of Beale's Treasure," The Virginia Magazine of History and Biography 90, no. 3 (July 1982), 310–24.

Nickell looked at punctuation, as well (table 9.5).

Table 9.5 Average Number of Commas and Semicolons per Sentence[a]

	Commas	Semicolons
Beale	2.6	0.06
Author	2.4	0.06
Marshall	0.32	0.02
Randolph	1.6	0.16
Tucker	2.96	0.18

[a]Nickell, Joe, "Discovered, The Secret of Beale's Treasure," The Virginia Magazine of History and Biography 90, no. 3 (July 1982), 310–24.

Lastly, Nickell presented some less obvious comparisons (table 9.6).

Table 9.6 Yet More Data Suggesting Beale = Author[a]

	Negatives	Negative Passives	Infinitives	Relative Clauses
Beale	24	6	44	30
Author	36	7	40	39
Marshall	15	0	21	8
Randolph	29[b]	0	18	9
Tucker	14	0	16	34

[a]Nickell, Joe, "Discovered, The Secret of Beale's Treasure," The Virginia Magazine of History and Biography 90, no. 3 (July 1982), 310–24. This test was carried out by Professor Jean G. Pival of the University of Kentucky.
[b]Nickell noted, "Ten of the negatives occur in one letter, in which Randolph tries to justify his participation in a duel."

All of these comparisons make a strong case for the material composed by the unknown author and the letters from Beale really having been written by a single individual. Comparisons could also be made between the writing styles of the unknown author and the material claimed to be from Morriss. If these match closely, it would further strengthen the hoax conjecture.

In addition to the five reasons for concluding that the ciphers are a hoax given above, I can make an "argument from authority." I don't really like such arguments, as they are often wrong. But for those who do like them, the authority William F. Friedman had Frank Rowlett (a man who played a large role in the breaking of the Japanese diplomatic cipher used during World War II) and his colleagues in the Signal Intelligence Service take on the Beale ciphers as a training exercise. Their conclusion? The ciphers were a hoax.

On the other hand, one bit of evidence has come to light in support of the story in *The Beale Ciphers* being genuine.

According to the pamphlet, Beale instructed Morriss to open the box if he didn't return within ten years. The pamphlet also reveals that Beale planned to have a friend mail Morriss the key at that same point in time. But, for some reason, Morriss never received this letter.

In 1984, an article appeared in *RUN*, a magazine for computer hobbyists, reporting on the missing letter. The author of the piece, Douglas Nicklow, reported that sometime in August 1832, the *St. Louis Beacon* printed a notice that a letter was being held for Robert Morriss. He credited this discovery to J. Solario of the Brookings Institution. A later researcher, Canadian computer analyst Wayne S. Chan, was apparently not thrilled with the level of detail provided by Nicklow. He investigated the claim for himself.

Chan learned that the *St. Louis Beacon* came out only on Thursdays and did regularly include a list of people who had mail being held for them at the post office. Such a list appeared in the issues of August 2, 9, and 16 of 1832. And Chan saw the name Robert Morriss on the lists. He asked, "Given the relatively rare double-"s" spelling of the Morriss surname, and the time frame, is this just an amazing coincidence?"[43]

He checked to see if there was someone living in the St. Louis area at the time with the name Robert Morriss, but the 1821 city directory and the 1840 city directory and census didn't show any Morriss in the area, much less a Robert Morriss.[44]

Chan checked out the September 1832 unclaimed mail lists and saw that these didn't include Robert Morriss, nor did the lists printed in July. Thus, the timing was right for the letter to have been mailed from St. Louis to Virginia when it was supposed to be, only to be brought back for some reason. Chan wasn't able to determine what happened to the letter after that.

So, is this just a coincidence, or is it significant?

I find it intriguing, but not enough to outweigh the strong evidence against the treasure being real. Although I reject the Beale story as a hoax, there's another tale of buried gold that I feel better about. This one may be real!

Forrest Fenn's Treasure

This challenge might be called "mini-Beale." Rather than enticing the would-be solver with tons of gold and silver, the lure is a mere twenty-some pounds of gold (and various jewels and other items). Is this one to be believed?

FIG. 9.26 Fenn's treasure. Photo by Addison Doty

The story appeared in a 2010 book, *The Thrill of the Chase*, authored by Forrest Fenn, the man claiming to have buried the treasure. He explained how the idea of burying the treasure traced back to 1988, when he was diagnosed with cancer at age fifty-eight and was given a twenty percent chance of surviving the next three years. He didn't live in denial, but rather thought he would die, as his doctor expected.[45] The story continues below in his own words.

Then one night, after the probability of my fate had finally hit bottom, I got an idea. It had been so much fun building my collection over the decades [Fenn collected all sorts of things], why not let others come searching for some of it while I'm still here, and maybe continue looking for it after I'm gone? So I decided to fill a treasure chest with gold and jewels, then secret it—leaving clues on how to find it for any searcher willing to try. It was a perfect match of mind and moment. Ha, I liked the idea but it would take some planning and the clock was ticking. No matter how far away a date is on the calendar, it always seems to arrive sooner in the face of unpleasant situations like mine.

Fortunately, I talked a museum friend into selling me his beautiful cast bronze chest that had three-dimensional female figures on its four sides and on the lid. I know I paid way too much for it but once in a while something comes along that is so special as to discount all logical rules of value. An excited antique scholar said the chest was probably a Romanesque Lock Box that dated to about 1150 AD. He also thought it might have once held a family bible or a Book of Days. Now it could hold my ancient jewelry and solid gold pieces. I was delighted. It was the perfect treasure chest.

I also wanted to include something personal with the treasure because maybe the lucky finder would want to know a little about the foolish person who abandoned such an opulent cache. So I placed a 20,000 word autobiography in the chest. It's in a small glass jar and the lid is covered with wax to protect the contents from moisture. The printed text is so small that a magnifying glass is needed to read the words. I tried to think of everything.

Then I started filling the chest with gold coins, mostly old American eagles and double eagles, along with lots of placer nuggets from Alaska. Two weigh more than a pound each and there are hundreds of smaller ones. Also included are pre-Columbian gold animal figures and ancient Chinese human faces carved from jade. The different objects in the cache are too numerous to mention one by one, but among them are a Spanish 17th century gold ring with a large emerald that was found with a metal detector, and an antique ladies gold dragon coat bracelet that contains 254 rubies, six emeralds, two Ceylon sapphires, and numerous small diamonds.

And with some reluctance I included a small silver bracelet that has twenty-two turquoise disc beads set side-by-side in a row. It fit snug to my arm and I loved it, but its history is what appealed to me the most. Richard Wetherell excavated the beads from a ruin in 1898, and a Navaho silversmith made the bracelet for him the same year. In 1901 Wetherell sold it to Fred Harvey, the hotel magnate. Sixty-four years later I won it playing pool with Byron Harvey, an heir of Fred's.

One of the prizes in my collection, a Tairona and Sinu Indian necklace from Columbia, is also part of the treasure. It contains thirty-nine animal fetishes carved from quartz crystal, carnelian, jadeite and other exotic stones. But special to the necklace are two cast gold objects—one, a jaguar claw and the other, a frog with bulbous eyes and legs cocked as if ready to spring. I held the 2,000-year-old piece of jewelry one last time and could almost feel its ancient power, its supremacy, before I finally lowered it into the chest and closed the lid. A little of me is also inside the box. There must be a few Indiana Jones types out there, like me, ready to throw a bedroll in the pickup and start searching, with a reasonable chance of discovering a treasure chest containing more than twenty troy pounds of gold. For me, it was always the thrill of the chase. What do you think?

I knew exactly where to hide the chest so it would be difficult to find but not impossible. It's in the mountains somewhere north of Santa Fe. Indecision is the key to flexibility and that's why I waited so long to secret my cache. George Burns was 100 years old when someone asked him how his health was. He replied, "My health's good, it's my age that's killing me." And like Eric Sloane, at age almost-eighty, I figured it was time to act. So I wrote a poem containing nine clues that if followed precisely, will lead to the end of my rainbow and the treasure:[46]

As I have gone alone in there
And with my treasures bold,
I can keep my secret where,
And hint of riches new and old.

Begin it where warm waters halt
And take it in the canyon down,
Not far, but too far to walk.
Put in below the home of Brown.

From there it's no place for the meek,
The end is drawing ever nigh;
There'll be no paddle up your creek,
Just heavy loads and water high.

If you've been wise and found the blaze,
Look quickly down, your quest to cease
But tarry scant with marvel gaze,
Just take the chest and go in peace.

So why is it that I must go
And leave my trove for all to seek?
The answers I already know
I've done it tired, and now I'm weak

So hear me all and listen good,
Your effort will be worth the cold.
If you are brave and in the wood
I give you title to the gold.

Well, these clues don't immediately pinpoint a location for me, but Fenn wrote, "There are also subtle clues sprinkled in the stories."[47] The stories were in the same book that was quoted from above, *The Thrill of the Chase*. It consisted almost entirely of anecdotes from Fenn's life, as did the sequel, *Too Far to Walk*.

It's interesting that Fenn wrote that the treasure is "somewhere north of Santa Fe." This roughly matches the location where it was claimed that Beale's group had found the gold and silver they later brought back to Virginia and hid. Is this a coincidence, or was Fenn making a nod to the older treasure hunt?

Having read Fenn's two volumes of autobiography, I got the sense that he was a pretty honest man. His anecdotes included only minor transgressions, such as having a fake driver's license as a teenager, peeing in someone's gas tank, going back to his childhood home and digging up toys he buried and keeping them (technically theft, as his family no longer owned the property).

The only incident Fenn wrote of that might possibly concern us came about when a writer and photographer for *National Geographic* magazine, working on a piece about Santa Fe, came to his gallery. He gave a quote to go with the pictures taken there. He said, "I'd much rather have a bad painting by a great artist than a great painting by an unknown."[48] He knew that he would be criticized for saying this, but he explained to his staff that "it was a business decision designed to create talk and spread the word about our gallery."[49]

Could Fenn have faked the treasure story to publicize what he believed was a greater cause—promoting people's love of nature, a love he felt deeply himself, as can be gathered from his books?

He opened his follow-up book, *Too Far to Walk*, with these words:

Oh, tell me wise sir,
Where are all the treasures?

And the wise man replied,

"Wherever you find them."[50]

If Fenn's plan was to promote hiking and camping, it may have backfired a bit, as a report from a television news program showed.

Although he is thrilled with the attention, Fenn said he is concerned that the treasure hunt has gotten "out of control." In an interview with *Action 7 News*, the millionaire said an estimated 30,000 people traveled to the Southwest to search for the buried treasure. He expects 50,000 to make the same journey this year.[51]

Fenn also told his interviewer that several treasure hunters came within two hundred feet of the buried chest, but were unaware of how close they really were.

The minor incidents of dishonesty mentioned above are far outweighed by Fenn's years of honorable military service. He joined the Air Force in 1950, during the Korean War, and continued on to the war in Vietnam, where he flew 328 sorties. Twice, he had to get back the hard way—without a plane. He later reflected,

I had fought in a war that was waged mostly for philosophical reasons designed by men who never allowed their sense of morality to get in the way of what they mistakenly thought was the right thing to do.[52]

This quote shows that Fenn has the courage to speak his mind on controversial and emotionally charged topics.

Later in his life, after he had become an art dealer, he might've shocked some readers of *National Geographic*, but he showed great integrity in his business dealings. An example of this is provided by what he did after he bought one hundred oil paintings by Elmyr de Hory, a famous forger of French Impressionist paintings. The fakes Fenn purchased were in the styles of Degas, Renoir, Gauguin, Van Gogh, Modigliani, Matisse, Picasso, and Monet, among others. He paid $225,000 for the lot, but could've easily gotten his money back from a single sale, if he passed off a painting as legitimate. Instead, he had each of his buyers sign a paper saying that he or she understood that the painting was by Elmyr de Hory, despite the signature that appeared on the canvas.

If the Fenn treasure is a hoax, then there should be some sort of motive. Could it be the profits from the book in which he describes the treasure? Well, he's been generous with his wealth in the past. For example, he donated *Hail to Peace*, a large painting by Edward Hopper, to the Scott & White Hospital in Temple, Texas. It wouldn't make much sense for a man who gave away so much to try to acquire more through dishonest means. And then there's the clincher:

> Fenn is donating all the proceeds from *The Thrill of the Chase* (as well as a substantial private donation) to help cancer patients who can't afford their treatments.[53]

So greed can be ruled out as motivation for a hoax. Maybe this story is real!

Fenn indicated that some subtle clues were included in his follow-up volume of autobiography, *Too Far to Walk*. I may have spotted one that I haven't seen described anywhere else. At the end of the book, Fenn provided a new map. As it includes a substantial portion of the United States, it isn't of much value in and of itself, but in presenting the map, Fenn wrote,

> We've recently been turned on to the map and atlas products of Benchmark Maps, and have had fun partnering with them to produce this treasure map. Their unique styles of cartography speak to our shared spirit of exploration. I declined their invitation to put an X on the map, but will admit that it is there in spirit.[54]

This intrigued me because earlier in the book there was a map marked with an X, in a manner. The relevant image is reproduced in fig. 9.27.

The bone crosses two prominent lines on the map, either one of which could be seen as completing an "X marks the spot."

Although it's possible to read too much into an image, and Fenn's books have likely led to the interpretation of thousands of "clues" that were not intended, we cannot rule out the "X marks the spot" possibility. If it's correct, then depending on which line on the map you think the bone is meant to cross, the treasure is indicated to be in either Colorado or Wyoming. In an interview, Fenn narrowed down the search area a little, and both of these states made the cut.

> Fenn also confirmed the chest is buried in Colorado, Montana, New Mexico, or Wyoming.[55]

Ultimately, this story may have a doubly happy ending. To start with, Fenn beat his cancer, despite his doctor's pessimistic assessment. He's still alive at almost eighty-five years of age, as of this writing. And he may just live to see the joy of whoever finds his treasure.

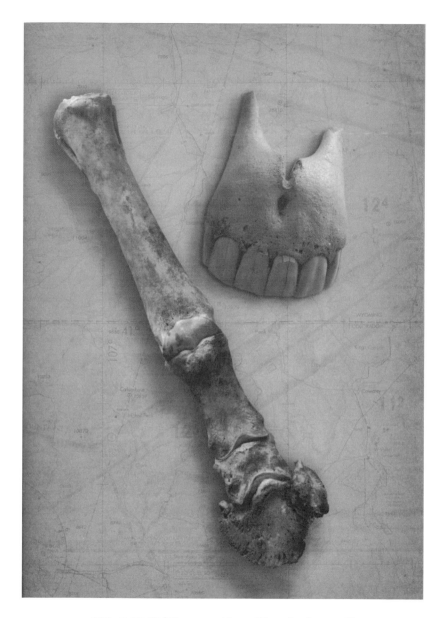

FIG. 9.27 Did Fenn provide an "X marks the spot"?

It should be noted that whether the treasures of Beale or Fenn are real or not, there do exist confirmed cases of treasures that have been found. A pair of examples are the following:

1. The Hoxne Hoard

In November 1992, a farmer in Suffolk, England, asked another man, Eric Lawes, to use his metal detector to search for a hammer he had lost on his property. Instead, Lawes found some coins, 14,780 of them, to be precise (565 gold, 14,191 silver, and 24 bronze). He also found hundreds of gold and silver spoons, jewelry, and statues, and all of it went back to the Roman Empire. Under British law, the treasure trove became government property, but the government was obligated to pay market value for it, so Lawes and the farmer got to split £1.75 million.

An even better find was made in America.

2. The Saddle Ridge Hoard.

In February 2013, a married couple, who wish to remain anonymous, noticed something shiny on the ground one day while walking their dog on their property in northern California. A bit of digging revealed eight metal cans with a total of more than 1,400 gold coins. The coins range in date from 1847 to 1894 and have a total face value of about $27,000. But because of the pristine condition of many of them, experts speculate that the collection could see more than $10 million at auction. One coin, an 1866-S No Motto Double Eagle, is worth almost a million dollars just by itself. How the coins came to be buried on the property remains a mystery.

There may be ciphers that lead to treasure, but at least once a treasure led to a cipher.

Chinese Gold Bar Ciphers

This tale begins with a set of seven gold bars weighing in at a total of 1.8 kilograms. The claim is that they were issued to a General Wang in Shanghai, China, in 1933, and represent certificates for a U.S. bank deposit. Chinese writing on the bars concerns a transaction of more than $300,000,000. The validity of the claim is disputed. A set of unsolved ciphers appearing on the bars may be the key to resolving the dispute.

An American museum curator brought the bars to the attention of the International Association for Cryptologic Research (IACR). No one in the group offered a decipherment, but the puzzle was placed on their website for anyone who wanted to take a crack at it. This website is the source of the scant information provided here. The same information appears elsewhere online, but little is added beyond speculation. The IACR website includes images of all of the bars and transcriptions of the ciphers. For easy reference, a few images and transcriptions are reproduced in figs. 9.28 through 9.30. The transcriptions are from the IACR website, which indicates that some of the letters are hard to read and may have been misidentified.

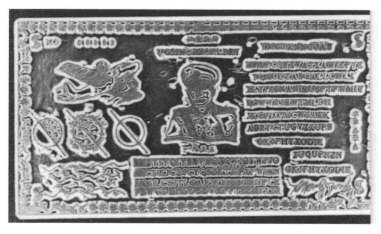

FIG. 9.28 A gold bar with a plane, a general, and ciphers?

Above the man, we have	UGMNCBXCFLDBEY.
Below him, we have	RHZVIYQIYSXVNQXQWIOVWPJO
	SKCDKJCDJCYQSZKTZJPXPWIRN
	MQOLCSJTLGAJOKBSSBOMUPCE
	FEWGDRHDDEEUMFFTEEMJXZR
And to his right, we have	VIOHIKNNGUAB
	HFXPCQYZVATXAWIZPVE
	YQHUDTABGALLOWLS
	XLYPISNANIRUSFTFWMIY
	KOWVRSRKWTMLDH
	JKGFIJPMCWSAEK
	ABRYCTUGVZXUPB
	GKJFHYXODIE
	ZUQUPNZN

GKJFHYXODIE

Another bar shows the figure on the one above, but also includes an obviously
different person.

FIG. 9.29 Another enigmatic gold bar

The ciphers above the man on the left are	ABRYCTUGVZXUPB
	XLYPISNANIRUSFTFWMIY
The ciphers below this man are	MVERZRLQDBHQ
	HLMTAHGBGFNIV
The ciphers at the top center are	MQOLCSJTLGAJOKBSSBOMUPCE
	FEWGDRHDDEEUMFFTEEMJXZR
The ciphers above the man on the right are	ZUQUPNZN
	VIOHIKNNGUAB
The ciphers below this man are	HFXPCQYZVATXAWIZPVE
	GKJFHYXODIE
	HLMTAHGBGFNIV

The back of one of the bars has both ciphers and Chinese text in alternating
lines. At the end of these, more lines of cipher appear, but they are inverted
relative to the longer texts.

FIG. 9.30 A gold bar with a mix of Chinese and ciphertext

The ciphers that alternate with the Chinese writing are

```
JKGFIJPMCWSAEK
SKCDKJCDJCYQSZKTZJPXPWIRN
MQOLCSJTLGAJOKBSSBOMUPCE
FEWGDRHDDEEUMFFTEEMJXZR
RHZVIYQIYSXVNQXQWIOVWPJO
MQOLCSJTLGAJOKBSSBOMUPCE
FEWGDRHDDEEUMFFTEEMJXZR
SKCDKJCDJCYQSZKTZJPXPWIRN
RHZVIYQIYSXVNQXQWIOVWPJO
MQOLCSJTLGAJOKBSSBOMUPCE
SKCDKJCDJCYQSZKTZJPXPWIRN
```

Notice that all but the first line repeat.

The ciphers that follow, inverted relative to those above, are

```
HLMTAHGBGFNIV
```

```
ZUQUPNZN ABRYCTUGVZXUPB
MVERZRLQDBHQ
GKJFHYXODIE UGMNCBXCFLDBEY
VIOHIKNNGUAB
HFXPCQYZVATXAWIZPVE
```

The IACR website for the gold bars includes a pair of names and addresses that questions can supposedly be directed to, but they're both out of date.

Before closing this chapter, there's still one more tale of treasure left to tell.

Pirate Plunder

The eighteenth-century Frenchman Olivier le Vasseur, also known as Olivier la Bouche, or simply La Buse (the Buzzard) brings together for us a cipher, another hoard of gold, and pirate adventure.

In April 1721, La Buse and an English pirate, John Taylor, working together scored an amazing haul, the Portuguese ship *Vierge du Cap*. This ship normally carried seventy guns, but had been storm damaged and tossed most of the guns overboard to save it from foundering.

As a result of her weakened state, La Buse and Taylor were able to acquire what's been described as "one of the richest prizes in the history of piracy." The prize included gold and silver bars, chests of gold guineas, pearls, casks of diamonds, silks, works of art, and sacred objects belonging to the Archbishop of Goa, who was on board. Each member of La Buse and Taylor's crew got five thousand gold guineas and forty-two diamonds as his share.[56]

La Buse's luck finally ran out in early 1730. Captain L'Ermitte, commanding the French ship *Méduse*, engaged him and, after a bloody battle, took him back to Bourbon in chains. La Buse was hanged for his crimes on July 17, 1730. His captured treasure was not found. Athol Thomas related the incident that brings La Buse into this book:

> The romanticists have it that, as the noose was being tightened round his throat, La Buse produced a piece of paper and flung it to the crowd, shouting: "Find my treasure who can!" The action was in keeping with the swashbuckling character le Vasseur, but so far it has been impossible to tell whether he was perpetrating a practical joke that would tease treasure-hunters for centuries or whether he was setting a genuine problem.[57]

A reproduction of what is claimed to be La Buse's cipher is in fig. 9.31.

FIG. 9.31 A pirate's cipher?

Any experienced codebreaker's response upon seeing a cipher like this would be one of sheer delight, for it appears to be one of the simplest methods of encryption. The sample key given in fig. 9.32 is used to show how it works.

As you can see, the letters were placed in the little cells or pens of a tic-tac-toe board and a large X. Because only half of the letters can be accommodated in this manner, the process was repeated, but this time each letter shared its pen with a dot. Now to encipher a message, we just replace each letter with a representation of its pen, including the dot, if one is present. So, for the message "THE CODE IS MORE LIKE GUIDELINES, REALLY," we have fig. 9.33.

FIG. 9.32 A sample key for a simple cipher

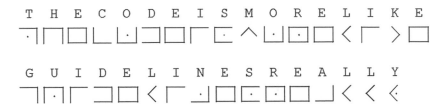

FIG. 9.33 A message enciphered with the pigpen cipher

Because of the use of "pens," this system is sometimes called the pigpen cipher. It was used extensively by the Society of Freemasons, so it's also referred to as the Masonic cipher.

The tic-tac-toe grids and the Xs aren't always used in the same order. For example, another user might start off with the tic-tac-toe grid, as we did, but then follow it by one with dots in it, then use an X, and finally an X with dots. Some versions make use of cells with two dots. Also, the alphabet can be placed into the cells in several different ways.

Still, no matter how it's done, the end result is a MASC, and we saw in chapter 1 how easily those are broken. But, for some reason, nobody's managed to turn La Buse's cipher into a nice plaintext. The best that's been found is one so riddled with errors that it's not useful for finding anything. Did La Buse put some extra twist on the cipher, or did he just mangle the encryption so badly that it can't quite be recovered? Or is it all a hoax?

The incredibly simple pigpen cipher has left us another example that long resisted solution. This time the message appears on a tombstone in Ohio.

Brent Morris, the thirty-third degree Mason and retired NSA mathematician quoted in chapter 5 on the Debosnys cipher, had this to say about the tombstone:

I suspect the message in the Masonic cipher is a listing of initials of offices. E.g., PM, PHP, PIM (Past Master, Past High Priest, Past Illustrious Master). However, it's too short to do much with.[58]

FIG. 9.34 A mysterious tombstone just south of Metamora, Ohio

While the publisher was copyediting this volume, I was lecturing at various locations on some of the unsolved ciphers covered in it. The tombstone cipher was one I mentioned to an audience that included German cipher expert Klaus Schmeh. He blogged about it, and his readers, in a group effort, found the solution.[59] You can follow the reference to see their result or attempt to solve it yourself. All I'll indicate here is that it wasn't what Morris thought!

Of greater importance than the meaning of this tombstone are the Scorpion ciphers of chapter 5. Flip back to the images of these and look for symbols that look like they arose from a pigpen cipher. If this is indeed what they are, then identifying those few symbols might be a lead that eventually unravels one of the ciphers completely. Remember, all it took to make the initial break into the Zodiac three-part cipher was suspecting that the message began with "I" and that the word "KILL" appeared in it.

I should confess that I felt a bit like Maxwell Smart as I presented the ciphers associated with large amounts of gold.

Me: At this very moment, there's more than a ton of gold buried in Buford County, Virginia. Would you believe it? Over a ton of gold!

You: I find that hard to believe.

Me: Would you believe twenty-some pounds somewhere north of Santa Fe?

You: No.

Me: How about two quarters and a wheat penny in my back yard?

10

Long Ciphers

In chapter 9, when investigating Cicada 3301, we saw how a book can provide the key to a cipher, and in chapter 1, we looked at a book that was a cipher itself, the Voynich manuscript. It turns out that the Voynich manuscript is not unique in that respect; there are many book-length ciphers. Some of them are unsolved. In this chapter, we look at some of these, as well as other long (if not quite book-length) ciphers.

It's somewhat surprising that these should resist decipherment. In general, the longer a cipher is, the easier it is to break. In fact, when intercepted messages are short, a common tactic of the codebreakers is to try to amass as many of them as possible. If they all make use of the same key, they are referred to as "messages in depth." Such sets of ciphers are similar in some respects to long ciphers or enciphered books. Viewed as a group, they are much easier to solve than the original individual ciphers.

German researcher Klaus Schmeh put together a list of encrypted books that is, as of this writing, still growing rapidly.[1] His list includes a numbering that is likely to become standard, so examples that are on his list are identified by number here.

Klaus began his list with the Voynich manuscript as number 00001. The leading zeros indicate that he was prepared for the list to become very long one day! Y2K would have been a nonissue if Klaus had done all of the programming. He continued with another old cipher, *Codex Rohonci* as number 00002.

Codex Rohonci (pre-1838)

This manuscript came to light in 1838, but its date of origin is unknown. It doesn't look as old as the Voynich manuscript, but at 450 pages, it's significantly longer. The little we know of its origins was summarized in a 2014 paper by Benedek Láng.

It was donated to the Library of the Hungarian Academy of Sciences together with the 30000-volume library of the late Hungarian count Gustav Batthyány in 1838. This library was earlier located at the family residence in the town of Rohonc (today: Rechnitz, Austria), hence the name of the codex.[2]

By 2005, three "solutions" had been put forth, but none was widely accepted.[3] A few pages from the book are reproduced as figs 10.1 and 10.2. It may be seen in its entirety at http://www.dacia.org/codex/original/original.html.

FIG. 10.1 The Rohonc Codex, pages 25–25a

As I indicated in the introduction to this book, some solved ciphers are included. This is, in my opinion, one of them. Images like the one above strongly indicate a Christian theme to the text, and indeed that is the result of a newly claimed decipherment. I've read an early draft of the paper that puts forth this decipherment and look forward to placing the final version in a future issue of *Cryptologia*, where you can read it and judge for yourself.

Ciphers are typically "stand-alone." They have context, but they don't come with a bunch of accessories. Our next example offers a strong exception. Like the Rohonc Codex, it was religiously inspired.

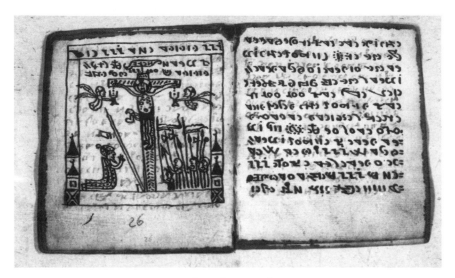

FIG. 10.2 The Rohonc Codex, pages 26–26a

James Hampton's Outsider Art

From 1950 to 1964, Meyer Wertlieb rented a garage in Washington, D.C., to a janitor named James Hampton. When the rent stopped being paid, Wertlieb went to the garage and discovered an elaborate artistic creation.

What had Hampton made and why? It was too late for Wertlieb to ask, for Hampton had died. That was the reason the rent had stopped arriving. To try to understand what Hampton was up to, we need to look at the few biographical details that have emerged.

Hampton was born in 1909 in Elloree, South Carolina. The 1910 census put the population of Elloree at only 540. Hampton's father was a self-ordained Baptist minister and gospel singer, who abandoned his wife and four children to be a traveling preacher. Almost nothing is known of Hampton's schooling. He later claimed on a job application to have had a tenth-grade education, but the school he listed has no record of him ever being there.

Hampton likely encountered a huge cultural shock in 1928, when he left rural South Carolina for Washington, D.C., to share an apartment with his older brother, Lee. Whether this was a contributing factor or not is unknown, but by 1931 he was having visions. He believed that God and his angels were physically visiting him. They told him to make a throne for the second coming of Jesus Christ. According to Hampton, such visits continued throughout his life. A record he made

THE THIRD HEAVEN
THE SECOND - CORINTHIANS
12 - 2 - 3

FIG. 10.3 James Hampton amid his creation

of one of these visits reads, "This is true that the great Moses the giver of the tenth commandment appeared in Washington, D.C., April 11, 1931."[4]

It's not known when he began work on the throne that was demanded. If he made much progress in the 1930s, it was to be interrupted by World War II. Hampton was with the 185th Aviation Squadron from 1942 until his honorable discharge in 1945. His duties took him to Texas, Seattle, Hawaii, Saipan, and Guam. He was tasked with carpentry and maintenance of airstrips, but luckily no combat.

In 1946, Hampton got a job in D.C. with the Government Services Administration (GSA) as a janitor. Living in a boardinghouse room, he found that he needed more space to build what he called *The Throne of the Third Heaven of the Nations' Millennium General Assembly*. This was the purpose for which he began renting the garage in 1950. It also seemed to be his life's purpose. He got off from his job at midnight and typically spent the next five or six hours working on the throne. He believed that God visited the unheated garage regularly to oversee the construction of the throne.

Hampton apparently had no other company at this point in his life, for he's been described as a "poor and friendless man."[5] His brother had died in 1948. He never found the "holy woman" he desired to help him with his work.

So how did he construct Jesus's throne? Well, he used a lot of aluminum foil. He was working on a tight budget. An account pieced together after his death shows how he gathered the materials he needed.

> He scavenged the neighborhood for gold and aluminum foil collected from store displays, cigarette boxes, and kitchen rolls; he even paid neighborhood vagrants for the foil on their wine bottles, and carried a sack wherever he went to hold any bits and pieces found on the streets. He also gathered light bulbs, desk blotters, sheets of plastic, insulation board, and kraft paper—all (apparently) from the trash of government buildings where he worked.[6]

Working in this manner, his installation grew to include 180 separate pieces.

According to a woman Hampton worked with, he wanted to become a minister after he retired. He didn't belong to any of the churches already in the area, but it's been speculated that he hoped to establish a ministry in a storefront church. In any case, he died before retiring, so he never had the chance.

Following his death, Hampton's sister came to D.C. to claim the body. This was the point at which she first saw the throne he obsessed over, and she did not want it.

For some reason, the National Collection of Fine Arts (which has since become the Smithsonian American Art Museum) did want it, agreed to pay the past due rent, and got all 180 pieces in 1970. They were first displayed publicly in the National Collection of Fine Arts exhibition "Hidden Aspects" in 1971.

What makes all of the above relevant to our purpose is that it provides context for possibly making sense of a strange notebook Hampton left behind. The notebook is some seventy pages long and was found by museum employees in the rented garage. A page from it, and a loose sheet, are reproduced as figs. 10.4 and 10.5.

Hampton believed that the odd script he used had been given to him by God. On every notebook page that he filled with it, he wrote "REVELATiON" at the bottom. He put his own name at the top of each page in the modified form "ST JAMES," outdoing his father's self-ordination.

Hampton used the same cipher on labels attached to the objects that make up his throne. But in these instances, the ciphertext usually follows one

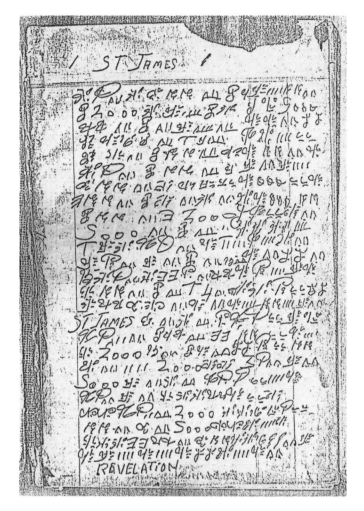

FIG. 10.4 A page from Hampton's enciphered notebook

or more words in plain English. If the ciphertexts are simply encryptions of the words they are paired with, then these pairings ought to offer valuable insight into the nature of the cipher. Still, nobody has been able to make sense of it yet! Some speculate that it's meaningless, but it certainly had meaning for Hampton. Is it possible to figure out exactly what it meant to him?

Stephen Jay Gould, who referred to himself as a "Jewish Agnostic,"[7] would certainly not have related to *The Throne* in the same way as Hampton, yet he gushed over it in print, calling it "one of the great works of American folk sculpture."[8] He elaborated

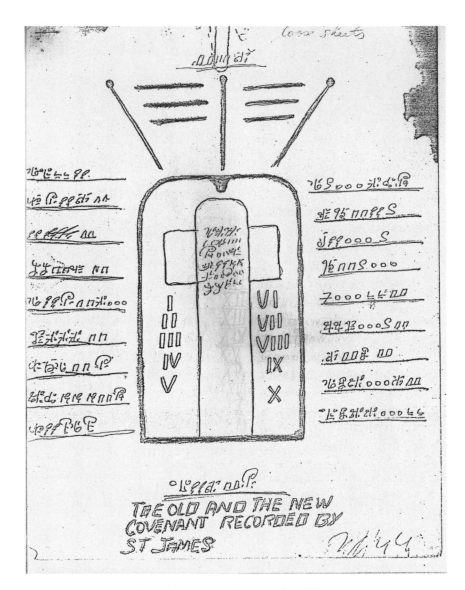

FIG. 10.5 A loose sheet containing samples of Hampton's cipher

I am not an art historian. I will make no aesthetic interpretation or judgment beyond a purely personal statement that Hampton's *Throne* stunned and delighted me when I happened upon it by accident during a coffee break from a meeting at the Smithsonian, and it has never failed, upon many subsequent and purposeful visits, to elicit the same pleasure and awe.[9]

I had a very different reaction. If I had inherited the work from a deceased relative, I would've put it out on the curb for trash. The ciphers are all that interest me.

But who am I to judge? A six-foot square piece of artwork I made from bottle caps, paper, magnet board, and wood dominates a wall in my home. See fig. 10.6.

FIG. 10.6 This is art!

The Penitentia Manuscript[10]

We already saw some of Gordon Rugg's work on an enciphered book, namely the Voynich manuscript, but he's connected with two more long unsolved ciphers—this time as the creator. He calls the puzzles The Penitentia Manuscript and The Ricardus Manuscript. Both have been available online since 2005.

Because Rugg is an atheist, we might expect the content of his unsolved ciphers to have little in common with the *Codex Rohonci* and the notebooks of Hampton, but Rugg has published in the area of biblical textual analysis, so who knows? Let's take a look at his ciphers and try to get inside his head a bit to figure out how he might have constructed them.

On his webpage devoted to The Penitentia Manuscript, Rugg wrote

Most modern codes are based on a shared set of underlying assumptions. I wondered what would happen if you deliberately ignored those assumptions. What sorts of code might that produce?

FIG. 10.7 Gordon Rugg (1955–)

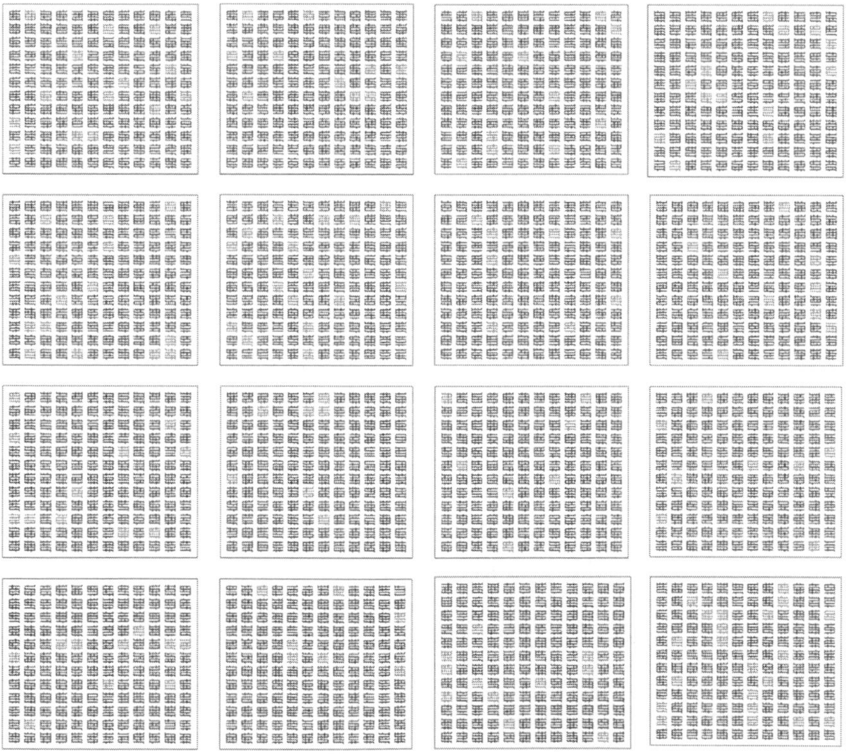

FIG. 10.8 The Penitentia Manuscript

One of the assumptions that he obviously ignored was that the ciphertext should consist of letters, numbers, or some combination of the two. He presented the puzzle in the format shown in fig. 10.8.

As this is too small to read, Rugg made it possible to enlarge any portion by clicking on it. Doing this for each individual position, we get the images like fig. 10.9.

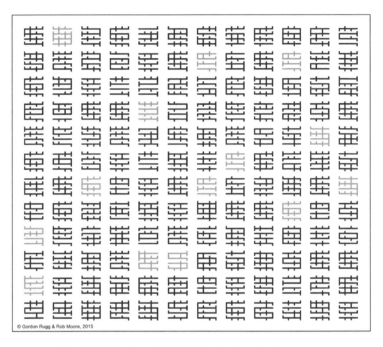

© Gordon Rugg & Rob Moore, 2015

FIG. 10.9 Position (1,1) of the Penitentia Manuscript

The symbols are far more intricate than is required to distinguish single letters. I think each represents instead two or more letters. Also, I think that Rugg may have ignored the assumption that text is to be read from left to right, top to bottom. Remember his work with the grille from chapter 1? Perhaps he filled the characters into the grid by following some path that can be generated by placing the grille in various positions, with possible rotations thrown in as well.

If all of this is correct, we have two separate problems to contend with:

1. Determining how the symbols correspond to combinations of characters, and
2. Determining the order in which the symbols are supposed to be read.

I believe that the first problem would have to be attacked first. Once the substitutions are known, statistics of the English language (assuming that it really is in English!) can help put everything in the correct order.

Of course, my guesses could be way off the mark! A quick look at Rugg's background may suggest other possibilities to you.

Most of the people publishing in the field of cryptology have degrees in mathematics or computer science. The notable exception is those publishing strictly on historical topics. However, Rugg doesn't match any of these profiles. His first degree, from Reading University, was in French and linguistics. He then went on to earn a Ph.D. in experimental psychology from the same institution. He also did postdoctoral work in artificial intelligence at Nottingham University and serves as the editor in chief of *Expert Systems: The International Journal of Knowledge Engineering and Neural Networks.*

With this background, he worked in fields as diverse as English lecturing, field archaeology, artificial intelligence, information retrieval, and human factors.

Describing his current position, Rugg wrote, "I am a Senior Lecturer in Computer Science at Keele University, and a visiting Senior Research Fellow in Computer Science at the Open University."[11]

For some, the world is an amazing intellectual playground where you can begin studying French and end up as a professor of computer science. Don't ever feel limited by your degrees or your current position. An excellent computer science professor, whom I had taken many classes with, once confessed to me that he had never taken a class in computer science himself. He was completely self-taught and had not too many years earlier been a tenured chemistry professor. He simply got bored with chemistry and needed a new intellectual challenge.

I asked Rugg for more details on his background, and the following paragraphs were among what I received.[12]

Linguistics showed me the difference between a humanities approach and a scientific approach to the same topic. That profoundly changed how I thought. It also showed me that intelligent people could study a topic for centuries and still miss a much more powerful way of handling it.

He mentioned that he served as an English lecturer in Nepal.

My time in Nepal showed me what real poverty looks like, and what life is like for a large proportion of the world's population. If I had been seriously ill, I would have been airlifted out, and given the best treatment available. My Nepali friends would have died. That didn't feel right. For me, the struggle against inequality and poverty is personal.

And he did work in field archaeology.

On one dig, I was part of a team that discovered one of the earliest Christian churches in Britain. It was built on the site of a Roman temple, which in turn was built on an Iron Age sacred site.

Back in the classroom, he noted,

My lectures have included using kites and a light sabre as teaching aids. The stories about my shooting an apple off a student's head with a crossbow are wildly exaggerated.

And he described some of his work with students and a colleague:

Several of my former undergraduate students have recently joined me in collaboration with Jason Dowdeswell, on the strength of the work they did in their undergraduate projects. Jason Dowdeswell is a leading visual effects specialist in Hollywood.[13]

The Search Visualizer software that I produced with my colleague Dr Ed de Quincey can do things that Google can't. Our work on identifying faulty reasoning helped us spot a set of possibilities that everyone else had missed.[14]

Does any of this give you insight into what he might have enciphered or how he did it?

The Ricardus Manuscript

Moving on to Rugg's second unsolved cipher, we once again have something that cannot immediately be plugged into a computer. This one was obviously inspired by the Voynich manuscript. It's called The Ricardus Manuscript, and a few images from it appear in figs. 10.10 through 10.14.[15]

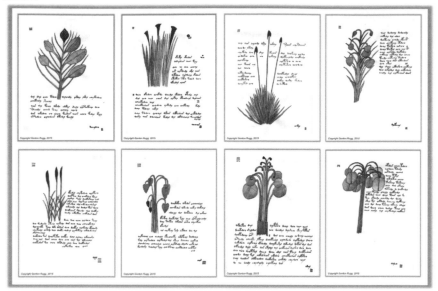

FIG. 10.10 The Ricardus Manuscript

Enlargements of these images can be found on Rugg's website for The Ricardus Manuscript.[16] Four of them appear on the following pages.

Rugg has promised that the first person to completely crack each code will receive (in addition to bragging rights!) an autographed canvas print of the code. Rugg's creations are indeed artistic, and our next contributor to the category of "Unsolved Ciphers, Long" also bridges the worlds of computer science and art.

The Book of Woo (2013)[17]

The online comic strip *Sandra and Woo* is drawn by Puri Powree and written by Oliver Knörzer. Oliver Knörzer is a German with a degree in computer science. So, in connection with our last two ciphers we saw someone with a degree in French working in computer science, and now we see someone with a degree in computer science engaged in the world of art. But, for the moment, Oliver's art is not his full-time gig. He explained, "I work as a software

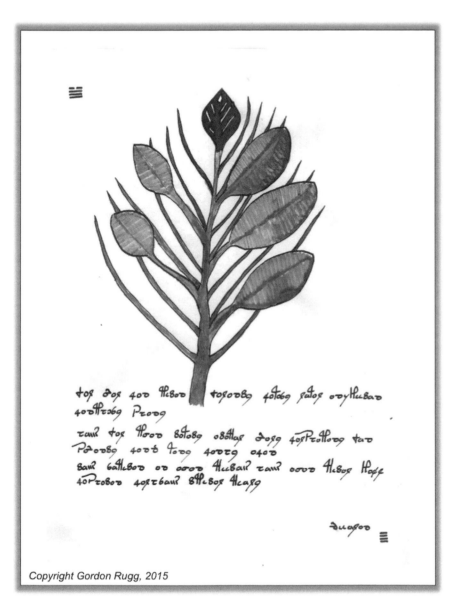

FIG. 10.11 The Ricardus Manuscript, Plant 1

FIG. 10.12 The Ricardus Manuscript, Plant 2

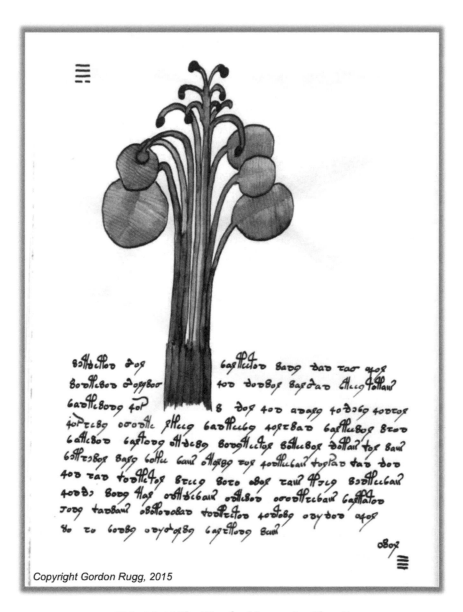

FIG. 10.13 The Ricardus Manuscript, Plant 7

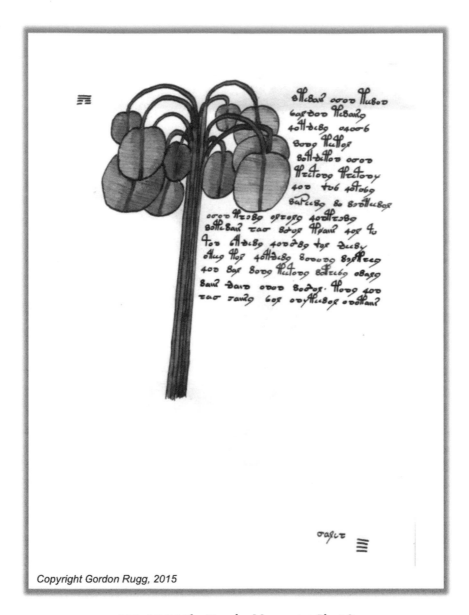

FIG. 10.14 The Ricardus Manuscript, Plant 8

developer for FARO Technologies. As part of our R&D department, I'm developing a webservice to view 3D data captured with laser scanners. Pretty exciting stuff!"[18]

The comic strip that introduced Knörzer's unsolved cipher is shown in fig. 10.15. It marked the five-hundredth installment and was titled *The Book of Woo.*

FIG. 10.15 Introduction to the *Book of Woo* cipher

In an e-mail to me, Knörzer noted that *Calvin and Hobbes* was one of his main influences on this strip, but that it was D. C. Simpson's *Ozy and Millie* that brought him into the world of webcomics. He also wrote, "I love raccoons and know a lot about them. I wrote almost all of the German and English Wikipedia articles about the species."[19] These articles could aid the cryptanalysts by revealing his preferred word choices.

The cipher follows over the next four pages (figs. 10.16–10.19).

Knörzer described the reaction to the cipher as "overwhelmingly positive." It set a record in terms of reader comments, and one reader, Foogod, put together a wiki (a website that allows collaborative editing of its content and structure by its users), so that those interested in breaking the cipher could easily work together.[20]

Among other useful information, the wiki now includes a hint from Knörzer. He wrote,

> I have decided to give you a little hint. The following word appears in the plain text:
>
> ENGLISH: Potbelly Hill | GERMAN: Bauchigen Hügel
>
> It should be quite easy to determine on which page. I hope this will generate some debate about the supposed content of that page.[21]

FIG. 10.16 *The Book of Woo* cipher, page 1

FIG. 10.17 *The Book of Woo* cipher, page 2

FIG. 10.18 *The Book of Woo* cipher, page 3

FIG. 10.19 *The Book of Woo* cipher, page 4

How a cipher works, is of course, influenced by the creator's background in the subject. Knörzer revealed,

> I was introduced to the basics of cryptography as part of several university courses. I found the topic to be particularly interesting, but decided against pursuing a career in it. Later, I found Nick Pelling's extensive blog "Cipher Mysteries" (http://www.ciphermysteries.com/) which renewed my interest in cryptography, and in particular in the Voynich Manuscript. I think there's something very fascinating about cipher texts that have eluded decipherment for decades or even centuries. And the Voynich Manuscript is the most mysterious of them all. My second favorite cipher is the Zodiac Killer's Z340.[22]

Knörzer promised to give the solution in strip number one thousand, if it isn't broken sooner; he anticipates that that strip will appear in late spring/ early summer 2018. But in the meanwhile, some serious progress has already been made. He wrote,

> Satsuoni was able to crack the last encipherment step. Since then, further analysis of the partially broken cipher text has yielded some interesting statistical results. But although some readers came up with seemingly promising interpretations of these results, no definitive breakthrough has been made yet. ... If you want to know more about the analytical results and theories, you should check out the wiki.[23]

> Many people must have put countless hours of their free time in solving the puzzle. This is certainly something that makes a writer happy.[24]

Why not visit the wiki yourself and see if you can make a contribution, and make him even happier!

By now, you may have noticed a common feature of all of the long unsolved ciphers presented thus far. None of them consist of the more traditional letters and/or numbers. So far, they are all made up of weird symbols.

Another thing that all of the ciphers in this chapter have in common is that they fail to disclose their method. It's likely that none of them could stay secure if *only* the key were unknown. In military contexts, it's always hoped that the method can be kept secret, but the assumption is made that it cannot. The enemy simply has too many ways to find out, especially if it's a system that's heavily used.

Claude Shannon, whose work with entropy, redundancy, and unicity points has already been described, put it like this, "The enemy knows the system." This is known as Shannon's maxim. But it's much older. Back in 1883, Auguste Kerckhoffs put forth the principle in the larger context of a set of six rules a good cipher ought to satisfy.[25]

K1. The system should be, if not theoretically unbreakable, unbreakable in practice.

K2. Compromise of the system should not inconvenience the correspondents. (This is Shannon's maxim. It means that if the enemy learns the system, no changes need to be made. Without the key, as well, the enemy ought not to be able to break the ciphers.)

K3. The method for choosing the particular member (key) of the cryptographic system to be used should be easy to memorize and change.

K4. Ciphertext should be transmittable by telegraph. (Today we would say computer.)

K5. The apparatus should be portable.

K6. Use of the system should not require a long list of rules or mental strain.

Looking back at the unsolved ciphers in this chapter, how many violate multiple rules in this list? Most of them? All of them? That's okay! None of them were intended to protect important military secrets. These ciphers were mostly just for fun (with the possible exception of the first two).

The last example in this chapter could have been placed in chapter 9, as it was part of the Cicada 3301 challenge in 2014. I put it here instead because it weighs in at fifty-eight pages. The first page is reproduced in fig. 10.20. The rest of it can be seen online.[26]

Cicada 3301 Rune Book

As has happened before, we have a cipher that looks much simpler than it really is. The appearance is that of a MASC, albeit one using an alternate alphabet. Rune rows have been used for enciphering everywhere, from *The Lord*

FIG. 10.20 First Page of the Cicada 3301 Rune Book from 2014

of the Rings dust jackets (J. R. R. Tolkien trained as a codebreaker before his writing career took off) to Ozzy Osbourne's *Speak of the Devil* album cover. This is something different.

The Cicada 3301 puzzle for 2014 had so many steps that a book could be devoted to it alone. Several puzzles involved pages of runes looking much like the one in fig. 10.20. They weren't MASCs either, but they were broken. The techniques that cracked those worked on only a single page of the current rune book, which appeared late in the challenge.

Once again, I'll leave it to you to explore this mystery. In closing this chapter, I point out how rapidly the catalogs of unsolved ciphers are growing. As of this writing (February 20, 2017) Klaus's list contains eighty-five encrypted books. Many are already broken, but the list will certainly be longer by the time this book sees print, and it's likely to contain at least one more that is unsolved. Check it out![27]

11

ET and RSA

The ultimate cipher could be a message originating from an extraterrestrial civilization, whether it be intentionally sent to us or simply intercepted. Messages believed to be just this, however briefly, have been encountered repeatedly.

The Martian Lights of December 1900

Subscribers to various newspapers around the world may have been surprised in December 1900 to see headlines indicating that Martians were sending messages to Earth. The method was via flashing light of various durations. Indeed, the astronomer to whom these stories trace back was also surprised to read of this!

The astronomer, Professor Edward C. Pickering, was director of Harvard's observatory at the time, and although he observed an interesting phenomenon, he didn't claim that it was an attempt at signaling. His explanation appeared in a newspaper article titled "Signal from Mars, Professor Pickering Saw Bright Lights Upon That Planet." Of course, the title of this article may have served to further the misinformation if the article itself wasn't read. For those who read beyond headlines, Pickering explained that he had received a telegram at the Harvard College Observatory on December 8, from the Lowell Observatory in Flagstaff, Arizona, with the following content

> Mr. A. E. Douglass, while observing the planet Mars on December 7, saw a projection on the north edge of Icarium Mare which lasted seventy minutes.

In this instance, "projection" meant "shaft of light."[1] Pickering passed word to Kiel, Germany, whence the information was distributed throughout Europe. Both Harvard and Kiel served as distribution hubs for astronomical news.

In Paris, France, Wilfrid Fonveille interpreted the phenomenon as an intentional signal and wrote of it in the European version of *The New York Herald*. From there, the altered story spread back to the United States.

Early reports further inspired Nikola Tesla, a highly eccentric genius in electrical engineering and related fields. Tesla was positive that radio communication with Mars was possible. He even claimed to have received radio signals from intelligent beings on the planet.

Pickering continued trying to set the record straight in the press, and in March 1901, *Scientific American* carried his lament, "In Europe it is stated that I have been in communication with Mars, and all sorts of exaggerations have sprung up. Whatever the light was, we have no means of knowing. Whether it had intelligence or not, no one can say. It is absolutely inexplicable."[2]

Nevertheless, Percival Lowell found an explanation in time to see print in the December 1901 issue of *Proceedings of the American Philosophical Society*. It is discussed in a moment.

Professor A. E. Douglass, the Arizona astronomer whose initial observations led to the eventual excitement, later wrote,

> The phenomenon on Mars which has given rise to the report of a message from that planet on December 7 of last year was really only a cloud on that planet lighted up by the setting sun. It was a true message, giving us knowledge of Martian climate, but not a message from any intelligent inhabitants.[3]

Despite the less exciting explanation being offered repeatedly, Charles Fort wrote about the incident in his 1923 book *New Lands* as if it might well have been a signal after all.

> It may have been a geyser of messages. It may be translated some day. If it were expressed in imagery befitting the salutation by a planet to its dominant, it may be known some day as the most heroic oration in the literature of the geo-system. See Lowell's account in *Popular Astronomy*, 10-187. Here are published several of the values in a possible code of long flashes and short flashes. Lowell takes a supposed normality for unity, and records variations of two thirds, one and one third, and one and a half. If there be, at Flagstaff, Arizona, records of all the long flashes and short flashes that were seen, for seventy minutes, upon this night of Dec. 7, 1900, it is either that the greetings of an island of space have been hopelessly addressed to

a continental stolidity, or there will have to be the descent, upon Flagstaff, Arizona, by all the amateur Champollions of this earth, to concentrate in one deafening buzz of attempted translation.[4]

The prestigious British science journal *Nature* reported on Lowell's explanation of the phenomenon in their May 1, 1902, issue. The journal reprinted the report in the "100 Years Ago" section of the May 2, 2002 issue. This reprint included the passage

> Mr. Lowell, in his concluding remarks, says that the surface marking, Icarium Mare, is undoubtedly a great tract of vegetation, and the observation of December is completely explained if it be assumed that a cloud was formed over this region and rose to a height of thirteen miles, and then, traveling east by north at about twenty-seven miles an hour, passed over the desert of Aeria and there was dissipated.[5]

The piece was perhaps reproduced to warn both reporters and scientists to be careful what conclusions they jump to and how much certainty they express in them. Lowell's conclusions were undoubtedly laughable a hundred years later.

On the Long Wavelength, 1920–1921

Before Charles Fort's *New Lands* appeared, another wave of excitement was generated by Guglielmo Marconi, a Nobel laureate in physics who was also credited with inventing radio (although there are many Tesla supporters who disagree).[6]

Once again, there was some assistance from the reporters, who misinterpreted some of his statements and sometimes got his first name wrong, resulting in paragraphs like the following, from *The Monthly Evening Sky Map*.[7]

> **Scientific World Stirred by Possibility of Communication from Mars**
> William Marconi, wireless inventor, recently announced that mysterious signals, dots and dashes like the letters of the Morse telegraphic code, had been recorded on radio instruments in both London and New York, which has aroused interest the world over, from the suggestion that they are signals from Mars or some other planet. Scientists generally doubt the

FIG. 11.1 Guglielmo Marconi (1874–1937)

possibility and believe the cause will be found due to terrestrial conditions or possibly the sun, while not a few believe it may be true.

Marconi says, in the *New York Journal*:

"I should be a bold speculator were I to declare positively that one or more of the planets are attempting to send us some kind of message, but it is equally futile to deny such a thing is possible, in view of our incomplete knowledge of such a profound subject.

"What is certain is that wireless stations are periodically receiving weird signals of uncertain import and indefinite origin." . . .

Resembles Morse Code

"What has actually happened in this throbbing mystery is that signals have been received which are apparently due to electro-magnetic waves of great length which are not merely stray signals.

"Occasionally such signals can be imagined to correspond with certain letters of the Morse code. They steal in at our stations irregularly at all seasons.

"What is their mysterious origin? The scientist who can answer this question will become immortal.

"At the same time, we must not overlook the fact that the possible seat of these signals may be the sun, where as is well known, electrical disturbances occur.

"Perhaps it may aid the inquiry if I say that we do not get the signals unless a minimum of a sixty-five mile wave length. Sometimes we hear the planetary or inter-planetary sounds twenty or thirty minutes after sending out a long wave. They do not interrupt traffic, *but when they occur they are very persistent.*"

Most Familiar Signal

"The most familiar signal received is curiously musical. It comes in the form of three short raps, which may be interpreted as the Morse letter "S," but there are other sounds which may stand for other letters.

"I must insist, however, that no connected letters fit in, and therefore nothing like a message intelligible to us has been received.

"The war prevented an investigation of the Hertzian mystery, but now our organization intends to undertake a thorough probe."

This piece was more sober than some and did present another possible explanation. A later piece, also in *The Monthly Evening Sky Map,*[8] was more dramatic.

Mars is Signaling to us, Says Marconi

William Marconi is now practically sure that he has received wireless messages from Mars, according to J. H. C. Macbeth, London manager of the Marconi Wireless Company, Ltd., who spoke at the Rotary Club in Manhattan recently. It was nearly two years ago that Signor Marconi first announced that he had received messages from the planet Mars. He was disputed by scientists, who said that his "messages" were caused by atmospheric disturbances. Marconi, however, by further testing, has found that his signals are interrupted regularly, regardless of other interference. Signor Marconi and other scientists now claim that the messages under discussion come from another planet, because the wave length is almost ten times that produced by our most powerful stations. According to Macbeth, Marconi has picked up waves estimated at

150,000 meters, when the largest waves our radio stations can produce are 17,000 meters.

Mr. Macbeth said that he believed it only a question of time before inventive genius, in deciphering unknown codes would result in the establishment of communication between this planet and Mars, if Signor Marconi's messages come from there.

Macbeths's call for an "inventive genius, in deciphering unknown codes" would be heeded a few years later when it was once again believed by some that a message had been received.

William Friedman Gets the Call

How would you like to be the go-to person for translating a possible alien message? This is exactly what happened to professional cryptologist William Friedman in 1924.

The story began with David Peck Todd, retired astronomy professor and former director of Amherst College's observatory. Todd had proposed in 1909 that Martians might be attempting to communicate with humans via radio and that a balloon at high altitude might be able to detect the signals. He never got to follow through with that particular experiment, but in August 1924, he was able to carry out another test. This month marked an opposition of Mars (i.e., a closest approach), so the timing was especially favorable for receiving signals if any were being sent from that planet.

Todd convinced the U.S. Army and Navy to observe radio silence in the Washington, D.C., area, as far as possible, to minimize interference. They also agreed to task their receiving stations with listening for communications from Mars during the opposition.

Todd decided to listen in at a wavelength of 6,000 meters (50 kHz). The signals were recorded, over a period of twenty-nine hours, on a film six inches wide and thirty feet long. The recording device was invented by C. Francis Jenkins, who called it a "radio photo message continuous transmission machine."[9] He explained the machine's functioning like so:

As the film unwinds an instrument passes over it from left to right 50 times every inch. An incoming sound causes a light to be flashed on the film, and this produces a black line.[10]

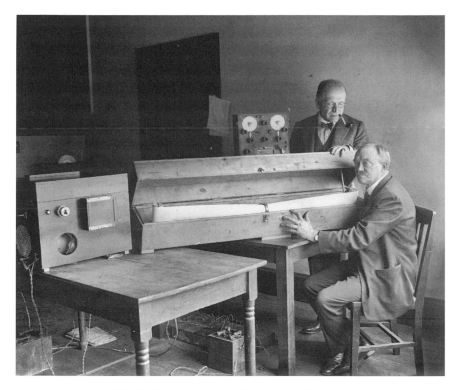

FIG. 11.2 Equipment for listening to Martians

With Todd's long interest in the subject, he was well aware of the history of previously received "messages." He remarked, "Three years ago Marconi was reported as saying he had heard signals from Mars. A few days ago he was quoted as saying he was too busy to listen to possible messages from Mars and that it was a ridiculous idea to do so. He changed his mind, and no one knows what he heard the first time. With our photograph, however, it is not a question of what one man heard. It is a permanent record, which all can study."[11]

The experiment was carried out in Jenkins's laboratory in Washington, D.C.

Apparently not very optimistic in regard to the chance of hearing from aliens, Jenkins said he expected the experiment to yield nothing more than a blank film. The result found after developing the film was quite different. According to a newspaper account,

The "messages" ranged from a fairly regular arrangement of dots and dashes, running the full length of the film down its left side, to pictures weirdly resembling a face, repeated at regular intervals on the right side of the film.[12]

A small sample of the developed film is reproduced in fig. 11.3.

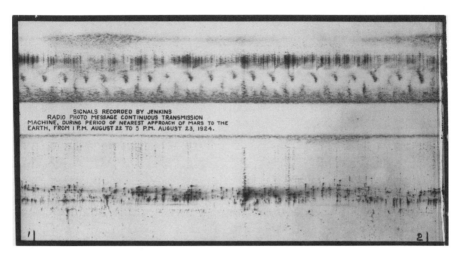

SIGNALS RECORDED BY JENKINS
RADIO PHOTO MESSAGE CONTINUOUS TRANSMISSION
MACHINE, DURING PERIOD OF NEAREST APPROACH OF MARS TO THE
EARTH, FROM 1 P.M. AUGUST 22 TO 5 P.M. AUGUST 23, 1924.

FIG. 11.3 A signal from Mars or another sort of learning opportunity?

Jenkins didn't give the appearance of someone interested in perpetrating a hoax. In fact, he resisted allowing the results to see print because he didn't want people to think that it was a hoax. After giving in to public interest, by allowing the results to be reproduced, he stressed that it was scientific curiosity that caused him to participate in the experiment and not a belief in intelligent life on Mars. In fact, he said,

> I don't think the results have anything to do with Mars. Quite likely the sounds recorded are the result of heterodyning, or interference of radio signals. The film shows a repetition, at intervals of about a half hour, of what appears to be a man's face. It's a freak which we can't explain.[13]

Todd was more open-minded (or deluded, depending on your perspective). He said,

> The Jenkins machine is perhaps the hypothetical Martians' best chance of making themselves known to earth. If they have, as well they may, a

machine that now is transmitting earthward their "close-up" of faces, scenes, buildings, landscapes and what not, their sunlight values having been converted into electric values before projection earthward, all these would surely register on the weirdly unique little mechanism.[14]

Ultimately, though, Todd was also unable to explain the results. In lieu of this, he stressed the value of having a record of the experiment. He provided William Friedman with about twelve feet of the record in the hope that the great codebreaker could find meaning in it. Another section was placed with Dr. J. H. Dellinger, chief of the radio division of the bureau of standards.

FIG. 11.4 Jenkins (left), Friedman (center), and Todd (right) examining the developed film © Bettmann / Getty Images

Friedman never made any sense of the signal.

However, when carrying out scientific investigations, there's sometimes an unexpected lesson to be learned. Friedman's training was not in astronomy or physics, so it's not surprising that he missed it in this instance. But Todd or Jenkins could conceivably have caught on to it. It's unclear why they chose to listen in at a wavelength of 6,000 meters, but the result was that they didn't stand a chance of picking up anything originating outside of the Earth's atmosphere, and what they did receive was therefore likely interference from terrestrial sources. However, if they had repeated the experiment (a good idea in general!) at different wavelengths, times, and directions, they would have stood a chance of discovering radio astronomy.

Today

Other potentially interesting signals continue to be found. One cause is due to the fact that computer memory is subject to ionization (charged particles hitting it), and too much of this activity sometimes causes double bit flip errors. So a device sent into space on whatever the current mission is can have the data it's collecting corrupted. There are correction algorithms built in, but they can't keep up with the errors when there's too much of this activity. The result is corrupted data on the solid-state memory device. When it's sent back and received on Earth, it can be misinterpreted as a signal.

Depending on the type of signal, the error is sometimes quickly noticed on the ground. Errors of this type occur about twice a year (per mission or device). Other times, it can take longer to recognize the error, generating a bit of excitement.

Much More!

I chose to focus here on old episodes that aren't as well-known today as more recent alarm raisers, such as the "Wow! signal" of 1977. Readers are encouraged to pursue the references for further tales of possible communications. This chapter must now move on by switching sides and looking at messages sent by Earthlings.

Role Reversal

Humans have not only made serious attempts to receive extraterrestrial signals; we've also (with incredible caution) sent out some of our own signals. The problem is that militaries tend to object to such activities. For if we manage to contact a civilization that is actually capable of coming here, then their technology would be far ahead of our own. So, if they chose to visit us and had ill intent, we would be completely at their mercy. They might be benevolent, but if not, we would be powerless to stop them from doing whatever they desired. For some strange reason that I don't understand, some people assume that an alien civilization with technology far beyond our own must be peaceful. I disagree. The worst excesses of humanity have not been mitigated by advances in technology. Indeed, these "advances" have often been applied to make the killing of members of our own species occur more easily and in greater numbers. Our technology has increased dramatically over the few thousand years of our recorded history, as have the numbers killed in our wars. I expect aliens to be much like us in this respect.

Anyway, when messages were sent out from Earth, briefly and with great care to make it extremely unlikely that anyone would ever receive them, here's how we did it.

The following is a message that was sent in 1974.

```
00000010101010000000000001010000010100000001001 00010
00100010010110010101010101010101001001000000000000000
00000000000000000000011000000000000000000001101000000
00000000000011010000000000000000001010100000000000000
0000001111100000000000000000000000000000000001100001110
0011000011000100000000000011001000011010001100011000
0110101111101111101111101111000000000000000000000000
00100000000000000000010000000000000000000000000001000
000000000000001111100000000000000111100000000000000
0000000011000011000011100011000100000001000000000100
0011010000110001110011010101111101111101111101111100000
00000000000000000000010000001100000000010000000000001
100000000000000010000011000000000011111000001100000
11111000000000011000000000000001000000010000000010000
01000000110000000100000001100001100000010000000000110
0010000110000000000000011001100000000000001100010000
11000000000011000011000001000000010000001000000001000
00100000001100000001000100000001100000001000100000
00001000000010000010000001000000100000001000000000
00110000000011000000011000000000100011101011000000
00001000000010000000000001000011111000000000000100
00101110100101101100000010011100100111111101110000111
0000011011100000000101000001110110010000001010000011
11110010000001010000011000001000011011000000000000
00000000000000000000011100000100000000000011101010
00101010101010011100000000010101010000000000000000101
```

```
00000000000000011111000000000000000001111111111000000000
00011100000000111000000000011000000000000110000000110100
00000010110000011001100000001100110001000101000010
10001000010001001000100100010000000010001010001000000
00000010000100001000000000000010000000001000000000000
01001010000000000011110011111010011111000
```

It was presumed that an alien intelligence receiving the signal would notice that the number of bits sent was 1,679, which is the product of two primes, namely 23 and 73. That is, $1,679 = 23 \times 73$. After noticing this, the aliens would hopefully write out the bits in the form of a rectangle with 23 bits in each of 73 rows, like so:

```
00000010101010000000000
00101000001010000000100
10001000100010010110010
10101010101010100100100
00000000000000000000000
00000000000011000000000
00000000001101000000000
00000000001101000000000
00000000010101000000000
00000000011111000000000
00000000000000000000000
11000011100011000011000
10000000000000110010000
11010001100011000011010
11111011111011111011111
00000000000000000000000
00010000000000000000010
00000000000000000000000
00001000000000000000001
11111000000000000011111
00000000000000000000000
11000011000011100011000
10000000100000000010000
11010000110001110011010
11111011111011111011111
00000000000000000000000
```

```
00010000000110000000010
00000000000110000000000
00001000001100000000001
11111000001100000011111
00000000000110000000000
00100000000100000000100
00010000001100000001000
00001100001100000010000
00000011000100001100000
00000000001100110000000
00000011000100001100000
00001100001100000010000
00010000001000000001000
00100000001100000000100
01000000001100000000100
01000000000100000001000
00100000001000000010000
00010000000000001100000
00001100000000110000000
00100011101011000000000
00100000001000000000000
00100000111110000000000
00100001011101001011011
00000010011100100111111
10111000011100000110111
00000000010100000111011
00100000010100000111111
00100000010100000110000
00100000110110000000000
00000000000000000000000
00111000001000000000000
00111010100010101010101
00111000000000101010100
00000000000000101000000
00000000111110000000000
00000011111111100000000
00001110000000111000000
```

```
000110000000000011000000
001101000000000010110000
011001100000000110011000
010001010000010100010000
010001001000100100010000
000001000101000100000000
000001000010000100000000
000001000000000100000000
000000010010100000000000
011110011111010011110000
```

The next step would be to replace the 0s and 1s with empty squares and filled in squares, respectively, yielding the image in fig. 11.5.

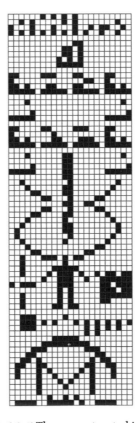

FIG. 11.5 The reconstructed image

Can you make much sense of the result? The idea is that an alien civilization with technology sufficiently advanced to receive the signal would not only be able to get to this stage of the decipherment, but would also succeed in grasping the information the image is intended to convey.

It's supposed to show (from top to bottom) the numbers 1 through 10 in binary; the components of DNA, via their atomic numbers; 1, 6, 7, 8, and 15, for hydrogen, carbon, nitrogen, oxygen, and phosphorus, respectively; molecular formulas for nucleotides; the DNA double helix and the number of nucleotides it's made up of for humans; a stick figure representing what we look like, along with the average height of an adult male and the total human population of the Earth; a map of our solar system roughly indicating the relative sizes of its members and the location of Earth; and finally the telescope at Arecibo, which sent the message, along with its diameter!

Frankly, I'm skeptical of an alien intelligence's ability to extract all of this information from fig. 11.5. A more honest attempt at communication would have used much larger primes to convey a higher resolution image that could convey information much more clearly and look less like a game on an Atari 2600. Another option is to send thousands of images that form an illustrated dictionary.[15]

As it was, the Arecibo message was sent just once, being "on the air" for less than three minutes, and was aimed at a globular star cluster, M13, which is somewhere around 21,000–25,000 light years away. Of course, this is a moving target, which will be missed by the time the message travels the immense distance. It should then stay free of any possible encounters until it arrives at another galaxy, an even more ridiculous distance away. As indicated previously, we were being careful not to actually contact anyone. It was just a publicity stunt.

The message was sent on a frequency of 2380 MHz. Arguments can be made for and against given frequencies, but I believe that we should consider how the intelligence community has been doing things for quite some time now. The concept is called "spread spectrum," and it involves spreading the signal over a wide *range* of frequencies. This method offers a number of advantages. First, if an enemy attempts to jam the signal with a powerful broadcast at a particular frequency, only the small portion of the message at that frequency is lost. You could say that spread spectrum makes the message more resilient. Also, a signal can be spread so widely that it drops down below the background noise. Imagine tuning through a radio dial and hearing nothing but static, never knowing that there's a meaningful signal underlying it all. It's a bit like steganography! The spread spectrum approach that works so well for secret earthly communications might be well-suited for nonsecret extraterrestrial communications as well. Perhaps attempts to listen in at specific frequencies are, in a sense, missing the forest for the trees. Could the messages be almost everywhere at once?

It will be interesting to see how efforts to listen in on possible extraterrestrial messages evolve over time, as well as how transmission attempts progress. It was almost twenty-five years after the Arecibo message before another such transmission was made. This second message was sent from a Ukrainian radio telescope in May 1999. By 2014, at least seven more messages had been sent, including a series of messages by a group of Russian teenagers working with astronomer Aleksandr Zaitsev in 2001. Military concerns persist, but groups wishing to send messages have many options in today's world.

The terse binary message sent from Arecibo was created by Frank Drake, Carl Sagan, and several others. It turns out that the development of the message connects with cryptology at several points. One connection came years later when Carl Sagan was invited to join the National Security Agency's Scientific Advisory Board. Agreeing to serve in this incredibly interesting position would have placed some restrictions on Sagan's work as an author. He would have had to go through a prepublication approval process to ensure that he wasn't accidentally revealing any classified material he encountered in his capacity as an advisory board member. Sagan turned the invitation down to avoid this restriction.

Another connection arose as I traced the idea of using factoring as a way of converting a string of bits into an image. World-class flutist (and also world-class cryptanalyst—he's in NSA's Hall of Honor) Lambros D. Callimahos discussed sending an Arecibo-type message in an article in the March 1966 *IEEE Spectrum*. Was he the one who came up with the idea in the first place? Apparently not, for he credited another. Callimahos wrote,

> After a 1961 conference at Green Bank W. Va., to discuss the possibility of communication with other planets, one of the participants, Bernard M. Oliver, made up a hypothetical message on the raster principle. The message, consisting of 1271 binary digits or "bits," is shown in Fig. 1. Since 1271 has but two prime factors, 31 and 41, we would naturally be led to write out the message in raster form in 41 lines of 31 bits each, or in 31 lines of 41 bits each; the latter case reveals a greater nonrandomness in the patterns disclosed, indicating these are the correct dimensions.[16]

Oliver's idea had appeared in a 1963 paper he wrote for a volume that had a mix of reprinted and new papers on the topic of interstellar communication. He discussed the idea explained above and wrote, "Using these principles, Frank Drake, of the National Radio Astronomy Observatory, constructed an imitation message which he mailed to those who had attended the meeting mentioned earlier (14)."[17]

So, the credit now seems to belong to Drake. The meeting was the same one referenced by Callimahos, but is more precisely given as being held at the National Radio Astronomy Observatory in November 1961.[18] More intriguing was reference 14, cited in this quotation. It referred to "Private communication to members of The Order of the Dolphin."[19]

It turns out that the Order of the Dolphin was a semisecret society whose goal was to establish communication with extraterrestrials. Its membership consisted of those who had attended the 1961 gathering, namely conference organizer J. P. T. Pearman; astronomers Carl Sagan, Frank Drake, Su-Shu Huang, and Otto Struve; inventor Bernard Oliver; neuroscientist John Lily; evolutionary biologist J. B. S. Haldane; physicist Philip Morrison; Dana Atchley (founder of the Fortune 500 company Microwave Associates); and chemist Melvin Calvin, who won a Nobel Prize while at the conference. The name for the society was inspired by Lily's work on dolphin communication. Not much information on this group has seen print. If someone can manage to get hold of a decent amount of their correspondence, it should make for an interesting paper. What other methods did they consider for composing easily interpreted messages?

A $200,000 Factoring Challenge

For the Arecibo message, if it's received a long time from now in a galaxy far, far away, the recipient will have to factor 1,679 to get at the image. This part is easy because the prime factors, 23 and 73, are small. But when the prime factors have hundreds of digits, it's another matter. For many years, RSA Laboratories offered prize money to anyone who could factor large challenge numbers and explain how they did it. The greatest challenge was for a $200,000 reward (!) and is reproduced here:[20]

RSA-2048

Status: Not Factored

2519590847565789349402718324004839857142928212620403202777713783604366202070759555626401852588078440069182906412495150821892985591491761845028084891200728449926873928072877767359714183472702618963750149718246911650776133798590957000973304597488084284017974291006424586918171951187461215151726546322822168699875491824224336372590851418654620435767984233871847744479207399342365848238242811981638150106748104516603773060562016196762561338441436038339044149526344321901146575444541784240209246165157233507787077498171257724679629263863563732899121548314381678998850404453640235273819513786365643912120103097122822120720357

If you want to plug this number into a computer and see if you can factor it, you should take a moment to make sure it was entered correctly—perhaps there's a typo in this book or on the website you try to copy and paste it from! As a way of checking, make sure that the number has 617 digits and that they sum to 2,738.

The number is known as RSA-2048 because when it's expressed in binary (base 2), it's 2,048 bits long.

The offer of a $200,000 reward has now expired, but the challenge remains. No one has been able to meet it. Though the solution in this case is simply a pair of numbers, and not a message, I feel that it fits the theme of the book because it's unsolved and connects strongly with cryptology. So, what is the connection, and why would a company offer such a big reward for the solution?

A Cipher System 337 Years in the Making

The story can be said to have begun in 1977, when three MIT professors, Ron Rivest, Adi Shamir, and Len Adleman, produced a paper that did what many thought was impossible. They found a way that enciphered messages could be securely exchanged between people who had not agreed on a secret key ahead of time. Their system became known as RSA, after their last initials.

The mathematical foundation for their approach originated far earlier than 1977, though. To see the true beginning, we need to roll the years back to 1640 and look at the work of French mathematician Pierre de Fermat (1601–1665). For some reason, Fermat was looking at what happens when numbers are raised to powers and then reduced by dividing by a prime and taking the remainder. This is called "exponentiation modulo p," where p is the prime that's used.

An example with actual numbers makes this clearer. We can tersely represent raising 7 to the 3rd power and then taking the remainder after dividing by 11, as 7^3 (mod 11). The abbreviation "mod" is often used for modulo. Because $7^3 = 343$ and dividing 343 by 11, we see that it goes in 31 times with a remainder of 2, our final answer is 2. That is, 7^3 (mod 11) = 2.

Mathematicians aren't content just making calculations, though. They want to find patterns and prove when they hold. So, Fermat kept looking at examples searching for some rule.

He found results like the following:

1^1 (mod 2) = 1	1^2 (mod 3) = 1	1^4 (mod 5) = 1	1^6 (mod 7) = 1
	2^2 (mod 3) = 1	2^4 (mod 5) = 1	2^6 (mod 7) = 1
		3^4 (mod 5) = 1	3^6 (mod 7) = 1
		4^4 (mod 5) = 1	4^6 (mod 7) = 1
			5^6 (mod 7) = 1
			6^6 (mod 7) = 1

Do you see the pattern? Can you predict what power you could raise numbers less than 11 to in order to get 1 out when you divide by 11 and take the remainder (i.e., mod 11)?

Fermat realized that given any prime p and a positive integer a less than p, we have $a^{p-1} = 1$ (mod p). This is known today as Fermat's little theorem, and it's an example of pure mathematics. One definition of pure mathematics is "Mathematics that gets only mathematicians excited." When a nonmathematician finds a particular bit of mathematics exciting, then it's likely to be applied mathematics.

Fermat's result was pure, but it was close to having an application. If we take $a^{p-1} = 1$ (mod p) and multiply both sides by a, we get $a^p = a$ (mod p). I'll show how this *might* be used as a cipher system, but first I need to discuss how to convert a message to a number.

There are many ways to do this, but for simplicity, we'll just replace every letter with its position in the alphabet. We don't want the confusion of wondering whether 17 is the pair AG (positions 1 and 7 in the alphabet) or the 17th letter, Q, so we'll always use a leading zero for the first nine letters. That is, we have A = 01, B = 02, C = 03, ..., Once we get to J = 10, we can write the numbers as we normally would. In this scheme, AG would by 0107, not 17.

For a long message, we'll get a big number, but that's okay for now. I'll talk about what to do if it's too big later. However big the number is, just imagine that the prime p that we're going to mod out by is even bigger.

We could take a message, convert it to a numerical value, call it a, and then raise it to some power less than p. Let's call our power e, for enciphering exponent. Raising a to the eth power and moding out by p gives us some completely different number, which has no obvious connection to our original number (message). This number can then be sent as a ciphertext. The person on the receiving end could raise this number to some other power, call it d for deciphering exponent, so that the grand total of the powers is p. By Fermat's little theorem, the end result is our original number, a, again. We can convert a back to the original message by using our substitutions 01 = A, 02 = B, etc., which (mod p) gives the original message once again.

The combined operations give us $(a^e)^d = a \pmod p$. However, this method does not actually work! Feel free to look back and try to see why if you haven't noticed the problem already.

The reason that $(a^e)^d = a \pmod p$ won't work out nicely is because $(a^e)^d = a^{ed}$, and if we want this to be guaranteed to be just a, then we must have $ed = p$. But this cannot be, because, as a prime, p cannot be nontrivially factored.

We can try to fix this problem by moding out by a number that isn't prime, but look at this:

$$2^{10} \pmod{10} = 4$$

We hoped to get 2, but we don't necessarily have $a^n = a \pmod n$ if n isn't prime.

So, although Fermat's result came close to offering us a cryptosystem, it doesn't work out nicely. But that's okay. Fermat wouldn't have cared. He wasn't trying to create a cipher system; he was simply pursuing pure mathematics. It's unlikely that he was even aware that his result could be investigated as a potential way to encipher messages.

The next big moment in this story came about in 1760 at the desk of the great Swiss mathematician Leonhard Euler. He was aware of Fermat's little theorem and sought a way to make it work modulo n for values of n that weren't prime. Above, I simply noted that it doesn't work. Euler wanted to find a way to modify it so that it would work. He didn't have any application in mind. Like Fermat, he was doing it for love of the game—pure mathematics!

What he found was that if n was not prime, he could find an exponent that would work, but only for certain bases. That is, the number that was being raised to a power had to have a certain property. If we call the base a (as we did above), then we need it to be true that a and the modulus n have no common divisors greater than 1. For example, $a = 4$ and $n = 9$ is okay. The positive divisors of a are 1, 2, and 4, and the positive divisors of 9 are 1, 3, and 9. The largest divisor they have in common is 1. Everything's okay in this case! On the other hand, a pair of numbers like $a = 7$ and $n = 21$ won't work because they have a common divisor of 7, which is greater than 1.

Given a pair of numbers that don't have any common divisors bigger than 1, we say that they are "relatively prime." This has nothing to do with if the individual numbers are prime or not. For example, 4 and 9 are relatively prime, although neither number is prime.

So, the result Euler found tells us that there's a number that a can be raised to modulo n to get 1 if a and n are relatively prime. Euler's result is a bit trickier than Fermat's, though. The power that a needs to be raised to isn't as simple as $n - 1$. It's a function of n, but one that's more easily expressed in words than in algebra. It's the number of positive integers less than n that are relatively prime to n. Euler used the Greek letter ϕ (phi) to represent the function that gives this number.

For example, if $n = 6$, then the positive integers less than 6 are 1, 2, 3, 4, and 5. However, these are not all relatively prime to 6. Looking at 2, we see that 2 and 6 have a common divisor of 2. Similarly, 4 and 6 have a common divisor. So, for the 5 numbers on our list, only 1, 3, and 5 are relatively prime to 6. That is, only three numbers work. Therefore, $\phi(6) = 3$.

Now consider a number that's a bit larger, $n = 15$. The positive integers less than 15 are 1, 2, 3, 4, 5, 6, 7, 8, 9, 10, 11, 12, 13, and 14. We want to count only the ones that are relatively prime to 15. We lose 5 and 10 because they and 15 are all divisible by 5. We also lose 3, 6, 9, and 12, because they and 15 are all divisible by 3. The numbers that remain are 1, 2, 4, 7, 8, 11, 13, and 14. This list has eight numbers, so $\phi(15) = 8$.

Putting it all together now, Euler's result was this:

If a and n are relatively prime, then $a^{\phi(n)} = 1 \pmod{n}$.

For the special case where n is prime, we have $\phi(n) = n - 1$. This is because every positive integer less than a given prime is relatively prime to that prime. So, if we take Euler's theorem and replace n with a prime p, the result reduces to Fermat's little theorem. That is, Euler's rule includes Fermat's as a special case. For this reason, it's known as "Euler's generalization of Fermat's little theorem."

Euler didn't use this result to create a cipher system. As was mentioned before, he was engaged in pure mathematics. However, more than two hundred years later, our heroes at MIT discovered the wonderful cryptologic application.

As we did with Fermat's little theorem, they multiplied both sides of Euler's formula by a, to get $a^{\phi(n)+1} = a \pmod{n}$.

Fermat's result failed to serve as a cryptosystem in the manner I proposed because the exponent needed to guarantee we'd get back to our starting point, a, was a prime and therefore not factorable. Here we don't have that problem. Though there are values of n that make $\phi(n) + 1$ prime, we don't have to worry about them. A few steps show why. First, we take the equality $a^{\phi(n)+1} = a \pmod{n}$ and multiply both sides by $a^{\phi(n)}$ to get $a^{2\phi(n)+1} = a^{\phi(n)+1} \pmod{n}$. Now we write this equation again, but change the way the right-hand side is expressed. We have $a^{2\phi(n)+1} = a^{\phi(n)} a^1 \pmod{n}$. But $a^{\phi(n)} = 1 \pmod{n}$. This means that the right-hand side of the equality can be simplified to give us $a^{2\phi(n)+1} = 1a^1 \pmod{n}$, but we don't need to write the 1s, so it becomes $a^{2\phi(n)+1} = a \pmod{n}$.

I showed a lot of steps, but basically we got this result by multiplying both sides by $a^{\phi(n)}$ and simplifying. If we do this again, we'll end up with $a^{3\phi(n)+1} = a \pmod{n}$. And we can keep going. Pick your favorite positive number k, and it will be true that $a^{k\phi(n)+1} = a \pmod{n}$.

So, though it's possible that $\phi(n) + 1$ will be prime, we won't have $k\phi(n) + 1$ prime for all positive integers k, and if we can factor $k\phi(n) + 1$ for some value of k, then we can find integers e and d that can be used to encipher and decipher messages.

But, we don't want to have to factor numbers of the form $k\phi(n) + 1$ to get e and d because factoring is hard! Fortunately, there's an old algorithm that helps us out. It's called the Euclidean algorithm. Yes, it is *that* old. This algorithm goes back to ancient Greece and provides a quick way to find e and d. All we have to do is pick an e that's relatively prime to $\phi(n)$ and then plug both e and $\phi(n)$ into the algorithm. After just a few steps, we have d.

I claimed earlier that the professors at MIT found a system that could be used by people who had not agreed on a secret key ahead of time. It might seem like we're not there yet because the users have to agree on e, d, and n. But this is not the case. If I want to have someone I haven't met be able to send me enciphered messages that I can actually decipher, I can do so easily. I start by selecting values for e, d, and n. I do this all by myself. I then put e and n on my website! These are referred to as my "public key" because anyone can look them up. They are made public! Now, the person I wish to communicate with can convert his or her message to a number, a, and then calculate a^e (mod n) and send it to me. I then raise this number to the power d, which I carefully keep secret, mod out by n and get a. Converting back from numbers to letters, I have the original message. The number d, which I keep secret, is known as my "private key."

Everyone in the world can do this. There are plenty of numbers to go around. Phonebooks could be published, in print or online, listing everyone's public keys. As long as the private keys are kept secret, messages can be sent securely.

Now, there's an important detail that needs to be addressed, and this is what connects this cipher with the $200,000 factoring challenge. The value for n needs to be chosen wisely. To be specific, it needs to be hard to calculate $\phi(n)$. For if someone can find $\phi(n)$, then he or she can plug $\phi(n)$ and e (part of my public key) into the Euclidean algorithm and quickly find d, my deciphering key. He or she can then read every enciphered message that's sent to me as easily as I can.

To make $\phi(n)$ hard to calculate, n should be big, but that's not enough. There's a simple formula that allows us to calculate $\phi(n)$ if we know how to factor n, so n should be hard to factor. The easiest way to make n big and hard to factor is to make it the product of two really big primes. People who use this system start with two primes, p and q, of more than a hundred digits each and multiply them together to get n. In order to use the system with people they haven't met, n needs to be made public.

This is why RSA Labs offered huge rewards to anyone who could factor products of two big primes. They were (and still are) selling encryption products based on the difficulty of factoring such numbers, and they wanted potential customers to know that this was an extremely challenging task. Offering cash rewards to anyone who could meet the challenge was a way of demonstrating confidence in their product and garnering publicity at the same time.

Of course, the person creating the system needs to know $\phi(n)$. Although this is hard for anyone else, it is easy for the creator because he or she knows the primes p and q such that $n = pq$. With this being the case, the creator applies the lovely formula $\phi(n) = (p-1)(q-1)$ to get the number he or she needs.

A few other points should be addressed to complete a basic understanding of RSA encryption.

First, the numbers we mod out by, our values for n, are big, but they wouldn't be used to encipher long messages all at once. Messages could be broken into pieces, with each piece enciphered separately.

Second, we don't even really do that. RSA is a great system, but it isn't a fast system. People are impatient, so RSA is usually combined with a much quicker system that doesn't offer its advantage of allowing people who didn't exchange keys in advance to communicate securely anyway. The way it typically works is that the person sending the message makes up a random key to use with some very fast system. He or she then enciphers the message with that key. The recipient has no idea what they key is, so it needs to be sent along with the message. But if the sender were to do that, then anyone could read it. To prevent this, he or she enciphers the key with the recipient's public key before sending it along with the enciphered message. This approach is known as a hybrid system, and it combines the best features of two different systems.

On Digital Signatures

In chapter 9, the Cicada 3301 challenges involved signed messages. At the time, I indicated that modern cryptology made it possible to do this, but I didn't provide any details. Now that RSA has been presented, signatures are easy to explain.

Above, I described how anyone can use my public enciphering power to send messages that only I can read, using my deciphering power d. I can turn this around and send out messages enciphered with my d. Now these will look like ciphers, but they have no security. Anyone can use my public key pair e and n to read such messages. However, because I'm the only one who knows d, I'm the only one who could have created these messages. You could say that I signed them because no one else can impersonate me in that way.

The ability to sign messages was considered so important that when Rivest, Shamir, and Adleman published their new system (soon to be known as RSA), they titled the paper "On Digital Signatures."[21]

It's important to understand that we're not limited to choosing between enciphering a message so that only one person can read it or signing a message so that everyone can read it and be sure it's genuine. We can do both. If I want to send you a message that is both enciphered and signed, I can apply both your public key and my private key.

RSA is a wonderful system, as long as factoring is hard! So, how hard is it? Let's look at some examples.

Factoring Challenges that Were Beaten

RSA encryption could be broken in every instance if a quick way could be found to factor. For small numbers, the human mind is sufficient. For example, when asked to factor 15, a child in elementary school could quickly recognize that the answer is 3 times 5. However, the technique he or she uses will certainly fail for a number such as 4,697,296,523, which factors as 37,657 times 124,739. Mathematicians and computer scientists have developed algorithms that can provide an answer in this case as quickly as a child, but when numbers reach larger sizes, the world's fastest computers armed with the best algorithms are stymied. That's why RSA-2048 hasn't been factored yet.

Although RSA-2048 still stands, smaller challenges have been overcome. An example is RSA-640:

```
31074182404900043721350750035888567930037346022842727545
72016194882320644051808150455634682967172328678243791627
28380334154710731085019195485290073377248227835257423864
54014691736602477652346609
```

This number was factored in 2005 by a German team,[22] earning them the smaller (but still nice!) prize of $20,000 from RSA Labs. They divided the computational labor among eighty computers with 2.2 GHz AMD Opteron CPUs, but it still took five months to determine the factors.

Another team, composed of researchers in France, Germany, Japan, the Netherlands, Switzerland, and the United States, factored RSA-768 in 2009. The time required to factor this number was "the equivalent of almost 2000 years of computing on a single core 2.2GHz AMD Opteron."[23] However, this was not how they did it! They divided the labor among "many hundreds of machines." Still, it took almost two and a half years.

They would've earned $50,000 for this effort, but the prizes were no longer being offered in 2009.

The divide-and-conquer approach worked (eventually) for these difficult factorization problems, but Rivest later issued a challenge for which such a method is not believed to offer any advantage. He wrote,

The puzzle is designed to foil attempts of a solver to exploit parallel or distributed computing to speed up the computation. The computation required to solve the puzzle is "intrinsically sequential".[24]

He put it more colorfully when he, along with Adi Shamir and David A. Wagner, first proposed the idea a few years earlier:

Solving the puzzle should be like having a baby: two women can't have a baby in 4.5 months.[25]

Time Capsule Crypto-Puzzle (1999)

Rivest gave his puzzle the unusual name shown above because he intentionally made it so that it would take thirty-five years to solve. This number was chosen to celebrate the thirty-fifth birthday of MIT's Laboratory for Computer Science (LCS), which occurred in 1999. The puzzle was associated with a physical "LCS Time Capsule of Innovations," a meter-tall container made out of lead, which would be opened in thirty-five years (2033), at about the time the puzzle would be solved if it was worked on that entire time. If the puzzle were to somehow be solved sooner, then the time capsule would be opened at that earlier date.

Along with the real puzzle, Rivest supplied a much easier scaled-down version. Looking at this version makes what's expected of the solver clearer. Then the actual puzzle is presented.

The puzzle has a piece (really a number) denoted by z. This number corresponds to an enciphered message. We need another piece (a numerical key) denoted by w to decipher it. How this is done is shown in a little bit. For now, we focus on how w can be found. It's actually just a straightforward computation:

$$w = 2^{2^t} \pmod{n}$$

where t and n are supplied by Rivest.

In his small example, $t = 10$ and $n = 253$. Notice that $253 = 11 \times 23$, the product of two primes.

We can calculate

$$w = 2^{2^t} = 2^{2^{10}} = 2^{1024} = \ldots$$

Well, this number is too big for my calculator to handle. An answer expressed in terms of scientific notation won't help, because we need every single digit. Remember, the next step is to divide by $n = 253$ and see what the remainder is. A computer program could easily be written to start with the number 2 and keep doubling it, 1,024 times, keeping track of all of the digits, and then divide by 253 and see what the remainder is. But this is a horribly inefficient way to do it. We have an alternative that gets the answer in just ten steps, instead of 1,024, and it can be handled easily by a computer (or even by hand, if you're careful).

The approach is called "repeated squaring," and it means just that. We start by squaring our given number (2 in this small example), and then square the result, and then square that result, and so on. We keep squaring the previous value. We get:

$2^2 = 4,$

$4^2 = 16,$

$16^2 = 256 = 3 \pmod{253}.$

Remember that we're calculating the power modulo 253.

$3^2 = 9,$

$9^2 = 81,$

$81^2 = 6,561 = 236 \pmod{253},$

$236^2 = 55,696 = 36 \pmod{253},$

$36^2 = 1,296 = 31 \pmod{253},$

$31^2 = 961 = 202 \pmod{253}.$

$202^2 = 40,804 = 71 \pmod{253}.$

And we can stop here! This work tells us that $2^{1024} \pmod{253} = 71.$

How did this happen? Why did it take just ten steps, instead of 1,024 steps?

To see, let's look at several steps at once. After we've squared four times, what we've calculated is $(((2^2)^2)^2)^2$, but when you raise a power to a power, you multiply the powers. So we have $(((2^2)^2)^2)^2 = 2^{16}$. That is, after four repeated squarings, we got up to the 16th power (the product of the four 2s that appear as exponents).

In the same way, ten repeated squarings gets us to a power that's the product of ten 2s, which is 1,024. That's why repeated squaring gave us the answer in just ten steps, instead of 1,024.

Another lovely feature of this method is that if you're looking for a final answer modulo 253 (or any other number), you can divide by this number and take the remainder at any time you like, as you go along, and you'll still get the right final answer. Moding out by the number at every opportunity stops the numbers from growing too large. So not only is this method incredibly fast, it's also a piece of cake with any calculator. We don't have to write a special program to handle large numbers.

Now that we have the key, $w = 71$, it's time to look at how Rivest instructed it should be used to figure out the message.

In his small example, z was given as "13 (hex)." The "hex" part refers to hexadecimal, a number system that uses base 16. That is, instead of the normal ten digits 0, 1, 2 , 3, 4, 5, 6, 7, 8, 9, it uses 16 symbols. They are 0, 1, 2, 3, 4, 5, 6, 7, 8, 9, A, B, C, D, E, F.

Table 11.1 shows numbers in base 10 (a.k.a. decimal, what we're used to) and their hexadecimal equivalents.

Table 11.1 Decimal and Hexadecimal Equivalents

Decimal	Hex	Decimal	Hex	Decimal	Hex
1	1	11	B	30	1E
2	2	12	C	40	28
3	3	13	D	50	32
4	4	14	E	60	3C
5	5	15	F	70	46
6	6	16	10	80	50
7	7	17	11	90	5A
8	8	18	12	100	64
9	9	19	13	500	1F4
10	A	20	14	1000	3E8

Hexadecimal notation is common in computer science, but rare elsewhere.

So we have $w = 71$ (decimal) and $z = 13$ (hex). The next step in Rivest's puzzle involves combining these numbers in a certain way, but first 71 needs to be converted to hexadecimal form, so the two numbers are expressed in the same way.

We then have $w = 47$ (hex) and $z = 13$ (hex).

The next step is to "XOR" these two numbers. I'll explain how it's done and then provide a handy shortcut, so that you can forget the explanation. We start by replacing the numbers with their representations in binary (base 2). The conversion from hex to binary goes like this:

0=0000	4=0100	8=1000	C=1100
1=0001	5=0101	9=1001	D=1101
2=0010	6=0110	A=1010	E=1110
3=0011	7=0111	B=1011	F=1111

So, 47 becomes 0100 0111 and 13 becomes 0001 0011. The spaces between the groups of 0s and 1s are not needed. They are simply there to make the substitutions I made clearer. We line up these numbers, as in an ordinary addition problem. We then add vertical pairs of numbers, but we do it in a slightly different way. We have 0 + 0 = 0, 0 + 1 = 1, 1 + 0 = 0 (as expected!), but we also have 1 + 1 = 0. While this may seem strange, it is actually a very useful way to add in many different contexts.

XOR is an abbreviation for exclusive-or. It is contrasted with inclusive-or, also written as OR. In everyday English the distinction is rarely made. Logicians often use True and False in place of 1 and 0. Doing so will help us understand the difference between OR and XOR. We have OR in, he's strong if he can bench press 300 lbs or squat 400 lbs. Clearly, we'd say he's strong if both are true! However, XOR doesn't allow both to be true. We encounter this kind of or in restaurants when the waiter asks, "Would you like french fries or a potato with your meal?" When I reply "Yes," trying to get both, the waiter frowns. We have True XOR True (I want both) = False (I don't get both). That is, 1 + 1 = 0. Applying the XOR rule we get

```
        0100  0111
   XOR  0001  0011
        0101  0100
```

Substituting back to hex values, using our little table above, gives 54. That is, in hex, we have

```
        47
   XOR  13
        54
```

To help you XOR numbers directly in hexadecimal (eliminating the conversion to binary and then back to hex again), table 11.2 is provided. The problem will be repeated using this table, which makes the calculation much easier.

Table 11.2 XOR Table for HEX

XOR	0	1	2	3	4	5	6	7	8	9	A	B	C	D	E	F
0	0	1	2	3	4	5	6	7	8	9	A	B	C	D	E	F
1	1	0	3	2	5	4	7	6	9	8	B	A	D	C	F	E
2	2	3	0	1	6	7	4	5	A	B	8	9	E	F	C	D
3	3	2	1	0	7	6	5	4	B	A	9	8	F	E	D	C
4	4	5	6	7	0	1	2	3	C	D	E	F	8	9	A	B
5	5	4	7	6	1	0	3	2	D	C	F	E	9	8	B	A
6	6	7	4	5	2	3	0	1	E	F	C	D	A	B	8	9
7	7	6	5	4	3	2	1	0	F	E	D	C	B	A	9	8
8	8	9	A	B	C	D	E	F	0	1	2	3	4	5	6	7
9	9	8	B	A	D	C	F	E	1	0	3	2	5	4	7	6
A	A	B	8	9	E	F	C	D	2	3	0	1	6	7	4	5
B	B	A	9	8	F	E	D	C	3	2	1	0	7	6	5	4
C	C	D	E	F	8	9	A	B	4	5	6	7	0	1	2	3
D	D	C	F	E	9	8	B	A	5	4	7	6	1	0	3	2
E	E	F	C	D	A	B	8	9	6	7	4	5	2	3	0	1
F	F	E	D	C	B	A	9	8	7	6	5	4	3	2	1	0

Using the table, we see that 7 XOR 3 = 4 and 4 XOR 1 = 5. So, we have

```
      47
XOR   13
      ──
      54
```

As a last step, Rivest tells us to look up this symbol in the ASCII code.[26] Hexadecimal is so common in computer science that most ASCII tables provide it along with the decimal number for each symbol, as in tables 11.3 and 11.4.

Using table 11.3, we see that 54 (hex) gives us the character T. Not a very exciting message, but this was just a sample to illustrate how the puzzle works.

The fact that Rivest used extended ASCII leads me to believe that the message for his real puzzle makes use of one or more of the characters in table 11.4 (i.e., in the base 10 range 128–255).

Now that it's clear how the puzzle works, it's time to look at the real deal.

Table 11.3 Extended ASCII (Smaller Values)

Dec Hex	Name	Char	Dec Hex	Char	Dec Hex	Char	Dec Hex	Char
0 0	Null	NUL	32 20	Space	64 40	@	96 60	`
1 1	Start of heading	SOH	33 21	!	65 41	A	97 61	a
2 2	Start of text	STX	34 22	"	66 42	B	98 62	b
3 3	End of text	ETX	35 23	#	67 43	C	99 63	c
4 4	End of xmit	EOT	36 24	$	68 44	D	100 64	d
5 5	Enquiry	ENQ	37 25	%	69 45	E	101 65	e
6 6	Acknowledge	ACK	38 26	&	70 46	F	102 66	f
7 7	Bell	BEL	39 27	'	71 47	G	103 67	g
8 8	Backspace	BS	40 28	(	72 48	H	104 68	h
9 9	Horizontal tab	HT	41 29	)	73 49	I	105 69	i
10 0A	Line feed	LF	42 2A	*	74 4A	J	106 6A	j
11 0B	Vertical tab	VT	43 2B	+	75 4B	K	107 6B	k
12 0C	Form feed	FF	44 2C	,	76 4C	L	108 6C	l
13 0D	Carriage feed	CR	45 2D	-	77 4D	M	109 6D	m
14 0E	Shift out	SO	46 2E	.	78 4E	N	110 6E	n
15 0F	Shift in	SI	47 2F	/	79 4F	O	111 6F	o
16 10	Data line escape	DLE	48 30	0	80 50	P	112 70	p
17 11	Device control 1	DC1	49 31	1	81 51	Q	113 71	q
18 12	Device control 2	DC2	50 32	2	82 52	R	114 72	r
19 13	Device control 3	DC3	51 33	3	83 53	S	115 73	s
20 14	Device control 4	DC4	52 34	4	84 54	T	116 74	t
21 15	Neg. acknowledge	NAK	53 35	5	85 55	U	117 75	u
22 16	Synchronous idle	SYN	54 36	6	86 56	V	118 76	v
23 17	End of xmit block	ETB	55 37	7	87 57	W	119 77	w
24 18	Cancel	CAN	56 38	8	88 58	X	120 78	x
25 19	End of medium	EM	57 39	9	89 59	Y	121 79	y
26 1A	Substitute	SUB	58 3A	:	90 5A	Z	122 7A	z
27 1B	Escape	ESC	59 3B	;	91 5B	[	123 7B	{
28 1C	File separator	FS	60 3C	<	92 5C	\	124 7C	|
29 1D	Group separator	GS	61 3D	=	93 5D	]	125 7D	}
30 1E	Record separator	RS	62 3E	>	94 5E	^	126 7E	~
31 1F	Unit separator	US	63 3F	?	95 5F	_	127 7F	DEL

Table 11.4 Extended ASCII (Larger Values)[a]

Dec	Hex	Char	Dec	Hex	Char	Dec	Hex	Char	Dec	Hex	Char
128	80	Ç	160	A0	á	192	C0	└	224	E0	α
129	81	ü	161	A1	í	193	C1	┴	225	E1	ß
130	82	é	162	A2	ó	194	C2	┬	226	E2	Γ
131	83	â	163	A3	ú	195	C3	├	227	E3	π
132	84	ä	164	A4	ñ	196	C4	─	228	E4	Σ
133	85	à	165	A5	Ñ	197	C5	┼	229	E5	σ
134	86	å	166	A6	a	198	C6	╞	230	E6	μ
135	87	ç	167	A7	o	199	C7	╟	231	E7	τ
136	88	ê	168	A8	¿	200	C8	╚	232	E8	Φ
137	89	ë	169	A9	⌐	201	C9	╔	233	E9	Θ
138	8A	è	170	AA	¬	202	CA	╩	234	EA	Ω
139	8B	ï	171	AB	½	203	CB	╦	235	EB	δ
140	8C	î	172	AC	¼	204	CC	╠	236	EC	∞
141	8D	ì	173	AD	¡	205	CD	═	237	ED	φ
142	8E	Ä	174	AE	«	206	CE	╬	238	EE	ε
143	8F	Å	175	AF	»	207	CF	╧	239	EF	∩
144	90	É	176	B0	░	208	D0	╨	240	F0	≡
145	91	æ	177	B1	▒	209	D1	╤	241	F1	±
146	92	Æ	178	B2	▓	210	D2	╥	242	F2	≥
147	93	ô	179	B3	│	211	D3	╙	243	F3	≤
148	94	ö	180	B4	┤	212	D4	╘	244	F4	⌠
149	95	ò	181	B5	╡	213	D5	╒	245	F5	⌡
150	96	û	182	B6	╢	214	D6	╓	246	F6	÷
151	97	ù	183	B7	╖	215	D7	╫	247	F7	≈
152	98	ÿ	184	B8	╕	216	D8	╪	248	F8	°
153	99	Ö	185	B9	╣	217	D9	┘	249	F9	·
154	9A	Ü	186	BA	║	218	DA	┌	250	FA	·
155	9B	¢	187	BB	╗	219	DB	█	251	FB	√
156	9C	£	188	BC	╝	220	DC	▄	252	FC	ⁿ
157	9D	¥	189	BD	╜	221	DD	▌	253	FD	2
158	9E	Pts	190	BE	╛	222	DE	▐	254	FE	■
159	9F	ƒ	191	BF	┐	223	DF	▀	255	FF	

[a]Some of the symbols in this table vary depending on the computer and font.

= =

```
Crypto-Puzzle for LCS35 Time Capsule.
Created by Ronald L. Rivest. April 2, 1999.
Puzzle parameters (all in decimal):
n = 6314466083072888893799357126131292332363298818330841375588990
    7727019571289248855473084460557532065136183466288489480886635
    0036848039658817136198766052189726781016228055747539383830826
    1759713218926668611776954526391570120690939973680089721274446
    6664233191878068305520679512530700820202412462339824107377537
    0512734449416950118097524189066796385875485631980550727370990
    4397119733614666701543905360152543373982524579313575317653646
    3319890646514021339852658003419919039821928447102124648874593
    8885358207031808428902320971090703239693491996277899532332018
    4064522476463966355937367009369212758092086293198727008292431
    243681
```

As in the small example, n is the product of two primes, but this time they're going to be hard to determine!

```
t = 79685186856218
```

Rivest chose t to be just big enough to require thirty-five years. Did he estimate correctly? He factored Moore's law into part of his estimate, assuming that internal chip speeds would increase by a factor of about 13 between 1999 and 2012, and then (less optimistically) another factor of 5 over the remaining years. It should be stressed that his time estimate was based on the fastest available computer being used, and being replaced every year with a new, faster model.

```
z = 4273385266812394147070994861525419078076239304748427595531276
    9957521280202136136722545165160035373394949568076023828487525
    8690199022379638588291839885522498545851997481849074579523880
    4226283637519132355620865854807750610249277739682050363696697
    8500226307631900353300045015777206708717225272801662783540046
    3807389033342175518988780339070669313124967596962087173533318
    1071167574435841870740398493890811235683625826527602500294010
    9087023128850957845498144088862975052260106933756431694036063
    1375375394366442662022050529545706707758321979377282989361374
    5614142047193712972117251792879310395477535810302267611143659
    071382
```

```
To solve the puzzle, first compute w = 2^(2^t) (mod n).
Then exclusive-or the result with z.
```

The two strings, w and z, won't be of the same length, which would make lining them up to perform the XOR operation confusing. Rivest explained that they should be lined up on the right. The extra values that stick out on the left and have nothing paired with them represent a number. This number, called b by Rivest, should be plugged into the expression $5^b \pmod{2^{1024}}$ and evaluated. This can be done quickly on a computer, even for large values of b. The result is just below one of the prime factors of n. Thus, this calculation allows the solver to find the prime and then factor n.

This is an important detail because it allows for the solution to be quickly verified. What would the LCS director of the future do if someone came claiming a solution and the only one who knew it (Rivest) was dead or could no longer remember what the message was? Rivest was born in 1947, so there's certainly no guarantee that he'll be around at about the time his puzzle should be solved.

The director wouldn't be able to ask the solver to do the calculation again under his or her observation to make sure that it was done right. It wouldn't take another thirty-five years, but it would be way too time-consuming nevertheless.

However, if the solution leads to a factorization of n, the solver could quickly convince the director that he or she has obtained that portion of the secret, for multiplying the numbers back together can be done rapidly on a computer, and the result can be checked to see if it matches n. As for the message itself, with the aforementioned factorization in hand, there's a quick way to do the calculation that provides the key to the message. It could be done again then and there, with an incredible gain in efficiency provided by the new information.

But factoring n isn't an easy backdoor for solving the puzzle. Rivest made the number of the same size as RSA-2048, so factoring it should be more time-consuming than solving the puzzle directly.

Rivest wrote,

Anyone who believes to have solved the puzzle should submit the resultant English sentence along with the factorization of n and relevant solution notes to the Director of LCS. Upon verification of the solution, the LCS Director will unseal the capsule at a ceremony set up for that purpose. If no solution is established by September 2033, then the LCS Director will unseal the capsule at the Laboratory's 70th anniversary celebration, or at a suitable alternate event. In the absence of an LCS Director, the President of MIT will designate another official or individual.[27]

Both the factoring challenges and the time capsule puzzles are ways of testing the state of the art in terms of attacks. There are many more challenges out there of this nature.

More to Come!

As I was writing this book, unsolved ciphers from decades, sometimes centuries, in the past were coming to light on a regular basis. New challenge ciphers were put forth regularly, as well. It would be a full-time job to thoroughly research each of them as they appear. But it would be a fun job!

So once you've solved all of the ciphers in this volume, know that there are plenty more out there to keep you busy. You might even see another book by me in a few years covering the best of them.

Notes

Chapter 1: A King's Quest

1   E-mail received by the author on November 21, 2014. Also see http://en.wikipedia.org/wiki
    /Jerusalem_artichoke.

2   Newbold included the section I labeled "cosmological" as part of the biological section. See New-
    bold, W. R., "The Voynich Roger Bacon Manuscript." *Transactions of the College of Physicians and
    Surgeons of Philadelphia* (1921): 431–74 (the relevant pages here are 461–63). Read April 20,
    1921. Also see Newbold, W. R., and Kent, R. G., *The Cipher of Roger Bacon*, 1928, pp. 44 and 46.
    Brumbaugh labels the sections as 1. Botanical, 2. Astrological, 3. Medical (perhaps dealing with
    human reproduction), 4. Medicinal, and 5. Appendix or Postscript. See Brumbaugh, Robert S.,
    "Science in Cipher," pp. 109–41, in Brumbaugh, Robert S. (ed.), *The Most Mysterious Manuscript*,
    1978, p. 110 cited here.

3   Six are suggested, for example, in Reddy, Sravana, and Kevin Knight, "What We Know about
    the Voynich Manuscript," in *Proceedings of the 5th ACL-HLT Workshop on Language Technology
    for Cultural Heritage, Social Sciences, and Humanities*, pp. 78–86, Portland, OR, June 24, 2011.
    Available online at http://www.aclweb.org/anthology/W11-1511 and https://www.aclweb
    .org/anthology/W/W11/W11-15.pdf.

4   Taken from Act 3, Scene 2 of William Shakespeare's play *Julius Caesar*. This line is delivered by
    Marc Antony.

5   The quote was taken from Evans, R. J. W., *Rudolf II and His World: A Study in Intellectual History
    1576–1612* (Oxford: Clarendon Press, 1973), 196.

6   Code penguin, http://tools.codepenguin.com/patterns/help#.

7   Evans, *Rudolf II and His World*, 58.

8   Evans, *Rudolf II and His World*, 198.

9   Evans, *Rudolf II and His World*, 230. Also see Mukherjee, Ashoke, "Giordano Bruno, An
    Ever-Burning Flame of Commitment to Truth and Reason," *Breakthrough* 8, no. 4 (2000): 1–13,
    available online at http://breakthrough-india.org/archives/bruno.pdf.

10 Evans, *Rudolf II and His World*, 279.

11 Stolfi, Jorge, Voynich Manuscript stuff, 2005, http://www.dcc.unicamp.br/stolfi/voynich/.

12 The graph is taken from Reddy and Knight, "What We Know," p. 4. Available online at http://
    www.isi.edu/natural-language/people/voynich-11.pdf.

13　Reddy and Knight, "What We Know," 4, footnote.

14　This translates to "On the Secret Workings of Art and Nature, and on the Vanity of Magic." Newbold, "The Voynich Roger Bacon Manuscript," p. 456 cited here.

15　Kennedy, Gerry, and Rob Churchill, *The Voynich Manuscript: The Unsolved Riddle of an Extraordinary Book Which Has Defied Interpretation for Centuries*, British hardcover edition, Orion Books, London, 2004, p. 20, which, in turn, took it from Bacon, Roger, *Compendium Studii Philosophiae*, 1271 or 1272.

16　Some sources give a death date two years later. See, for example, Kennedy and Churchill, *The Voynich Manuscript*, p. 21. At http://www.nndb.com/people/582/000114240/ a death date of 1294 is conjectured, although uncertainty is expressed. Also see Newbold, "The Voynich Roger Bacon Manuscript," 445–46, for the 1292 death date.

17　Lieutenant Colonel H. W. L. Hime, Royal Artillery, "Roger Bacon and Gunpowder," pp. 321–36 in *Roger Bacon Essays Contributed by Various Writers on the Occasion of the Commemoration of the Seventh Centenary of His Birth*, collected and edited by A. G. Little, Clarendon Press, Oxford, 1914, available online at http://archive.org/stream/rogerbaconessays00litt /rogerbaconessays00litt_djvu.txt. Also see Goldstone, Lawrence, and Nancy Goldstone, *The Friar and the Cipher: Roger Bacon and the Unsolved Mystery of the Most Unusual Manuscript in the World* (New York: Doubleday, 2005), 107–8.

18　Thorndike's argument does not get into the details of the cipher itself. See Thorndike, Lynn, "Roger Bacon and Gunpowder," *Science*, New Series 42 , no. 1092 (December 3, 1915): 799–800. This piece was reprinted as Appendix II (pp. 688–91) of Thorndike's *A History of Magic and Experimental Science during the First Thirteen Centuries of our Era*, Vol. II (New York: Columbia University Press, 1923), which is available online at https://archive.org/details /historyofmagicex02thor. The controversy is also mentioned in Goldstone and Goldstone, *The Friar and the Cipher*, 251.

19　Newbold, "The Voynich Roger Bacon Manuscript," p. 457 cited here. Newbold noted that this isn't what Ethicus actually did, at least not in the ciphers of his that are known today.

20　Newbold, "The Voynich Roger Bacon Manuscript," p. 456 cited here.

21　Newbold, "The Voynich Roger Bacon Manuscript," p. 457 cited here.

22　Roberts, R. J., and A. G. Watson (eds.), *John Dee's Library Catalogue* (London: The Bibliographical Society, 1990).

23　Kennedy and Churchill, *The Voynich Manuscript*, 2004, p. 66, citing Prinke's online article. Images comparing known samples of Dee's numbers to those on the manuscript pages are provided on p. 67 of Kennedy and Churchill. Also see Pelling, Nick, *The Curse of the Voynich: The Secret History of the World's Most Mysterious Manuscript* (Surbiton, Surrey, UK: Compelling Press, 2006), 11–13.

24　Goldstone and Goldstone, *The Friar and the Cipher*, pp. 187–88. When Sir Francis Walsingham received intercepted messages penned by the imprisoned Mary, Queen of Scots, and conspirators planning to set her free and assassinate Queen Elizabeth, he likely had much confidence that the plaintexts could be recovered. It was his cryptanalyst Thomas Phelippes who succeeded in breaking them, and the rest is history. Simon Singh's *The Code Book* (New York: Doubleday, 1999) does an excellent job of recounting this tale and the consequences for Mary (see chapter 1). Thomas Phelippes was also at a meeting of Walsingham and Dee. See Gold-stone and Goldstone, *The Friar and the Cipher,* 204–5.

25 Fell-Smith, Charlotte, *John Dee (1527–1608),* (London: Constable & Co. Ltd., 1909), 311. Available online at https://archive.org/details/cu31924028928327.

26 Kelley was blackmailed by Sir Francis Walsingham into becoming a spy.

27 Brumbaugh, Robert S., editor and contributor, *The Most Mysterious Manuscript: The Voynich "Roger Bacon" Cipher Manuscript* (Carbondale, IL: Southern Illinois University Press, 1978), 131.

28 I learned of this in Pelling, *The Curse of the Voynich,* p. 158, where credit for the discovery of the numbering system is given to Allan Wechsler and Ivan Derzhanski, who came upon it independently.

29 Shelfmark APUG 557, fol. 353.

30 Translation taken from http://www.voynich.net/neal/barschius_translation.html, which incorrectly dates it as 1637, the date of a previous letter that we do not have. The original Latin may be found at http://www.voynich.nu/letters.html#gb39, along with other letters connected with the manuscript.

31 The text of this letter appears in Goldstone and Goldstone, *The Friar and the Cipher,* pp. 237–39 and elsewhere.

32 Identified as "Dr Raphael Missowsky, Attorney General to Ferdinand III" in Kennedy and Churchill, *The Voynich Manuscript,* 73.

33 Kircher, Athanasius, *The Volcanos: Or Burning and Fire-Vomiting Mountains, Famous in the World* (London: J. Darby,1669): 35.

34 Goldstone and Goldstone, *The Friar and the Cipher,* 239.

35 Goldstone and Goldstone, *The Friar and the Cipher,* 239–40.

36 Voynich, Wilfrid M., "A Preliminary Sketch of the History of the Roger Bacon Cipher Manuscript." *Transactions of the College of Physicians of Philadelphia* 43, ser. 3 (1921): 415–30. Read April 20, 1921.

37 Voynich, "A Preliminary Sketch."

38 Voynich, "A Preliminary Sketch."

39 Voynich, "A Preliminary Sketch."

40 Voynich, "A Preliminary Sketch."

41 Voynich, "A Preliminary Sketch."

42 Sowerby, E. Millicent, *Rare People and Rare Books* (London: Constable & Co., Ltd, 1967).

43 Sowerby, *Rare People and Rare Books.*

44 Taratuta, Evgeniya, "Our Friend Ethel Lilian Boole/Voynich." Moscow, 1964, translated from the Russian by Séamus Ó Coigligh with additional notes, 2008.

45 Orioli, Giuseppe, *Adventures of a Bookseller* (New York: Robert M. McBride & Co., 1938).

46 Sowerby, *Rare People and Rare Books.*

47 Taratuta, "Our Friend Ethel Lilian Boole/Voynich."

48 Manly, John M., "The Most Mysterious Manuscript in the World." *Harper's Monthly Magazine,* 143 (July 1921): 186–97.

49 Newbold, "The Voynich Roger Bacon Manuscript."

50 Newbold, "The Voynich Roger Bacon Manuscript."

51   Newbold, "The Voynich Roger Bacon Manuscript."

52   Newbold, "The Voynich Roger Bacon Manuscript."

53   Newbold, "The Voynich Roger Bacon Manuscript."

54   Newbold, "The Voynich Roger Bacon Manuscript."

55   Newbold, "The Voynich Roger Bacon Manuscript."

56   Newbold, "The Voynich Roger Bacon Manuscript."

57   Newbold, "The Voynich Roger Bacon Manuscript."

58   Newbold, "The Voynich Roger Bacon Manuscript."

59   Newbold, "The Voynich Roger Bacon Manuscript."

60   Newbold, "The Voynich Roger Bacon Manuscript."

61   Newbold, William Romaine. *The Cipher of Roger Bacon,* edited with foreword and notes by Roland Grubb Kent. (Philadelphia: University of Pennsylvania Press; London, H. Milford, Oxford University Press, 1928).

62   Newbold, "The Voynich Roger Bacon Manuscript."

63   Bird, J. Malcolm, "The Roger Bacon Manuscript Investigations into Its History, and the Efforts to Decipher It." *Scientific American Monthly,* June 1921, 492–96.

64   Anonymous, "The Roger Bacon Manuscript, What It Looks Like, and a Discussion of the Probabilities of Decipherment." *Scientific American,* May 28, 1921, 432, 439–40.

65   Anonymous, "The Roger Bacon Manuscript."

66   Thorndike, Lynn, "Roger Bacon." *The American Historical Review* 34, no. 2 (January 1929): 317–19.

67   Manly, John Matthews., "Roger Bacon and the Voynich Manuscript." *Speculum* 6 (July 1931): 345–91.

68   Manly, "Roger Bacon and the Voynich Manuscript."

69   Manly, "Roger Bacon and the Voynich Manuscript."

70   Manly, "Roger Bacon and the Voynich Manuscript."

71   Manly, "Roger Bacon and the Voynich Manuscript."

72   This name is hard to make out in the letter and may not have been reproduced correctly here.

73   "The Quirinal" refers to the Italian government of the time, so-called in reference to Quirinal Palace, which was used by kings, as well as being one of the homes of the presidents who followed them.

74   Also known as Peter and Pieter.

75   Kraus, H. P., *A Rare Book Saga* (New York: G. P. Putnam's Sons, 1978), 222.

76   Thanks to Klaus Schmeh for pointing many of these out. He wrote, "Regarding the Voynich manuscript alone, more than 15 'solutions' are known to me, and all have turned out to be false." See Schmeh, Klaus, "The Pathology of Cryptology—A Current Survey," *Cryptologia,* 36, no. 1 (January 2012): 14–45. By the end of 2015, he was aware of more than thirty false solutions.

77   Strong, Leonell C., "Anthony Askham, the Author of the Voynich Manuscript." *Science,* New Series 101, no. 2633 (June 15, 1945): 608–9.

78   Goldstone and Goldstone, *The Friar and the Cipher,* p. 276. The decipherment also included a "recipe for an herbal contraceptive."

79   Brumbaugh, R. S. "The Solution of the Voynich 'Roger Bacon' Cipher." *Yale University Library Gazette* 49, no. 4 (April 1975): 347–55. Brumbaugh later claimed that he found the solution

in 1972, but he didn't explain why no results were published until 1974. See Brumbaugh, R. S., "The Voynich 'Roger Bacon' Cipher Manuscript: Deciphered Maps of Stars." *Journal of the Warburg and Courtauld Institutes* 39 (1976): 139–50.

80 Brumbaugh, "The Solution of the Voynich 'Roger Bacon' Cipher."

81 Child, James R. "The Voynich Manuscript Revisited." *NSA Technical Journal* 21, no. 3 (Summer 1976): 1–4.

82 Child, J. R., "Again, the Voynich Manuscript." 2007. http://web.archive.org/web/20090616 205410/http://voynichmanuscript.net/voynichpaper.pdf.

83 Goldstone and Goldstone, *The Friar and the Cipher*, 285.

84 Finn, James E., "The Voynich Manuscript. Extraterrestrial Contact during the Middle Ages?" January 14, 2001. http://www.bibliotecapleyades.net/ciencia/esp_ciencia_manuscrito02.htm. Accessed March 23, 2015.

85 Zbigniew Banasik's Manchu theory can be found online at http://www.ic.unicamp.br/~stolfi /voynich/04-05-20-manchu-theo/ and at the links contained therein.

86 Kennedy and Churchill, *The Voynich Manuscript*, 242.

87 Geheimnisvollstes Manuskript der Welt entschlüsselt, press release, 2005. Available online at http://www.ms408.de/downloads/Pressemitteilung.pdf.

88 http://www.ms408.de/context_d.htm.

89 http://www.edithsherwood.com/voynich_decoded/.

90 https://web.archive.org/web/20090609075537/http://www.voynich.nl/.

91 http://www.ciphermysteries.com/2009/11/12/richard-rogers-voynich-theory.

92 Watson, Leon, "Prophet of God" claims mysterious manuscript's code has been cracked, December 3, 2011, http://www.dailymail.co.uk/news/article-2069481/Prophet-God-claims-mysterious -manuscripts-code-cracked.html.

93 Watson, "Prophet of God."

94 Watson, "Prophet of God."

95 https://stephenbax.net/?page_id=11.

96 http://www.livescience.com/43542-voynich-manuscript-10-words-cracked.html.

97 http://theworldsfairest.com/writtenintongues/Written.in.Tongues-The.Voynich.Manuscript .Solved.pdf.

98 http://www.proza.ru/2015/01/11/2343.

99 Foster, Caxton C., "A Comparison of Vowel Identification Methods," *Cryptologia* 16, no. 3 (July 1992): 282–86.

100 An executable version of this algorithm can be downloaded from http://sun1.bham.ac.uk /G.Landini/evmt/evmt.htm. It's called VFQ and was written by Jacques Guy. The details of the algorithm are taken here from G. T. Sassoon, "The Application of Sukhotin's Algorithm to Certain Non-English Languages," *Cryptologia* 16, no. 2 (April 1992): 165–73.

101 This matrix is symmetric— although it mistakenly isn't in the example given in G. T. Sassoon, "The Application of Sukhotin's Algorithm to Certain Non-English Languages," *Cryptologia* 16, no. 2 (April 1992) 165–73.

102 Jacques Guy, "Statistical Properties of Two Folios of the Voynich Manuscript," *Cryptologia* 15, no. 3 (1991). The previous issue of *Cryptologia* (Jacques Guy, "Voynich Revisited" (letter to the editor), *Cryptologia* 15, no. 2, (April 1991)) contains a letter to the editor from Guy in which

he claims that Sukhotin's algorithm resulted in only four vowels, but he did qualify that "not having kept my notes, [I] am quoting mostly from memory."

103　This course was actually offered by Sanjoy Mahajan at MIT in 2008. See http://ocw.mit.edu /courses/mathematics/18-098-street-fighting-mathematics-january-iap-2008/.

104　Guy, "Voynich Revisited" (letter to the editor), p. 162 quoted here.

105　Values taken from William Ralph Bennett, Jr., *Scientific and Engineering Problem-Solving with the Computer*, (Englewood Cliffs, NJ: Prentice-Hall, Inc., 1976), 140, 193, and 194. Missing values in the third column simply hadn't been computed at the time this book was published.

106　Gerry Kennedy and Rob Churchill, *The Voynich Manuscript: The Unsolved Riddle of an Extraordinary Book Which Has Defied Interpretation for Centuries*, British hardcover edition, (London: Orion Books, 2004), 142. Pictures of seven different handwritings used in the manuscript appear on pp. 144–45. Currier originally estimated somewhere between six and eight different scribes.

107　John F. Clabby, *Brigadier John Tiltman: A Giant among Cryptanalysts* (Fort Meade, MD: Center for Cryptologic History, National Security Agency, 2007), 34.

108　Andreas Schinner, "The Voynich Manuscript: Evidence of the Hoax Hypothesis," *Cryptologia* 31, no. 2 (April 2007): 95–107.

109　D'Imperio, M. E. *The Voynich Manuscript—An Elegant Enigma* (Fort Meade, MD: National Security Agency, 1976). Reprinted by Aegean Park Press, Laguna Hills, CA, 1978.

110　Clark also published a biography of Edison that same year.

111　https://www.nsa.gov/public_info/declass/friedman_documents/.

112　W. F. Friedman and E. S. Friedman, "Acrostics, Anagrams, and Chaucer," *Philological Quarterly* 38 (1959): 1–20.

113　C. A. Zimansky, "William F. Friedman and the Voynich Manuscript," *Philological Quarterly* 49, no. 4 (1970): 433–42.

114　Lawrence Goldstone and Nancy Goldstone, *The Friar and the Cipher: Roger Bacon and the Unsolved Mystery of the Most Unusual Manuscript in the World* (New York: Doubleday, 2005), 284.

115　E-mail from David Hatch to the author, received July 6, 2015.

116　http://www.ciphermysteries.com/.

117　G. W. L. Hodgins, "2011 Forensic Investigation of the Voynich Manuscript," presented at the Voynich 100 Conference, Frascati, Italy, May 2011. Also reported on in A. J. T. Jull, "Some Interesting and Exotic Applications of Carbon-14 Dating by Accelerator Mass Spectrometry," *10th International Conference on Clustering Aspects of Nuclear Structure and Dynamics, Journal of Physics: Conference Series* 436 (2013): 012083.

118　Sravana Reddy and Kevin Knight, "What We Know about the Voynich Manuscript," *Proceedings of the 5th ACL-HLT Workshop on Language Technology for Cultural Heritage, Social Sciences, and Humanities*, Portland, OR (June 24, 2011), (Madison, WI: Omnipress, Inc.), 78–86.

119　David Kahn, *The Codebreakers* (New York: Macmillan, 1967), 863.

Chapter 2: Ancient Ciphers

1　Thomas Hoving, *False Impression: The Hunt for Big-Time Art Fakes* (New York: Simon & Schuster, 1996), 52–53.

2　Mary-Ann Russon, "Tiny Egyptian Mummy Confirmed by CT Scan to be Baby's Remains," *International Business Times*, May 8, 2014.

3   W. M. Flinders Petrie, *Ten Years' Digging in Egypt 1881–1891*, 2nd ed. (London: The Religious Tract Society, 1893), 124–25.

4   W. M. Flinders Petrie, *Illahun Kahun and Gurob, 1889–90* (London: David Nutt, 1891).

5   Alan R. Schulman, "The Ossimo Scarab Reconsidered," *Journal of the American Research Center in Egypt* 12 (1975): 15–18, p. 16 quoted here.

6   John Coleman Darnell, *The Enigmatic Netherworld Books of the Solar-Osirian Unity, Cryptographic Compositions in the Tombs of Tutankhamun, Ramesses VI and Ramesses IX*, (Göttingen, Germany: Academic Press Fribourg, Vandenhoek & Ruprecht, 2004), 1–2.

7   Étienne Drioton, "Les principes de la cryptographie égyptienne," *Comptes rendus des séances de l'Académie des Inscriptions et Belles-Lettres*, 97e année, no. 3 (1953): 355–64.

8   John Coleman Darnell, *The Enigmatic Netherworld Books of the Solar-Osirian Unity, Cryptographic Compositions in the Tombs of Tutankhamun, Ramesses VI and Ramesses IX*, (Göttingen, Germany: Academic Press Fribourg, Vandenhoek & Ruprecht, 2004), 5.

9   Alan R. Schulman, "The Ossimo Scarab Reconsidered," *Journal of the American Research Center in Egypt* 12 (1975): 15–18, p. 16 quoted here.

10   Plutarch's *Life of Lysander*, sec. 19.

11   Albert C. Leighton, "Secret Communication among the Greeks and Romans," *Technology and Culture* 10, no. 2 (April 1969): 139–54, p. 154 cited here. Leighton's citation for the letter was Ausonius' *Epistles*, 28.

12   Leighton, "Secret Communication," 153, footnote 61, cited here.

13   A dissenting view is offered by Henry R. Immerwahr in *Attic Script, A Survey* (New York: Oxford University Press, 1990). On p. 15, he wrote, "Before the time of Draco Athens was a cultural backwater—not a suitable location for the composition of the Homeric poems, as some have thought." And on p. 20 he wrote, "Corinth clearly took the lead in the custom of applying inscriptions to vases." He views the work in Athens as imitative of the work in Corinth.

14   Immerwahr, *Attic Script*, 45.

15   John Boardman, *Athenian Black Figure Vases* (New York: Oxford University Press, 1974), 201.

16   Dan Vergano, "Amazon Warriors' Names Revealed Amid 'Gibberish' on Ancient Greek Vases," *National Geographic* (September 22, 2014), http://news.nationalgeographic.com/news/2014/09/140923-amazon-greek-vase-translations-science/.

17   Umberto Eco, *Foucault's Pendulum*, trans. William Weaver (San Diego, New York, London: Harcourt Brace Jovanovich, 1989), 24.

18   *Big Bang Theory* transcripts, https://bigbangtrans.wordpress.com/series-4-episode-09-the-boyfriend-complexity/.

19   Adrienne Mayor, John Colarusso, and David Saunders, "Making Sense of Nonsense Inscriptions Associated with Amazons and Scythians on Athenian Vases," *Hesperia: The Journal of the American School of Classical Studies at Athens* 83, no. 3 (July–September 2014): 447–93, p. 466 cited here.

20   Adrienne Mayor, John Colarusso, and David Saunders, "Making Sense of Nonsense Inscriptions Associated with Amazons and Scythians on Athenian Vases," *Hesperia: The Journal of the American School of Classical Studies at Athens* 83, no. 3 (July–September 2014): 447–93, pp. 486–87 cited here.

21   Dan Vergano, "Amazon Warriors' Names Revealed Amid 'Gibberish' on Ancient Greek Vases," *National Geographic*, September 22, 2014.

22   Adrienne Mayor, *The Amazons: Lives and Legends of Warrior Women across the Ancient World* (Princeton, NJ: Princeton University Press, 2014), 11.

23   E-mail from Adrienne Mayor to the author, received June 25, 2015.

24   Adrienne Mayor, John Colarusso, and David Saunders, "Making Sense of Nonsense Inscriptions Associated with Amazons and Scythians on Athenian Vases," Version 2.0, Princeton/Stanford Working Papers in Classics, July 2012, 27–28.

25   E-mail from Adrienne Mayor to the author, received June 25, 2015.

26   E-mail from Adrienne Mayor to the author, received June 25, 2015.

27   Adrienne Mayor, John Colarusso, and David Saunders, "Making Sense of Nonsense Inscriptions Associated with Amazons and Scythians on Athenian Vases," *Hesperia: The Journal of the American School of Classical Studies at Athens* 83, no. 3 (July–September 2014): 447–93, p. 489 cited here.

28   Adrienne Mayor, John Colarusso, and David Saunders, "Making Sense of Nonsense Inscriptions Associated with Amazons and Scythians on Athenian Vases," *Hesperia: The Journal of the American School of Classical Studies at Athens* 83, no. 3 (July–September 2014): 447–93, p. 465 cited here.

29   E-mail from David Saunders to the author, received June 26, 2015.

30   Ole Franksen, *Mr. Babbage's Secret: The Tale of a Cypher and APL* (Englewood Cliffs, NJ: Prentice-Hall, 1985), 86.

31   Stig Eliasson, "When 'Chance' Explanations Turn Tenuous: Basque Verbal Morphology, Argument Marking, and an Allegedly Nonsensical Danish Runic Inscription."

32   Eliasson, "When 'Chance' Explanations Turn Tenuous."

33   Michael P. Barnes, *Runes: A Handbook* (Suffolk, U.K.: Boydell Press, 2012), 167.

34   Barnes, *Runes*, 202.

35   Barnes, *Runes*, 24.

Chapter 3: Dorabella

1   This is actually a series of six marches titled *Pomp and Circumstance Military Marches*, of which the *Trio* section, "Land of Hope and Glory," of the first is played at many American graduations.

2   Edward Elgar, *My Friends Pictured Within* (London: Novello and Company Ltd, 1946).

3   John C. Tibbetts, "John C. Tibbetts Interviews Eric Sams," available online at http://www.ericsams.org/index.php/interview-by-john-c-tibbetts.

4   Ron Rosenbaum, "A Visit with an Avenging Angel," *The Shakespeare Wars* (New York: Random House, 2008), 66–75, available online at http://www.ericsams.org/index.php/ron-rosenbaum-the-avenging-angel.

5   John C. Tibbetts, "John C. Tibbetts Interviews Eric Sams," available online at http://www.ericsams.org/index.php/interview-by-john-c-tibbetts.

6   Solution #04, revised as #36, posted at http://unsolvedproblems.org/index_files/Solutions.htm.

7   E-mail received by author on June 19, 2015.

8   Found in an odd book by one of the people who claimed a solution to the Voynich manuscript: Joseph Martin Feely, *Electrograms from Elysium: A Study on the Probabilities in Postmortuary Communication through the Electronics of Telepathy and Extrasensory Perception, Including the Code of Anagrams in the Purported Sender's Name* (New York, 1954), 246.

9   See https://www.cqu.edu.au/about-us/staff-directory/profiles/higher-education-division/higher
-education-division-honadjunct-appoint/robertst and http://unsolvedproblems.org/S12x.pdf.

10   Jerrold Northrup Moore, *Edward Elgar: A Creative Life* (Oxford, U.K.: Oxford University Press,
1984), 114. The passage also appears online at http://www.benzedrine.ch/dorabella.html.

11   John Holt Schooling, "Secrets in Cipher IV. From the Time of George II. to the Present Day," *The
Pall Mall Magazine* 8, no. 36 (April 1896): 608–18, p. 618 cited here.

12   http://www.ciphermysteries.com/2013/10/09/elgars-little-cipher.

13   Robert J. Buckley, *Sir Edward Elgar* (London: Ballantyne, Hanson & Co., 1905).

14   The Elgar Birthplace Museum, L189.

15   The Elgar Birthplace Museum, L296.

16   This was short for "Darling Chuck," where Chuck was Elgar's daughter Carice.

17   Mrs. Richard Powell, *Edward Elgar: Memories of a Variation*, 4th ed., rev. and ed. Claud Powell,
addendum Jerrold Northrop Moore (Aldershot, Hants, U.K.: Scolar Press, 1994), 36.

18   Powell, *Edward Elgar*, 4th ed., 90–91.

19   Powell, *Edward Elgar*, 4th ed., 93.

20   R.B.T. was Richard Baxter Townshend. Elgar's variation for Townshend ridiculed Townshend's
inability to control his voice in the amateur theatre productions he performed in.

21   Powell, *Edward Elgar*, 4th ed., 132–33.

22   Ian Parrott, *Elgar* (London: J. M. Dent and Sons Ltd., and New York: Farrar, Straus and Giroux,
Inc., 1971), 122.

23   Kevin Jones, "The Puzzling Mr. Elgar," *New Scientist* 184, no. 2479 (December 25, 2004): 56.

24   Mrs. Richard Powell, *Edward Elgar: Memories of a Variation* (London: Oxford University Press,
1937), 35.

25   Rosa Burley and Frank Carruthers, *Edward Elgar: The Record of a Friendship* (London: Barrie
and Jenkins, 1972), 132.

26   Elgar Birthplace Museum, L1104.

27   Martin Bird, ed., *Darling Chuck, The Carice Letters*, Edward Elgar: Collected Correspondence,
ser. II, vol. 1, Elgar Works (Chippenham, Wiltshire, U.K.: Antony Rowe Ltd., 2014). Examples
from pp. 145, 159, 160, 162, 163, 164, 164, 164, 168, 182, 192, 201, 204, 237, 272, 381, 381,
and 381, respectively.

Chapter 4: Zodiac

1   Robert Graysmith, *Zodiac* (New York: St. Martin's/Marek, 1986), 30.

2   Robert Graysmith, *Zodiac* (New York: St. Martin's/Marek, 1986), 32–33.

3   Images from Robert Graysmith, *Zodiac* (New York: St. Martin's/Marek, 1986), 50–51.

4   Robert Graysmith, *Zodiac* (New York: St. Martin's/Marek, 1986), 58 for some "solutions."

5   Robert Graysmith, *Zodiac* (New York: St. Martin's/Marek, 1986), 67.

6   Robert Graysmith, *Zodiac* (New York: St. Martin's/Marek, 1986), 69.

7   Robert Graysmith, *Zodiac* (New York: St. Martin's/Marek, 1986), 71.

8   Robert Graysmith, *Zodiac* (New York: St. Martin's/Marek, 1986), 75.

9   Robert Graysmith, *Zodiac* (New York: St. Martin's/Marek, 1986), 83.

10   Robert Graysmith, *Zodiac* (New York: St. Martin's/Marek, 1986), 85 and 91.

11   Robert Graysmith, *Zodiac* (New York: St. Martin's/Marek, 1986), 91.

12   Robert Graysmith, *Zodiac* (New York: St. Martin's/Marek, 1986), 102.

13   Robert Graysmith, *Zodiac* (New York: St. Martin's/Marek, 1986), 107.

14   Michael D. Kelleher and David van Nuys, *"This Is the Zodiac Speaking": Into the Mind of a Serial Killer* (Westport, CT: Praeger, 2001).

15   Robert Graysmith, *Zodiac* (New York: St. Martin's/Marek, 1986), 122.

16   Robert Graysmith, *Zodiac* (New York: St. Martin's/Marek, 1986), 243.

17   Robert Graysmith, *Zodiac* (New York: St. Martin's/Marek, 1986), 136–37.

18   Robert Graysmith, *Zodiac Unmasked: The Identity of America's Most Elusive Serial Killer Revealed* (New York: Berkley Books, 2002), 128.

19   Robert Graysmith, *Zodiac* (New York: St. Martin's/Marek, 1986), 148.

20   Gareth Penn, "The Calculus of Evil," *MENSA Bulletin*, no. 288 (July/August 1985): 5–8, 25, p. 5 quoted here.

21   Gareth Penn, "The Calculus of Evil," *MENSA Bulletin*, no. 288 (July/August 1985), 5–8, 25, p. 5 quoted here.

22   Gareth Penn, "The Calculus of Evil," *MENSA Bulletin*, no. 288 (July/August 1985), 5–8, 25, p. 25 quoted here. Also see Gareth Penn, *Times 17: The Amazing Story of the Zodiac Murders in California and Massachusetts 1966–1981* (New York: Foxglove Press, 1987), 26.

23   http://www.zodiackillerfacts.com/Ray%20Grant.htm.

24   Fletcher Pratt, *Secret and Urgent* (New York: Bobbs-Merrill, 1939), 258. This is a book used by Harden!

25   See the killer's letter of January 29, 1974, for the claim of thirty-seven victims. Available online at http://www.zodiackillerfacts.com/gallery/thumbnails.php?album=19.

26   Robert Graysmith, *Zodiac* (New York: St. Martin's/Marek, 1986), 308–11.

27   I refer the interested reader to Kelleher and Van Nuys, *"This Is the Zodiac Speaking"* again. Their argument against Zodiac being connected with Bates is persuasive.

28   Robert Graysmith, *Zodiac* (New York: St. Martin's/Marek, 1986), 308.

29   http://www.imdb.com/title/tt0469999/.

30   http://www.imdb.com/title/tt0371739/.

31   http://www.imdb.com/title/tt0443706/.

Chapter 5: More Killer Ciphers

1   Cheri L. Farnsworth, *Adirondack Enigma: The Depraved Intellect and Mysterious Life of North Country Wife Killer Henry Debosnys* (Charleston, SC: The History Press, 2010).

2   Farnsworth, *Adirondack Enigma*, p. 26, quoting a reporter who visited Henry in jail.

3   Farnsworth, *Adirondack Enigma*, 69.

4   Farnsworth, *Adirondack Enigma*, 27.

5   Farnsworth, *Adirondack Enigma*, 69.

6   Farnsworth, *Adirondack Enigma*, 69–71.

7   Farnsworth, *Adirondack Enigma*, 71–72.

8   Farnsworth, *Adirondack Enigma*, 72.

9   Farnsworth, *Adirondack Enigma*, 72.

10   Farnsworth, *Adirondack Enigma*, 72.

11 Farnsworth, *Adirondack Enigma*, 72–73.

12 Farnsworth, *Adirondack Enigma*, 73.

13 Farnsworth, *Adirondack Enigma*, 73.

14 Farnsworth, *Adirondack Enigma*, 73.

15 Farnsworth, *Adirondack Enigma*, 73.

16 Farnsworth, *Adirondack Enigma*, 43.

17 Farnsworth, *Adirondack Enigma*, 76.

18 Farnsworth, *Adirondack Enigma*, 76.

19 Farnsworth, *Adirondack Enigma*, 77–78.

20 Farnsworth, *Adirondack Enigma*, 79–80.

21 Farnsworth, *Adirondack Enigma*, 80.

22 Farnsworth, *Adirondack Enigma*, 80–81.

23 *Philadelphia Times*, August 11, 1882.

24 Farnsworth, *Adirondack Enigma*, 27 and 63.

25 See http://www.isi.edu/~knight/ to get an idea of the lively, creative, eccentric way he views the world.

26 E-mail received by the author on May 8, 2015.

27 Farnsworth, *Adirondack Enigma*, 107.

28 American Cryptogram Association, http://cryptogram.org/.

29 Taken from Tom Mahon and James J. Gillogly, *Decoding the IRA* (Cork, Ireland: Mercier Press, 2008), 266, fig. 27, pt. 9.

30 Tom Mahon and James J. Gillogly, *Decoding the IRA* (Cork, Ireland: Mercier Press, 2008), 48.

31 Mahon and Gillogly, 33.

32 Mahon and Gillogly, 53.

33 Brainy Quotes, John Walsh quotes, http://www.brainyquote.com/quotes/authors/j/john_walsh.html.

34 *America's Most Wanted*, http://en.wikipedia.org/wiki/America's_Most_Wanted#Profiling_missing_persons. The man was captured in April 2008.

35 That would give us 156 distinct symbols instead of the 155 we actually observed. That's why I used the word "approximately."

36 Nick Pelling, Cipher Mysteries blog, http://www.ciphermysteries.com/2014/05/19/cracking-the-scorpion-ciphers.

37 Successories website, John Walsh Quotes, http://www.successories.com/iquote/author/6223/john-walsh-quotes/1.

38 IMDb, *America's Most Wanted*, http://www.imdb.com/title/tt0094415/quotes.

39 It was actually February 2, 2004. The bodies were found on the third.

40 Tom Farmer, "Suicide Note Leaves Motive for Saugus Killings a Mystery," *The Boston Herald*, February 5, 2004, p. 14.

41 Mac Daniel and Diane Allen, "Killings, Suicide Baffle Authorities. Husband Left Note; Motive a Mystery," *The Boston Globe*, February 5, 2004, 3rd ed., p. B1.

42 Posted January 30, 2006, 11:16 a.m.

43 Posted January 30, 2006, 11:42 a.m.

44 Posted January 31, 2006, 11:58 a.m.

45   E-mail received by the author on August 3, 2015.

46   Posted February 18, 2006, 1:29 p.m.

47   Posted February 19, 2006, 7:02 a.m.

48   Successories, John Walsh Quotes, http://www.successories.com/iquote/author/6223/john
      -walsh-quotes/8.

49   Greg Olsen and Rebecca Morris, *If I Can't Have You: Susan Powell, Her Mysterious Disappear-
      ance, and the Murder of Her Children* (New York: St. Martin's Press, 2014), 102. Also see p. 228.

50   Anonymous, "Police Close Susan Powell Case, Offer New Details," *USA Today*, May 20, 2013,
      available online at http://www.usatoday.com/story/news/nation/2013/05/20/susan-powell
      -case/2344681/.

51   David Lohr, "Susan Powell's Parents Search For Clues in Father-in-Law's Former Home," *Huff-
      ington Post*, May 27, 2014, available online at http://www.huffingtonpost.com/2014/05/27
      /susan-powell-missing-clues_n_5398147.html.

52   Communication with the author.

53   Ingo Wagner, "Verschlüsselt und nicht knackbar, Polizei scheitert an Festplatten des 'Masken
      manns,' " *Focus Magazine*, January 23, 2012. Available online at http://www.focus.de/panorama
      /welt/verschluesselt-und-nicht-knackbar-polizei-scheitert-an-festplatten-des-maskenmanns
      _aid_705621.html. Quoted portion translated by the author.

Chapter 6: From the Victims

1   Anthony Boucher, "QL696.C9." *Ellery Queen's Mystery Magazine*, May 1943.

2   Eye Doctor Guide.com, http://www.eyedoctorguide.com/Eye-Color/hazel-eyes-eye-color.html.

3   "Inquest into the Death of a Body Located at Somerton on 1.12.48." [refers to Dec. 1, 1948],
      1949, p. 69. Hands can tell us a lot about a person. In this instance, it was also noted that the
      hands were "hard, but were not rough from performing manual work."

4   Gerald (Gerry) Michael Feltus, *The Unknown Man: A Suspicious Death at Somerton Park* (Rich-
      mond, Australia: Hyde Park Press, 2010), 143.

5   Feltus, *The Unknown Man*, 39–40.

6   Kerry Greenwood, *Tamam Shud: The Somerton Man Mystery* (Sydney, Australia: NewSouth
      Publishing, University of New South Wales Press Ltd., 2012), 28.

7   Descriptions vary slightly.

8   "Inquest into the Death of a Body," 79.

9   All of the marks were on the pocket label of the same pair of pants.

10   "Inquest into the Death of a Body," 81.

11   Feltus claims that the paper was rolled up. Greenwood described it as folded.

12   "Inquest into the Death of a Body," 37.

13   "Inquest into the Death of a Body," 39.

14   "Inquest into the Death of a Body," 39.

15   "Inquest into the Death of a Body," 41.

16   "Inquest into the Death of a Body," 45.

17   "Inquest into the Death of a Body," 47.

18   "Inquest into the Death of a Body," 81.

19   "Inquest into the Death of a Body," 85.

20   "Inquest into the Death of a Body," 85.

21   "Inquest into the Death of a Body," 85.

22   "Inquest into the Death of a Body," 95.

23   "Inquest into the Death of a Body," 97.

24   "Inquest into the Death of a Body," 97.

25   "Inquest into the Death of a Body," 97.

26   "Inquest into the Death of a Body," 9. Note: There are two Clelands involved with this case, Thomas and John. They were cousins.

27   "Inquest into the Death of a Body," 9.

28   "Inquest into the Death of a Body," 7.

29   Greenwood, *Tamam Shud*, 80.

30   The suitcase can be seen in the 1978 documentary at http://www.youtube.com/watch?v =nnPqlYPQ9lY, but it was destroyed in 1986.

31   "Inquest into the Death of a Body," 73.

32   Feltus, *The Unknown Man*, 17.

33   Feltus, *The Unknown Man*, 171.

34   He wished to remain anonymous. His name is given as Ronald Francis in the books by Feltus and Greenwood, but this is a pseudonym. A list of common errors made by people writing about this case states, "Reports sometimes refer to the man who found the copy of *The Rubaiyat,* tossed in the back of his car, as a 'doctor' or 'chemist'. This is not certain. In fact both his name and real occupation have been withheld. Also the reason why his details are withheld is also suppressed. All we can assume is that he was some kind of professional." From https://www.eleceng.adelaide.edu.au/personal/dabbott/wiki/index.php/List_of_facts_on_the _Taman_Shud_Case_that_are_often_misreported.

35   *The Advertiser* (Adelaide, South Australia), "Police Test Book for Somerton Body Clue," July 26, 1949, p. 3.

36   Maybe. The 1949 inquest, p. 5, claims that the "Tamam Shud" paper came from a second edition. In any case, by "first edition," I mean the first from this publisher. It is not the first appearance of the book in print.

37   Some reports say "in the back of the book," whereas others say "on the back of the book." One official police document says that the back cover was missing.

38   If the hypothesis that follows is correct, this is exactly what we would expect. If the cipher consists of just initial letters, we should have a close match with initial frequencies, less of a match with overall frequencies (but not a *total* disconnect, since initial letters are included in determining overall frequencies), and a strong mismatch when comparing with terminal frequencies, since almost none of the initial letters are counted when compiling terminal frequencies—the only exceptions arise from the one-letter words A and I.

39   I redacted most of the editor's phone number. Please e-mail him if you have the solution!

40   Simon Singh, *The Codebook* (New York: Doubleday, 1999).

41   Feltus believes that he found the entire correct solution, but he doesn't reveal all of it in his book. Instead, he refers the reader to the back cover, which bears an image of the cipher. He also included a phone number alongside the cipher, but it's a fake. That is, it doesn't match the phone number that played an important role in the case.

42 Some proposed solutions, displayed on pages 119 and 120, are of the type I propose (taking cipher letters as the first letters of words), but there is no reason to favor them over many other potential solutions.

43 There are claims that more than one telephone number was present, but nothing authoritative enough to warrant inclusion here. This is another instance where we are greatly hampered because most of the police file is no longer available.

44 The term "nurse" may not have been technically accurate for the woman at this point in her life. See https://www.eleceng.adelaide.edu.au/personal/dabbott/wiki/index.php/List_of _facts_on_the_Taman_Shud_Case_that_are_often_misreported, item no. 4.

45 Boxall's copy was a 1924 Sydney edition.

46 Greenwood, *Tamam Shud*, 60.

47 "Inquest into the Death of a Body," 7.

48 "Inquest into the Death of a Body," 53.

49 Feltus, *The Unknown Man*, 151.

50 Feltus, *The Unknown Man*, 108.

51 Feltus, *The Unknown Man*, 178–79.

52 For more on early intelligence sharing with Australia, and a slightly different version of the story above, see Tom Johnson, *American Cryptology during the Cold War: 1945–1989, Book I: The Struggle for Centralization 1945–1960* (Ft. Meade, MD: Center for Cryptologic History, National Security Agency, 1995), 18–19. Discrepancies are minor, such as labeling the crypt-analyst as being from SIS (Signals Intelligence Service) instead of ASA. There were frequent name changes during these years. The evolution led to the National Security Agency (NSA) in 1952.

53 Wikipedia, http://en.wikipedia.org/wiki/Taman_Shud_Case, citing Phillips, John Harber, "So When That Angel of the Darker Drink," *Criminal Law Journal* 18, no. 2 (April 1994): 110.

54 Wikipedia, http://en.wikipedia.org/wiki/Taman_Shud_Case.

55 Victor Navasky, ed., *The Nation*, New York, Commentary, "Harry Dexter White," letter to the editor, April 1, 1988, available online at http://www.commentarymagazine.com/article/harry-dexter -white/.

56 On November 3, 1948, the day after Harry S. Truman defeated Thomas E. Dewey in the presidential election, the *Chicago Daily Tribune* ran the headline "Dewey Defeats Truman." Actually, the Dewey presidency has more sources to back it up than the digitalis-induced death of White. *The Journal of Commerce* ran the headline "Dewey Victory Seen as Mandate to Open New Era of Government-Business Harmony, Public Confidence" on November 3.

57 Trove digitized newspapers, http://trove.nla.gov.au/ndp/del/article/76017408. Thanks to James Ramm for pointing this article out to me!

58 Greenwood, *Tamam Shud*, 122–23.

59 https://web.archive.org/web/20110612061900/http://www.weirld.com/Paranormal/UFOs -Aliens/The-Strange-Case-of-the-Somerton-Man-Time-Traveller-Human/Alien-Hybrid-or -Secret-Agent-Spy.html.

60 "Weirld" is not a typo but rather a portmanteau of "weird" and "world."

61 *America's Most Wanted*, http://en.wikipedia.org/wiki/America's_Most_Wanted#Profiling –missing–persons. The man was captured in April 2008.

62 *Unsolved Mysteries*, https://web.archive.org/web/20140422010517/http://unsolved.com/about.html.

63 Greenwood, *Tamam Shud*, 129.

64 Available online at http://www.youtube.com/watch?v=iy5s50F3uB8 ("Somerton Man" segment). This is the source of the quotes that follow.

65 Some secondary sources say "formaldehyde," but that is technically wrong, as it refers to a gas. However, "formaldehyde solution" is equivalent to formalin, which is a mix of formaldehyde gas and water.

66 Renato Castello, "New Twist in Somerton Man Mystery as Fresh Claims Emerge," *Sunday Mail* (SA), November 23, 2013.

67 Castello, "New twist."

68 Castello, "New twist."

69 Gerry Feltus, "Second Application to Exhume Refused - 4 Jun 2014," http://theunknownman.com/comments.html This opinion also appeared at Pete Bowes, "the dead will not arise ... Rau, his second ruling," June 26, 2014, http://tomsbytwo.com/2014/06/26/the-dead-will-not-arise-rau-his-second-ruling/.

70 Kynton Grace, "South Australia's *X-Files*: Part 2—The Somerton Man Mystery and the Secrets of Adelaide's Tunnels," *The Advertiser*, June 9, 2014.

71 "Dad Identifies Poisoned Youth as N.Y. Student," *Washington Times-Herald*, January 22, 1953, p. 3.

72 "Body Found at Airport, Coded Note on Abdomen," January 21, 1953.

73 "Code Note Taped to Body in Ditch," *The Evening Bulletin*, Philadelphia, 1, column 2, Sports Final edition, January 20, 1953.

74 "Cyanide Caused Death at Airport," *The Evening Bulletin*, Philadelphia, p.1, column 3, Postscript edition, January 21, 1953.

75 "Cyanide Caused Death at Airport."

76 "Mystery Man at Airport Is Found Dead of Poison," *The Philadelphia Inquirer* January 21, 1953, p. 1.

77 "Mystery Man Cyanide Clue Hints." Murder Most accounts just say "cyanide." Potassium cyanide is specified in "Man Poisoned, Paper in Code Taped to Body; FBI Called In," *Washington Star*, p. A-26, and in some other articles. It is "potassium or sodium syanide [sic]" in "Code Message Found Taped to Dead Man," *Washington Times-Herald*, January 21, 1953, p. 35.

78 "Mystery Man Cyanide Clue Hints Murder."

79 "Mystery Man Cyanide Clue Hints Murder."

80 "Code Note Taped to Body in Ditch," *The Evening Bulletin*, Philadelphia, January 20, 1953, p. 1.

81 "Body with Code Note Found Here," *Daily News*, Philadelphia, January 20, 1953, p. 2.

82 " 'Dulles' Name Taped to Body of 'Suicide,'" *Washington Post*.

83 "Code Message Found Taped to Dead Man," *Washington Times-Herald*, January 21, 1953, p. 35. The quote is attributed to Dr. Edward Burke of the police crime laboratory, in "Mystery Man Cyanide Clue Hints Murder."

84 We also have the quote "Police said they found no cyanide container near the body." in "Cyanide Caused Death at Airport."

85   "Test Tube Found at Site of Airport Ditch Death," *The Philadelphia Inquirer*, January 23, 1953, p. 28. Available online through http://fultonhistory.com/Fulton.html.

86   "'Dulles' Name Taped to Body of 'Suicide.'" Also stated in "Student Identified as Cyanide Victim," *New York Times*, January 22, 1953.

87   "Mystery Man Cyanide Clue Hints Murder."

88   "Code Note Taped to Body in Ditch," *The Evening Bulletin*, Philadelphia, January 20, 1953, p. 1. Described as eight inches by three inches in "Code Message Found Taped to Dead Man," *Washington Times-Herald*, January 21, 1953, p. 35. It was described as only five inches in "'Dulles' Name Taped to Body of 'Suicide.'"

89   E-mail from David Hatch to the author, October 3, 2012.

90   "Mystery Man at Airport Is Found Dead of Poison." Another source claims that there was no wallet! See "Code Message Found Taped to Dead Man."

91   "Body Found at Airport, Coded Note on Abdomen."

92   "Cyanide Caused Death at Airport." Some sources only mention "a car key." Were there four keys or just one?

93   "Code Message Found Taped to Dead Man."

94   "Codes Seen Key in Airport Death," January 22, 1953.

95   For false report, see "Codes Seen Key in Airport Death."

96   "Mystery Man Cyanide Clue Hints Murder."

97   Rubin's age was typically reported as eighteen, but some articles gave his age as nineteen. See "Body at Airport Is Identified as NYU Student," Philadelphia, January 21, 1953, and "Airport Death Called Murder by Ominsky," Philadelphia, January 21, 1953. In this second article, Rubin's age is later stated as eighteen.

98   "Youth Dabbled in Cryptography," *Philadelphia Evening Bulletin*, January 22, 1953.

99   In most accounts, it was forty-seven cents. For example, see "Code Note Taped to Body in Ditch." But it was forty-six cents in "Body with code note found here."

100 "Poison Victim Identified as N.Y. Youth," *The Philadelphia Inquirer*, January 22, 1953, p.1, illustrated on p. 3 and continued on p. 13, quote from p. 13.

101 "Poison Victim Identified as N.Y. Youth," quote from p. 13.

102 "FBI Investigates Possibilities of Murder in Mysterious Cyanide-Death of Student," *Schenectady Gazette,* January 22, 1953.

103 Jay Apt, "Note Baffles Decoders in Poison Death," *Daily News*, January 22, 1953, p. 2.

104 "FBI Investigates Possibilities of Murder in Mysterious Cyanide-Death of Student."

105 "Hazing Studied in Airport Death," *The Philadelphia Inquirer*, January 28, 1953, p. 5.

106 "Student Identified as Cyanide Victim," *New York Times*, January 22, 1953.

107 E-mail from Claire Ashley Wolford, May 2, 2012.

108 "Life on the Newsfronts of the World, Student Leaves a Coded Clue to His Death, a Soldier Plants a Flag and Flu Virus Sweeps the World," *Life*, February 2, 1953, p. 30.

109 "Youth Dabbled in Cryptography."

110 "Poison Victim Identified as N.Y. Youth," quote from p. 13.

111 "Mystery Shrouds Death by Cyanide," *The Philadelphia Inquirer*, March 25, 1953, p. 19.

112 "Mystery Shrouds Death by Cyanide."

113 "Mystery Shrouds Death by Cyanide." The names Edwin and Edward both appear in newspaper accounts.

114   "Body at Airport."

115   "Body at Airport."

116   "Gay rights campaigner calls for fresh inquiry into death of Alan Turing," *Manchester Evening News*, December 24, 2013, http://www.manchestereveningnews.co.uk/news/greater-manchester-news/codebreaker-alan-turing-gay-rights-6445286, accessed July 11, 2014.

117   Langen, Harry [*sic*—it should say Henry], *My Life Between the Lines*.

118   Langen, *My Life Between the Lines*.

119   I found this paper as a clipping in an uncatalogued portion of the National Cryptologic Museum's collection. It is not clear from the clipping which paper it came from!

120   Wolfram Alpha, http://www.wolframalpha.com/input/?i=average+english+word+length. You may find much larger estimates, but these simply average the lengths of all English words from some list. They do not weight the average by frequency. That is, the value above looks at actual usage.

121   Shane Anthony, "From 1999: Body found in field puzzles police," *St. Louis Post-Dispatch*, July 2, 1999, but Christpher Tritto implied that the sentence was about two years in his article.

122   According to Christopher Tritto. Shane Anthony reported that he served 11 months.

123   Christopher Tritto, "Code Dead: Do the encrypted writings of Ricky McCormick hold the key to his mysterious death?" *Riverfront Times* (St. Louis), June 14, 2012.

124   Tritto, "Code Dead." This article reports that "At least one gas-station employee" claims to have seen McCormick there that day. There's no elaboration concerning the uncertainty expressed in the number of witnesses.

125   Anthony, "From 1999." This article actually says "Highway 367," but it's a typo, one that has spread to the Wikipedia page for the McCormick cipher.

126   Still available online at the FBI site, http://www.fbi.gov/news/stories/2011/march/cryptanalysis_032911.

127   Jennifer Mann, "FBI wants public to help crack code in St. Charles County cold case," *St. Louis Post-Dispatch*, March 31, 2011.

128   Mann, "FBI wants public to help."

129   Tritto, "Code Dead."

130   Tritto, "Code Dead."

131   Tritto, "Code Dead."

Chapter 7: From Beyond the Grave?

1   R. H. Thouless, "A Test of Survival," *Proceedings of the Society for Psychical Research* 48 (July 1948): 253–63.

2   Thouless, "A Test of Survival."

3   Thouless, "A Test of Survival."

4   Thouless, "A Test of Survival."

5   Cambridge University Library, "The Society for Psychical Research [3000] collection," SPR.MS 63/2/113. Thanks to Klaus Schmeh for sharing this.

6   *Oxford English Dictionary*, 2nd ed. (Oxford, U.K.: Oxford University Press), http://public.oed.com/history-of-the-oed/dictionary-facts/.

7   Thouless, "A Test of Survival."

8   Thouless, "A Test of Survival."

9   Thouless, "A Test of Survival."

10   Thouless, "A Test of Survival."

11   Thouless, "A Test of Survival."

12   Thouless, "A Test of Survival."

13   Thouless, "A Test of Survival," p. 262.

14   There are several different ways to carry out "double Playfair" encipherment. This is just one example.

15   Robert H. Thouless, "Additional Notes on a Test of Survival," *Proceedings of the American Society for Psychical Research* 48 (1948): 342–43.

16   James J. Gillogly and Larry Harnisch, "Cryptograms from the Crypt," *Cryptologia* 20, no. 4 (October 1996): 325–29.

17   Gillogly and Harnisch, "Cryptograms from the Crypt."

18   He apparently didn't want much attention and published his paper under the name T. E. Wood. No point of contact was given in the paper. I found his full name in a document from Cambridge University Library, "The Society for Psychical Research [3000] collection," SPR.MS 63/1/1. Thanks to Klaus Schmeh for sharing this. A fuller biography also appears in this file and more from Wood follows. The biography might help suggest what languages he used.

19   Originally, he had a normal Vigenère cipher in mind, but he learned of its weakness before going to print. See Cambridge University Library, "The Society for Psychical Research [3000] collection," SPR.MS 63/2/4. Thanks to Klaus Schmeh for sharing this.

20   T. E. Wood, "A Further Test for Survival," *Proceedings of the Society for Psychical Research* 49, (1950): 105–6, p. 105 quoted here.

21   Wood, "A Further Test for Survival," p. 105 quoted here.

22   Cambridge University Library, "The Society for Psychical Research [3000] collection," SPR.MS 63/2/46. Thanks to Klaus Schmeh for sharing this.

23   Cambridge University Library, "The Society for Psychical Research [3000] collection," SPR.MS 63/2/97. Thanks to Klaus Schmeh for sharing this.

24   Frank C. Tribbe, "The Tribbe/Mulders Code," *Journal of the Academy of Religion and Psychical Research* 3, no. 1 (January 1980): 44–46.

25   Arthur S. Berger, *Aristocracy of the Dead: New Findings in Postmortem Survival* (London and Jefferson, NC: McFarland & Co., 1987), p. 152 cited here.

26   Berger, *Aristocracy of the Dead*, 165.

27   Berger, *Aristocracy of the Dead*, 163.

28   Berger, *Aristocracy of the Dead*, 165.

29   Arthur S. Berger, "Reincarnation by the Numbers: A Criticism and Suggestion," *Reincarnation: Fact or Fable?* Arthur S. Berger and Joyce Berger, eds. (London: The Aquarian Press, 1991), 221–33, p. 229 cited here.

30   Berger, "Reincarnation by the Numbers," 231.

31   For details of the lock, see p. 248 of Ian Stevenson, "The Combination Lock Test for Survival," *Journal of the American Society for Psychical Research* 62 (1968): 246–54.

32   See, for example, Matt Blaze, "Safecracking for the computer scientist," 2004 draft. Available online at http://www.crypto.com/papers/safelocks.pdf (accessed May 17, 2016).

33 Stevenson, "The Combination Lock Test for Survival."

34 Joseph Martin Feely, *Electrograms from Elysium: A Study on the Probabilities in Postmortuary Communication through the Electronics of Telepathy and Extrasensory Perception, Including the Code of Anagrams in the Purported Sender's Name* (New York, 1954), 202.

35 Feely, *Electrograms from Elysium*, 338–39.

36 Feely, *Electrograms from Elysium*, 236.

37 Feely, *Electrograms from Elysium*, 274.

38 Feely, *Electrograms from Elysium*, 240, 247.

39 Feely, *Electrograms from Elysium*, 329.

40 Feely, *Electrograms from Elysium*, 331.

41 Michael Levin, "Encryption Algorithms in Survival Evidence," *The Journal of Parapsychology* 58, no. 2 (June 1994): 189–95.

42 Levin, "Encryption Algorithms in Survival Evidence."

43 Levin, "Encryption Algorithms in Survival Evidence."

44 Levin was much more successful with his other papers, which earned a total of 7,214 citations as of this writing.

45 Susy Smith (pen name of Ethel Elizabeth Smith). Introduction by Gary E. R. Schwartz and Linda G. S. Russek. *The Afterlife Codes: Searching for Evidence of the Survival of the Human Soul* (Charlottesville, VA: Hampton Roads Publishing Company, Inc., 2000).

46 The Internet Archive Wayback Machine, https://archive.org/web/.

47 Afterlife Codes, https://web.archive.org/web/20010517033428/http://www.afterlifecodes .com/.

48 The linked page was archived from June 14, 2001.

49 It was just a MASC. See the section on Frank C. Tribbe earlier in this chapter or, for more detail, read Frank C. Tribbe, "The Tribbe/Mulders Code," *Journal of the Academy of Religion and Psychical Research* 3, no. 1 (January 1980): 44–46.

50 Elsewhere on the website, Smith wrote, "Naturally, we also thought of the possibility that someone might try to break my code by fraud, but we discounted that idea because we felt sure people weren't mean and unprincipled enough to do such a thing. But then came the '90s and an unfortunate moral decline. Thus we are having to advise those who register now not to offer rewards, which might be tempting. After considering the idea that I should apply this to myself as well, I have decided not to revoke my offer. It's such a part of my history now that a few more years won't make any difference. Except that although my body complains constantly with arthritis and such, my mind still enjoys hanging around keeping in touch. I could go on living until my funds are all used up. Not to worry. In that case Gary and Linda have vowed to provide the reward, should I be able to send my message after my demise. Fortunately also, the amount of $10,000 has been left with the University of Arizona Foundation for the Susy Smith Project by Robert Bigelow's National Institute for Discovery Science of Las Vegas. This, plus future contributions, should guarantee the continuation of our computer's code activity over the years."

51 Klaus Schmeh, *Nicht zu knacken* (Munich, Germany: Carl Hanser Verlag, 2012).

Chapter 8: A Challenge Cipher

1   In 2011, Steven Bellovin uncovered the fact that the one-time pad had, in fact, only been rediscovered by Mauborgne and Vernam. It was previously known to Frank Miller and was published by him in an 1882 commercial code book. For details, see Steven M. Bellovin, "Frank Miller: Inventor of the One-Time Pad," *Cryptologia* 35, no. 3 (July 2011): 203–22.

2   A line from the film *Fight Club*.

3   A line from the film *Saw*.

4   A line from the film *Freddy Got Fingered*.

5   It was re-presented by William F. Friedman, who had already solved it but delayed revealing his solution until the next issue to give readers a chance to attack it for themselves.

6   Taken here from Louis Kruh, "A 77-Year Old Challenge Cipher," *Cryptologia* 17, no. 2 (April 1993): 172–74.

7   Kruh, "A 77-Year Old Challenge Cipher," 172.

8   Louis Kruh, "Riverbank Laboratory Correspondence, 1919 (SRH-50)," *Cryptologia* 19, no. 3 (July 1995): 236–46.

9   A line from the film π.

10   Thanks to Moshe Rubin for helping me identify this individual.

11   TRIO (ACA pen name for Fenwick Wesencraft), "Solutions of the M-94 Test Messages," *The Cryptogram* 48, no. 8 (November–December 1982): 6–7.

12   Byrne promoted the system most famously in John F. Byrne, *Silent Years: An Autobiography with Memoirs of James Joyce and Our Ireland* (New York: Farrar, Straus and Young, 1953).

13   The error was pointed out in Jeff Calof, Jeff Hill, and Moshe Rubin, "Chaocipher Exhibit 5: History, Analysis, and Solution of *Cryptologia*'s 1990 Challenge," *Cryptologia* 38, no. 1 (January 2014): 1–25.

Chapter 9: More Challenge Ciphers

1   For a fuller account of this episode of government censorship, see Craig Bauer and Joel Burkholder, "From the Archives: Reading Stimson's Mail," *Cryptologia* 31, no. 2 (April 2007): 179–84.

2   Alexander d'Agapeyeff, *Codes and Ciphers* (London: Oxford University Press, 1939), front inside dust jacket.

3   d'Agapeyeff, *Codes and Ciphers*, 62–63.

4   J 77/2621/1445, Divorce Court File: 1445. Appellant: Josephine Christian Lilian Passy d'Agapeyeff. Respondent: Alexander d'Agapeyeff. Type: Wife's petition for divorce [WD], 1929, The National Archives, Kew, England.

5   Ralph Erskine and John Gallehawk located and photographed d'Agapeyeff's SOE file for me, but it didn't indicate why he was turned down. The reference is Special Operations Executive: Personnel Files (PF Series). Alexander d'Agapeyeff, Collection: Records of Special Operations Executive, 01 January 1939–31 December 1946, HS 9/9/5, The National Archives, Kew, England. Erskine thought it was doubtful that d'Agapeyeff was turned down because of the adultery. He pointed out, "Sir Stewart Menzies, head of the UK SIS, had at least 1 mistress, and was also divorced. Hinsley said privately that Godfrey, the Director of Naval Intelligence had a number of women." On the other hand, when interviewing potential new hires, NSA

polygraphers ask questions like "Have you ever cheated on your wife?" Beyond the trust issue, they don't want to hire people who may be subject to blackmail.

6   d'Agapeyeff, *Codes and Ciphers*, 131.

7   Image from d'Agapeyeff, *Codes and Ciphers*, 103.

8   The series begins with Gordon Rugg and Gavin Taylor, "A very British mystery: The case of the D'Agapeyeff Cipher," https://hydeandrugg.wordpress.com/2013/07/02/a-very-british-mystery-the-case-of-the-dagapeyeff-cipher/. As of this writing, part 6, in which Van Zandt's work is discussed, is unpublished. Rugg shared a draft version of it with me. I expect that a combined and revised version of these articles will see print in *Cryptologia*.

9   Gordon Rugg and Gavin Taylor, "A very British mystery, part 6: A possible solution? Probably not ... ," draft version.

10 Richard Feynman, *Surely You're Joking, Mr. Feynman* (New York: W. W. Norton and Co., 1985), 148–49.

11 Feynman's cipher, https://groups.google.com/forum/#!topic/sci.crypt/u1O1W1mL7Fw.

12 Feynman ciphers, https://groups.google.com/forum/#!searchin/sci.crypt/Jack$20C.$20Morrison$20/sci.crypt/RAxvau5mxJ4/VWLRaSbb-xMJ.

13 The last word didn't appear in its entirety in the deciphered text. That's why it was completed in brackets. http://www.vanderbilt.edu/usac/documents/BoTPresentation.pdf.

14 *Central Intelligence Agency Employee Bulletin*, June 1, 1988, released in response to a Freedom of Information Act (FOIA) request.

15 Remarks by William H. Webster, Director of Central Intelligence, at the dedication of the sculpture for the New Headquarters Building, November 5, 1990, released in response to a FOIA request.

16 Webster remarks.

17 Robert M. Andrews, "Sculpture with Code Poses Mystery at the CIA," *Los Angeles Times*, April 28, 1991, p. A34.

18 Part of a petition sent to the CIA Fine Arts Commission on August 2, 1989, released under a FOIA request. The sixteen signatures were primarily from the Office of Soviet Analysis (SOVA).

19 This method can be found at https://kryptosfan.wordpress.com/k3/k3-solution-3/.

20 John Markoff, "C.I.A.'s Artistic Enigma Yields All but Final Clue," *New York Times*, June 16, 1999, p. A24.

21 Markoff, "C.I.A.'s Artistic Enigma Yields All but Final Clue."

22 Markoff, "C.I.A.'s Artistic Enigma Yields All but Final Clue."

23 Markoff, "C.I.A.'s Artistic Enigma Yields All but Final Clue."

24 Lester S. Hill, "Cryptography in an Algebraic Alphabet," *The American Mathematical Monthly* 36, no. 6 (1929): 306–12.

25 These figures actually include non-invertible matrices. It was easier to just test all matrices than to eliminate those that weren't desirable from a cryptologic perspective.

26 Markoff, "C.I.A.'s Artistic Enigma Yields All but Final Clue."

27 Andrews, "Sculpture With Code Poses Mystery at the CIA."

28 Wikipedia, http://en.wikipedia.org/wiki/Cicada_3301.

29 Quote taken from *Anonymous: How Hackers Changed the World*, BBC documentary, https://www.youtube.com/watch?v=jwA4tSwmS-U.

30   This message can now be found at this site: https://www.reddit.com/r/a2e7j6ic78h0j /comments/o2rbx/ukbn_txltbz_nal_hh_uoxelmgox_wdvg_akw_hvu_ogl_rsm/.

31   You can hear the message at this site: https://www.youtube.com/watch?v=k24ZrFR2IUQ.

32   http://pgp.mit.edu:11371/pks/lookup?search=0x7A35090F&op=index.

33   See http://www.clevcode.org/cicada-3301/ for Eriksson's interpretation of the rest of this clue.

34   And still is, as of this writing. See http://www.williamgibsonbooks.com/source/agrippa.asp.

35   "Norway Terror Attacks Fast Facts," CNN Library, July 17, 2015.

36   "Norway Terror Attacks Fast Facts," CNN Library, July 17, 2015. But according to Bjoern Amland and Sarah DiLorenzo, "Lawyer: Norway Suspect Wanted a Revolution," July 24, 2011, he killed at least ninety-two and wounded ninety-seven others. It's unclear where this number for the deaths came from. Much later accounts don't support it. See, for example, Olivia Herstein, "Still Standing—As One," *Viking* 108, no. 12 (December 2011): 16–20, 22, 24. Wikipedia says that Breivik killed 77 and injured 241.

37   Amland and DiLorenzo, "Lawyer."

38   Dave Ramsden, *Unveiling the Mystic Ciphers: Thomas Anson and the Shepherd's Monument Inscription* (CreateSpace, 2014).

39   Archived at https://web.archive.org/web/20110919074847/http://jahbulon.com/.

40   https://web.archive.org/web/20121210124921/http://jahbulon.com/about/.

41   List of errors taken from Todd D. Mateer, "Cryptanalysis of Beale Cipher Number Two," *Cryptologia* 37, no. 3 (July 2013): 225–26.

42   Letter from M. E. Ohaver to John Ingles, August 8, 1933. This letter is discussed in Louis Kruh, "A Basic Probe of the Beale Cipher as a Bamboozlement," *Cryptologia* 6, no. 4 (October 1982): 378–82.

43   Wayne S. Chan, "Key Enclosed: Examining the Evidence for the Missing Key Letter of the Beale Cipher," *Cryptologia* 32, no. 1 (January 2008): 33–36.

44   Chan was unable to get any listings closer to 1832 for St. Louis.

45   Forrest Fenn, *The Thrill of the Chase, A Memoir* (Santa Fe, NM: One Horse Land & Cattle Co., 2010), 128.

46   Fenn, *The Thrill of the Chase*, 129–32.

47   Fenn, *The Thrill of the Chase*, 133.

48   Forrest Fenn, *Too Far to Walk* (Santa Fe, NM: One Horse Land & Cattle Co., 2013), 164.

49   Fenn, *Too Far to Walk*, 164.

50   Fenn, *Too Far to Walk*.

51   Jennifer Deutschmann, "Forrest Fenn: Hunt for Buried Treasure Is 'Out of Control,' " *Inquisitr*, April 28, 2015, available online at http://www.inquisitr.com/2050321/forrest-fenn/.

52   Fenn, *Too Far to Walk*, 86.

53   Margie Goldsmith, "Well Over $1 Million in Buried Treasure: Find It!" *Huffington Post*, February 18, 2011, updated February 27, 2013.

54   Fenn, *Too Far to Walk*, 265.

55   Deutschmann, "Forrest Fenn."

56   Athol Thomas, *Forgotten Eden: A View of the Seychelles Islands in the Indian Ocean* (London and Harlow, U.K.: Longmans, Green and Co. Ltd., 1968), 133–34.

57   Thomas, *Forgotten Eden*, 134.

58   E-mail sent to the author on July 13, 2015.

59   Klausis Krypto Kolumne, http://scienceblogs.de/klausis-krypto-kolumne/2016/03/27/wer
      -kann-diesen-grabstein-entschluesseln-und-wo-befindet-er-sich-ueberhaupt/.

Chapter 10: Long Ciphers

1   Klausis Krypto Kolumne, http://scienceblogs.de/klausis-krypto-kolumne/klaus-schmehs-list
      -of-encrypted-books/.

2   Benedek Láng, "Why Don't We Decipher an Outdated Cipher System? The Codex of Rohonc,"
      Cryptologia 34, no. 2 (April 2010): 115–44.

3   The solutions were pointed out by Klaus Schmeh in "The Pathology of Cryptology—A Current
      Survey," *Cryptologia* 36, no. 1 (January 2012): 14–45, and are included in the references at the
      end of this book.

4   Lynda Roscoe Hartigan, *James Hampton: The Throne of the Third Heaven of the Nations' Mil-
      lennium General Assembly* (Boston: Museum of Fine Arts, October 19 (1975?)–February 13,
      1976). *This essay was originally published for a 1976 exhibition of James Hampton's work.*

5   Stephen Jay Gould, "James Hampton's Throne and the Dual Nature of Time," *Smithsonian Stud-
      ies in American Art* 1, no. 1 (Spring 1987): 46–57.

6   Gould, "James Hampton's Throne."

7   Stephen Jay Gould, "Nonoverlapping Magisteria," *Natural History* 106 (March 1997): 16–22.

8   Gould, "James Hampton's Throne."

9   Gould, "James Hampton's Throne."

10   This is 00011 in Klaus Schmeh's list. See http://www.scm.keele.ac.uk/research/knowledge
      _modelling/km/people/gordon_rugg/cryptography/penitentia/index.html.

11   Keele University, School of Computing and Mathematics, http://www.keele.ac.uk/scm/staff
      /academic/drgordonrugg/.

12   E-mail from Gordon Rugg to the author on June 29, 2015.

13   IMDb, http://www.imdb.com/name/nm0235710/.

14   Search Visualizer, http://www.searchvisualizer.com/.

15   This is 00012 in Klaus Schmeh's list.

16   Keele University, Knowledge Modelling, http://www.scm.keele.ac.uk/research/knowledge
      _modelling/km/people/gordon_rugg/cryptography/ricardus_manuscript.html.

17   Sandra and Woo, http://www.sandraandwoo.com/2013/07/29/0500-the-book-of-woo/. This is
      number 00022 in Klaus Schmeh's list.

18   E-mail from Oliver Knörzer to the author, July 16, 2015.

19   E-mail from Oliver Knörzer to the author, July 16, 2015.

20   E-mail from Oliver Knörzer to the author, July 16, 2015. The wiki is at http://bookofwoo
      .foogod.com/wiki/Book_of_Woo_Wiki.

21   Book of Woo Wiki, http://bookofwoo.foogod.com/wiki/What_We_Know. The hint was given
      on July 21, 2014.

22   E-mail from Oliver Knörzer to the author, July 16, 2015.

23   E-mail from Oliver Knörzer to the author, July 16, 2015.

24   E-mail from Oliver Knörzer to the author, July 16, 2015.

25   Auguste Kerckhoffs, "La Cryptographie Militaire." *Journal des science militaires* (Paris: Baudoin,
      1883), vol. IX, 5–83.

26   See imgur, http://imgur.com/a/8xnWx/all.

27   Klausis Krypto Kolumne, http://scienceblogs.de/klausis-krypto-kolumne/klaus-schmehs-list
      -of-encrypted-books/.

Chapter 11: ET and RSA

1   "Science Notes," *Scientific American* 84, no. 12 (March 23, 1901), 179.

2   "Science Notes," *Scientific American*.

3   "Annual Report of the Board of Regents of the Smithsonian Institution," 1900, p. 169 (taken
      from *Boston Transcript*, February 2, 1901). Available online at http://www.biodiversitylibrary
      .org/item/53390#page/291/mode/1up. Note: This is not to be confused with a different vol-
      ume of almost identical title, but specifying "1900 Incl. Rept. US Natl Mus."

4   Charles Fort, *New Lands*, in The Books of Charles Fort, with an introduction by Tiffany Thayer
      (omnibus ed.) (New York: Henry Holt and Co., 1941), ch. 32, p. 494. Available online in
      single-volume edition at http://www.resologist.net/landsei.htm. See part 2, chapter 20.

5   "100 Years Ago," *Nature* 417, no. 6884 (May 2, 2002): 31.

6   Charles Fort also wrote about this episode. See Fort, *New Lands*, ch. 36, pp. 526–27. Available
      online in single-volume edition at http://www.resologist.net/landsei.htm. See part 2, chap. 24.

7   *The Monthly Evening Sky Map* XIV, no. 159 (March 1920).

8   *The Monthly Evening Sky Map* XV, no. 179 (November 1921). Available online at Google Books.
      More detail can be found in "Marconi Believes He Received Wireless Messages From Mars,"
      New York Tribune, Sept. 2, 1921, p. 3.

9   "Weird 'Radio Signal' Film Deepens Mystery of Mars," *Washington Post*, August 27, 1924.

10   "Weird 'Radio Signal' Film."

11   "Weird 'Radio Signal' Film."

12   "Weird 'Radio Signal' Film."

13   "Weird 'Radio Signal' Film."

14   "Weird 'Radio Signal' Film."

15   This idea was mentioned by Ronald Bracewell in "What to Say to the Space Probe when it
      Arrives," *Horizon*, January 1977. Bracewell preferred the idea of sending animated cartoons
      because they can convey words that can't be made clear by single images.

16   Harold Wooster (moderator), Paul J. Garvin, Lambros D. Callimahos, John C. Lilly, William
      O. Davis, and Francis J. Heyden, "Communication with Extraterrestrial Intelligence," *IEEE Spec-
      trum* 3, no. 3 (March 1966): 156 and 158.

17   Bernard M. Oliver, "Interstellar Communication," *Interstellar Communication;—A Collection of
      Reprints and Original Contributions*, A. G. W Cameron, ed. (New York: W.A. Benjamin, 1963),
      294–305.

18   This conference is also where Frank Drake first put forth what's become known as the Drake
      equation.

19   Oliver, "Interstellar Communication," 305.

20   Reproduced from http://www.rsa.com/rsalabs/node.asp?id=2093.

21   R. L Rivest, A. Shamir, and L. Adleman, *On Digital Signatures and Public-Key Cryptosystems*
      (There was soon a title change to *A Method for Obtaining Digital Signatures and Public-key
      Cryptosystems*. The date is the same for both.), M.I.T. Laboratory for Computer Science Report

MIT/LCS/TM-82, April 1977. This report later appeared in *Communications of the ACM* 21, no. 2 (February 1978): 120–26, with the second title.

22   The team consisted of F. Bahr, M. Boehm, Jens Franke, and Thorsten Kleinjung.

23   Thorsten Kleinjung, Kazumaro Aoki, Jens Franke, Arjen K. Lenstra, Emmanuel Thomé, Joppe W. Bos, Picrrick Gaudry, Alexander Kruppa, Peter L. Montgomery, Dag Arne Osvik, Herman te Riele, Andrey Timofeev, and Paul Zimmermann, "Factorization of a 768-bit RSA modulus," version 1.4, February 18, 2010, 14, available online at http://eprint.iacr.org/2010/006.pdf.

24   Ronald L. Rivest, "Description of the LCS35 Time Capsule Crypto-Puzzle," April 4, 1999, available online at http://people.csail.mit.edu/rivest/lcs35-puzzle-description.txt.

25   Ronald L. Rivest, Adi Shamir, and David A. Wagner, "Time-lock puzzles and timed-release Crypto," Revised March 10, 1996, p. 2, available online at http://theory.lcs.mit.edu/~rivest /RivestShamirWagner-timelock.ps.

26   Rivest specified "ascii at 8 bits per character," meaning that extended ASCII is intended. The 8 bits offer 256 possibilities.

27   Ronal L. Rivest, "Desription of the LCS35 Time Capsule Crypto-Puzzle," http://people.csail.mit .edu/rivest/lcs35-puzzle-description.txt.

References and Further Reading

Chapter 1

Jim Reeds's bibliography follows, with my additions (and some deletions). Permission to use Jim's work as a basis for my own bibliography was granted by Jim at the Cryptologic History Symposium in October 2011. Jim got some references from voynich@rand.org, Rene Zandbergen, Gabriel Landini, Denis Mardle, Dennis Stallings, and especially from Brian Smith.

MANUSCRIPTS

Items with WFF numbers are in the William F. Friedman collection of the George Marshall Library in Lexington, Virginia.

"The Voynich 'Roger Bacon' Cipher Manuscript." Beinecke Rare Book Library, Yale University: MS 408. Supplementary material in folders and boxes labelled A–N.

Positive photocopies of ff. 1–56 of Voynich MS. British Library MS Facs 461.

Positive photocopies of miscellaneous folios of Voynich MS, misc. VMS correspondence of R. Steele, 1921, and misc. VMS articles. (Contains: 68r1/68r2, 65v/66r, 78v/79r, 107v/108r, 108v/111r, 111v/112r, 112v/113r, 1r, and 116v. Correspondence includes letters to Steele from W. Voynich, W. Newbold, and from A. W. Pollard. Articles include a clipping from *Morning Post* newspaper, 26.9.21, "Astrological Anagrams: the Diary of Roger Bacon," clipping from *Daily Chronicle* newspaper, n.d., "Key to Cypher in historic MS./America's new light on Roger Bacon/600 years' mystery," penciled draft of article by Steele, photostat of typed lecture notes by Newbold, copy of J. Manly's *Harper's* article, copy of Louis Cons February 4, 1922, article.) British Library MS Facs 439.

Brumbaugh, Robert S. "Voynich Newsletter." Various issues: February 1978, 6-p typescript. November 1978, 7-p typescript. January 1980, 7-p typescript.

Carter, Albert H. "Some Impressions of the Voynich Manuscript." Unpublished notes, September 10, 1946. WFF 1614.

Carton, Raoul. "The Cipher of Roger Bacon." 55-p typescript, translated from French by E. L. Voynich. In N.Y. Academy of Medicine Library. 1930.

Currier, Prescott. "Voynich MS Transcription Alphabet; Plans for Computer Studies; Transcribed Text of Herbal A and B Material; Notes and Observations." Unpublished communications to John H. Tiltman and M. D'Imperio. Darimascotta, ME.

D'Imperio, M. E. "Structure of Voynich Text Groups: A Statistical Model." 2-p typescript, 1978.

D'Imperio, M. E. "An Application of Cluster Analysis and Multiple Scaling to the Question of 'Hands' and 'Languages' in the Voynich Manuscript." January 28, 1992. [Confirms Currier's findings. In Gillogly collection. From abstract: "This paper is an extensively revised and updated version of an earlier paper in an in-house technical journal, dated 20 June 1978. It includes corrected and expanded data sets."]

D'Imperio, M. E. "Odd Repetitions or Near Repetitions in the Text." January 1992.

D'Imperio, M. E. "Some Ideas on the Construction of the Voynich Script." 3-p typescript, January 1992.

Firth, Robert. "Notes on the Voynich Manuscript." Numbered series of essays: 1–24, 1991–1995.

Friedman, William F. Two "First Study Group" transcription alphabet sheets. WFF 1609.1.

Friedman, William F. "First Study Group" transcription alphabet sheet. WFF 1609.2.

Friedman, William F., Mark Rhoads, et al. Minutes of the "Voynich Manuscript Research Group," 1944–45. National Cryptologic Museum, VF 10-8.

Friedman, William F., et al. Printout of transcription, onto 131 printout sheets, ca. 1944–46. WFF 1609.

Friedman, William F., et al. Printout of partial transcription, Second Study Group, ca. 1963. WFF 1609.3.

Friedman, William F., et al. Correspondence with RCA Corp. about activities of "Second Study Group," 1963. WFF 1609.4.

Friedman, William F., et al. Printout of partial transcription, ca. 1944–46 Unnumbered item in Friedman Collection.

Friedman, William F., et al. Printout of partial transcription, ca. 1944–46. Unnumbered item in NSA Historical Records Collection.

Guy, Jacques B. M. "The distribution of letters c and o in the Voynich Manuscript: Evidence for a real language?" April 1994.

Krischer, Jeffrey P. *The Voynich Manuscript*, Harvard University term paper, 1969.

Mervyn, Tim, unpublished manuscript. [This manuscript is mentioned briefly on p. ix and then in much more detail on pp. 216–20 of Kennedy, Gerry, and Rob Churchill, *The Voynich Manuscript: The Unsolved Riddle of an Extraordinary Book Which Has Defied Interpretation for Centuries*, British hardcover edition (London: Orion Books, 2004).]

Panofsky, Erwin. "Answers to Questions for Prof. E. Panofsky." Letter to William F. Friedman, March 19, 1954. Correspondence between Friedman, Panofsky, and J. v. Neumann. Letters from Richard Salomon to Erwin Panofsky and Gertrud Bing. WFF 1614.

Petersen, Theodore C. "Notes to Mr. Tiltman's [1951] Observations on the Voynich Cipher MS." Unpublished. April 23, 1953.

Petersen, Theodore C. Hand Transcript and Concordance of the Voynich Manuscript and Other Working Papers. In the Friedman Collection, George Marshall Library, Lexington, Virginia, dated 1966, but this is simply when the library acquired it (following Petersen's death).

Puckett, Frances M. Partial transcription of ff. 111v–114r, in WFF 1613.

Strong, Leonell C. Collection of letters, notebook entries, and worksheets. In a private collection.

Tiltman, John. "Interim Report on the Voynich MS." Personal communication to William F. Friedman, May 5, 1951. 2-p typescript. In NSA Historical Records Collection. Copy in WFF 1615.13.

Tiltman, John. Partial transcription, 1951. In NSA Historical Records Collection.

Tiltman, John. Biography of T. C. Petersen. WFF 1615.

Tiltman, John. "The Voynich MS" Script of an address presented at the Baltimore Bibliophiles. March 4, 1951. [The date looks wrong. Cited by D'Imperio. Almost certainly the 1967 Baltimore Bibliophiles address.]

Voynich, Wilfrid, Ethel Voynich, and A. M. Nill. Notes concerning the history of the cipher manuscript. Voynich Archives, Library of the Grollier Club of New York, 1917–196? [Possibly identical with Beinecke MS 408 B.]

PRINTED BOOKS AND ARTICLES

Altick, Richard D. *The Scholar Adventurers.* (New York: The Free Press, 1966). [First ed. in 1950. Has discussion of VMS and Manly's refutation on pp. 200–6. A revised reprint appeared in 1987.]

Anonymous. "Art Works Worth $1,500,000 Arrive to Escape War." *Chicago Daily Tribune.* Oct. 9, 1915, p. 1, col. 2. [Exhibit of WMV's books at the Art Institute of Chicago, which includes "a work by Roger Bacon in cipher to which the key has never been discovered."]

Anonymous. "Antique Books Worth $500,000." *Chicago Sunday Tribune.* Oct. 10, 1915, sec. 2, p. 1, col. 5. [Exhibit of WMV's books at the Art Institute of Chicago. A few paragraphs describe the VMS. "The manuscript is from the hand of Roger Bacon . . . was bought by Emperor Rudolf . . . and at the end of the sixteenth century passed into the hands of King Ferdinand of Bohemia."]

Anonymous. "Review of Brumbaugh's Most Mysterious Manuscript." *Choice* 15 (October 1978), 1080.

Anonymous. "The Roger Bacon Manuscript." *Scientific American* 124 (May 7, 1921): 362.

Anonymous. "The Roger Bacon Manuscript: What It Looks Like, and a Discussion of the Possibilities of Decipherment." *Scientific American* 124, no. 22 (May 28, 1921): 432, 439, 440.

Ashbrook, Joseph. "Roger Bacon and the Voynich Manuscript." *Sky and Telescope* (April 1966): 218–19.

Barlow, Michael. "The Voynich Manuscript—By Voynich?" *Cryptologia* 10, no. 4 (October 1986): 210–16.

Barlow, Michael. "Voynich Solved?" (Review of Levitov), pp. 47–48 in "Reviews and Things Cryptologic." *Cryptologia* 12, no. 1 (January 1988): 37–51.

Barthélemy, Pierre. "L'indéchiffrable manuscrit Voynich résiste toujours au décryptage." *Le Monde,* December 20, 2000.

Barthélemy, Pierre (introduction only). *Le Code Voynich.* (Paris: Jean-Claude Gawsewitch, 2005). Color facsimile edition of the full text of the Voynich ms.

Bauer, Craig P. *Secret History: The Story of Cryptology* (Boca Raton, FL: Chapman and Hall/CRC, 2013).

Bax, Stephen. Voynich—a provisional, partial decoding of the Voynich script, https://www.youtube.com/watch?v=fpZD_3D8_WQ.

Bennett, William Ralph. *Scientific and Engineering Problem Solving with the Computer* (Englewood Cliffs, NJ: Prentice-Hall, 1976). [Contains a chapter on VMS.]

Bird, J. Malcom. "The Roger Bacon Manuscript: Investigation into its History, and the Efforts to Decipher It." *Scientific American Monthly* 3 (June 1921): 492–96.

Black, John. *Codex Gigas* (the Devil's Bible), the largest manuscript in the world, http://www.ancient-origins.net/myths-legends-europe/codex-gigas-devil-s-bible-largest-manuscript-world-001276, January 27, 2014.

Blunt, Wilfrid, and Sandra Raphael. *The Illustrated Herbal* (London: Thames and Hudson, in association with the Metropolitan Museum of Art, 1979). [Provides two-color reproductions of folios from the Voynich manuscript, along with many other pictures from medieval herbals, allowing comparison.]

Bolton, Henry Carrington. *The Follies of Science at the Court of Rudolph II 1576–1612* (Milwaukee: Pharmaceutical Review Publishing Co., 1904), available online at https://archive.org/details/folliessciencea00boltgoog.

Boston Transcript. Review of Newbold's Cipher of Roger Bacon, June 30, 1928, p. 2.

Brooke, Tucker. "Doctor Mirabilis." *Yale Review* 19 (1929): 207–8. [Review of Newbold.]

Brumbaugh, Robert S. "Botany and the Voynich 'Roger Bacon' Manuscript Once More." *Speculum* 49 (1974): 546–48.

Brumbaugh, Robert S. "The Solution of the Voynich 'Roger Bacon' Cipher." *Yale Library Gazette* 49, no. 4 (April 1975): 347–55.

Brumbaugh, Robert S. "The Voynich 'Roger Bacon' Cipher Manuscript: Deciphered Maps of Stars." *Journal of the Warburg and Courtauld Institutes* 39 (1976): 139–50.

Brumbaugh, Robert S., ed. and contrib. *The Most Mysterious Manuscript.* (Carbondale, IL: Southern Illinois University Press, 1978, and London: Weidenfeld and Nicholson, 1977).

Bühler, Markus. *Der Fährtenleser 4: Das Voynich-Manuskript: Ein kryptobotanisches Rätsel* (German ed.), Feb. 4, 2009. This is a magazine for cryptozoology. The Voynich ms qualifies for inclusion as the supposed nonexistence of the plants it depicts makes it a case of cryptobotany, a related field.

Bull. Art Inst. Chicago IX (1915): 100.

Carton, Raoul. "Le Chiffre de Roger Bacon." *Révue d'Histoire de la Philosophie* 3 (1929): 31–66, 165–79.

Casanova, Antoine. *Méthodes d'analyse du langage crypte: une contribution a l'étude du manuscrit de Voynich: These pour obtenir le grade de Docteur de l'Universite Paris 8*, March 19, 1999.

Child, James R. "The Voynich Manuscript Revisited." *NSA Technical Journal* 21, no. 3 (Summer 1976): 1–4, available online at https://www.nsa.gov/public_info/_files/tech_journals/voynich_manuscript_revisited.pdf, p. 1.

Child, James R. "Again, The Voynich Manuscript," 2007, available online at http://web.archive.org/web/20090616205410/http://voynichmanuscript.net/voynichpaper.pdf.

Clabby, John F. *Brigadier John Tiltman: A Giant among Cryptanalysts* (Fort Meade, MD: Center for Cryptologic History, National Security Agency, 2007).

Clark, Ronald W. *The Man Who Broke Purple: The Life of Colonel William F. Friedman, Who Deciphered the Japanese Code in World War II* (Boston: Little Brown & Co., 1977).

Cons, Louis. "Un manuscrit mystérieux: Un traité scientifique du treizième siècle, attribué à Roger Bacon." *L'Illustration* 159, no. 4118 (February 4, 1922): 112.

Cons, Louis. "Newbold's Trail." *Saturday Review of Literature* 5 (October 27, 1928): 292.

Corrales, Scot. "The Books of the Damned: Fact or Fiction?" *FATE* (July 2000).

Currier, Captain Prescott H. "Some Important New Statistical Findings." *Proceedings of a Seminar Held on 30th November 1976 in Washington, D.C.*, available online at http://www.ic.unicamp.br/~stolfi/voynich/mirror/gillogly/currier.paper.

Daiger, Michael. "The World's Most Unusual Manuscript." *Occult* (January 1976).

"Devil's Bible." season 1, episode 5 of the National Geographic Channel series *The Truth Behind*, 2008, aired July 30, 2010. [On Codex Gigas].

Dilas, Jonathan. Das Voynich-Manuskript, http://www.jonathan-dilas.de/Texte/voynich.html.

D'Imperio, M. E., ed. *New Research on the Voynich Manuscript: Proceedings of a Seminar* (Washington, DC: Privately printed pamphlet, November 30, 1976).
Partial contents:
James Child. "A Linguistic Approach to the Voynich Text."
Capt. Prescott H. Currier. "Some Important New Statistical Findings."
Dr. Sydney Fairbanks. "Suggestions Toward a Decipherment of the 'Key.' "
M. E. D'Imperio. "The Solution Claim of Dr. Robert S. Brumbaugh."

Capt. Prescott Currier. "Further Details of New Statistical Findings."

Capt. Prescott Currier. "The Voynich Manuscript, Some Notes and Observations."

Stuart H. Buck. "What Constitutes Proof?"

D'Imperio, M. E. *The Voynich Manuscript—An Elegant Enigma* (Fort Meade, MD: National Security Agency, 1976). Reprinted by Aegean Park Press, Laguna Hills, CA, 1978.

D'Imperio, M. E. "The Voynich Manuscript: A Scholarly Mystery." *Manuscripts* 29, no. 2 (Spring 1977): 85–93 and vol. 29, no. 3 (Summer 1977): 161–73 and vol. 30, no. 1 (Winter 1978): 34–48. [Three-part article. Parts about physical history of the manuscript, about attempts at decipherment, and about Brumbaugh's, Currier's, and Child's work, respectively.]

D'Imperio, M. E. "An Application of Cluster Analysis and Multidimensional Scaling to the Question of 'Hands' and 'Languages' in the Voynich Manuscript." *National Security Agency Technical Journal* 23, no. 3 (Summer 1978): 59–75.

D'Imperio, M. E. "An Application of PTAH to the Voynich Manuscript." *National Security Agency Technical Journal* 24, no. 2 (Spring 1979): 65–91.

Drucker, Johanna. *The Alphabetic Labyrinth* (New York: Thames and Hudson, 1995).

Ephron, H. (Pseud. "DENDAI"). "A burning question in re the Voynich MS (slightly revised)." *The Cryptogram* 43 (March–April 1977): 22, 46–48; (May–June 1977): 49, 51–52, 72.

Evans, R. J. W. *Rudolf II and His World: A Study in Intellectual History 1576–1612* (Oxford, U.K.: Oxford University Press, 1973).

Feely, Joseph M. *Roger Bacon's Cipher: The Right Key Found* (Rochester, NY: self-published, 1943).

Fell-Smith, Charlotte. *John Dee (1527–1608)* (London: Constable & Company Ltd., 1909). Available online at https://archive.org/details/cu31924028928327.

Fischer, Elliot. "Language Redundancy and Cryptanalysis." *Cryptologia* 3, no. 4 (October 1979): 233–35.

Fleischaker, Julia. "Has a botanist solved the mystery of the Voynich Manuscript?" http://www.mhpbooks.com/has-a-botanist-solved-the-mystery-of-the-voynich-manuscript/. [On Dr. Arthur Tucker's "solution."]

Fletcher, John Edward. *A Study of the Life and Works of Athanasius Kircher, 'Germanus Incredibilis'* (Leiden, Netherlands, and Boston: Brill, 2011).

Foster, Caxton C., "A Comparison of Vowel Identification Methods." *Cryptologia* 16, no. 3 (July 1992): 282–86.

Friedman, Elizebeth S. " 'The Most Mysterious Manuscript' Still an Enigma." *The Washington Post,* August 5, 1962, pp. E1, E5.

Friedman, W. F., and E. S. Friedman. "Acrostics, Anagrams, and Chaucer." *Philological Quarterly* 38 (1959): 1–20.

Garland, Herbert. "The Mystery of the Roger Bacon Cipher MS." *Bookman's Journal and Print Collector* 5 (October 1921): 11–16.

Garland, Herbert. "A Literary Puzzle Solved?" *Illustrated London News* 160 (May 20, 1922): 740–42.

Garland, Herbert. "Notes on the Firm of W. M. Voynich." *Library World* 34 (April 1932): 225–28.

Gilson, Étienne. Review of Newbold's *The Cipher of Roger Bacon*. *Révue critique d'histoire et de littérature,* Paris 62 (August 1928): 378–83.

Goldstone, Lawrence, and Nancy Goldstone. *The Friar and the Cipher: Roger Bacon and the Unsolved Mystery of the Most Unusual Manuscript in the World* (New York: Doubleday, 2005).

Grossman, Lev. "When Words Fail: The Struggle to Decipher the World's Most Difficult Book." *Lingua Franca* 9, no. 3 (April 1999): 9–15.

Gutmann, J. *Athanasius Kircher (1602–1680) und das Schöpfungs—und Entwicklungsproblem* (Fulda, Germany: Druck Parzeller & Co., 1938). [This is a doctoral dissertation from the University of Würzburg that discusses Kircher's anticipation of evolution.]

Guy, Jacques. "Voynich Revisited." Letter to the editor. *Cryptologia* 15, no. 2 (April 1991): 161–66.

Guy, Jacques B. M. "Statistical Properties of Two Folios of the Voynich Manuscript." *Cryptologia* 15, no. 3 (July 1991): 207–18. [Applies Sukhotin's vowel-finding algorithm to VMS text.]

Guy, Jacques B. M. "On Levitov's Decipherment of the Voynich Manuscript." December 9, 1991. Available online at http://www.necronomi.com/magic/hermeticism/levitov+voynich.txt.

Guzman, Gregory S. Review of Brumbaugh's "Most Mysterious Manuscript." *Historian* 42 (November 1979): 120–21.

Harnisch, Larry (Pseud. "AR-MYR"). "The Voynich Manuscript." *The Cryptogram* 43 (May–June 1976): 45, 62–63; (July–August 1976): 69, 74–77.

Holzer, Hans. *The Alchemist: The Secret Magical Life of Rudolf von Habsburg* (New York: Stein and Day, 1974).

James, Peter J., and Nick Thorpe. *Ancient Inventions* (New York: Ballantine Books, 1994).

Jay, Mike. "Maze of Madness." *Fortean Times* (*UK*), no. 130 (January 2000).

Johnson, Charles. Review of Newbold's "Cipher of Roger Bacon." *English Historical Review* 44 (October 1929): 677–78.

Joven, Enrique. *The Book of God and Physics: A Novel of the Voynich Mystery* (New York: William Morrow, 2009).

Kahn, David. "The Secret Book." *Newsday*, June 26, 1962.

Kahn, David. *The Codebreakers* (New York: Macmillan, 1967). [VMS discussed on pp. 863–72, 1120–21.]

Kennedy, Gerry, and Rob Churchill. *The Voynich Manuscript: The Unsolved Riddle of an Extraordinary Book Which Has Defied Interpretation for Centuries*, British hardcover ed. (London: Orion Books, 2004).

Kennedy, Gerry, and Rob Churchill. *The Voynich Manuscript: The Mysterious Code That Has Defied Interpretation for Centuries*, American ed. (Rochester, VT: Inner Traditions, 2006). [Note: The pagination differs from the British edition.]

Kent, Roland G. "Deciphers Roger Bacon Manuscripts." *Pennsylvania Gazette* 19 (May 27, 1921): 851–53.

Kircher, Athanasius. *Polygraphia Nova et Universalis ex Combinatoria Arte Detecta* (Rome: Varese, 1663).

Kircher, Athanasius. *Ars Magna Sciendi Sive Combinatoria* (Amsterdam: Johannes Janssonius a Waesberge, 1669).

Knobloch, Eberhard. "Renaissance Combinatorics." *Combinatorics: Ancient and Modern*, Robin Wilson and John J. Watkins, eds. (Oxford, U.K.: Oxford University Press, 2013). [Pp. 123–46 and 135–37 are relevant here for more on Kircher's combinatorial work.]

Knox, Sanka. "700-Year-Old Book for Sale; Contents, In Code, Still Mystery." *New York Times,* July 18, 1962, p. 27.

Knuth, Donald E. "Two Thousand Years of Combinatorics." *Combinatorics: Ancient and Modern*, Robin Wilson and John J. Watkins, eds. (Oxford, U.K.: Oxford University Press, 2013), 7–37.

Kraus, H. P. *Catalogue 100. Thirty-Five Manuscripts: Including the St. Blasien Psalter, the Llangattock Hours, the Gotha Missal, the Roger Bacon (Voynich) Cipher Ms.* (New York: H.P. Kraus, 1962).

Kraus, H. P. *A Rare Book Saga* (New York: G. P. Putnam's Sons, 1978).

Kullback, Solomon. *Statistical Methods in Cryptanalysis* (Laguna, CA: Aegean Park Press, 1976).

Landini, G., and R. Zandbergen. "A Well-kept Secret of Medieval Science: The Voynich Manuscript." *Aesculapius*, no. 18 (July 1998): 77–82.

Landmann, E. "Das sogenannte Voynich-Manuskript." *Magazin 2000plus*, Alte Kulturen Spezial, 2007.

"Law Report, May 10. High Court of Justice. King's Bench Division. Cathedral Library Thefts: Old Volume Traced. The Dean and Chapter of the Cathedral Church of Lincoln v. Voynich." *The Times*, London, May 11, 1916, p. 4.

Levitov, Leo. *Solution of the Voynich Manuscript: A Liturgical Manual for the Endura Rite of the Cathari Heresy, the Cult of Isis* (Laguna Hills, CA: Aegean Park Press, 1987).

Liebert, Herman W. "The Beinecke Library Accessions 1969." *Yale University Library Gazette* 44 (April 1970). [Describes Kraus's gift of VMS on pp. 192–93.]

Loeser, R. "Roger Bacon's Chiffremanuskript." *Die Umschau* 26 (1922): 115–17.

"Lovecraft and the Voynich Manuscript." *INFO Journal*, no. 48 (ca. 1984). [The International Fortean Organization's *INFO Journal*.]

Manly, John M. "Roger Bacon's Cypher Manuscript." *American Review of Reviews* 64 (July 1921): 105–6.

Manly, John M. "The Most Mysterious Manuscript in the World." *Harper's Monthly Magazine* 143 (July 1921): 186–97.

Manly, John M. "Roger Bacon and the Voynich Manuscript." *Speculum* 6 (July 1931): 345–91.

Marshall, Peter. *The Magic Circle of Rudolf II: Alchemy and Astrology in Renaissance Prague* (New York: Walker & Company, 2006).

McKaig, Betty. "The Voynich Manuscript—Cipher of the Secret Book." (Interview with Leonell Strong). *North County Independent*, Oct. 7, 1970. Reprinted courtesy Independent Newspapers, Inc., San Diego.

McKenna, Terence K. "Has the World's Most Mysterious Manuscript Been Read at Last?" *Gnosis Magazine* 7 (Summer 1988): 48–51.

McKenna, Terence K. *The Archaic Revival: Speculations on Psychedelic Mushrooms, the Amazon, Virtual Reality, UFOs, Evolution, Shamanism, the Rebirth of the Goddess, and the End of History* (San Francisco: HarperSanFrancisco, 1991).

McKeon, Richard. "Roger Bacon." *The Nation* 127 (August 29, 1928): 205–6.

Moses, Montrose J. "A Cinderella on Parchment: The Romance of the New 600 Year-Old Bacon Manuscript." *Hearst's International* (1921): 16–17, 75.

Nadis, Steve. "Look Who's Talking." *New Scientist* 179, no. 2403 (July 12, 2003).

New York Times. March 26, 1921, p. 6; March 27, 1921, sec 2, p. 1; April 21, 1921, p. 3; April 22, 1921, p. 13. [All on Newbold's findings.]

New York Times. "Roger Bacon's Formula Yields Copper Salts, Proving Newbold Secret Cipher Translation." December 2, 1926, p. 5, with follow-up articles December 3, p. 22, and December 12, sec. 20, p. 12.

New York Times. "Will Orders Sale of Bacon Cipher." April 15, 1930, p. 40.

Newbold, William R. "The Voynich Roger Bacon Manuscript." *Transactions of the College of Physicians and Surgeons of Philadelphia* (1921) 431–74. Read April 20, 1921.

Newbold, William R. "The Eagle and the Basket on the Chalice of Antioch." *American Journal of Archaeology* 29, no. 4 (October–December 1925): 357–80. [A paper on the Holy Grail.]

Newbold, William Romaine. *The Cipher of Roger Bacon,* edited with foreword and notes by Roland Grubb Kent. (Philadelphia: University of Pennsylvania Press; London, H. Milford, Oxford University Press, 1928).

Newsom, Eugene. *A Split in the Mystery Curtain,* 20-p pamphlet held at Central Arkansas Library System, Butler Center for Arkansas Studies, Little Rock, AR, 1994.

O'Neill, Hugh. "Botanical Remarks on the Voynich MS." *Speculum* 19 (1944): 126.

Orioli, Giuseppe. *Adventures of a Bookseller* (New York: Robert M. McBride & Co., 1938).

Pelling, Nick. *The Curse of the Voynich: The Secret History of the World's Most Mysterious Manuscript* (Surbiton, Surrey, U.K.: Compelling Press, October 2006).

Pesic, Peter. "François Viète, Father of Modern Cryptanalysis—Two New Manuscripts." *Cryptologia* 21, no. 1 (January 1997). [Viète may be the father of vowel recognition algorithms.]

Pollak, Michael. "Can't Read It? You Can Look at the Pictures." *New York Times,* September 16, 1999, p. G11. Available online at https://partners.nytimes.com/library/tech/99/09/circuits/articles/16voyn.html.

Poundstone, William. *Labyrinths of Reason* (New York: Doubleday, 1988).

Pratt, Fletcher. *Secret and Urgent* (New York: Bobbs Merrill, 1939). [Discussion of VMS on pp. 30–38.]

Prinke, Rafał T. in "Did John Dee *Really* Sell the Voynich MS to Rudolf II?" available online at http://main2.amu.edu.pl/~rafalp/WWW/HERM/VMS/dee.htm.

Publishers Weekly. "Kraus Marks Anniversary with Catalog of Treasures." 181 (June 25, 1962): 39–40.

Reddy, Sravana, and Kevin Knight. "What We Know About the Voynich Manuscript." *Proceedings of the 5th ACL–HLT Workshop on Language Technology for Cultural Heritage, Social Sciences, and Humanities,* Portland, OR (Madison, WI: Omnipress, Inc., 2011), 78–86. Available online at http://www.aclweb.org/anthology/W11-1511 and https://www.aclweb.org/anthology/W/W11/W11-15.pdf.

Reeds, James. "Entropy Calculations and Particular Methods of Cryptanalysis." *Cryptologia* 1, no. 3 (July 1977): 235–54.

Reeds, Jim. "William F. Friedman's Transcription of the Voynich Manuscript." *Cryptologia* 19, no. 1 (January 1995): 1–23.

Review of "The Cipher of Roger Bacon (Newbold)." *Quarterly Review of Biology* 3 (December 1928): 595–96.

Ricci, Seymour de. *Census of Medieval and Renaissance Mss in the United States and Canada,* 2 vols. (1937) (Kraus reprint, 1961) [VMS: vol. 2, pp. 1845–47.]

Roberts, R. J., and Andrew G. Watson, eds. *John Dee's Library Catalogue* (London: The Bibliographical Society, 1990). [Claims the folio numbers in the VMS are in John Dee's hand.]

Roitzsch, E. H. Peter. *Das Voynich-Manuskript: ein ungelöstes Rätsel der Vergangenheit* (Münster, Germany: MV-Verlag, 2008). There's also a second edition from 2010.

Rugg, Gordon. "An Elegant Hoax? A Possible Solution to the Voynich Manuscript." *Cryptologia* 28, no. 1 (January 2004): 31–46.

Rugg, Gordon. "The Mystery of the Voynich Manuscript." *Scientific American* (July 2004): 104–9.

Rugg, Gordon. *Blind Spot: Why We Fail to See the Solution Right in Front of Us* (New York: HarperCollins, 2013).

Ruysschaert, Jose. *Codices Vaticani Latini 11414–11709* (Rome: Biblioteca Apostolica Vaticana, 1959). [Describes the MSs acquired by the Vatican from the Collegium Romanum and mentions that

W. Voynich bought a number of them, which have been transferred to various American libraries, including the VMS.]

Salomon, Richard. Review of Manly's Critique of Newbold's Decipherment, *Bibliotek Warburg, Kulturwissenschaftliche Bibliographie zum Nachleben der Antike* 1 (1934): 96.

Sarton, George. Review of Manly's Critique of Newbold's Decipherment, in "Eleventh Critical Bibliography of the History and Philosophy of Science and of the History of Civilization. (To October 1921)." *Isis* 4, no. 2 (October 1921): 390–453, p. 404 relevant here.

Sarton, George. Review of Newbold's *Cipher of Roger Bacon, Isis* 11 (1928): 141–45.

Sassoon, George T. "The Application of Sukhotin's Algorithm to Certain Non-English Languages." *Cryptologia* 16, no. 2 (April 1992): 165–73.

Schaefer, Bradley E. "The Most Mysterious Astronomical Manuscript: Baffled Researchers Are Looking for Astronomical Clues to Help Decipher a Medieval Manuscript." *Sky & Telescope* 100, no. 5 (November 2000).

Schinner, Andreas. "The Voynich Manuscript: Evidence of the Hoax Hypothesis." *Cryptologia* 31, no. 2 (April 2007): 95–107.

Schuster, John. *Haunting Museums* (New York: Tom Doherty Associates, 2009).

Sebastian, Wencelas. "The Voynich Manuscript; Its History and Cipher." *Nos Cahiers,* 2 (1937): 47–69 (Montréal: Studium Franciscain).

Serafini, Luigi. *Codex Seraphinianus* (Milan, Italy: Franco Maria Ricci, 1981) (in two volumes). [There have also been later editions, at least one of which contains some new material.]

Seymour, Ian. "Thirteenth Century Magic Glass." *Astronomy Now* (June 1992): 59.

Shepherdson, Nancy. "Mystery Codes." *Boys' Life* 87, no. 11 (November 1997): 42.

Shuker, Karl P. N. *The Unexplained: An Illustrated Guide to the World's Natural and Paranormal Mysteries* (London: Carlton Books Ltd., 1996).

Smolka, J., and R. Zandbergen, eds. "Athanasius Kircher und seine erste Prager Korrespondenz." *Bohemia Jesuitica 1556–2006, Festschrift zum 450. Jahrestag des Ordens in Böhmen* (Prague, Czech Republic: V. Praze, 2010), 677–705.

Smyth, Frank. "A Script Full of Secrets." *The Unexplained* 6, no. 70 (1982): 1381–85.

Smyth, Frank. "The Uncrackable Code." *The Unexplained* 6, no. 71 (1982): 1418–20.

Smyth, Frank. "A Script Full of Secrets" and "The Uncrackable Code," reprinted in *Mysteries of Mind, Space & Time: The Unexplained* (Westport, CT: H. S. Stuttman, Inc., 1992), pp. 3062–69. [Originally published in *The Unexplained* in the U.K.]

Sowerby, E. Millicent. *Rare People and Rare Books* (London: Constable & Co. Ltd., 1967).

Stallings, Dennis, J. "Catharism, Levitov, and the Voynich Manuscript." October 10, 1998, available online at http://www.bibliotecapleyades.net/ciencia/esp_ciencia_manuscrito04.htm.

Steele, Robert. "Luru Vopo Vir Can Utriet." *Nature* 121 (February 11, 1928): 208–9. [About Bacon "gunpowder cipher," not VMS.]

Steele, Robert. "Science in Medieval Cipher." *Nature* 122 (October 13, 1928): 563–65.

Stojko, John. *Letters to God's Eye: The Voynich Manuscript for the First Time Deciphered and Translated into English* (New York: Vantage Press, 1978).

Stokley, James. "Did Roger Bacon Have a Telescope?" *Science News Letter* 14 (September 1, 1928): 125–26, 133–34.

Strong, Leonell C. "Anthony Askham, the Author of the Voynich Manuscript." *Science*, New Series 101, no. 2633 (June 15, 1945): 608–9.

Strong, Leonell C., and E. L. McCawley. "A Verification of a Hitherto Unknown Prescription of the 16th Century." *Bulletin of the History of Medicine* (Baltimore, Md.) 21 (November–December 1947): 898–904.

Sypher, F. J. *Eric Sams, Cryptography and the Voynich Manuscript*, a pamphlet (New York, 2011), available online at http://www.ericsams.org/index.php/a-portrait-of-f-j-sypher/278-eric-sams-cryptography-and-the-voynich-manuscript.

Taratuta, Evgeniya. *Our Friend Ethel Lilian Boole/Voynich* (Moscow: Izdatel'stvo, 1957, translated from the Russian by Séamus Ó Coigligh with additional notes, 2008), available online at http://www.corkcitylibraries.ie/media/SOCoiglighwebversion171.pdf.

Taylor, Frances Grandy. "The Mystery of Manuscript 408." *The Hartford Courant,* October 12, 1999, available online at http://articles.courant.com/1999-10-12/features/9910120627_1_manuscript-library-rare-documents-mystery.

Theroux, Michael. "Deciphering 'The Most Mysterious Manuscript in the World' The Final Word?" *Borderlands* 50 (1994): 36–43.

Thorndike, Lynn. "The "Bacon" Manuscript," letter to *Scientific American* 124 (June 25, 1921): 509.

Thorndike, Lynn. "Review of Newbold's Cipher of Roger Bacon." *American Historical Review* 34 (1929): 317–19.

Tiltman, John. "The Voynich Manuscript, 'The Most Mysterious Manuscript in the World.'" *NSA Technical Journal* 12 (July 1967): 41–85.

Times [Newspaper of London]. "Mr. W. M. Voynich." March 22, 1930, p. 17; March 25, 1930, p. 21; March 26, 1930, p. 18. [Obituary.]

Toresella, Sergio. "Gli erbari degli alchimisti." [Alchemical herbals.] *Arte farmaceutica e piante medicinali—erbari, vasi, strumenti e testi dalle raccolte liguri* [Pharmaceutical arts and medicinal plants—herbals, jars, instruments and texts of the Ligurian collections.] Liana Saginati, ed. (Pisa, Italy: Pacini Editore, 1996), 31–70. [Profusely illustrated. Fits the VMS into an "alchemical herbal" tradition.]

Von Schleinitz, Otto. "Die Bibliophilen W. M. Voynich." *Zeitschrift für Bücherfreunde* 10 (1906–1907): 481–87. [Contains information about Voynich's life.]

Voynich, Wilfrid M. "A Preliminary Sketch of the History of the Roger Bacon Cipher Manuscript." *Transactions of the College of Physicians and Surgeons of Philadelphia* (1921): 415–30. Read April 20, 1921. Available online at https://archive.org/stream/s3transactionsstud43coll/s3transactionsstud43coll_djvu.txt.

"Voynich Manuscript. Botanical Clue, Evidence indicating Roger Bacon could not have written the Voynich manuscript ..." *Science News Letter* (July 29, 1944): 69.

"Voynich Manuscript Translated." *INFO Journal* no. 56 (ca. 1988). [The International Fortean Organization's *INFO Journal.*]

Way, Peter. *Codes and Ciphers* (London: Aldus, 1974).

Weekly World News [newspaper]. March 7, 2000, 4–7. [Interview with Mike Jay, author of VMS article in *Fortean Times.*]

Werner, Alfred. "The Most Mysterious Manuscript." *Horizon* 5 (January 1963): 4–9.

Westacott, Evalyn. *Roger Bacon in Life and Legend* (New York: Philosophical Library, 1953).

Wickware, Francis Sill. "The Secret Language of War." *Life* 19 (November 26, 1945): 63–70. [Only one sentence mentions the VMS, calling it "possibly the only unbreakable code," which provoked Strong to write an angry letter to the editor.]

Williams, Robert L. "A Note on the Voynich Manuscript." *Cryptologia* 23, no. 4 (October 1999): 305–9.

Wilson, Colin. *The Encyclopedia of Unsolved Mysteries* (Chicago: Contemporary Books, 1988).

"The World's Most Baffling Manuscript." *Parade Magazine*, February 21, 1982.

Wrixon, Fred B. *Codes, Ciphers, and Other Cryptic and Clandestine Communication* (New York: Black Dog and Leventhal, 1998).

Zandonella, Catherine. "Book of riddles." *New Scientist* 172, no. 2317 (November 17, 2001): 36–39.

Zimansky, Curt A. "William F. Friedman and the Voynich Manuscript." *Philological Quarterly* 49 (1970): 433–42.

WEBSITES

Anonymous. http://hurontaria.baf.cz/CVM/ao.htm. [Many articles on the Voynich manuscript are linked from this page.]

Landini, G., and René Zandbergen. The European Voynich Manuscript Transcription Project Home Page, http://web.bham.ac.uk/G.Landini/evmt/evmt.htm.

Prinke, Rafał T. Some facts, thoughts and speculations related to the Voynich manuscript, http://main2.amu.edu.pl/~rafalp/WWW/HERM/VMS/vms.htm.

Voynich Manuscript Mailing List HQ, www.voynich.net. [This is the website for the mailing list begun by Jim Gillogly and Jim Reeds in 1991.]

Yale University. Beinecke Rare Book and Manuscript Library, General Collection of Rare Books and Manuscripts, Medieval and Renaissance Manuscripts, http://brbl-net.library.yale.edu/pre1600ms/docs/pre1600.ms408.htm. [This is Yale's website for the Voynich manuscript.]

Zandbergen, René. The Voynich Manuscript, www.voynich.nu.

Chapter 2

ON ANCIENT EGYPT

Black, Jeremy, Graham Cunningham, Eleanor Robson, and Gábor Zólyomi. *The Literature of Ancient Sumer* (Oxford, U.K.: Oxford University Press, 2004). The introduction by Robson is relevant here.

Bohleke, Briant. "Amenemopet Panehsi, Direct Successor of the Chief Treasurer Maya." *Journal of the American Research Center in Egypt* 39 (2002): 157–72.

Bosticco, S. "Scarabei egiziani della Necropoli die Pithecusa nell'Isola di Ischia." *La Parola del Passato* 54 (1957): 215–29.

Charles, Robert P. "Les scarabées Égyptiens et Égyptisants de Pyrga district de Larnaca (Chypre)." *ASAE* 58 (1964): 3–36.

Charles, Robert P. "Remarques sur une maxime religieuse à propos d'un scarabée égyptien à Kyrenia (Chypre)." In *Melanges de Carthage offerts à Charles Saumagne* (Louis Poinssot, Maurice Pinard, Paris, Libraire Orientaliste, Paul Geuthner 1964–1965), 11–20.

Darnell, John Coleman. *The Enigmatic Netherworld Books of the Solar-Osirian Unity, Cryptographic Compositions in the Tombs of Tutankhamun, Ramesses VI and Ramesses IX* (Fribourg, Germany:

Academic Press and Göttingen, Germany: Vandenhoek & Ruprecht, 2004). This book is a slight reworking of Darnell's doctoral dissertation, which is available through University Microfilms International.

Deveria, T. "L'écriture secrète dans les textes hiéroglyphiques des anciens Égyptiens." *Bibliothèque Égyptologique* 5 (1897): 49–90.

Drioton, Étienne. "Essai sur la cryptographie privée de la fin de la XVIIIe dynastie." *Revue d'Égyptologie* I (1933): 1–50.

Drioton, Étienne. "Une figuration cryptographique sur une stèle du Moyen Empire." *Revue d'Égyptologie* I (1933): 203–29.

Drioton, Étienne. "La cryptographie égyptienne." *Revue Lorraine d'Anthropologie* VI (1933–1934): 5–28.

Drioton, Étienne. "La cryptographie égyptienne." *Chronique d'Egypte* IX, (1934): 192–206.

Drioton, Étienne. "Les jeux d'écriture et les rébus de l'Egypte antique." *Rayon d'Egypte* VIII (1935): 173–75.

Drioton, Étienne. "Notes sur le cryptogramme de Montouemhêt." *Université libre—de Bruxelles, Annuaire de l'Institut de philologie et d'histoire orientales* III. Volume offert à Jean Capart, Bruxelles, 1935, 133–40.

Drioton, Étienne. "Les protocoles ornementaux d'Abydos." *Revue d'Égyptologie* II (1936): 1–20.

Drioton, Étienne. "Le cryptogramme de Montou de Médamoud." *Revue d'Égyptologie* II (1936): 22–33.

Drioton, Étienne. "Un rébus de l'Ancien Empire." *Mémoires de l'Institut français d'Archéologie orientale du Caire* LXVI (Cairo, Egypt: Mélanges Gaston Maspero, I, 1935–1938): 697–704.

Drioton, Étienne. "Note sur un cryptogramme récemment découvert à Athribis." *Les Annales du Service des Antiquités de l'Egypte* XXXVIII (1938): 109–16.

Drioton, Étienne. "Deux cryptogrammes de Senenmout." *Les Annales du Service des Antiquités de l'Egypte* XXXVIII (1938): 231–46.

Drioton, Étienne. "Senenmout cryptographe." *Atti del XIX Congresso internazionale degli Orientalisti* September 23–29, 1935, XIII (Rome, Tipografia del Senato, G. Bardi, 1938), 132–38.

Drioton, Étienne. "Cryptogrammes de la reine Nefertari." *Les Annales du Service des Antiquités de l'Egypte* XXXIX (1939): 133–44.

Drioton, Étienne. "Recueil de cryptographie monumentale." *Les Annales du Service des Antiquités de l'Egypte* XL (1940): 305–427.

Drioton, Étienne. *Recueil de cryptographie monumentale* (Cairo, Egypt: L'institut français d'archéologie orientale, 1940). This book is available online at http://www.cfeetk.cnrs.fr/fichiers/Documents /Ressources-PDF/documents/K1234-DRIOTON.pdf.

Drioton, Étienne. "L'écriture énigmatique du *Livre du Jour et de la Nuit*." A. Piankoff, *Le Livre du Jour et de La Nuit* (Cairo, Egypt: L'institut français d'archéologie orientale, 1942), 83–121.

Drioton, Étienne. "La cryptographie du Papyrus Sait 825." *Les Annales du Service des Antiquités de l'Egypte* XLI (1942): 199–234.

Drioton, Étienne. "A propos du cryptogramme de Montouemhêt." *Les Annales du Service des Antiquités de l'Egypte* XLII (1943): 177–81.

Drioton, Étienne. "Procédé acrophonique ou principe consonantal?" *Les Annales du Service des Antiquités de l'Egypte* XLIII (1943): 319–49.

Drioton, Étienne. "La cryptographie par perturbation." *Les Annales du Service des Antiquités de l'Egypte* XLVI (1944): 17–33.

Drioton, Étienne. "Notes diverses. 9. Le cynocéphale et l'écriture du nom de Thot. — 10. Chawabtiou à inscriptions cryptographiques." *Les Annales du Service des Antiquités de l'Egypte* XLV (1945): 17–29.

Drioton, Étienne. "Plaques bilingues de Ptolémée IV." *Discovery of the Famous Temple and Enclosure of Serapis at Alexandria*, Alan Rowe, *Supplément aux Annales du Service des Antiquités de l'Egypte* no. 2, (1946), 97–112.

Drioton, Étienne. "La cryptographie de la chapelle de Toutânkhamon." *The Journal of Egyptian Archaeology* XXXV (1949): 117–22.

Drioton, Étienne. "Inscription énigmatique du tombeau de Chéchanq III à Tanis." *Kêmi* XII (1952): 24–33.

Drioton, Étienne. "Les principes de la cryptographie égyptienne." *Comptes rendus des séances de l'Académie des Inscriptions et Belles-Lettres* 97e année, no. 3 (1953): 355–64, available online at http://www.persee.fr/web/revues/home/prescript/article/crai_0065-0536_1953_num_97_3_10159.

Fairman, H. W. "Notes on the Alphabetic Signs Employed in the Hieroglyphic Inscriptions in the Temple of Edfu." *ASAE* 43 (1943): 191–310.

Fairman, H. W. "An Introduction to the Study of Ptolemaic Signs and their Values." *BIFAO* 43 (1945): 52–138.

Graves-Brown, Carolyn. "A Mummified Baby?" Egypt Centre, Swansea, May 1, 2014, available online at http://egyptcentre.blogspot.co.uk/2014/05/a-mummified-baby.html. This is the most scholarly account available. Graves-Brown is the Egypt Centre curator.

Halévy, Joseph. "Observations critiques sur les prétendus Touraniens de la Babylonie." *Journal Asiatique* 3 (1874): 461–536. This puts forth the incorrect argument that Sumerian was a form of secret writing used by priests.

Hornung, E., and E. Staehelin. *Skarabäen und andere Siegelamulette aus Basler Sammlungen* (Mainz, Germany: P. von Zabern, 1976), 173–80. [For information on cryptographic scarabs and seals.]

Hoving, Thomas. *False Impression: The Hunt for Big-Time Art Fakes* (New York: Simon & Schuster, 1996).

Junker, H. *Über das Schriftsystem im Tempel der Hathor in Dendera* (Berlin: August Schaefer, 1903).

Jurman, Claus. "Ein Siegelring mit kryptographischer Inschrift in Bonn." *Ägypten und Levante XX/ Egypt and the Levant* 20 (2010): 227–42.

Kahn, David. *The Codebreakers* (New York: Macmillan, 1967). This classic book on the history of cryptology includes not only a couple of pages on ancient Egyptian cryptology, but also an accessible account of how Egyptian hieroglyphs came to be deciphered using techniques similar to those of the cryptanalysts.

Leibevitch, J. "Un écho posthume du Chanoine Étienne Drioton." *BSFÉ* 36 (1963): 34–36.

Livius. "Mummified Fetus Found in Egyptian Sarcophagus." The History Blog, http://www.thehistory blog.com/archives/30530.

Lorenzi, Rossella. "Mummified Fetus Found in Tiny Ancient Egyptian Sarcophagus." May 13, 2014, http://news.discovery.com/history/archaeology/mummified-fetus-found-in-tiny-ancient -egyptian-sarcophagus-140512-140512.htm.

Morenz, Ludwig D. "Tomb Inscriptions: The Case of the I Versus Autobiography in Ancient Egypt." *Human Affairs* 13 (2003): 179–96. On pp. 190–91, Morenz wrote

A stela from Naga ed Deir provides an interesting example of illiterate imitation of writing (Un of Cal N 3993). The execution of the monument is rather crude. Tile proportions of the sitting man are not well done. As in high culture the picture is provided with a framing line. On top of it we see a line with signs imitating hieroglyphs. They are probably inspired by the offering-formula. On this stelae writing is imitated. Unfortunately we know nothing more about this man. He remains even nameless. On the other hand he becomes a representative of a social group we know next to nothing about except ceramics. It is quite important to see that this stelae imitates the monuments in the style of high culture. It suggests a broader acceptance and imitation of this code of the elite. From Gebelein we have no such examples, but Un of Cal. (6-19911) comes fairly close. The layout of the inscription also seems rather confused but still is intelligible.

Morenz, Ludwig D. *Sinn und Spiel der Zeichen: Visuelle Poesie im Alten Ägypten* (Cologne, Germany: Böhlau Verlag, 2008).

Petrie, W. M. Flinders. *Illahun Kahun and Gurob. 1889–90* (London: David Nutt, 1891), 27–28, available online at http://www.lib.uchicago.edu/cgi-bin/eos/eos_title.pl?callnum=DT73.I3P5_cop2 and https://archive.org/details/cu31924086199514.

Petrie, W. M. Flinders. *Ten Years' Digging in Egypt 1881–1891*, 2nd ed., rev. (London: The Religious Tract Society, 1893), 124–25.

Pinches, T. "Sumerian or Cryptography." *Journal of the Royal Asiatic Society* 32, no. 1 (January 1900): 75–96. The idea of Sumerian as a form of secret writing only used by priests, and never spoken, is finally rejected definitively.

Russon, Mary-Ann. "Tiny Egyptian Mummy Confirmed by CT Scan to be Baby's Remains." *International Business Times*, May 8, 2014, available online at http://www.ibtimes.co.uk/tiny-egyptian-mummy-confirmed-by-ct-scan-be-babys-remains-1447717.

Sauneron, S. "Le papyrus magique illustré de Brooklyn [Brooklyn Museum 47.218.156]." *Wilbur Monographs* 3 (New York: The Brooklyn Museum, 1970).

Schulman, Alan R. "The Ossimo Scarab Reconsidered." *Journal of the American Research Center in Egypt* 12 (1975): 15–18.

Schulman, Alan Richard. "Two scarab impressions from Tel Michal." *Tel Aviv* 5, no. 3–4 (1978): 148–51. In the acknowledgments section of this paper, Schulman wrote, "I should like to thank Dr. Z. Herzog for reminding me that I had looked at these items 8 years ago and for allowing me to revise what I had written then, since it in retrospect, was nonsense." I like people who can admit this sort of thing!

Woollaston, Victoria. "The mummified FOETUS: Scans reveal tiny ancient Egyptian sarcophagus contains the remains of a 16-week-old embryo," *Daily Mail*, May 9, 2014, available online at http://www.dailymail.co.uk/sciencetech/article-2624136/The-mummified-FOETUS-Scans-reveal-tiny-ancient-Egyptian-sarcophagus-contains-remains-16-week-old-baby.html.

ON ANCIENT GREECE

Bakker, Egbert J. *A companion to the Ancient Greek Language* (Chichester, U.K.: Wiley-Blackwell, 2010).

Boardman, John. " 'Reading' Greek Vases?" *Oxford Journal of Archaeology* 22, no. 1 (2003): 109–14. This paper argues that the inscriptions on the vases were meant to be read aloud. But how to read gibberish? Boardman's explanation is this:

> As for the many nonsense inscriptions on vases; were these simply an invitation to improvisation on the part of the owner displaying them, a form of Hellenic *karaoke*? All seem to be either decorative or making a false pretense to literacy. They are part of the look of the vase, nothing to do with reading.

Clark, A. J., Maya Elston, and Mary Louise Hart. *Understanding Greek Vases: A Guide to Terms, Styles, and Techniques* (Los Angeles: Getty Publications, 2002).

Clifford, Kathleen Elizabeth. "Lingering Words: A Study of Ancient Greek Inscriptions on Attic Vases." (master's thesis, Florida State University Electronic Theses, Treatises and Dissertations, Paper 3580, 2007). This thesis is a very nice (gentle) introduction to the topic of inscriptions on Attic vases. It can be read and enjoyed by a layperson. I suggest that any nonspecialist wishing to delve deeper begin here, even though the possibility of encryption is not addressed.

Cook, R. M. *Greek Painted Pottery* (London: Routledge, 1997).

Gardthausen, Viktor Emil. *Griechische Palaeographie* (Leipzig, Germany: Veit, 1913).

Grammenos, Dimitris. "Abba-Dabba-Ooga-Booga-Hoojee-Goojee-Yabba-Dabba-Doo: Stupidity, Ignorance & Nonsense as Tools for Nurturing Creative Thinking." *Proceedings CHI EA '14, CHI '14, Extended Abstracts on Human Factors in Computing Systems* (New York: ACM, 2014), 695–706, available online at https://www.ics.forth.gr/_publications/dgrammenos_sin_FINAL.pdf. This paper includes an image of the Memnon pieta as an example of a nonsense inscription. The author wrote:

> In Ancient Greek pottery inscriptions are sometimes seemingly meaningless combinations of letters. Up to now, the prevailing explanation was that these were made by illiterate vase-painters either to imitate the decorative effect of literate inscriptions or, to give the impression that they were literate. But, "nonsense inscriptions" often coexist with others that do make sense. Furthermore, there are too many of them (about 1/3 of vases in the Corpus of Attic Vase Inscriptions). Recently, Mayor et al. came to a groundbreaking conclusion. There is evidence that (at least some of them) constitute names and words of "barbarian" tongues transliterated into Greek. Sometimes experts label what they cannot understand as nonsense, while in reality the distance between nonsense and sense is just a matter of standpoint. After all, our everyday lives are full of nonsense and there is so much of it that we rarely even notice. For example, in June 2010, almost half of the earth's population spent at least one minute watching 22 grown-up guys kicking around an inflated piece of plastic (i.e., FIFA World Cup South Africa). In 2006, "no. 5, 1948" a painting by Jackson Pollock showing colored paint drizzles was sold for $140m, the highest sum ever been paid for a painting. And so on.

Hunt, A. S. "A Greek Cryptogram." *Proceedings of the British Academy* 15 (1929): 1–10. This easy-to-read paper presents a Greek ciphertext in which the letters were turned half over or modified in other small ways to disguise the writing. No knowledge of Greek is needed to follow this ten-page paper, as an English translation of the ciphertext is provided. Unfortunately, an approximate date for the ciphertext examined (Michigan cryptographic papyrus) is not given.

Hyman, Malcolm D. "Of Glyphs and Glottography." *Language & Communication* 26 (2006): 231–49.

Immerwahr, Henry R. *Attic Script: A Survey*, Oxford Monographs on Classical Archaeology (Oxford, U.K.: Clarendon Press, 1990).

Immerwahr, Henry R. "Observations on Writing Practices in the Athenian Cera-micus." *Studies in Greek Epigraphy and History in Honor of Stephen V. Tracy* (Ausonius Éditions Études 26), eds. G. Reger, F. X. Ryan, and T. F. Winters (Pessac, France: Ausonius Éditions, 2010), 107–22.

Immerwahr, Henry R. Corpus of Attic Vase Inscriptions, http://www2.lib.unc.edu/dc/attic/about .html.

This website contains the following quotes:

> I should add at this point that we are dealing with a total of over 100,000 known Attic vases (a rough guess), of which (again roughly) one in ten has some inscription. There are certainly over 10,000 inscribed Attic vases. Rudolf Wachter, who is working on a project based on this corpus, conjectures that the number is about 4,000; my guess is that he did not include nonsense inscriptions and perhaps not graffiti made by users (as against in-scriptions written in the workshop). Four thousand is probably about right for meaningful decorative inscriptions. In fact, Sir John Beazley, in 1947, advised me not to undertake this collection because I would not finish it. He was right, in the sense that my collection is not complete, although with well over 8,000 items it is extensive.

> I made the decision that the so-called nonsense inscriptions (of which there are thou-sands) had to be included, for the corpus was to be of service not only to linguists but also to archaeologists, and the nonsense inscriptions are a part of the ornamentation of the vase, which varies from workshop to workshop.

Note his reluctance to include them—it sounds like he'd really rather not.

> The next form this project takes will be Wachter's *AVI* (note that it is no longer called a corpus), and it will also be in electronic form which will make it possible to update it con-tinuously. This is especially important for vases, for new material is constantly produced by excavation and tomb robbery.

"Inscriptions," University of Oxford, Classical Art Research Centre and the Beazley Archive, http://www.beazley.ox.ac.uk/tools/pottery/inscriptions/.

Leighton, Albert C. "Secret Communication among the Greeks and Romans." *Technology and Culture* 10, no. 2 (April 1969): 139–54.

Lorber, Fritz. *Inschriften auf Korinthischen Vasen. Archaologisch-Epigraphische Untersuchungen zur Korinthischen Vasenmalerei im 7 end 6 JH. V. CHR.*, Deutsches Archäologisches Institut, Archäol-ogische Forschungen 6 (Berlin: Gebr. Mann Verlag, 1979). This catalog includes nonsense inscrip-tions.

Mayor, Adrienne, John Colarusso, and David Saunders. "Making Sense of Nonsense Inscriptions Asso-ciated with Amazons and Scythians on Athenian Vases," Version 2.0, Princeton/Stanford Work-ing Papers in Classics, July 2012.

Mayor, Adrienne, John Colarusso, and David Saunders. "Making Sense of Nonsense Inscriptions Asso-ciated with Amazons and Scythians on Athenian Vases." *Hesperia: The Journal of the American School of Classical Studies at Athens* 83, no. 3 (July–September 2014): 447–93.

Nock, A. D. "A Greek Cryptogram by Arthur S. Hunt." *The Classical Review* 43, no. 6 (December 1929): 238.

Pappas, A. "More than Meets the Eye: The Aesthetics of (Non)sense in the Ancient Greek Symposium." *Aesthetic Value in Classical Antiquity* (*Mnemosyne* Suppl. 305), eds. I. Sluiter and R. M. Rosen (Leiden, Netherlands: Brill, 2012), 71–111. Note: The Symposia were celebratory drinking parties for men. The author gives an odd explanation for the purpose of nonsense inscriptions.

Plutarch's Life of Lysander, sec. 19, available online at http://www.gutenberg.org/files/14114/14114-h /14114-h.htm#LIFE_OF_LYSANDER.

Polybius. *Histories*, Book 10 is relevant here; see sec. 45–47. Available online at http://www.gutenberg .org/files/44126/44126-h/44126-h.htm.

Sironen, Erkki. "Edict of Diocletian and a Theodosian Regulation at Corinth." *Hesperia: The Journal of the American School of Classical Studies at Athens* 61, no. 2 (April–June 1992): 223–26.

Steiner, Ann. *Reading Greek Vases* (Cambridge, U.K.: Cambridge University Press, 2007). Steiner argues that some nonsense inscriptions could be ridiculing the accents of foreigners.

Suess, Wilhelm. "Ueber antike Geheimschreibemethoden und ihr Nachleben." *Philologus* LXXVIII (June 1922): 142–75.

Sulzer, A. B. "Making Sense of Nonsense Inscriptions: Orality, Literacy and the Greek Dipinti on Four Vases in the Arthur M. Sackler Museum" (senior thesis, Harvard University, 2003).

Thomsen, Megan Lynne. "Herakles Iconography on Tyrrhenian Amphorae" (master's thesis, University of Missouri–Columbia, 2005).

Vanderpool, E. "An Unusual Black-Figured Cup." *American Journal of Archaeology* 49, no. 4 (October 1945): 436–40.

Vergano, Dan. "Amazon Warriors' Names Revealed Amid 'Gibberish' on Ancient Greek Vases." *National Geographic*, September 22, 2014, available online at http://news.nationalgeographic.com /news/2014/09/140923-amazon-greek-vase-translations-science/. This brief article presents the work of Mayor et al. in a very entertaining way.

ON THE VIKINGS

I've included some references that deal with controversial decipherments. I'm simply providing the titles; it's up to you to decide whether or not the claims are correct.

Barnes, Michael P. *Runes: A Handbook* (Suffolk, U.K. and Rochester, NY: Boydell Press, 2012). Despite having a chapter devoted to "Cryptic inscriptions and cryptic runes," Barnes is dismissive of the so-called nonsense inscriptions.

Braunmüller, Kurt. *Der Maltstein: Versuch einer Deutung* (Berlin: Walter de Gruyter, 1992).

Eliasson, Stig. " 'The letters make no sense at all . . .': språklig struktur i en 'obegriplig' dansk runinskrift?" *Nya perspektiv inom nordisk språkhistoria. Föredrag hållna vid ett symposium i Uppsala 20–22 januari 2006*, ed. Lennart Elmevik, Acta Academiae Regiae Gustavi Adolphi, 97, 45-80 (Uppsala, Sweden: Kungl. Gustav Adolfs Akademien för svensk folkkultur, 2007). This paper describes the Sørup runestone, Rundata ID DR 187, in Denmark, which Eliasson interprets as possibly representing Basque in his other paper cited here.

Eliasson, Stig. "When 'chance' explanations turn tenuous: Basque verbal morphology, argument marking, and an allegedly nonsensical Danish runic inscription," available online at http:// www.orientalistik.uni-mainz.de/robbeets/verbalmorphv8/_Media/abs-eliasson2.pdf.

Franksen, Ole. *Mr. Babbage's Secret: The Tale of a Cypher and APL* (Englewood Cliffs, NJ: Prentice-Hall, 1985).

Gordon, Cyrus, and Roy Bongartz. "Stone Inscription Found in Tennessee Proves that America was Discovered 1500 Years before Columbus." *Argosy* (January 1971): 23–27. Gordon was a World War II cryptanalyst.

Landsverk, O. G. *Ancient Norse Messages on American Stones* (Glendale, CA: Norseman Press, 1969).

Landsverk, O. G. "Cryptography in runic inscriptions." *Cryptologia* 8, no. 4 (1984): 302–19.

Ljosland, Ragnhild. "Pondering Orkney's runic inscriptions." *The Orcadian*, June 4, 2015, p. 21, available online at http://www.uhi.ac.uk/en/research-enterprise/cultural/centre-for-nordic-studies/mimirs-well-articles/pondering-orkneys-runic-inscriptions. Ljosland wrote:

> New light was also shed on the metal pendant found by James Barrett's excavation team on the Brough of Deerness. The runes on this pendant look very strange, a bit like Christmas trees with wild-growing branches. Could they be in the tree-like cipher code, like some of the Maeshowe inscriptions, which are also found in Anglo-Saxon manuscripts (as shown by Aya Van Renterghem at the conference)? Yet, interpreting the Deerness runes as such didn't bring up any sensible text.
>
> But resulting from a closer study of the Deerness pendant, Sonia Pereswetoff-Morath discovered that the runes are actually complex ligatures made from combinations of ordinary runes. She has not yet interpreted the whole text, but was able to give us a few words, involving talk of a "large payment" or "big secret", and a command to "go away". It was great to hear that the runes are actually meaningful, and not just nonsense-runes as we first thought. It's also exciting to have another runic object whose function seems to have been as an amulet, in addition to the bear's or seal's tooth amulet from the Brough of Birsay, which was also discussed at the conference. Birsay itself was identified in Jan Ragnar Hagland's talk as a centre of runic literacy, along with Orphir.

Meijer, Jan. "Corrections in Viking Age Rune-Stone Inscriptions." *Arkiv för nordisk filologi* 110 (January 1995): 77–83, available online at http://journals.lub.lu.se/ojs/index.php/anf/article/viewFile/11543/10639.

Mongé, A., and O. G. Landsverk. *Norse Medieval Cryptography in Runic Carvings* (Glendale, CA: Norseman Press, 1967). Mongé's skills as a cryptanalyst are discussed briefly in chapter 7.

Page, Raymond Ian. *Runes* (*Reading the Past*) (Berkeley, CA, and Los Angeles: University of California Press/British Museum, 1987). Examples of gibberish appear on p. 29 (the Lindholm "amulet," listed as DR 261 in Rundata) and p. 58 (the Hunterston brooch).

Stevenson, Robert B. K. "The Hunterston Brooch and Its Significance." *Medieval Archaeology* 18 (1974): 16–42. The odd inscription is dismissed in this paper.

Syversen, Earl. *Norse Runic Inscriptions with Their Long-Forgotten Cryptography* (Sebastopol, CA: Vine Hill Press, 1979). This book has the lovely dedication "In memory of my true friend ALF MONGE who solved the enigma of Medieval Norse Runic Cryptography."

Thompson, Claiborne W. "Nonsense Inscriptions in Swedish Uppland." *Studies for Einar Haugen*, ed. E. S. Firchow et al. (The Hague, Netherlands: Mouton, 1972), 522–34.

Whittaker, Helène. "Social and Symbolic Aspects of Minoan Writing." *European Journal of Archaeology* 8, no. 1 (2005): 29–41. This wide-ranging paper talks about the Phaistos disc and the Arkalochori axe, among other challenges, but also includes Scandinavian bracteates bearing unintelligible runic inscriptions. Whittaker concludes, "It seems obvious that these bracteates were made by illiterate goldsmiths for customers who were themselves illiterate and unable to check the genuineness of the writing." I have no idea why this is supposed to seem obvious.

Chapter 3

Adams, Byron, ed. *Edward Elgar and His World*, Bard Music Festival (Princeton, NJ: Princeton University Press, 2007).

All About Elgar, A comprehensive guide to the man, his music and the organisations which support him, available online at http://www.elgar.org/welcome.htm.

Anderson, Martin. Code-breaker, civil servant, musicologist and Shakespeare scholar, available online at http://www.ericsams.org/index.php/eric-sams/100-code-breaker-civil-servant-musicologist-and-shakespeare-scholar.

Anderson, Robert. *Elgar*, The Master Musicians (New York: Schirmer Books, 1993).

Buckley, Robert J. *Sir Edward Elgar* (London: Ballantyne, Hanson & Co., 1905).

Centro Studi Eric Sams—the Eric Sams Archive, http://www.ericsams.org/.

Centro Studi Eric Sams, Eric Sams: on Cryptography, Solutions to ciphers, esssays, articles, reviews, http://www.ericsams.org/index.php/on-cryptography has seventeen pieces on crypto by Eric Sams.

Elgar, Edward. *My Friends Pictured Within* (London: Novello & Co Ltd., 1946).

Elgar, Edward, and Jerrold Northrup Moore. *Edward Elgar: Letters of a Lifetime* (Oxford, U.K.: Oxford University Press, 1990).

Fardon, Michael. *"Dear Carice . . ." Postcards from Edward Elgar to His Daughter* (Worcester, U.K.: Osbourne Books, 1997).

Fiske, Roger. "The Enigma: A Solution." *Musical Times* 110, no. 1521 (November 1969): 1124–26.

Grimley, Daniel M., and Julian Rushton, eds. *The Cambridge Companion to Elgar*, Cambridge Companions to Music (Cambridge, U.K.: Cambridge University Press, 2004).

Jones, Kevin. "The Puzzling Mr. Elgar." *New Scientist* 184, no. 2479 (December 25, 2004): 56.

Kennedy, Michael. *Portrait of Elgar*, 3rd ed. (Oxford, U.K.: Clarendon Press, 1987), Clarendon paperback edition 1993.

Kennedy, Michael. *The Life of Elgar*, Musical Lives (New York: Cambridge University Press, 2004).

Kent, Christopher. *Edward Elgar: A Thematic Catalogue and Research Guide,* Routledge Music Bibliographies, 2nd ed. (New York: Routledge, 2013).

Kenyon, Nicholas, introduction. *Elgar: An Anniversary Portrait* (London: Continuum, 2007).

Kolodin, Irving. "What is Enigma?" *Saturday Review*, February 2, 1953, pp. 53, 55, and 71.

Kruh, Louis. "Still Waiting to be Solved: Elgar's 1897 Cipher Message." *Cryptologia* 22, no. 2 (April 1998): 97–98.

Macnamara, Mark. "The Artist of the Unbreakable Code, Composer Edward Elgar still has cryptographers playing his tune." *Nautilus* no. 6, October 17, 2013, http://nautil.us/issue/6/secret-codes/the-artist-of-the-unbreakable-code.

McVeagh, Diana. *Elgar the Music Maker* (Woodbridge, U.K.: Boydell Press, 2007).

Messenger, Michael. *Edward Elgar: An Illustrated Life of Sir Edward Elgar (1857–1934)* (Princes Risborough, U.K.: Shire Library, 2005).

Moore, Jerrold Northrup. *Edward Elgar: A Creative Life* (Oxford, U.K.: Oxford University Press, 1984).

Palmer, Jean. *The Agony Column Codes & Ciphers* (Bedfordshire, U.K.: Bright Pen, 2005).

Parrott, Ian. "Music and Cipher." *The Musical Times* 109, no. 1508 (October 1968): 920–21.

Parrott, Ian. *Elgar* (London: J. M. Dent and Sons Ltd., and New York: Farrar, Straus and Giroux, Inc., 1971).

Powell, Mrs. Richard. *Edward Elgar: Memories of a Variation* (London: Oxford University Press, 1937).

Powell, Mrs. Richard. *Edward Elgar: Memories of a Variation*, 4th ed., rev. and ed. Claud Powell, with an addendum by Jerrold Northrop Moore (Aldershot, Hants, U.K.: Scolar Press, 1994).

Rushton, Julian. *Elgar: Enigma Variations*, Cambridge Music Handbooks (Cambridge, U.K.: Cambridge University Press, 1999).

Sams, Eric. "Elgar's Cipher Letter to Dorabella." *The Musical Times* 111, no. 1524 (February 1970): 151–54.

Sams, Eric. "Musical Cryptography." *Cryptologia* 3, no. 4 (October 1979): 193–201.

Sams, Eric. "Cracking the Historical Codes." *Times Literary Supplement*, February 8, 1980, p. 154.

Sams, Eric. "Elgar and Cryptology," letter to the editor, *The Musical Times* 125, no. 1695 (May 1984): 251.

Sams, Eric. "Cryptanalysis and Historical Research." *Archivaria* 21 (Winter 1985–1986).

Sams, Eric. "Elgar's Enigmas." *Music & Letters* 78, no. 3 (August 1997): 410–15.

Santa, Charles Richard, and Matthew Santa. "Solving Elgar's Enigma." *Current Musicology* 89 (2010).

Schooling, John Holt. "Secrets in Cipher IV. From the Time of George II to the Present Day." *The Pall Mall Magazine* 8, no. 36 (April 1896): 608–18. This article contains the cipher that Elgar broke on p. 618.

Schridde, Christian. "The Dorabella Cipher (Part 3)," July 31, 2013, available online at http://numberworld.blogspot.co.uk/2013/07/the-dorabella-cipher-part-3.html.

Sypher, F. J. "Eric Sams, Cryptography and the Voynich Manuscript," a pamphlet, New York, 2011, available online at http://www.ericsams.org/index.php/a-portrait-of-f-j-sypher/278-eric-sams-cryptography-and-the-voynich-manuscript.

Tatlow, Ruth. *Bach and the Riddle of the Number Alphabet* (Cambridge, U.K.: Cambridge University Press, 1991). This volume investigates claims of Bach's use of cryptography.

Tibbetts, John C. "John C. Tibbetts Interviews Eric Sams," available online at http://www.ericsams.org/index.php/interview-by-john-c-tibbetts.

Trowell, B. "Elgar's Marginalia." *The Musical Times* 125, no. 1693 (March 1984): 139–41, 143. Trowell looked at the notebook pages where Elgar played with a cipher and noted

> The presence of the sentence 'DO YOU GO TO LONDON TOMORROW', which contains no E (the commonest letter in English by far), with the figure '23' (for the total of letters) and the remark '9 Os', strongly suggests that the Dorabella message has been deliberately constructed in order to defeat the solver's normal resource of analyzing the frequency-count of the letters of the alphabet in English.

> Eric Sams rejected this notion, saying that it would still have been solved quickly. See Sams, Eric. "Elgar and Cryptology," letter to the editor, *The Musical Times* 125, no. 1695 (May 1984): 251.

Young, Percy M. *Elgar O. M. A Study of a Musician* (London: Collins, 1955).

Chapter 4

America's Most Wanted. December 1998. Also in the companion magazine, *Manhunter.*

Anonymous. "Are They Closing in on Zodiac?" *Detective Cases* (April 1974).

Butterfield, Michael. *Zodiac Death Machine: The Untold Story of the Unsolved Zodiac,* to appear. For now, there's the website http://zodiackillerfacts.com/.

Covino, Joseph, Jr. *San Francisco's Finest: Gunning for the Zodiac* (Walnut Creek, CA: Epic Press, 2012).

Crowley, Kieran. *Sleep My Little Dead: The True Story of the Zodiac Killer* (New York: St. Martin's Paperbacks, 1997). This is about a copycat killer in New York, not the original Zodiac. The author signed my copy "To Craig—What's Your Sign?" It gave me a chill—thanks!

Dell, Jessica. The Zodiac Killer Annotated Bibliography, http://jsscdell7.wordpress.com/2011/04/22 /the-zodiac-killer-annotated-bibliography-project-3/ (mostly websites).

Doss, Diane. *NEXUS,* date unknown, but not long before August 20, 1985. Doss found Penn's evidence "voluminous and overwhelming."

Fraley, Craig. *Zodiac Killer Final Thoughts,* CreateSpace Independent Publishing Platform, 2014.

Francis, Carmen. "Zodiac Casts a Strangler's Shadow." *Startling Detective* (March 1970): 18–21, 62–64.

Graysmith, Robert. Letter to Waltz, May 25, 1979, National Cryptologic Museum files.

Graysmith, Robert. *Zodiac* (New York: St. Martin's/Marek, 1986). This is the best book on the Zodiac case. It's creepy and was made into a movie that's also creepy.

Graysmith, Robert. *Zodiac Unmasked: The Identity of America's Most Elusive Serial Killer Revealed* (New York: Berkley Books, 2002).

Haugen, Brenda. *The Zodiac Killer: Terror and Mystery* (Mankato, MN: Capstone Point Books, 2010)— True Crime—96 pages. This is a young adult book.

Hodel, Steve, with Ralph Pezzullo. *Most Evil: Avenger, Zodiac, and the Further Serial Murders of Dr. George Hill Hodel* (New York: Dutton, 2009).

Holt, Tim. "The Men Who Stalk the Zodiac Killer." *San Francisco Magazine,* April 1974.

Jordan, John Robert. Hunter among the Stars: A Critical Look at the Zodiac Killer as Serial Killer, Occultist, and Speller (CreateSpace Independent Publishing Platform, 2011).

Kelleher, Michael D., and David Van Nuys. *"This Is the Zodiac Speaking": Into the Mind of a Serial Killer* (Westport, CT: Praeger Publishers, 2002).

Lafferty, Lyndon E. *The Zodiac Killer Cover-Up: AKA The Silenced Badge* (Vallejo, CA: MANDAMUS, 2012).

Lowall, Gene. "Zodiac California's Blood-Thirsty Phantom." *Argosy,* September 1970.

Montgomery, John. "Your Daughter May Be Next." *Inside Detective,* January 1969.

O'Hare, Michael. "Confessions of a Non-Serial Killer. Conspiracy theories are all fun and games until you become the subject of one." *Washington Monthly,* May/June 2009, available online at http:// www.washingtonmonthly.com/features/2009/0905.ohare.html.

Oranchak, David. http://www.zodiackillerciphers.com/ (my favorite Zodiac website).

Oswell, Douglas. *The Unabomber and the Zodiac* (Douglas Evander Oswell, 2007).

Penn, Gareth. "The Calculus of Evil." *MENSA Bulletin,* no. 288 (July/August 1985): 5–8, 25. At the end of this piece, it was noted that "A different form of this story appeared serially in *The Ecphorizer,* then edited by Richard Amyx." *MENSA Bulletin* chose not to include references that specifically identified Penn's suspect, although Penn had already gone public with his accusation, according to the editor's note. Penn's address was provided for readers who wanted more information. In response to such a query, the writer would not receive a name, but rather enough information so that the suspect's identity could be found with the help of a library with a "half-decent reference collection", according to Gareth Penn, from a response to queries, August 20, 1985. As I stated in the main text of this chapter, I think Penn's suspect is innocent and has, in fact, been victimized by Penn's accusation.

Penn, Gareth. Form response to queries, August 20, 1985, 8 pp.

Penn, Gareth. *Times 17: The Amazing Story of the Zodiac Murders in California and Massachusetts 1966–1981* (New York: Foxglove Press, 1987). Penn seems to have put much more thought into the killings than Zodiac himself. It seems to me that almost all of the patterns he picked up on are coincidental, but I do like the radian theory.

Schillemat, Brandon. "Massachusetts Man Says He's Cracked Zodiac Killer Code," July 21, 2011, available online at http://belmont-ca.patch.com/articles/massachusetts-man-says-hes-cracked-zodiac-killer-code; comments can be found at http://www.fark.com/comments/6408574/705 65655#c70565655.

Smith, Dave. "Zodiac Killer: Is He Still at Large?" *Coronet*, October 1973.

Stephens, Hugh. "Has the Zodiac Killer Trapped Himself?" *Front Page Detective*, February 1970.

Stephens, Hugh. "He Wants Slave Girls Waiting for Him in Paradise." *Front Page Detective*, September 1975.

Stewart, Gary L., and Susan Mustafa. *The Most Dangerous Animal of All: Searching for My Father . . . and Finding the Zodiac Killer* (New York: Harper, 2014).

Symons, C. *Solving the Zodiac: The Zodiac Killer Case Files* (CreateSpace Independent Publishing Platform, 2009).

Weissman, Jerry. *Zodiac Killer* (Los Angeles: Pinnacle Books, 1979).

Williams, Bryan. "The Zodiac Killings: California's No. 1 Murder Mystery." *True Detective* 95, no. 4 (August 1971): 12–17, 65–67.

Yancey, Diane. *The Case of the Zodiac Killer*, Crime Scene Investigations (Detroit: Lucent Books, 2008).

Chapter 5

Anderson, Jeanne. "Breaking the BTK Killer's Cipher." *Cryptologia* 37, no. 3 (July 2013). This is a nice example of a killer's cipher for which the solution was found.

Anderson, Jeanne. "Kaczynski's Ciphers." *Cryptologia* 39, no. 3 (July 2015). Again, killer ciphers that were solved. This time, because the key was kept with them.

Anonymous. "Police Close Susan Powell Case, Offer New Details." *USA Today*, May 20, 2013, available online at http://www.usatoday.com/story/news/nation/2013/05/20/susan-powell-case/2344681/.

Beattie, Robert. *Nightmare in Wichita: The Hunt for the BTK Strangler* (New York: New American Library, 2005). This book covers the noncryptologic aspects of this case.

Bennett, Donald H. "An Unsolved Puzzle Solved." *Cryptologic Spectrum* (Spring 1980). Available online at https://www.nsa.gov/public_info/_files/cryptologic_spectrum/unsolved_puzzle.pdf.

Bennett, Donald H. "An Unsolved Puzzle Solved." *Cryptologia* 7, no. 3 (July 1983): 218–34.

Carlisle, Nate. "Susan Powell Case Closed, Files Are Opened." *The Salt Lake Tribune*, May 21, 2013, available online at http://www.sltrib.com/sltrib/news/56338157-78/powell-susan-police-valley.html.csp.

Daniel, Mac, and Diane Allen. "Killings, Suicide Baffle Authorities. Husband Left Note; Motive a Mystery." *The Boston Globe*, February 5, 2004, 3rd ed., p. B1. This article was posted to Bruce Schneier's blog.

Davis, William F. "The Debosnys Murder Case." Undated, but after 1961. Davis served as deputy sheriff in Essex County. The report on the case was a joint effort carried out by Davis and his wife, according to Cheri L. Farnsworth, *Adirondack Enigma: The Depraved Intellect and Mysterious Life of North Country Wife Killer Henry Debosnys* (Charleston, SC: The History Press, 2010), 51.

"Debosnys Suspected of Another Murder." *Elizabethtown Post*, August 10, 1882.

Farmer, Tom. "Suicide Note Leaves Motive for Saugus Killings a Mystery." *The Boston Herald*, February 5, 2004, p. 14.

Farnsworth, Cheri L. *Adirondack Enigma: The Depraved Intellect and Mysterious Life of North Country Wife Killer Henry Debosnys* (Charleston, SC: The History Press, 2010).

Hodgkins, James D. "The Copiale Cipher: An Early German Masonic Ritual Unveiled." *The Scottish Rite Journal* (March/April 2012): 4–8, available online at http://scottishrite.org/about/media -publications/journal/article/the-copiale-cipher-an-early-german-masonic-ritual-unveiled/.

Knight, Kevin, Beáta Megyesi, and Christiane Schaefer. "The Copiale Cipher." *Proceedings of the 4th Workshop on Building and Using Comparable Corpora*, 49th Annual Meeting of the Association for Computational Linguistics, Portland, OR, June 24, 2011 (Stroudsburg, PA: Association for Computational Linguistics, 2011) 2–9.

Kourofsky, Niki. *Adirondack Outlaws: Bad Boys and Lawless Ladies*, Bedside Readers Series (Helena, MT: Farcountry Press, 2015). The Henry Debosnys case is detailed on pp. 71–78.

Lohr, David. "Susan Powell's Parents Search for Clues in Father-in-Law's Former Home." May 27, 2014, *Huffington Post*, available online at http://www.huffingtonpost.com/2014/05/27/susan-powell -missing-clues_n_5398147.html.

Mahon, Tom, and James J. Gillogly. *Decoding the IRA* (Cork, Ireland: Mercier Press, 2008).

Meguid, Halia. Best known for her singing on television's *Doctor Who*, Halia indicated online that she's at work on a novel about the Henry Debosnys case. See Halia Meguid, http://haliameguid .tumblr.com/post/91617594118/pls-read-if-you-are-an-ingenious-codebreaker-or.

Morris, Brent. "Fraternal Cryptography." *Cryptologic Spectrum* (Summer 1978).

Morris, Brent. "Fraternal Cryptography, Cryptographic Practices of American Fraternal Organiza-tions." *Cryptologia* 7, no. 1 (January 1983): 27–36.

Morris, S. Brent. *The Folger Manuscript: The Cryptanalysis and Interpretation of an American Masonic Manuscript* (Ft. Meade, MD, 1992).

Morris, S. Brent. The Folger Manuscript, a lecture, available online at http://www.themasonictrowel .com/articles/manuscripts/manuscripts/folger_manuscript/the_folger_manuscript_lecture.htm.

Olsen, Greg, and Rebecca Morris. *If I Can't Have You: Susan Powell, Her Mysterious Disappearance, and the Murder of Her Children* (New York: St. Martin's Press, 2014).

Rule, Ann. *Fatal Friends, Deadly Neighbors and Other True Cases*, Ann Rule's Crime Files: Vol. 16 (New York: Pocket Books, 2012). Pages 1–156 of this book are devoted to the Powell case. No mention is made of encryption.

Ryba, Jim. "What Is He Saying?" *The Phoenician*, Summer 2006, p. 26.

Schmeh, Klaus. Codeknacker auf Verbrecherjagd, Folge 4: Der Maskenmann, http://scienceblogs .de/klausis-krypto-kolumne/2014/03/12/codeknacker-auf-verbrecherjagd-folge-4-der -maskenmann/, March 12, 2014.

Schmeh, Klaus. http://scienceblogs.de/klausis-krypto-kolumne/2015/02/23/die-ungeloesten-codes
-des-mutmasslichen-frauenmoerders-henry-debosnys-teil-1/, February 23, 2015.

Schmeh, Klaus. Die ungelösten Codes des mutmaßlichen Frauenmörders Henry Debosnys (Teil 2),
http://scienceblogs.de/klausis-krypto-kolumne/2015/02/25/die-ungeloesten-codes-des
-mutmasslichen-frauenmoerders-henry-debosnys-teil-1-2/, February 25, 2015.

Schneier, Bruce. Handwritten Real-World Cryptogram, Schneier on Security, https://www.schneier
.com/blog/archives/2006/01/handwritten_rea.html.

Wagner, Ingo."Verschlüsselt und nicht knackbar, Polizei scheitert an Festplatten des 'Maskenmanns.' " *Focus
Magazine*, January 23, 2012. Available online at http://www.focus.de/panorama/welt/verschluesselt
-und-nicht-knackbar-polizei-scheitert-an-festplatten-des-maskenmanns_aid_705621.html.

Chapter 6

ON SOMERTON MAN

Abbott, Derek. Cipher Cracking, https://www.eleceng.adelaide.edu.au/personal/dabbott/wiki/index
.php/Cipher_Cracking. Scroll down to the section titled "See Also." This section has a number
of useful links, including "Primary source material on the Taman Shud Case" and "Secondary
source material on the Taman Shud Case." Each of these leads to many useful documents.

Andrew, Christopher. *Defend the Realm: The Authorized History of MI5* (New York: Alfred A. Knopf,
2009). See this for background on possibly relevant Cold War spy cases.

Anonymous. "Dead Man Found Lying on Somerton Beach." *The News*, Adelaide, December 1, 1948, p. 1.
Available online at http://trove.nla.gov.au/ndp/del/article/129897161.

Anonymous. "Body Found on Beach." *The Advertiser*, Adelaide, December 2, 1948, p. 3.

Anonymous. " 'Dead' Man Walks into Police HQ." *The News*, Adelaide, December 2, 1948.

Anonymous. "Somerton Beach Body Mystery." *The Advertiser*, Adelaide, December 4, 1948.

Anonymous. "Luggage as Clue to Beach Body." *The News*, Adelaide, January 12, 1949.

Anonymous. "Attempts to Solve Somerton Mystery Deepen Mystery." *Adelaide Truth*, June 25, 1949, p. 6.

Anonymous. "Police Test Book for Somerton Body Clue." *The Advertiser,* Adelaide, July 26, 1949, p. 3.
Available online at http://trove.nla.gov.au/ndp/del/article/36677872.

Ashton, Lucy F. "Logical thinking," Letter to the Editor, *The Advertiser*, November 15, 1995, p. 14.

Balint, Ruth. "The Somerton Man an Unsolved History." *Cultural Studies Review* 16, no. 2 (September
2010): 159–78, http://epress.lib.uts.edu.au/journals/index.php/csrj/index.

The Body on the Beach: The Somerton Man—Taman Shud Case, http://brokenmeadows.hubpages.
com/hub/The-Mystery-of-the-Somerton-Man-Taman-Shud-Case This is a great article with
some rare pictures.

Bouda, Simon. *Crimes that Shocked Australia* (Sydney, Australia: Bantam Books, 1991). This book contains an incorrect version of Somerton Man's cipher.

Campbell, MacGregor. "Unbreakable: Somerton Man's Poetic Mystery." *New Scientist*, May 26, 2011.

Castello, Renato. "New Twist in Somerton Man Mystery as Fresh Claims Emerge." *The Advertiser, Sunday Mail*, South Australia, November 23, 2013, available online at http://www.adelaidenow.com
.au/news/south-australia/new-twist-in-somerton-man-mystery-as-fresh-claims-emerge/story
-fni6uo1m-1226766905157.

Clegg, Edward (His Honour Judge Clegg QC). *Famous Australian Murders* (Sydney, Australia: Angus and Robertson, 1975.

Clemow, Matt. " 'Poisoned in SA'—Was He a Red Spy?" *Sunday Mail*, November 7, 2004, 76–77. Portions reproduced in Feltus's book, pp. 186–87.

Coupe, Stuart, and Julie Ogden, eds. *Case Reopened* (Sydney, Australia: Allen and Unwin, 1993). This anthology contains a Kerry Greenwood short story, where an unknown man is smuggling for Ireland. It makes use of an incorrect version of the cipher from Simon Bouda's *Crimes that Shocked Australia* (Sydney, Australia: Bantam Books, 1991). The short story was reprinted in Greenwood's nonfiction *Tamam Shud: The Somerton Man Mystery* on pp. 177–216; p. 176 offers an introduction.

Dash, Mike. "The Body on Somerton Beach." Smithsonian.com, August 12, 2011, available online at http://www.smithsonianmag.com/history/the-body-on-somerton-beach-50795611/.

Debelle, Penelope. "A Body, a Secret Pocket and a Mysterious Code. Can the Riddle be solved?" *The Advertiser, Weekend Magazine,* August 1, 2009, pp. W14–W16.

Feltus, Gerald (Gerry) Michael. *The Unknown Man: A suspicious death at Somerton Beach* (Richmond, Australia: Hyde Park Press, 2010). Also see www.theunknownman.com. Feltus was born in 1943 and grew up near the area where the unknown man was found. He joined the South Australian police force in 1964, but knew about the case before becoming a police officer. In 1975, as a member of the Major Crimes Squad, the old unsolved case came under his purview. It wasn't as cold of a case as the intervening decades might make it seem, as periodic media attention led to new leads. Feltus left the Major Crimes Squad in 1979, but returned in 1992 (the name had changed to the Major Crime Task Force). He began devoting some of his spare time to the case (just like Graysmith did with Zodiac). He retired from the police force in 2004 (bio taken from pp. 10–11).

Feltus, Gerry. The Unknown Man, http://theunknownman.com/. This is the website for the book, and one can find useful updates and other information not in the book here, as well.

Ferguson, John. "After Discovery of Body on Somerton Beach in 1948 . . . Mystery Remains on the 'Unknown Man.' " *The News,* May 15, 1987, p. 5.

Fife-Yeomans, Janet. "The Man with No Name." *The Weekend Australian Magazine*, September 15–16, 2001. Portions of this article were reproduced in Feltus's book, pp. 185–86.

Gibson, Candy. "Students Aim to Crack 60-Year-Old Mystery." *Adelaidean* 18, no. 3 (May 2009): 9.

Grace, Kynton. "South Australia's *X-Files*: Part 2—The Somerton Man Mystery and the Secrets of Adelaide's Tunnels." *The Advertiser*, June 9, 2014, available online at http://www.theaustralian.com.au/news/south-australias-xfiles-part-2-the-somerton-man-mystery-and-the-secrets-of-adelaides-tunnels/story-e6frg6n6-1226941307664.

Greenwood, Kerry. *Tamam Shud: The Somerton Man Mystery* (Sydney, Australia: NewSouth Publishing, University of New South Wales Press Ltd., 2012).

Greenwood, Kerry. "Riddle on the Sands." *The Sydney Morning Herald*, November 28, 2012, edited extract from *Tamam Shud: The Somerton Man Mystery* by Kerry Greenwood, available online at http://www.smh.com.au/national/riddle-on-the-sands-20121119-29kwz.html.

"Inquest into the death of a body located at Somerton on 1.12.48" [refers to December 1, 1948] 1949, pdf available online at http://www.eleceng.adelaide.edu.au/personal/dabbott/tamanshud/inquest1949ocr.pdf.

"Inquest into the death of a body located at Somerton on 1.12.48" [refers to December 1, 1948] 1958, pdf available online at http://www.eleceng.adelaide.edu.au/personal/dabbott/tamanshud/inquest1958.pdf.

Jory, Rex. "The Dead Man Who Sparked Many Tales." *The Advertiser*, December 1, 2000, p. 18.

King, Stephen. *The Colorado Kid* (New York: Hard Case Crime, 2005). Although a reader might get the sense that this book was inspired by the Somerton Man case, King said that it was not. This novel inspired the TV series *Haven* (2010).

Lewes, Jacqueline Lee. "30-Year-Old Death Riddle Probed in New Series." *TV Times,* August 19, 1978, pp. 20–21.

Loftus, Tom. "The Somerton Body Mystery." *The News*, December 1–2, 1982. Portions reproduced in Feltus's book, pp. 197–200.

MacGregor Campbell. "Killer Codes." *New Scientist,* May 21, 2011, pp. 40–45.

Maguire, Shane. "Death Riddle of a Man with No Name." *The Advertiser*, March 9, 2005, p. 28.

Orr, Stephen. "Riddle of the End." *The Sunday Mail,* January 11, 2009, pp. 71–76.

Pedley, Derek. "Detective Still on Trail 47 Years On." *The Advertiser*, November 4, 1995, p. 14.

Pelling, Nick. "Sorry, the unknown man is (very probably) not H. C. Reynolds . . .." March 15, 2013, http://www.ciphermysteries.com/2013/03/15/sorry-but-the-unknown-man-is-almost-certainly-not-h-c-reynolds.

Phillips, John Harber. "So When That Angel of the Darker Drink." *Criminal Law Journal* 18, no. 2 (April 1994): 108–10.

Schmeh, Klaus. *Nicht zu Knacken* [Impossible to Crack] (Munich, Germany: Carl Hanser Verlag, 2012). This German language book is a survey of unsolved ciphers. Chapter 7 is devoted to Somerton Man. No English edition is available.

Steadwell. Somerton Man: A True Mystery "Down Under" with *Casablanca* Intrigue, August 23, 2011, http://blogcritics.org/culture/article/somerton-man-a-true-mystery-down/#ixzz1esR3uG12.

Taman Shud Case, http://en.wikipedia.org/wiki/Taman_Shud_Case.

Trove, digitised newspapers and more, http://trove.nla.gov.au/newspaper. A large number of Australian newspaper articles on Somerton Man are available here.

Turner, Jeff. "Beach Keeps Its Grim Secret." *The Advertiser*, December 7, 1998, p. 19. Portion reproduced in Feltus's book, p. 200.

Watkins, Emily. "Is British Seaman's Identity Card Clue to Solving 63-Year-Old Beach Body Mystery?" *The Advertiser, Sunday Mail*, Adelaide, November 20, 2011, pp. 4–5, available online at http://www.adelaidenow.com.au/is-british-seamans-identity-card-clue-to-solving-63-year-old-beach-body-mystery/story-e6frea6u-1226200076344.

Videos

All of the videos can be found at the indicated YouTube pages, but the first three immediately below are also linked from the webpage https://www.eleceng.adelaide.edu.au/personal/dabbott/wiki/index.php/Primary_source_material_on_the_Taman_Shud_Case.

The Somerton Beach Mystery, *Inside Story*, Australian Broadcasting Corporation, 1978. Online in three parts: http://www.youtube.com/watch?v=nnPqlYPQ9lY, http://www.youtube.com/watch?v=605V1-03r1Y, and http://www.youtube.com/watch?v=ieczsZRQnu8.

Tamam Shud—The Time Traveler that Died Before He Was Born! http://www.youtube.com/watch?v=zV3tfU0Jyr4. No time travel is actually suggested in this video!

Taman Shud Case, *Stateline Report*, Australian Broadcasting Corporation, March 27, 2009, and May 1, 2009. Online in two parts: http://www.youtube.com/watch?v=GIUP-wVw6ok and http://www.youtube.com/watch?v=GNgsA1aHNHA.

The segment "The Somerton Man" from *60 Minutes* (Australia) is available at http://sixtyminutes .ninemsn.com.au/article.aspx?id=8759245. If you want to watch the entire program (including unrelated segments) go to http://www.youtube.com/watch?v=iy5s50F3uB8. This program aired on November 24, 2013. For "Extra Minutes," providing an extended interview with Roma and Rachel Egan discussing the reasons they want the Somerton Man exhumed, go to http://www .youtube.com/watch?v=pJKMnX4WHSA.

Derek Abbott and Matthew Berryman have led several student projects on Somerton Man. The You-Tube videos associated with this work are listed here:

Somerton Man Code Investigation. http://www.youtube.com/watch?v=YrjOvbk6QVI (2009) Students Andrew Turnbull and Denley Bihari drew the same conclusion that I did.

Somerton Man Code Investigation. http://www.youtube.com/watch?v=rFsFpSBGhQw (2010) Students Kevin Ramirez and Michael Lewis-Vassallo.

Project Video. http://www.youtube.com/watch?v=K0DEF-FDLso (2011) Students Steven Maxwell and Patrick Johnson. The audio portion of this one doesn't work.

Code Cracking: Who Murdered the Somerton Man? http://www.youtube.com/watch?v =jWE7xl9LiOw&feature=youtu.be (2012) Students Aidan Duffy and Thomas Stratfold.

Code Cracking: Who Murdered the Somerton Man? http://www.youtube.com/watch?v=JX4bt7VuJQs (2013) Students Lucy Griffith and Peter Varsos. These students tried other languages, but found that English had the best match. They also conducted mass spectral analysis on one of Somerton Man's hairs and learned that the amount of lead in his system decreased toward his time of death.

ON PAUL RUBIN

Some of the newspaper articles cited here were found in clipped form in a scrapbook in the uncat-aloged portion of the National Cryptologic Museum's library. Others were obtained as photocopied clippings from an FBI file on the case held at the National Archives and Records Administration. Hence, full bibliographic details were not always apparent. Journalists were hardly ever credited by name in these articles.

"Airport Death Called Murder by Ominsky." Philadelphia, January 21, 1953.

Apt, Jay. "Note Baffles Decoders in Poison Death." *Daily News*, Philadelphia, January 22, 1953, p. 2. Story continues on page 19, but retains the headline.

"Body at Airport Is Identified as NYU Student." Philadelphia, January 21, 1953.

"Body Found at Airport, Coded Note on Abdomen." *The Philadelphia Inquirer*, January 21, 1953, p. 1. This story continues on p. 18, col. 3, under the headline "Mystery Death at City Airport."

"Body with Code Note Found Here." *Daily News*, Philadelphia, p. 2, January 20, 1953.

"Code Death Mystery Still Baffles FBI." January 23, 1953.

"Code Message Found Taped to Dead Man." *Washington Times-Herald*, January 21, 1953, p. 35.

"Code Note Taped to Body in Ditch." *The Evening Bulletin*, Philadelphia, January 20, 1953, p. 1, col. 2.

"Coded Note, Poison Deepen Mysteries of Airport Death." *Philadelphia Daily News* 28, no. 253, January 21, 1953.

"Codes Seen Key in Airport Death." January 22, 1953.

"Coroner Hints Poison Slaying of Student." *The Knickerbocker News*, Albany, NY, January 22, 1953, p. 14-B.

"Cyanide Caused Death at Airport." *The Evening Bulletin*, Philadelphia, January 21, 1953, p. 1.

"Cyanide Killed Cloak-and-Dagger Airport Victim." (big headline, references story on p. 3)

"Dad Identifies Poisoned Youth as N.Y. Student." *Washington Times-Herald*, January 22, 1953, p. 3.

"Dead Youth with Message on Body was NY Pupil." *Buffalo Courier-Express*, January 22, 1953, p. 2.

"Death of Student." *Shamokin News-Dispatch*, Shamokin, PA, January 23, 1953, p. 2.

" 'Dulles' Name Taped to Body of 'Suicide.' " *Washington Post*.

"FBI Investigates Possibilities of Murder in Mysterious Cyanide-Death of Student." *Schenectady Gazette*, January 22, 1953.

"Hazing Studied in Airport Death." *The Philadelphia Inquirer*, January 28, 1953, p. 5.

"Life on the Newsfronts of the World, Student Leaves a Coded Clue to his Death, a Soldier Plants a Flag and Flu Virus Sweeps the World." *Life*, February 2, 1953, p. 30.

"Man Poisoned, Paper in Code Taped to Body; FBI Called In." *Washington Star*, p. A-26.

"Murder Hinted in Cyanide Death." *The Washington Daily News*, January 22, 1953.

"Murder Hinted in Youth's Death." *Long Island Star-Journal*, January 22, 1953, p. 4.

"Mystery Man at Airport Is Found Dead of Poison." *The Philadelphia Inquirer* 248, no. 21, January 21, 1953, p. 1. Story continues on p. 37.

"Mystery Man Cyanide Clue Hints Murder."

"Mystery of Body, Code Note Deepens in Airport Death." January 21, 1953.

"Mystery Shrouds Death by Cyanide." *The Philadelphia Inquirer*, March 25, 1953, p. 19.

"Note Taped to Body at Airport Bears Symbols, Word 'Dulles.' " January 20, 1953.

"Philly Jury Finds Boro Student Died from Cyanide Poisoning." *Brooklyn Eagle*, March 25, 1953, p. 4.

"Poison Victim Identified as N.Y. Youth." *The Philadelphia Inquirer*, January 22, 1953, final city ed., p.1, illustrated on p. 3 and continued on p. 13.

"Sift Cyanide Death of Boro Collegian." *Brooklyn Eagle*, January 22, 1953, 7 star ed., p. 1.

"Student Identified as Cyanide Victim." *New York Times*, January 22, 1953.

"Test Tube Found at Site of Airport Ditch Death." *The Philadelphia Inquirer*, January 23, 1953, p. 28.

"Test Tube Near Code Body Yields no Clue." January 23, 1953.

"Verdict Remains Open in Mysterious Death Philadelphia." *The Evening Sun*, Hanover, PA, March 25, 1953, p. 9.

"Youth Dabbled in Cryptography." *Philadelphia Evening Bulletin*, January 22, 1953.

Many of the newspaper articles cited above are available here as pdfs: http://fultonhistory.com/Fulton.html.

ON RICKY McCORMICK

Anthony, Shane. "From 1999: Body Found in Field Puzzles Police." *St. Louis Post-Dispatch*, July 2, 1999, available online at http://www.stltoday.com/news/article_bcc02074-5b1a-11e0-b199-0017a4a78c22.html.

Castigliola, Angelo. "Ricky McCormick FBI Letters Decoded." *Angelo on Security*, March 30, 2011, http://www.castigliola.com/index.php?option=com_content&task=view&id=123&Itemid=1 (a proposed solution).

Cryptanalysts Part 2: Help Solve an Open Murder Case, http://www.fbi.gov/news/stories/2011/march/cryptanalysis_032911.

Help Break the Code, https://forms.fbi.gov/code.

Howard, Trisha L. "Store Manager Gets 38-Year Sentence in Killing of Customer." *St Louis Post-Dispatch*, September 5, 2002. This is the place to start if you want to follow the continuing adventures of Baha Hamdallah.

Kessler, Ronald. *The Secrets of the FBI* (New York: Crown Publishers, 2011). This book has some useful general information beginning on p. 257.

Mann, Jennifer. "FBI Wants Public to Help Crack Code in St. Charles County Cold Case." *St. Louis Post-Dispatch*, March 31, 2011, available online at http://www.stltoday.com/news/local /crime-and-courts/fbi-wants-public-to-help-crack-code-in-st-charles/article_10af026c -e2d1-59f1-92e5-0576d36e2ffc.html.

Ricky McCormick's encrypted notes. http://en.wikipedia.org/wiki/Ricky_McCormick's_encrypted_notes.

Schmeh, Klaus. Top-25 der ungelösten Verschlüsselungen—Platz 7: Der Mord an Ricky McCormick, Klausis Krypto Kolumne, http://scienceblogs.de/klausis-krypto-kolumne/2013/08/29/top-25 -der-ungelosten-verschlusselungen-platz-7-der-mord-an-ricky-mccormick/. This piece is in German, although some use comments appear in English.

Tritto, Christopher. "Code Dead: Do the Encrypted Writings of Ricky McCormick Hold the Key to His Mysterious Death?" *Riverfront Times*, St. Louis, June 14, 2012, available online at http://www .riverfronttimes.com/2012-06-14/news/ricky-mccormick-code-mysterious-death-st-louis/.

Tweedie, Neil. "Calling All Codebreakers . . . ." *The Telegraph*, April 7, 2011, available online at http:// www.telegraph.co.uk/news/uknews/crime/8432893/Calling-all-codebreakers....html. This arti-cle provides various experts' opinions on the cipher.

Chapter 7

Anonymous. "£20 to Solve Riddle." *Psychic News*, July 15, 1950, p. 4. This piece is about Wood's cipher.

Anonymous. "£20—If You Can Beat the Spirits to It!" *Two Worlds* 63, no. 3277 (September 16, 1950): 923. This piece is about Wood's cipher.

Anonymous. "S. P. R. Member's Survival Test Amplified." *Two Worlds* (December 2, 1950). In this piece, Wood referred to the "substantial prize" he offered.

Anonymous. "After Death." *Daily Express*, August 15, 2015. This piece is about Wood's cipher.

Augustine, Keith. "The Case against Immortality." *Skeptic Magazine* 5, no. 2 (1997), available online in an expanded form at http://infidels.org/library/modern/keith_augustine/immortality.html.

Berger, Arthur S. "Better than a Gold Watch: The Work of the Survival Research Foundation." *Theta* 10 (1982): 82–84.

Berger, Arthur S. "Death Comes Alive." *Journal of Religion and Psychical Research* 5 (1982): 139–47.

Berger, Arthur S. "Project: Unrecorded Information." *Christian Parapsychologist* 4 (1982): 159–61.

Berger, Arthur S. "Letter to the Editor." *Journal of the Society for Psychical Research* 52 (1983): 156–57.

Berger, Arthur S. "Experiments with False Keys." *Journal of the American Society for Psychical Research* 78, no. 1 (January 1984): 41–54.

Berger, Arthur S. "The Development and Replication of Tests for Survival." *Parapsychological Journal of South Africa* 5 (1984): 24–35.

Berger, Arthur S. *Aristocracy of the Dead: New Findings in Postmortem Survival* (London and Jefferson, NC: McFarland & Co., 1987).

Berger, Arthur S. "Reincarnation by the Numbers: A Criticism and Suggestion." *Reincarnation: Fact or Fable?* Arthur S. Berger and Joyce Berger, eds. (London: The Aquarian Press, 1991), 221–33.

Carroll, R. T. "How Not to Conduct Scientific Research." (review of *The Afterlife Experiments* by Gary Schwartz), The Skeptic's Dictionary, available online at http://skepdic.com/refuge/afterlife.html.

Feely, Joseph Martin. *Electrograms from Elysium: A Study on the Probabilities in Postmortuary Communication through the Electronics of Telepathy and Extrasensory Perception, Including the Code of Anagrams in the Purported Sender's Name* (New York, 1954).

Fox, Margalit. "Ian Stevenson Dies at 88; Studied Claims of Past Lives." *New York Times*, February 18, 2007, available online at http://www.nytimes.com/2007/02/18/health/psychology/18stevenson .html?_r=0.

Gillogly, James J., and Larry Harnisch. "Cryptograms from the Crypt." *Cryptologia* 20, no. 4 (October 1996): 325–29.

Kellock, Harold. *Houdini: His Life-story, from the Recollections and Documents of Beatrice Houdini* (New York: William Heinemann, 1928).

Leighton, A. C. "Has Dr. Thouless Survived Death?" *Cryptologia* 10, no. 2 (April 1986): 108–9.

Levin, Michael. "Encryption Algorithms in Survival Evidence." *The Journal of Parapsychology* 58, no. 2 (June 1994): 189–95.

MacDonald, Leo. "Rebuttal to Keith Augustine's Look at the Tests Done by Ian Stevenson," February 3, 2009, available online at http://paranormalandlifeafterdeath.blogspot.com/2009/02/rebuttal -to-keith-augustines-look-at.html.

Martin, Michael, and Keith Augustine. *The Myth of an Afterlife: The Case against Life After Death* (Lanham, MD: Rowman & Littlefield Publishers, 2015).

Mongé, A. "Solution of a Playfair Cipher." *Signal Corps Bulletin* 93 (November–December 1936). Reprinted in W. F. Friedman. *Cryptography and Cryptanalysis Articles* 1 (Laguna Hills, CA: Aegean Park Press, 1976) and in Brian J. Winkel. "A Tribute to Alf Mongé." *Cryptologia* 2, no. 2 (1978): 178–85.

Polidoro, Massimo. "The Day Houdini (Almost) Came Back from the Dead." *Skeptical Inquirer* 36, no. 2 (March/April 2012), available online at http://www.csicop.org/si/show/the_day_houdini _almost_came_back_from_the_dead.

Roggo, D. Scott. "Parapsychology—Its Contributions to the Study of Death." *Omega Journal of Death and Dying* 5, no. 2 (July 1974): 99–113.

Salter, W. H. "F.W.H. Myers' Posthumous Message." Proceedings *of the Society for Psychical Research* 52 (1958): 1–32 (a sealed envelope message).

Schmeh, Klaus. *Nicht zu knacken* [Impossible to Crack] (Munich, Germany: Carl Hanser Verlag, 2012). Chapter 10 of Schmeh's book covers Thouless's ciphers.

Schmeh, Klaus. "Parapsychologische Verschlüsselungsexperimente, Tote verraten keine Geheimwörter—bisher jedenfalls." *Skeptiker* 3 (2012): 2–9.

Schmeh, Klaus. "Skurriles Experiment: Verraten Tote Geheimwörter?" *Telepolis*, May 13, 2013, available online at http://www.heise.de/tp/artikel/38/38934/1.html.

Schwartz, Gary E. R., and L. G. S. Russek. "Testing the Survival of Consciousness Hypothesis: The Goal of the Codes." *Journal of Scientific Exploration* 11, no. 1 (1997): 79–88.

Schwartz, Gary E. R., and Linda G. S. Russek. "Celebrating Susy Smith's Soul: Preliminary Evidence for the Continuance of Smiths Consciousness after Her Physical Death." *Journal of Religion and*

Psychical Research 24, no. 2 (April 2001): 82–91, available online at http://www.innerknowing .net/research.html.

Schwartz, Gary E., with William L. Simon, foreword by Deepak Chopra. *The Afterlife Experiments, Breakthrough Scientific Evidence of Life After Death* (New York: Pocket Books, 2002).

Smith, Susy (pen name of Ethel Elizabeth Smith). Introduction by Gary E. R. Schwartz and Linda G. S. Russek. *The Afterlife Codes: Searching for Evidence of the Survival of the Human Soul* (Charlottesville, VA: Hampton Roads Publishing Company, Inc., 2000).

Stevenson, Ian. "The Combination Lock Test for Survival." *Journal of the American Society for Psychical Research* 62 (1968): 246–54.

Stevenson, Ian. "Further Observations on the Combination Lock Test for Survival." *Journal of the American Society for Psychical Research* 70 (1976): 219–29.

Stevenson, Ian, Arthur T. Oram, and Betty Markwick. "Two Tests of Survival After Death: Report on Negative Results." *Journal of the Society for Psychical Research* 55, no. 815 (April 1989): 329–36.

Thouless, Robert H. "A Test of Survival." *Proceedings of the Society for Psychical Research* 48 (July 1948): 253–63.

Thouless, Robert H. "Additional Notes on a Test of Survival." *Proceedings of the American Society for Psychical Research* 48 (1948): 342–43.

Tribbe, Frank C. "The Tribbe/Mulders Code." *Journal of the Academy of Religion and Psychical Research* 3, no. 1 (January 1980): 44–46.

Winkel, Brian J. "A Tribute to Alf Mongé." *Cryptologia* 2, no. 2 (April 1978): 178–85.

Wood, T. E. "A Further Test for Survival." *Proceedings of the Society for Psychical Research* 49 (1950): 105–6.

Chapter 8

Bauer, Craig. *Secret History: The Story of Cryptology* (Boca Raton, FL: Chapman and Hall/CRC, 2013).

Bauer, Craig, and Elliott Gottloeb. "Results of an Automated Attack on the Running Key Cipher." *Cryptologia* 29, no. 3 (July 2005): 248–54.

Bauer, Craig, and Christian N. S. Tate. "A Statistical Attack on the Running Key Cipher." *Cryptologia* 26, no. 4 (October 2002): 274–82.

Bellovin, Steven M. "Frank Miller: Inventor of the One-Time Pad." *Cryptologia* 35, no. 3 (July 2011): 203–22.

Byrne, John F. *Silent Years: An Autobiography with Memoirs of James Joyce and Our Ireland* (New York: Farrar, Straus and Young, 1953).

Byrne, John, Cipher A. Deavours, and Louis Kruh. "Chaocipher Enters the Computer Age When Its Method Is Disclosed to *Cryptologia* Editors." *Cryptologia* 14, no. 3 (July 1990): 193–98.

Calof, Jeff, Jeff Hill, and Moshe Rubin. "Chaocipher Exhibit 5: History, Analysis, and Solution of *Cryptologia*'s 1990 Challenge." *Cryptologia* 38, no. 1 (January 2014): 1–25.

Friedman, William F. "The Cryptanalyst Accepts a Challenge." *The Signal Corps Bulletin* 103 (January–March 1939).

Griffing, Alexander. "Solving the Running Key Cipher with the Viterbi Algorithm." *Cryptologia* 30, no. 4 (October 2006): 361–67. This is the best paper on the topic of breaking running key ciphers.

"Historical Survey of Strip Cipher Systems." This is available from NARA; NSA Historical Collections 190/37/7/1, NR 3525 CBRK24 12957A 19450000.

"History of Army Strip Cipher, SRH-366." This is available from NARA; RG 0457: NSA/CSS Finding Aid A1, 9020 U.S. Navy Records Relating to Cryptology 1918–1950 Stack 190 Begin Loc 36/12/04 Location 1-19.

Kruh, Louis. "The Genesis of the Jefferson/Bazeries Cipher Devices." *Cryptologia* 5, no. 4 (October 1981): 193–208.

Kruh, Louis (under his ACA pen name, MEROKE). "The M-94 Test Messages." *The Cryptogram* XLVIII, no. 6 (July–August 1982): 4–5, available online at http://www.prc68.com/I/M94TM.htm.

Kruh, Louis (under his ACA pen name, MEROKE). "A 77-Year-Old Challenge Cipher." *The Cryptogram* (March/April 1991): 7.

Kruh, Louis. "A 77-Year-Old Challenge Cipher." *Cryptologia* 17, no. 2 (April 1993): 172–74. Note: Mauborgne is misspelled in this paper, and the reference to the paper in *The Cryptogram* leads one to look in early 1992 issues, when it is actually March/April 1991.

Kruh, Louis. "Riverbank Laboratory Correspondence, 1919 (SRH-50)." *Cryptologia* 19, no. 3 (July 1995): 236–46.

Mauborgne, Joseph O. *Practical Uses of the Wave Meter in Wireless Telegraphy* (New York: McGraw-Hill Book Co., 1913).

Mauborgne, Joseph O. *An Advanced Problem in Cryptography and Its Solution* (Fort Leavenworth, KS: Press of the Army Services Schools, 1914). A second edition appeared in 1918.

Mauborgne, Joseph O. *Data for the Solution of German Ciphers: Also a Diagram of Cipher Analysis* (Fort Leavenworth, KS: Army Service School Press, 1917).

Mauborgne, Joseph O. "One Method of Solution of the Schooling 'Absolutely Indecipherable' Cryptogram." *The Signal Corps Bulletin* 104 (April–June 1939): 27–40.

Mauborgne, Joseph O. "Reminiscences of Joseph Oswald Mauborgne: Oral History." 1971 (held in Columbia University Library's rare book collection).

Smoot, Betsy Rohaly. "Parker Hitt's First Cylinder Device and the Genesis of U.S. Army Cylinder and Strip Devices." *Cryptologia* 39, no. 4 (October 2015): 315–21. It was long believed that Hitt came up with his version of the cipher wheel in 1913. This paper shows that it was 1912. So, it shouldn't surprise anyone if Mauborgne's version turns out to be a little older than believed as well!

Wesencraft, Fenwick (under his ACA pen name TRIO). "Solutions of the M-94 Test Messages." *The Cryptogram* XLVIII no. 8 (November–December 1982): 6–7, available online at http://www.prc68.com/I/M94S.htm.

Chapter 9

ON D'AGAPEYEFF'S CIPHER

Barker, Wayne G. "The Unsolved d'Agapeyeff Cipher." *Cryptologia* 2, no. 2 (April 1978): 144–47.

d'Agapeyeff, Alexander. *Codes and Ciphers* (London: Oxford University Press, 1939).

d'Agapeyeff, Alexander. *Maps* (London: Oxford University Press, 1942). Some online sources claim that this book appeared before d'Agapeyeff's *Codes and Ciphers*, but that's not correct.

d'Agapeyeff cryptogram revisited, http://www.rodinbook.nl/dagapeyeff.html.

Lann, Jew-Lee Irena. D'Agapeyeff's Code: A New Breakthrough Leads to A New Paradigm, May 25, 2009, available online at www.thekryptosproject.com/tjp/release.doc.

Rugg, Gordon, and Gavin Taylor. A very British mystery: The case of the D'Agapeyeff Cipher, https://hydeandrugg.wordpress.com/2013/07/02/a-very-british-mystery-the-case-of-the-dagapeyeff-cipher/ July 2, 2013, last accessed May 19, 2015.

Rugg, Gordon, and Gavin Taylor. A very British mystery, part 2: The D'Agapeyeff Cipher and the first edition, https://hydeandrugg.wordpress.com/2013/07/17/a-very-british-mystery-part-2-the-dagapeyeff-cipher-and-the-first-edition/.

Rugg, Gordon, and Gavin Taylor. A very British mystery, part 3: The D'Agapeyeff Cipher's Table of Contents, https://hydeandrugg.wordpress.com/2013/07/25/a-very-british-mystery-part-3-the-dagapeyeff-ciphers-table-of-contents/, July 25, 2013, last accessed May 19, 2015.

Rugg, Gordon, and Gavin Taylor. A very British mystery, part 4: Quiet Bodies, https://hydeandrugg.wordpress.com/2013/08/12/a-very-british-mystery-part-4-quiet-bodies/, August 12, 2013, last accessed May 19, 2015.

Rugg, Gordon, and Gavin Taylor. A very British mystery, part 5: Gavin Finds a Typo, https://hydeandrugg.wordpress.com/2013/08/16/a-very-british-mystery-part-5-gavin-finds-a-typo/, August 16, 2013, last accessed May 19, 2015.

Rugg, Gordon, and Gavin Taylor. A very British mystery, part 6: A possible solution? Probably not . . .. As of this writing, this installment of Rugg and Taylor's series has not yet appeared online. Rugg shared a draft version of it with me.

Shulman, David (under his ACA pen name AB STRUSE). "The D'Agapeyeff Cryptogram: A Challenge." *The Cryptogram* (April/May 1952): 39–40, 46.

Shulman, David (under his ACA pen name AB STRUSE). "D'Agapeyeff Cipher: Postscript." *The Cryptogram* (March/April 1959): 80–81.

Van Zandt, Armand. His proposed solution has been archived at https://web.archive.org/web/20131207173351/http://www.gather.com/viewArticle.action?articleId=281474981054022.

ON KRYPTOS

Andrews, Robert M. "Sculpture with Code Poses Mystery at the CIA." *Los Angeles Times*, April 28, 1991, p. A34, available online at http://articles.latimes.com/1991-04-28/news/mn-1468_1_mystery-sculpture.

Bauer, Craig P. *Secret History: The Story of Cryptology* (Boca Raton, FL: Chapman and Hall/CRC, 2013).

Bauer, Craig, Gregory Link, and Dante Molle. "James Sanborn's *Kryptos* and the Matrix Encryption Conjecture." *Cryptologia* 40, no. 6 (November 2016), 541–52. .

Bauer, Craig, and Katherine Millward. "Cracking Matrix Encryption Row by Row." *Cryptologia* 31, no. 1 (January 2007). A good attack on matrix encryption is described in this paper, but it requires that the numerical assignments for the letters be known. Since publication, other authors have improved this attack.

Brown, Dan. *The Da Vinci Code* (New York: Doubleday, 2003).

Brown, Dan. *The Lost Symbol* (New York: Doubleday, 2009).

Central Intelligence Agency. *Central Intelligence Agency Employee Bulletin*. June 1, 1988, released in response to a Freedom of Information Act (FOIA) request.

Central Intelligence Agency. *Central Intelligence Agency Employee Bulletin*. May 17, 1989, released in response to a Freedom of Information Act (FOIA) request.

Central Intelligence Agency. "The Puzzle at CIA Headquarters: Cracking the Courtyard Crypto." *Studies in Intelligence* 43, no. 1 (1999): 11 pages. Available online at http://www2.gwu.edu/~nsarchiv/NSAEBB /NSAEBB431/docs/intell_ebb_010.PDF.

Dunin, Elonka. *Kryptos*, http://elonka.com/kryptos/.

Ellis, David, reported by Daniel S. Levy. "The Spooks' Secret Garden." *Time*, March 18, 1991.

Gillogly, James (under his ACA pen name, Scryer). "The Kryptos Sculpture Cipher: A Partial Solution." *The Cryptogram* 65, no. 5 (September–October 1999): 1–7.

Gillogly, James (under his ACA pen name, Scryer). "Kryptos Clue." *The Cryptogram* (January–February 2011): 11.

Hill, Lester S. "Cryptography in an Algebraic Alphabet." *The American Mathematical Monthly* 36, no. 6 (1929): 306–12.

Markoff, John. "C.I.A.'s Artistic Enigma Yields All but Final Clue." *New York Times*, June 16, 1999, p. A24. Available online at http://www.nytimes.com/library/tech/99/06/biztech/articles/16code.html. This piece gets the artist's name wrong in a caption, calling him David Sanborn, instead of James Sanborn (he actually prefers Jim). The article also errs in attributing the sculpture's dedication to October 1990 instead of November 1990.

National Security Agency. "CIA KRYPTOS Sculpture—Challenge and Resolution." *United States Government—Memorandum*, June 9, 1993. Available online at https://docs.google.com/file/d /0B7G1aFZQuZtXRmRkcmhkNGtqQ2c/edit.

Overbey, J., W. Traves, and J. Wojdylo. "On the Keyspace of the Hill Cipher." *Cryptologia* 29, no. 1 (January 2005): 59–72.

Schwartz, John. "Clues to Stubborn Secret in C.I.A.'s Backyard." *New York Times*, November 20, 2010.

ON CICADA 3301

Anonymous. *How Hackers Changed the World* (BBC Documentary) https://www.youtube.com /watch?v=jwA4tSwmS-U , last accessed March 3, 2015.

Bell, Chris. "The Internet Mystery That Has the World Baffled." *The Telegraph* (November 25, 2013), http://www.telegraph.co.uk/technology/internet/10468112/The-internet-mystery-that-has-the -world-baffled.html.

Bell, Chris. "Cicada 3301 Update: The Baffling Internet Mystery Is Back." *The Telegraph* (January 7, 2014), http://www.telegraph.co.uk/technology/internet/10555088/Cicada-3301-update-the-baffling -internet-mystery-is-back.html.

Dailey, Timothy. *The Paranormal Conspiracy: The Truth about Ghosts, Aliens, and Mysterious Beings* (Minneapolis: Chosen Books, 2015). Chapter 10 is titled "The Mystery of Cicada 3301."

Eriksson, Joel. "Cicada 3301," *ClevCode*, http://www.clevcode.org/cicada-3301/. Scroll down to get to a clear explanation of the beginning of the 2012 Cicada 3301 puzzle.

Ernst, Douglas. "Secret Society Seeks World's Brightest: Recruits Navigate 'Darknet' Filled with Terrorism, Drugs." *The Washington Times*, November 26, 2013, available online at http://www .washingtontimes.com/news/2013/nov/26/secret-society-seeks-worlds-smartest-cicada-3301-r/.

Ethliel. The Forum > Technical Corner > Hidden message in image, Two Cans and String, http:// twocansandstring.com/forum/technical/4123/.

Grothaus, Michael. Meet the man who solved the mysterious cicada 3301 puzzle, http://www .fastcompany.com/3025785/meet-the-man-who-solved-the-mysterious-cicada-3301-puzzle.

Kushner, David. "Cicada: Solving the Web's Deepest Mystery." *Rolling Stone,* no. 1227 (January 15, 2015): 52–59, available online at http://www.rollingstone.com/culture/features/cicada-solving-the-webs-deepest-mystery-20150115.

NPR Staff. "The Internet's Cicada: A Mystery without an Answer." *All Things Considered,* National Public Radio, January 5, 2014. Text and audio recording also available at link, http://www.npr.org/2014/01/05/259959632/the-internets-cicada-a-mystery-without-an-answer.

Nursall, Kim. "Cicada 3301—The hunt continues in 2014." *Toronto Star,* January 10, 2014, available online at http://www.thestar.com/life/2014/01/10/cicada_3301_the_hunt_continues_in_2014.html.

Øverlier, Lasse, and Paul Syverson. "Locating Hidden Servers." *IEEE Symposium on Security and Privacy* (May 2006).

Uncovering Cicada, http://uncovering-cicada.wikia.com/wiki/Uncovering_Cicada_Wiki. This is an extensive site that covers all of the challenges to date.

Vincent, James. "Masonic Conspiracy or MI6 Recruitment Tool? Internet Mystery Cicada 3301 Starts Up Again." *Belfast Telegraph,* July 1, 2014, available online at http://www.belfasttelegraph.co.uk/technology/masonic-conspiracy-or-mi6-recruitment-tool-internet-mystery-cicada-3301-starts-up-again-29896340.html.

ON JAHBULONIAN

Amland, Bjoern, and Sarah DiLorenzo. "Lawyer: Norway Suspect Wanted a Revolution." July 24, 2011, available online at http://news.yahoo.com/lawyer-norway-suspect-wanted-revolution-100757635.html.

Belfield, Richard. *The Six Unsolved Ciphers: Inside the Mysterious Codes that Have Confounded the World's Greatest Cryptographers* (Berkeley, CA: Ulysses Press, 2007). Chapter 4 is titled "Shugborough: The Shepherd's Monument." I chose not to include an in-depth treatment of this cipher (and many others) to prevent the present volume from growing to thousands of pages.

Defalcouss, Liath. "Jahbulonian 'Sons of Fallen' Oddity Data." *Cybercomopolotian,* https://cybercosmopolitan.wordpress.com/p-c-c-t-jahbulonian-website/jahbulonian-sons-of-fallen-oddity-data/.

Herstein, Olivia. "Still Standing—As One." *Viking* 108, no. 12 (December 2011): 16–20, 22, 24. This article details Breivik's attack and the aftermath. It includes the visit of Norway's prime minister, Jens Stoltenberg, to a mosque and his quote, "We will be one community, across religion, ethnicity, gender, and rank."

Jahbulon, https://en.wikipedia.org/wiki/Jahbulon.

Norway Terror Attacks Fast Facts, CNN Library, July 17, 2015, available online at http://www.cnn.com/2013/09/26/world/europe/norway-terror-attacks/.

Ramsden, Dave. *Unveiling the Mystic Ciphers: Thomas Anson and the Shepherd's Monument Inscription* (CreateSpace, 2014).

"The True Sons of the Fallen Are Back—Mysterious Website." Before It's News, December 6, 2013, http://beforeitsnews.com/strange/2013/12/the-true-sons-of-the-fallen-are-back-mysterious-website-2453196.html.

ON BEALE

Aaron, Frank H. "Historical Facts Supporting Beale." *Proceedings of the Fourth Beale Cypher Association,* 1986 (Warrington, PA: Beale Cipher Association, 1987).

Anonymous. "Believers Still Searching for the Beale Treasure." *Bedford Bulletin-Democrat*, August 31, 1967.

Anonymous. " 'Buford County' Still Attracts Buried Treasure Hunters." *Bedford Bulletin-Democrat*, August 1, 1968.

Anonymous. "Newspaper Reports Cipher of Bedford Treasure Broken." *Roanoke Times*, April 20, 1972.

Anonymous. "Beale's Treasure Tale Revived." *The News*, April 21, 1972.

Anonymous. "Using Computers to Hunt Beale Treasure." *Bedford Bulletin-Democrat*, May 4, 1972.

Anonymous. "Beale Treasure Termed Hoax." *Staunton Leader*, February 5, 1974.

Anonymous. "The Second Beale Cipher Sympoisum—Call for Papers." *Cryptologia* 3, no. 3 (1979): 191.

Anonymous. "Treasure Hunter Freed on Bond." *Roanoke World News*, January 13, 1983.

Anonymous. "Retrial Is Ordered for Woman Charged with Disturbing Graves." *Richmond Times Dispatch*, February 24, 1983.

Anonymous. "Follow Guidelines When Treasure Hunting." *Lynchburg News*, July 14, 1985.

Anonymous. "Many Still Seek Beale Treasure." *Lynchburg News*, July 14, 1985.

Atwell, Albert. "The Mystery of Beale's Treasure Solved." (Ridgeway, VA: A. L. Atwell, 1990), 36 pp.

Barnes, Raymond. "Famed 1822 Beale Treasure Led Roanoke Brothers to Futile Hunt 66 Years Ago." *Roanoke World News*, June 1, 1963.

Bauman, Ken Andrew. *The National (Beale) Treasure at Red Knee* (Pittsburgh: RoseDog Books, 2007).

Beale Cypher Association. *Proceedings of the Third Beale Cipher Symposium* (Medfield, MA: Beale Cypher Association, 1981).

Beale Cypher Association. *The Beale Ciphers in the News* (Medfield, MA: Beale Cypher Association, 1983).

Bechtel, Stefan. "Solid Gold Mystery." *Southern World* (July–August 1980).

Boegli, Jacques S. "Madison-Beale-Hite Connection." *Proceedings of the Fourth Beale Cypher Association,* 1986 (Warrington, PA: Beale Cipher Association, 1987).

Brooks, Dorothy S. "Story of Buried Treasure has Disappointing Ending." *Lynchburg Daily Advance*, May 8, 1970.

Burchard, Hank. "Legendary Treasure Quests." *The Washington Post*, October 5, 1984.

Burleson, Bill. "Trio Hunts Famed Beale Treasure." *Roanoke Times & World News*, May 8, 1962.

Burleson, Bill. "Bedford County's Buried Treasure Lures Hunters." *Lynchburg Daily Advance*, May 19, 1962.

Chan, Wayne S. "Key Enclosed: Examining the Evidence for the Missing Key Letter of the Beale Cipher." *Cryptologia* 32, no. 1 (January 2008): 33–36.

Clayton, Stan. *Beale Treasure Map to Cipher Success* (Peterborough, U.K.: FastPrint Publishing, 2012).

Daniloff, Ruth. "A Cipher's the Key to the Treasure in Them Thar Hills." *Smithsonian Magazine* (April 1981).

"Death of James B. Ward." *The Lynchburg News*, May 17, 1907.

Easterling, E. J. *In Search of a Golden Vault: The Beale Treasure Mystery* (Roanoke, VA: Avenel, 1995).

Gillogly, James. "The Beale Cipher: A Dissenting Opinion." *Cryptologia* 4, no. 2 (April 1980): 116–19.

Greaves, Richard H. "One Letter, One Enclosure Subject: The Beale Treasure." private publication, 1986.

Greaves, Richard H. "Subject: The Beale Treasure," advertisement in *Lynchburg News & Daily Advance*, September 21, 1986.

Hart, George L., Sr. The Beale Papers, Manuscript to Roanoke Library, 1952, 1964.

Hinson, Larry C. *Secret Mission of Thomas Jefferson Beale: Intrigue and Hidden Treasure—With Beale Code 3 Solved* (Denver: Outskirts Press, Inc., 2011).

Hohmann, Robert E. "Beale Code No. 3 Deciphered." *True Treasure Mag.* (March–April 1973).

Holst, Per A. *Handbook of the Beale Ciphers* (Medfield, MA: Beale Cypher Association, 1980).

Innis, Pauline B. "The Beale Fortune." *Argosy Magazine* (August 1964).

Innis, Pauline B. and Walter Dean Innis. *Gold in the Blue Ridge, the True Story of the Beale Treasure* (Washington, DC: R. B. Luce, 1973).

Jolley, Boyd M. "Has Beale's Fabulous Treasure Been Found?" *Treasure Mag.* (August 1982).

Jolley, Boyd M. "Circle Tightens Further on Beale Treasure." *Treasure Magazine* (December 1982).

Kendall, Ray. *Solved: The T. J. Beale Treasure Code of 1822* (Birmingham, AL: Colonial Press, 1998).

Kennedy, Joe. "And the Treasure Hunt Continues." *Roanoke Times*, August 11, 1982.

Kenny, Tom. 30 Million Dollar Beale Treasure Hoax, Private, 1990.

King, John C. "A Reconstruction of the Key to Beale Cipher Number Two." *Cryptologia* 17, no. 3 (July 1993): 305–18. In this paper, King gave a more thorough analysis of the unlikely patterns that Gillogly found arose in Cipher 1, when the Declaration of Independence was applied as the key.

Kruh, Louis. "Reminiscences of a Master Cryptologist." *Cryptologia* 4, no. 1 (January 1980): 45–50. The master cryptologist was Frank Rowlett. These reminiscences include a paragraph where Rowlett talks about the Beale and Swift ciphers, as well as another similar one of which he couldn't recall many details. He believed all three to be hoaxes.

Kruh, Louis. "Beale Society Material: Book Reviews." *Cryptologia* 6, no. 1 (January 1982): 39.

Kruh, Louis. "A Basic Probe of the Beale Cipher as a Bamboozlement." *Cryptologia* 6, no. 4 (October 1982): 378–82. [This is a transcript of a slide-illustrated talk delivered by the author at the Third Beale Cipher Symposium, held September 12, 1981, in Arlington, VA.]

Kruh, Louis. "The Beale Cipher as a Bamboozlement—Part II." *Cryptologia* 12, no. 4 (October 1988): 241–46.

Leighton, Albert C., and Stephen M. Matyas. "Search for the Key Book to Nicholas Trist's Book Ciphers." *Cryptologia* 7, no. 4 (October 1983).

Masters, Al. "Has the Beale Treasure Code Been Solved?" *True Treasure Mag.* (September–October 1968).

Mateer, Todd D. "Cryptanalysis of Beale Cipher Number Two." *Cryptologia* 37, no. 3 (July 2013): 215–32.

Matyas, Stephen M. *The Beale Ciphers: Containing Research Findings, and Documents and Data, As Well As Several Predictions about How The Ciphers Were Constructed* (Poughkeepsie, NY: Published privately, 1966).

McCartney, Sean. *Breaking the Beale Code: The Treasure Hunters Club Book 2* (Ogden, UT: Mountainland Publishing, Inc., 2011).

Nelson, Carl W., Jr. Historical & Analytical Studies in Relation to the Beale Cyphers, Proprietary, 1970.

Nickell, Joe. "Uncovered—The Fabulous Silver Mines of Swift and Filson." *Filson Club History Quarterly* LIV (1980): 325–45. This paper details an earlier fictional story that has much in common with the Beale Papers. Did it inspire the unknown author?

Nickell, Joe. "Discovered, The Secret of Beale's Treasure." *The Virginia Magazine of History and Biography* 90, no. 3 (July 1982): 310–24.

Nickell, Joe, and John F. Fischer. *Mysterious Realms: Probing Paranormal, Historical, and Forensic Enigmas* (Buffalo, NY: Prometheus Books, 1992), 53–67.

Nicklow, Douglas. "Beale's Buried Treasure." *RUN* (August 1984): 48–57.

Ostler, Reinhold. "100 Millionen! Das Gold in der Höhle." *Bild am Sonntag* (Berlin: Axel Springer AG, December 1, 1985).

Poe, Edgar Allan. "The Gold Bug." 1843. This short story has some features in common with the Beale papers. In particular, consider the style of the beginnings. "The Gold Bug" may have served as inspiration for a hoax. First saw print in two installments in *Philadelphia Dollar Newspaper*, June 21 and June 28, 1843.

Price, Steve. "Bedford Treasure Hunt Goes On and On." *Lynchburg News*, August 20, 1967.

Ray, Richard. "Silent for Years, Famous THer Reveals New Clues to Famous Cache." *Treasure Search Mag.* (January–February 1987).

Rubin, Robert. "Jail Is at the End of Treasure-Hunter's Rainbow." *Roanoke Times and World News*, January 15, 1983.

Smith, Becca C. *Alexis Tappendorf and the Search for Beale's Treasure*, The Alexis Tappendorf Series, Book 1, Kindle ed. (Red Frog Publishing, 2012).

Timm, John W. *Mystery Treasure* (Bedford, VA: Tracy Book Co., 1973).

Viemeister, Peter. *The Beale Treasure: New History of a Mystery* (Bedford, VA: Hamilton's, 1997).

Ward, James B. *The Beale Papers* (Lynchburg, VA: Virginian Book and Print Job, 1885).

Williams, Clarence R. The Beale Papers, Library of Congress, Legislative Reference Service Memorandum, April 26, 1934.

Yancey, Dwayne. "Buried Treasure in the Blue Ridge." *Commonwealth* (September 1980).

ON FENN AND OTHER TREASURE CIPHERS

Charroux, Robert. *Trésors du monde, trésors de France, trésors de Paris: Enterrés, emmurés, engloutis* (Paris: Fayard, 1972).

de La Roncière, Charles. *Le Flibustier mystérieux, histoire d'un trésor cache* (Paris: Masque, *1934*).

Deutschmann, Jennifer. "Forrest Fenn: Hunt for Buried Treasure Is 'Out of Control.' " *Inquisitr* (April 28, 2015), available online at http://www.inquisitr.com/2050321/forrest-fenn/.

Fenn, Forrest. *The Thrill of the Chase, A Memoir* (Santa Fe, NM: One Horse Land & Cattle Co., 2010). The relevant excerpt is available online at https://www.oldsantafetradingco.com/assets/book-previews/thrill-of-the-chase.pdf.

Fenn, Forrest. *Too Far to Walk* (Santa Fe, NM: One Horse Land & Cattle Co., 2013).

Ford, Dana. "California Couple Strikes Gold after Finding $10 Million in Rare Coins." CNN, February 26, 2014, available online at http://www.cnn.com/2014/02/25/us/california-gold-discovery/.

Goldsmith, Margie. "Well Over $1 Million in Buried Treasure: Find It!" *Huffington Post*, February 18, 2011, updated February 27, 2013, available online at http://www.huffingtonpost.com/margie-goldsmith/over-1-million-in-buried-_b_822894.html.

Lammle, Rob. Get Rich Quick: 6 People Who Accidentally Found a Fortune, *mental_floss*, February 26, 2014, available online at http://mentalfloss.com/article/22449/get-rich-quick-6-people-who-accidentally-found-fortune.

McCurley, Kevin. Cryptograms on Gold Bars from China, International Association for Cryptologic Research, http://www.iacr.org/misc/china/. This is, sadly, the best reference for this story. It has pictures of all of the bars, but very little information beyond that.

Thomas, Athol. *Forgotten Eden: A View of the Seychelles Islands in the Indian Ocean* (London and Harlow, U.K.: Longmans, Green and Co. Ltd., 1968).

Chapter 10

Enăchiuc, V. *Rohonczy Codex* (Bucharest, Romania: Editura Alcor, 2002). A claimed solution to the Rohonc Codex.

Gould, Stephen Jay. *Time's Arrow, Time's Cycle: Myth and Metaphor in the Discovery of Geological Time* (Cambridge, MA: Harvard University Press, 1987).

Gould, Stephen Jay. "James Hampton's Throne and the Dual Nature of Time." *Smithsonian Studies in American Art* 1, no. 1 (Spring 1987): 46–57.

Gould, Stephen Jay. "Nonoverlapping Magisteria." *Natural History* 106 (March 1997): 16–22.

Hampton, James, http://fortean.wikidot.com/james-hampton.

Hampton, James writings [ca. 1950–1964], Archives of American Art, http://www.aaa.si.edu /collections/james-hampton-writings-7162.

Hartigan, Lynda Roscoe. *James Hampton: The Throne of the Third Heaven of the Nations' Millennium General Assembly* (Boston: Museum of Fine Arts, October 19 (1975?)–February 13, 1976). This essay was originally published for a 1976 exhibition of James Hampton's work.

Hartigan, Lynda Roscoe. *The Throne of the Third Heaven of the Nations' Millennium General Assembly* (Montgomery, AL: Montgomery Museum of Fine Arts, 1977).

Ingalls, Helen. "James Hampton's Throne of the Third Heaven of the Nations' Millennium General Assembly," https://www.youtube.com/watch?v=0abZVM-02IQ. This is a video of a lecture titled "James Hampton's *Throne*: All That Glitters Is Not Gold," delivered by Helen Ingalls, of the Lunder Conservation Center of the Smithsonian American Art Museum, under whose care the objects came in 1988.

Kerckhoffs, Auguste. "La Cryptographie Militaire." *Journal des science militaires* (Paris: Baudoin, 1883), vol. IX, 5–83.

Láng, Benedek. "Why Don't We Decipher an Outdated Cipher System? The Codex of Rohonc." *Cryptologia* 34, no. 2 (April 2010): 115–44.

Marshall, Steve. "St James the Janitor." *Fortean Times* no. 150 (2001). Available online at http://web .archive.org/web/20020125163459/http://www.forteantimes.com/articles/150_jamesjanitor.shtml.

Nyíri, A. "Megszólal 150 év után a Rohonci-kódex? [After 150 Years, the Rohonc Codex Starts to Speak?]." *Theologiai Szemle* 39 (1996): 91–98. A claimed solution to the Rohonc Codex.

Rugg, Gordon. Home page at Keele University, http://www.keele.ac.uk/scm/staff/academic /drgordonrugg/.

Rugg, Gordon. The Penitentia Manuscript, http://www.scm.keele.ac.uk/research/knowledge_modelling /km/people/gordon_rugg/cryptography/penitentia/index.html.

Rugg, Gordon. The Ricardus Manuscript, http://www.scm.keele.ac.uk/research/knowledge_modelling /km/people/gordon_rugg/cryptography/ricardus_manuscript.html.

Rugg, Gordon. "Visualising Structures in Ancient Texts." *Search Visualizer*, Nov. 16, 2012, https:// searchvisualizer.wordpress.com/2012/11/16/visualising-structures-in-ancient-texts/.

Rugg, Gordon. "The 'Genesis Death Sandwich' Story." *Search Visualizer*, Feb. 21, 2013, https:// searchvisualizer.wordpress.com/2013/02/21/the-genesis-death-sandwich-story/.

Schmeh, Klaus. Klaus Schmeh's List of Encrypted Books, http://scienceblogs.de/klausis-krypto -kolumne/klaus-schmehs-list-of-encrypted-books/.

Schmeh, Klaus. "The Voynich Manuscript: The Book Nobody Can Read." *Skeptical Inquirer* 35.1 (January/February 2011), available online at http://www.csicop.org/si/show/the_voynich _manuscript_the_book_nobody_can_read/_1.

Schmeh, Klaus. "The Pathology of Cryptology—A Current Survey." *Cryptologia* 36, no. 1 (January 2012): 14–45.

Schmeh, Klaus. "Neue Scans zeigen den Codex Rohonci in seiner ganzen Schönheit," April 25, 2015, http://scienceblogs.de/klausis-krypto-kolumne/2015/04/25/neu-scans-zeigen-den-codex -rohonci-in-seiner-ganzen-schoenheit/.

Schmeh, Klaus. "Encrypted Books: Mysteries that Fill Hundreds of Pages." *Cryptologia* 39, no. 4 (October 2015): 1–20.

Singh, Mahesh Kumar. "Rohonci Kódex [The Codex of Rohonc]." *Turán* 6 (2004): 9–40. Also, there is more in 1 (2005). A claimed solution to the Rohonc Codex.

Stallings, Dennis J. The Secret Writing of James Hampton, African American Sculptor, Outsider Artist, Visionary, http://ixoloxi.com/hampton/index.html.

Stamp, Mark. Hamptonese, http://www.cs.sjsu.edu/faculty/stamp/Hampton/hampton.html.

Walsh, Mike. The Miracle of St. James Hampton, *Expresso Tilt*, available online at http://www .missioncreep.com/tilt/hampton.html.

Chapter 11

Anonymous. "Signal from Mars. Professor Pickering Saw Bright Lights Upon That Planet." *Toledo Weekly Blade*, January 17, 1901, p. 5. Available online at https://news.google.com/newspapers ?nid=1350&dat=19010117&id=uO8SAAAAIBAJ&sjid=cP4DAAAAIBAJ&pg=5259,940727&hl=en.

Anonymous. "Scientific World Stirred by Possibility of Communication from Mars." *The Monthly Evening Sky Map* XIV, no. 159 (March 1920). Available online at Google Books.

Anonymous. "Mars Is Signaling to Us, Says Marconi." *The Monthly Evening Sky Map* XV, no. 179 (November 1921). Available online at Google Books.

Anonymous. "Asks Air Silence When Mars is Near. Prof. Todd Obtains Official Aid from Washington Despite Doubts of Its Efficacy." *New York Times*, August 21, 1924, available online at http:// theartpart.jonathanmorse.net/tag/david-peck-todd/.

Anonymous. "Weird 'Radio Signal' Film Deepens Mystery of Mars." *Washington Post*, August 27, 1924, available online at http://www.shorpy.com/node/12482.

Anonymous. "Seeks Sign from Mars in 30-foot Radio Film, Dr. Todd Will Study Photograph of Mysterious Dots and Dashes Recently Recorded." *New York Times*, August 28, 1924, available online at http://theartpart.jonathanmorse.net/tag/david-peck-todd/.

Bracewell, Ronald. "What to Say to the Space Probe When It Arrives." *Horizon*, January 1977.

Callimahos, Lambros D. "Communication with Extraterrestrial Intelligence." Published by the National Security Agency, pp. 79–86 and 109. I haven't been able to find the name of the journal, but a copy of the paper is available online at https://www.nsa.gov/public_info/_files/ufo /communication_with_et.pdf. NSA also published it somewhere with the page numbers 107–15. See https://www.nsa.gov/public_info/_files/tech_journals/communications_extraterrestrial _intelligence.pdf. The paper was reprinted in *Cryptologic Spectrum* 5, no. 2 (Spring 1975): 4–10, available online at https://www.nsa.gov/public_info/_files/cryptologic_spectrum/communications _with_extraterrestrial.pdf. Few papers get published three times!

Cameron, A. G. W. *Interstellar communication: A Collection of Reprints and Original Contributions* (New York: W.A. Benjamin, 1963).

Campaigne, Howard. "Key to the Extraterrestrial Messages." *NSA Technical Journal* XIV, no. 1 (Winter 1969): 13–23, available online at https://www.nsa.gov/public_info/declass/tech_journals.shtml.

Campbell, MacGregor. "Unbreakable: The MIT Time-Lock Puzzle." *New Scientist* no. 2813 (May 27, 2011).

Darling, David. "Green Bank Conference (1961)." *Encyclopedia of Science*, http://www.daviddarling .info/encyclopedia/G/GreenBankconf.html.

Darling, David. "Todd, David Peck (1855–1939)." *Encyclopedia of Science*, http://www.daviddarling .info/encyclopedia/T/Todd.html.

Ehman, Jerry R. "The Big Ear Wow! Signal What We Know and Don't Know about It after 20 Years." original draft completed: September 1, 1997. Last revision: February 3, 1998, available online at http://www.bigear.org/wow20th.htm.

Fort, Charles. *New Lands* In *The Complete Books of Charles Fort*, with an introduction by Tiffany Thayer (Omnibus ed.) (New York: Henry Holt and Co., 1941), chap. 32, p. 494. Available online in single-volume edition at http://www.resologist.net/landsei.htm—see part 2, chap. 20.

Gardner, Martin. Mathematical Games, "Thoughts on the Task of Communication with Intelligent Organisms on Other Worlds." *Scientific American* 213, no. 2 (August 1, 1965): 96–100. This paper also appears in Martin Gardner, *Martin Gardner's 6th Book of Mathematical Diversions from Scientific American* (Chicago: The University of Chicago Press, 1971) as chap. 25, "Extraterrestrial Communication." pp. 253–62. As Gardner typically did with his collected articles, an addendum is provided at the end of this piece. In this instance, it takes the form of a pair of letters related to the piece that appeared in the January 1966 issue of *Scientific American*. However, all they discuss is the reality, or not, of the canals on Mars. Not relevant to our interest here. In the body of the paper, an Arecibo-style communication scheme is discussed, but instead of primes it uses 100 bits to form a 10 × 10 square. Gardner points out, "Indeed, this is the technique by which pictures are now transmitted by radio as well as the basis of television-screen scanning." Of course, that was done at a much higher resolution.

Gregg, Justin. "Dolphins Aren't as Smart as You Think." *The Wall Street Journal*, December 18, 2013. Available online at http://www.wsj.com/articles/SB10001424052702304866904579266183573854204.

Kraus, John. "We Wait and Wonder." *Cosmic Search* 1, no. 3 (Summer 1979), available online at http:// www.bigear.org/CSMO/PDF/CS03/cs03p31.pdf. This paper is on the Wow! signal.

Lowell, Percival. *Proceedings of the American Philosophical Society* 40, no. 167 (December 1901): 166– 76. This was reprinted, with the addition of many illustrations, in the journal *Popular Astronomy* 10 (April 1902): 185–94. Available online through Google Books and http://adsabs.harvard.edu /full/1902PA.....10..185L.

Lunan, Duncan. "Spaceprobe from Epsilon Bootes." *Spaceflight*, British Interplanetary Society, 1973. This paper presented another claim of a signal being received. The Lunan references that follow continue the story.

Lunan, Duncan. "Long-Delayed Echoes and the Extraterrestrial Hypothesis." *Journal of the Society of Electronic and Radio Technicians* 10, no. 8 (September 1976).

Lunan, Duncan. *The Mysterious Signals from Outer Space* (New York: Bantam Books, 1977).

Lunan, Duncan. "Epsilon Boötis Revisited." *Analog Science Fiction and Fact* 118, no. 3 (March 1998).

Morton, Ella. "Messages to the Universe: A Short History of Interstellar Communication." *Slate*,

November 14, 2014, available online at http://www.slate.com/blogs/atlas_obscura/2014/11/14 /the_arecibo_message_and_other_interstellar_communication_attempts.html.

Oliver, Bernard M. "Interstellar Communication." *Interstellar Communication: A Collection of Reprints and Original Contributions*, ed., A. G. W. Cameron (New York: W.A. Benjamin, 1963), 294–305.

Piper, H. Beam. "Omnilingual." *Astounding Science Fiction*, February 1957, available online at http:// www.gutenberg.org/files/19445/19445-h/19445-h.htm. "To translate writings, you need a key to the code—and if the last writer of Martian died forty thousand years before the first writer of Earth was born . . . how could the Martian be translated . . .?" I won't give away the punchline, but I will reveal that it isn't mathematics that provides the first break.

Rivest, Ron, Adi Shamir, and Len Adleman. "On Digital Signatures and Public-Key Cryptosystems." MIT/LCS/TM-82, Massachusetts Institute of Technology, Laboratory for Computer Science, Cambridge, MA, 1977. There was soon a title change to "A Method for Obtaining Digital Signatures and Public-Key Cryptosystems," but the date is the same for both. This report later appeared in *Communications of the ACM* 21, no. 2 (1978): 120–26, with the latter title.

Rivest, Ronald L. Description of the LCS35 Time Capsule Crypto-Puzzle, April 4, 1999, available online at http://people.csail.mit.edu/rivest/lcs35-puzzle-description.txt.

Rivest, Ronald L., Adi Shamir, and David A. Wagner. "Time-Lock Puzzles and Timed-Release Crypto." Revised March 10, 1996, available online at http://theory.lcs.mit.edu/~rivest /RivestShamirWagner-timelock.ps.

RSA Secret-Key Challenge, http://en.wikipedia.org/wiki/RSA_Secret-Key_Challenge. This page details other challenges put forth by RSA Labs between 1997 and 2007.

Steele, Bill. "It's the 25th Anniversary of Earth's First (and Only) Attempt to Phone E.T." *Cornell News* (November 12, 1999), available online at http://web.archive.org/web/20080802005337/http:// www.news.cornell.edu/releases/Nov99/Arecibo.message.ws.html.

Wooster, Harold (moderator), Paul J. Garvin, Lambros D. Callimahos, John C. Lilly, William O. Davis, and Francis J. Heyden. "Communication with Extraterrestrial Intelligence." *IEEE Spectrum* 3, no. 3 (March 1966).

Photo and Illustration Credits

Fig. 1.19   From Newbold, William Romaine. *The Cipher of Roger Bacon*, edited with foreword and notes by Roland Grubb Kent, Philadelphia: University of Pennsylvania Press; London, H. Milford, Oxford UniversityPress, 1928)

Fig. 1.20   From Yale, Beinecke Rare Book & Manuscript Library, with permission

Fig. 1.21   From Yale, Beinecke Rare Book & Manuscript Library, with permission

Fig. 1.22   From "Something More Than a Secretary," *Christian Science Monitor*, Sept. 30, 1924

Fig. 1.23   From Kraus, Hans Peter, *A Rare Book Saga*, New York: G.P. Putnam's Sons, 1978. Courtesy of Mary Ann Kraus Folter

Fig. 1.24   From D'Imperio, M. E., *The Voynich Manuscript—An Elegant Enigma*, National Security Agency, 1976

Fig. 1.25   Courtesy of the National Cryptologic Museum

Fig. 1.26   From D'Imperio, M. E., *The Voynich Manuscript—An Elegant Enigma*, National Security Agency, 1976

Fig. 1.27   Courtesy of the National Cryptologic Museum

Fig. 1.28   D'Imperio, M. E., *The Voynich Manuscript—An Elegant Enigma*, National Security Agency, 1976

Fig. 1.29   From Rugg, Gordon, "An Elegant Hoax? A Possible Solution to the Voynich Manuscript," *Cryptologia*, Vol. 28, No. 1, January 2004, with permission

Fig. 1.30   Courtesy of the National Cryptologic Museum

Fig. 2.1   © The Egypt Centre, Swansea University, with permission

Fig. 2.3   © Musée du Louvre, Dist. RMN-Grand Palais / Hervé Lewandowski / Art Resource, NY

Fig. 2.4   © Musée du Louvre, Dist. RMN-Grand Palais / Hervé Lewandowski / Art Resource, NY

Fig. 2.5   © Musée du Louvre, Dist. RMN-Grand Palais / Hervé Lewandowski / Art Resource, NY

Fig. 2.6   Courtesy of Adrienne Mayor

Fig. 2.7   © Musée du Louvre, Dist. RMN-Grand Palais / Hervé Lewandowski / Art Resource, NY

Fig. 2.8   From Vanderpool, E., "An Unusual Black-Figured Cup," *American Journal of Archaeology*, Vol. 49, No. 4, pp. 436–440 (fig. 1), October 1945. Courtesy of the Archaeological Institute of America and the *American Journal of Archaeology*

Fig. 2.9   From Vanderpool, E., "An Unusual Black-Figured Cup," *American Journal of Archaeology*, Vol. 49, No. 4, pp. 436–440 (fig. 3), October 1945. Courtesy of the Archaeological Institute of America and the *American Journal of Archaeology*

Fig. 2.10   From Riksantikvarieämbetet / Swedish National Heritage Board, with permission

Fig. 2.11   From Franksen, O. I., *Mr. Babbage's Secret: The Tale of a Cypher—and APL*, Englewood Cliffs, NJ: Prentice Hall, 1984

Fig. 2.12   From Franksen, O. I., *Mr. Babbage's Secret: The Tale of a Cypher—and APL*, Englewood Cliffs, NJ: Prentice Hall, 1984

Fig. 2.13   From Riksantikvarieämbetet / Swedish National Heritage Board, with permission

Fig. 2.15   From Riksantikvarieämbetet / Swedish National Heritage Board, with permission

Fig. 2.18   From Franksen, O. I., *Mr. Babbage's Secret: The Tale of a Cypher—and APL*, Englewood Cliffs, NJ: Prentice Hall, 1984

Fig. 2.19   From Wimmer, L. F. A. *De danske runemindesmærker*, København: Gyldendal, 1914.

Fig. 2.20   From Riksantikvarieämbetet / Swedish National Heritage Board, with permission

Fig. 2.21   From Riksantikvarieämbetet / Swedish National Heritage Board, with permission

Fig. 2.22	From Riksantikvarieämbetet / Swedish National Heritage Board, with permission
Fig. 2.23	From Riksantikvarieämbetet / Swedish National Heritage Board, with permission
Fig. 2.24	From Riksantikvarieämbetet / Swedish National Heritage Board, with permission
Fig. 2.25	From Riksantikvarieämbetet / Swedish National Heritage Board, with permission
Fig. 2.26	From Riksantikvarieämbetet / Swedish National Heritage Board, with permission
Fig. 2.27	From Riksantikvarieämbetet / Swedish National Heritage Board, with permission
Fig. 2.28	From Riksantikvarieämbetet / Swedish National Heritage Board, with permission
Fig. 2.29	From Riksantikvarieämbetet / Swedish National Heritage Board, with permission
Fig. 2.30	Based on Sverigekarta-Landskap.svg by Lapplänning
Fig. 3.1	From Elgar, Edward, *My Friends Pictured Within*, Novello & Co Ltd, 1946
Fig. 3.2	From Elgar, Edward, *My Friends Pictured Within*, Novello & Co Ltd, 1946
Fig. 3.3	From Powell, Mrs. Richard, *Edward Elgar: Memories of a Variation*, London: Oxford University Press, 1937
Fig. 3.4	Redrawn from Elgar, Edward and Jerrold Northrup Moore, *Edward Elgar: Letters of a Lifetime*, Oxford University Press, 1990
Fig. 3.5	Reproduced by kind permission of the Elgar Will Trust and the Elgar Birthplace Museum
Fig. 3.6	Reproduced by kind permission of the Elgar Will Trust and the Elgar Birthplace Museum
Fig. 3.7	Reproduced by kind permission of the Elgar Will Trust and the Elgar Birthplace Museum
Fig. 3.8	Reproduced by kind permission of the Elgar Will Trust and the Elgar Birthplace Museum
Fig. 3.9	From Schooling, John Holt, "Secrets in Cipher IV. Form the time of George II to the present day," *The Pall Mall Magazine*, Vol. 8, No. 36, April 1896
Fig. 3.10	From Schooling, John Holt, "Secrets in Cipher IV. Form the time of George II to the present day," *The Pall Mall Magazine*, Vol. 8, No. 36, April 1896
Fig. 3.11	From Schooling, John Holt, "Secrets in Cipher IV. Form the time of George II to the present day," *The Pall Mall Magazine*, Vol. 8, No. 36, April 1896
Fig. 3.12	Reproduced by kind permission of the Elgar Will Trust and the Elgar Birthplace Museum
Fig. 3.13	Reproduced by kind permission of the Elgar Will Trust and the Elgar Birthplace Museum
Fig. 3.14	Reproduced by kind permission of the Elgar Will Trust and the Elgar Birthplace Museum
Fig. 3.15	Reproduced by kind permission of the Elgar Will Trust and the Elgar Birthplace Museum
Fig. 3.16	Reproduced by kind permission of the Elgar Will Trust and the Elgar Birthplace Museum
Fig. 3.17	Reproduced by kind permission of the Elgar Will Trust and the Elgar Birthplace Museum
Fig. 3.18	Reproduced by kind permission of the Elgar Will Trust and the Elgar Birthplace Museum
Fig. 3.19	Reproduced by kind permission of the Elgar Will Trust and the Elgar Birthplace Museum
Fig. 3.20	Reproduced by kind permission of the Elgar Will Trust and the Elgar Birthplace Museum
Fig. 3.21	Reproduced by kind permission of the Elgar Will Trust and the Elgar Birthplace Museum
Fig. 3.22	Reproduced by kind permission of the Elgar Will Trust and the Elgar Birthplace Museum
Fig. 3.23	Reproduced by kind permission of the Elgar Will Trust and the Elgar Birthplace Museum
Fig. 3.24	From Powell, Mrs. Richard, *Edward Elgar: Memories of a Variation*, London: Oxford University Press, 1937
Fig. 3.25	From the collection of the author
Fig. 4.1	© Bettmann / Getty Images
Fig. 4.2	© Bettmann / Getty Images
Fig. 4.3	© Bettmann / Getty Images

Fig. 4.4   From http://www.dajia.info/ikeymview-michael-mageau.html

Fig. 4.5   From Graysmith, Robert, *Zodiac*, New York: St. Martin's / Marek, 1986

Fig. 4.6   From Graysmith, Robert, *Zodiac*, New York: St. Martin's / Marek, 1986

Fig. 4.7   From Graysmith, Robert, *Zodiac*, New York: St. Martin's / Marek, 1986

Fig. 4.8   © Bettmann / Getty Images

Fig. 4.9   © Associated Press

Fig. 4.10   © Associated Press

Fig. 4.11   © Bettmann / Getty Images

Fig. 4.12   The cover of *Argosy*, September 1970

Fig. 4.13   Retrieved from http://zodiackillerfacts.com/feed/

Fig. 4.14   © Associated Press

Fig. 4.15   © *San Francisco Chronical* / Polari

Fig. 4.16   Courtesy of the National Cryptologic Museum

Fig. 4.17   From DC Entertainment

Fig. 4.18   From www.zodiackillerfacts.com/radian.htm

Fig. 4.19   From www.zodiackillerfacts.com/radian.htm

Fig. 4.20   From www.zodiackillerfacts.com/radian.htm

Fig. 4.22   Modified from www.zodiackillerfacts.com/radian.htm

Fig. 5.1   From the Collection of the Adirondack History Museum/Essex County Historical Society, with permission

Fig. 5.2   From the Collection of the Adirondack History Museum/Essex County Historical Society, with permission

Fig. 5.3   From the Collection of the Adirondack History Museum/Essex County Historical Society, with permission

Fig. 5.4   From the Collection of the Adirondack History Museum/Essex County Historical Society, with permission

Fig. 5.5   From http://stp.lingfil.uu.se/~bea/copiale/. Courtesy of Beáta Megyesi

Fig. 5.6   From http://stp.lingfil.uu.se/~bea/copiale/. Courtesy of Beáta Megyesi

Fig. 5.7   From Morris, S. Brent, *The Folger Manuscript: The Cryptanalysis and Interpretation of an American Masonic Manuscript*, Ft. Meade, MD, 1992, with permission

Fig. 5.8   From Morris, S. Brent, *The Folger Manuscript: The Cryptanalysis and Interpretation of an American Masonic Manuscript*, Ft. Meade, MD, 1992, with permission

Fig. 5.9   From the Collection of the Adirondack History Museum/Essex County Historical Society, with permission

Fig. 5.10   From the Collection of the Adirondack History Museum/Essex County Historical Society, with permission

Fig. 5.11   From Mahon, Tom and Gillogly, James J., *Decoding the IRA*, Cork, Ireland: Mercier Press, 2008. Courtesy of Jim Gillogly

Fig. 5.12   From http://coldcasecameron.com/wp-content/uploads/2014/02/John-Walsh -Zodiac-Killer-Letter.pdf

Fig. 5.13   From http://cipherfoundation.org/modern-ciphers/scorpion-ciphers

Fig. 5.14   From http://cipherfoundation.org/modern-ciphers/scorpion-ciphers

Fig. 5.15   From http://coldcasecameron.com/wp-content/uploads/2014/02/John-Walsh
             -Zodiac-Killer-Letter.pdf

Fig. 5.16   From http://coldcasecameron.com/wp-content/uploads/2014/02/John-Walsh
             -Zodiac-Killer-Letter.pdf

Fig. 5.17   From http://coldcasecameron.com/wp-content/uploads/2014/02/John-Walsh
             -Zodiac-Killer-Letter.pdf

Fig. 5.18   From *The Phoenician*, Summer 2006, p. 26

Fig. 5.19   From http://www.thephoenixsociety.org/puzzles/special_puzzle.htm

Fig. 5.20   From http://scienceblogs.de/klausis-krypto-kolumne/2014/03/12/codeknacker
             -auf-verbrecherjagd-folge-4-der-maskenmann/

Fig. 6.1   Redrawn from Feltus, Gerald (Gerry) Michael, *The Unknown Man: A Suspicious Death at*
             Somerton Park, Richmond, Australia: Hyde Park Press, 2010, with permission

Fig. 6.2   From Kerry Greenwood, *Tamam Shud: The Somerton Man Mystery*, Sydney, Australia:
             NewSouth Publishing, University of New South Wales Press Ltd., 2012, with permission

Fig. 6.3   From Feltus, Gerald (Gerry) Michael, *The Unknown Man: A Suspicious Death at Somer-*
             ton Park, Richmond, Australia: Hyde Park Press, 2010, with permission

Fig. 6.4   Retrieved from https://www.eleceng.adelaide.edu.au/personal/dabbott
             /wiki/index.php/The_Taman_Shud_Case_Coronial_Inquest#Exhibit_C.2.

Fig. 6.5   From Feltus, Gerald (Gerry) Michael, *The Unknown Man: A Suspicious Death at Somer-*
             ton Park, Richmond, Australia: Hyde Park Press, 2010, with permission

Fig. 6.6   From the South Australia Police Historical Society

Fig. 6.7   From Feltus, Gerald (Gerry) Michael, *The Unknown Man: A Suspicious Death at Somerton*
             Park, Richmond, Australia: Hyde Park Press, 2010, with permission

Fig. 6.8   From Feltus, Gerald (Gerry) Michael, *The Unknown Man: A Suspicious Death at Somerton*
             Park, Richmond, Australia: Hyde Park Press, 2010, with permission

Fig. 6.9   From the Australian police

Fig. 6.14   From *The Phoencian*, Winter 2012–13

Fig. 6.15   From "The Body on the Beach: The Somerton Man–Taman Shud Case," http://hubpages.
             com/education/The-Mystery-of-the-Somerton-Man-Taman
             -Shud-Case

Fig. 6.16   Redrawn from "The Body on the Beach: The Somerton Man–Taman Shud Case," http://
             hubpages.com/education/The-Mystery-of-the-Somerton-Man
             -Taman-Shud-Case

Fig. 6.17   From the Australian police

Fig. 6.18   Retrieved from https://en.wikipedia.org/wiki/File:SomertonManEars.jpg

Fig. 6.19   From "Perth Poet-Suicide Chose Omar Verse as His Epitaph," *Mirror*, (Perth, WA), August 25,
             1945, http://trove.nla.gov.au/ndp/del/article/76017408

Fig. 6.20   From "Perth Poet-Suicide Chose Omar Verse as His Epitaph," *Mirror*, (Perth, WA), August 25,
             1945, http://trove.nla.gov.au/ndp/del/article/76017408

Fig. 6.21   Redrawn from "The Body on the Beach: The Somerton Man–Taman Shud Case," http://
             hubpages.com/education/The-Mystery-of-the-Somerton-Man
             -Taman-Shud-Case

Fig. 6.22   From the FBI

Fig. 6.23   From collection of the author

Fig. 6.24   From *Information Bulletin, the Monthly Magazine of the Office of US High Commissioner for Germany*, March 1953, https://commons.wikimedia.org /wiki/File:Dr._James_B_Conant_1953_Berlin.jpeg

Fig. 6.25   From *Galaxy Science Fiction*, February 1953

Fig. 6.26   Courtesy of the National Cryptologic Museum

Fig. 6.28   From http://www.investigatingcrimes.com/unsolved-ciphers-hiding-murder -mysteries/, accessed February 27, 2016.

Fig. 6.29   From the FBI

Fig. 6.30   From the FBI

Fig. 7.1   From the Master and Fellows of Corpus Christi College, Cambridge, with permission

Fig. 7.2   Courtesy of Paragon House

Fig. 7.4   From Berger, Arthur S., *Aristocracy of the Dead: New Findings in Postmortem Survival*, London and Jefferson, NC: McFarland & Co., 1987

Fig. 7.5   Courtesy of Jim Tucker

Fig. 8.1   Courtesy of Dirk Rijmenants

Fig. 8.2   Courtesy of the National Cryptologic Museum

Fig. 8.3   Courtesy of Nicholas Gessler

Fig. 8.4   Courtesy of the National Cryptologic Museum

Fig. 8.5   Courtesy of the National Cryptologic Museum

Fig. 9.1   From d'Agapeyeff, Alexander, *Codes and Ciphers*, London: Oxford University Press, 1939

Fig. 9.2   From d'Agapeyeff, Alexander, *Codes and Ciphers*, London: Oxford University Press, 1939

Fig. 9.3   From d'Agapeyeff, Alexander, *Codes and Ciphers*, London: Oxford University Press, 1939

Fig. 9.4   From the collection of the author

Fig. 9.5   Courtesy of James Sanborn

Fig. 9.6   From https://www.cia.gov/about-cia/headquarters-tour/kryptos/Kryptos Print.pdf

Fig. 9.7   From https://www.cia.gov/about-cia/headquarters-tour/kryptos/Kryptos Print.pdf

Fig. 9.8   From https://stilljane7.files.wordpress.com/2015/02/kryptos_transcript.jpg

Fig. 9.9   From the secretive group Cicada 3301, retrieved from http://uncovering -cicada.wikia.com/wiki/What_Happened_Part_1_(2012)

Fig. 9.10   From the secretive group Cicada 3301, retrieved from http://i.imgur.com /m9sYK.jpg

Fig. 9.11   From the secretive group Cicada 3301, retrieved from http://www.reddit .com/r/a2e7j6ic78h0j/

Fig. 9.12   From the secretive group Cicada 3301, retrieved from http://uncovering -cicada.wikia.com/wiki/What_Happened_Part_1_(2012)

Fig. 9.13   From the secretive group Cicada 3301, retrieved from http://uncovering -cicada.wikia.com/wiki/What_Happened_Part_1_(2012)

Fig. 9.14   Retrieved from http://www.davidkushner.com/article/cicada-solving-the -webs-deepest-mystery/

Fig. 9.15	From the secretive group Cicada 3301, retrieved from http://i.imgur.com /vjuNp.jpg
Fig. 9.16	Retrieved from https://web.archive.org/web/20101020055419/http://pccts.com/
Fig. 9.17	Retrieved from http://jahbulonian.byethost7.com/
Fig. 9.18	Retrieved from http://jahbulonian.byethost7.com/
Fig. 9.19	Courtesy of Andrew Baker
Fig. 9.20	Courtesy of Andrew Baker
Fig. 9.21	Retrieved from http://jahbulonian.byethost7.com/
Fig. 9.23	Retrieved from http://jahbulonian.byethost7.com/666partsofhell.html
Fig. 9.24	Modified from http://jahbulonian.byethost7.com/666partsofhell.html
Fig. 9.25	From Ward, James B. *The Beale Papers*, Lynchburg, VA: Virginian Book and Print Job, 1885
Fig. 9.26	Courtesy of Addison Doty
Fig. 9.27	From Fenn, Forrest, *Too Far to Walk*, Santa Fe, NM: One Horse Land & Cattle Co., 2013, with permission
Fig. 9.28	Courtesy of the IACR
Fig. 9.29	Courtesy of the IACR
Fig. 9.30	Courtesy of the IACR
Fig. 9.31	From de la Roncière, Charles, *Le Flibustier mystérieux, histoire d'un trésor cache*, Paris, 1934
Fig. 9.34	From http://www.findagrave.com/cgi-bin/fg.cgi?page=gr&GRid=8179530
Fig. 10.1	From Codex Rohonci [16th–19th c.?], Library and Information Centre of the Hungarian Academy of Sciences, Department of Manuscripts and Rare Books (Budapest, Hungary), K 114. Published with the permission of the Library and Information Centre of the Hungarian Academy of Sciences (Budapest, Hungary)
Fig. 10.2	From Codex Rohonci [16th–19th c.?], Library and Information Centre of the Hungarian Academy of Sciences, Department of Manuscripts and Rare Books (Budapest, Hungary), K 114. Published with the permission of the Library and Information Centre of the Hungarian Academy of Sciences (Budapest, Hungary)
Fig. 10.3	Retrieved from http://www.missioncreep.com/tilt/hampton.html
Fig. 10.4	From the Smithsonian Institute
Fig. 10.5	From the Smithsonian Institute
Fig. 10.7	Courtesy of Gordon Rugg
Fig. 10.8	Courtesy of Gordon Rugg
Fig. 10.9	Courtesy of Gordon Rugg
Fig. 10.10	Courtesy of Gordon Rugg
Fig. 10.11	Courtesy of Gordon Rugg
Fig. 10.12	Courtesy of Gordon Rugg
Fig. 10.13	Courtesy of Gordon Rugg
Fig. 10.14	Courtesy of Gordon Rugg
Fig. 10.15	Courtesy of Oliver Knörzer
Fig. 10.16	Courtesy of Oliver Knörzer
Fig. 10.17	Courtesy of Oliver Knörzer
Fig. 10.18	Courtesy of Oliver Knörzer

Fig. 10.19   Courtesy of Oliver Knörzer

Fig. 10.20   From the secretive group Cicada 3301, retrieved from http://uncovering
-cicada.wikia.com/wiki/File:0.jpg

Fig. 11.1   Photograph by Pach Bros, 1908, from the United States Library of Congress, Prints and
Photographs division.

Fig. 11.2   Photograph by the National Photo Co. / Shorpy, Inc.

Fig. 11.3   From the David Peck Todd papers, 1862–1939 (inclusive), Manuscripts and Archives,
Yale University, with permission

Fig. 11.4   © Bettmann / Getty Images

Fig. 11.5   Retrieved from http://www.americaspace.com/2014/06/03/of-alien-life
-and-intelligence-are-we-ready-for-contact-part-3/

Index